Sustainable Civil Infrastructures

Sustainable Civil Infrastructures (SUCI) is a series of peer-reviewed books and proceedings based on the best studies on emerging research from all fields related to sustainable infrastructures and aiming at improving our well-being and day-to-day lives. The infrastructures we are building today will shape our lives tomorrow. The complex and diverse nature of the impacts due to weather extremes on transportation and civil infrastructures can be seen in our roadways, bridges, and buildings. Extreme summer temperatures, droughts, flash floods, and rising numbers of freeze-thaw cycles pose challenges for civil infrastructure and can endanger public safety. We constantly hear how civil infrastructures need constant attention, preservation, and upgrading. Such improvements and developments would obviously benefit from our desired book series that provide sustainable engineering materials and designs. The economic impact is huge and much research has been conducted worldwide. The future holds many opportunities, not only for researchers in a given country, but also for the worldwide field engineers who apply and implement these technologies. We believe that no approach can succeed if it does not unite the efforts of various engineering disciplines from all over the world under one umbrella to offer a beacon of modern solutions to the global infrastructure. Experts from the various engineering disciplines around the globe will participate in this series, including: Geotechnical, Geological, Geoscience, Petroleum, Structural, Transportation, Bridge, Infrastructure, Energy, Architectural, Chemical and Materials, and other related Engineering disciplines.

SUCI series is now indexed in SCOPUS and EI Compendex.

Bingxiang Yuan · Hüseyin Bilgin · Qingzi Luo ·
Zejun Han · Xueqiang Yang
Editors

Structural Safety and Ground Improvement on Bridge

Proceedings of 2024 8th International Conference on Civil Architecture and Structural Engineering

Springer

Editors
Bingxiang Yuan
School of Civil and Transportation Engineering
Guangdong University of Technology
Guangzhou, Guangdong, China

Qingzi Luo
Guangdong University of Technology
Guangzhou, Guangdong, China

Xueqiang Yang
Guangdong University of Technology
Guangzhou, Guangdong, China

Hüseyin Bilgin
EPOKA University
Tirana, Albania

Zejun Han
Guangdong University of Technology
Guangzhou, Guangdong, China

ISSN 2366-3405 ISSN 2366-3413 (electronic)
Sustainable Civil Infrastructures
ISBN 978-3-031-76101-0 ISBN 978-3-031-76102-7 (eBook)
https://doi.org/10.1007/978-3-031-76102-7

This Springer imprint is published by the registered company Springer Nature Switzerland AG
The registered company address is: Gewerbestrasse 11, 6330 Cham, Switzerland

Preface

As we gather to commemorate the momentous occasion of the 2024 8th International Conference on Civil Architecture and Structural Engineering (ICCASE 2024), held from April 19–21, 2024, in Guangzhou, China, it is with immense pride and gratitude that we present this proceedings, a compilation of the ideas, insights, and innovations shared by our esteemed participants.

The ICCASE 2024 attracted a diverse and esteemed group of participants, boasting a robust attendance of about 200 scholars, researchers, and professionals from various fields, including a significant contingent of international delegates hailing from diverse countries across the globe. This diversity of perspectives and backgrounds added immeasurable depth and breadth to the discussions and collaborations that took place during the conference.

The conference agenda was designed to cover a wide range of topics within the fields of civil architecture and structural engineering. It commenced with the inauguration ceremony, in which Xiaowen Chen (Secretary-General of the Guangdong Geotechnical Mechanics and Engineering Society, China) and Huiyou Ye (Secretary of the College of Water Conservancy and Civil Engineering, South China Agricultural University, China) extended cordial welcome to all the leaders, experts, and guests attending the conference. Then, a series of plenary sessions, speeches, and academic discussions were held. One of the highlights was the series of keynote speeches delivered by leading figures in relevant domains, including Jun Teng (Professor of Harbin Institute of Technology, Shenzhen, China), Tugen Feng (Professor of Jiangxi University of Science and Technology/Hohai University, China), Zhenyu Yin (Professor of The Hong Kong Polytechnic University, China), Xuehua Chen (Chairman of Guangzhou Jintu Geotechnical Engineering Technology Co., Ltd., China), among others. These speakers, representing both academia and industry, provided unique perspectives on the current trends, challenges, and future directions of research in their respective areas. Their insights were met with keen interest and lively discussions, fostering a vibrant intellectual atmosphere throughout the conference.

The call for papers generated a robust response from the academic community. The proceedings of ICCASE 2024 is a testament to the quality and diversity of the submissions received. These papers cover a wide range of topics, including but not limited to bridge engineering, structural safety of bridges, high-performance concrete for bridges, pile foundation, ground improvement, etc. The authors, representing a diverse array of institutions, have presented their researches in a clear and concise manner, making the proceedings a valuable resource for both scholars and practitioners alike.

In conclusion, the 2024 8th International Conference on Civil Architecture and Structural Engineering was a resounding success. We are grateful to all the participants, speakers, and authors for their contributions to this momentous event, and we look forward to the continued growth and development of these important fields in the coming years.

The Committee of ICCASE 2024

Organization

Conference Chair

Xiangsheng Chen	Shenzhen University, China/Chinese Academy of Engineering, China/Russian Academy of Engineering (RAE), Russia

Technical Program Committee

Feng Liu	Guangdong University of Technology, China
Jinyuan Liu	Toronto Metropolitan University, Canada

Technical Program Committee Members

Linwang Su	CCCC Fourth Harbor Engineering Institute Co., Ltd., China
Pingshan Chen	CCCC Fourth Harbor Engineering Institute Co., Ltd., China
Zhijun Liu	CCCC Fourth Harbor Engineering Institute Co., Ltd., China
Zongjie An	Shenzhen Modern City Architectural Design Co. Ltd., China
Bing Bai	Beijing Jiaotong University, China
Xiaoyu Bai	Qingdao University of Technology, China
Yulei Bai	Beijing University of Technology, China
Xiaohua Bao	Shenzhen University, China
Jun Bi	Nanjing Tech University, China
Wengui Cao	Hunan University, China
Zhangnv Zeng	Henan University of Technology, China
Fusheng Zha	Hefei University of Technology, China
Jinjian Chen	Shanghai Jiao Tong University, China
Keping Chen	Central South University, China
Rui Chen	Harbin Institute of Technology, Shenzhen, China
Xiaoping Chen	Jinan University, China
Xingzhou Chen	Xi'an University of Science and Technology, China

Wenwu Chen	Lanzhou University, China
Chunyi Cui	Dalian Maritime University, China
Beibing Dai	Sun Yat-sen University, China
Guoliang Dai	Southeast University, China
Xuanming Ding	Chongqing University, China
Youkou Dong	China University of Geosciences (Wuhan), China
Qinwen Du	Chang'an University, China
Yanjun Du	Southeast University, China
Yumin Du	Northwestern Polytechnical University, China
Tianhui Fan	South China University of Technology, China
Deluan Feng	Guangdong University of Technology, China
Yong Feng	Henan University of Technology, China
Hongmei Gao	Nanjing Tech University, China
Weiming Gong	Southeast University, China
Xing Gong	Guangdong University of Technology, China
Kai Gu	Nanjing University, China
Wei Guo	Tianjin University, China
Yongchang Guo	Guangdong University of Technology, China
Yuancheng Guo	Zhengzhou University, China
Zejun Han	Guangdong University of Technology, China
Huan He	Southeast University, China
Jian Ji	Hohai University, China
Tong Jiang	North China University of Water Resources and Electric Power, China
Yan Jiang	Guangdong Research Institute of Water Resources and Hydropower, China
Chong Jiang	Central South University, China
Mingjie Jiang	Guangxi University, China
Gangqiang Kong	Hohai University, China
Liang Kong	Qingdao University of Technology, China
Hailei Kou	Ocean University of China, China
Diyuan Li	Central South University, China
Jiale Li	Hebei University of Technology, China
Jinhui Li	Harbin Institute of Technology, Shenzhen, China
Lihua Li	Hubei University of Technology, China
Lianxiang Li	Shandong University, China
Weichao Li	Tongji University, China
Xia Li	Southeast University, China
Yurun Li	Hebei University of Technology, China
Yuanhai Li	China University of Mining and Technology (Xuzhou), China
Shihua Liang	Guangdong University of Technology, China

Hang Lin	Central South University, China
Peiyuan Lin	Sun Yat-sen University, China
Yuliang Lin	Central South University, China
Chao Liu	Guangzhou University, China
Donghai Liu	Tianjin University, China
Hai Liu	Guangzhou University, China
Junwei Liu	Qingdao University of Technology, China
Kaiwen Liu	Southwest Jiaotong University, China
Qingbing Liu	China University of Geosciences (Wuhan), China
Run Liu	Tianjin University, China
Xianshan Liu	Chongqing University, China
Yongjian Liu	Guangdong University of Technology, China
Mengmeng Lu	China University of Mining and Technology (Xuzhou), China
Zhao Lu	The Hong Kong University of Science and Technology, China
Congshuang Luo	Henan University of Urban Construction, China
Qingzi Luo	Guangdong University of Technology, China
Jianbing Lv	Guangdong University of Technology, China
Guoxiong Mei	Hohai University, China
Linlong Mu	Tongji University, China
Huafu Pei	Dalian University of Technology, China
Feifan Ren	Tongji University, China
Xiang Ren	Xi'an University of Science and Technology, China
Xiusong Shi	Hohai University, China
Hongyan Shi	Guangdong University of Technology, China
Jiangwei Shi	Hohai University, China
Dong Su	Shenzhen University, China
Liqiang Sun	Tianjin University, China
Liang Tang	Harbin Institute of Technology, China
Lei Su	Qingdao University of Technology, China
Liyuan Tong	Southeast University, China
Yixian Wang	Hefei University of Technology, China
Chenghua Wang	Tianjin University, China
Dongxing Wang	Wuhan University, China
Dong Wang	Ocean University of China, China
Lei Wang	Anhui University of Science & Technology, China
Wanying Wang	Guangdong University of Technology, China
Yonghong Wang	Qingdao University of Technology, China
Zhongwei Wang	Shenzhen Modern City Architectural Design Co. Ltd., China

Wengang Zhang	Chongqing University, China
Heng Zhao	Hunan University, China
Yanru Zhao	Shenzhen University, China
Gang Zheng	Tianjin University, China
Dequan Zhou	Changsha University of Science & Technology, China
Feng Zhou	Nanjing Tech University, China
Haizuo Zhou	Tianjin University, China
Hang Zhou	Chongqing University, China
Qizhi Zhu	Hohai University, China
Haiyang Zhuang	Nanjing Tech University, China
Yan Zhuang	Southeast University, China
Xinjun Zou	Hunan University, China
Fadi Hage Chehade	Lebanese University, Lebanon
Haran Pragalath	British Applied College, Umm Al Quwain, UAE
Ramadhansyah Putra Jaya	Universiti Malaysia Pahang, Malaysia
Mohammad Arif Kamal	Architecture Section Aligarh Muslim University, India
Abdollah Tabaroei	Eshragh Institute of Higher Education, Iran
S. M. Anas	Jamia Millia Islamia (A Central University), India
Gunalan Vasudevan	Tunku Abdul Rahman University of Management and Technology, Malaysia
Huzaifa bin Hashim	University of Malaya, Malaysia
Ng Chiew Teng	Universiti Teknologi Malaysia, Malaysia
Chin Chee Loong	Universiti Teknologi Malaysia, Malaysia
Onn Chiu Chuen	University of Malaya, Malaysia
Suhana Binti Koting	University of Malaya, Malaysia
Yap Soon Poh	University of Malaya, Malaysia
Abhijith R.	Federal Institute of Science and Technology, India

Organizing Committee Chairs

Yongchang Guo	Guangdong University of Technology, China
Bingxiang Yuan	Guangdong University of Technology, China
Wei Zhang	South China Agricultural University, China

Organizing Committee Members

Linwang Su	CCCC Fourth Harbor Engineering Institute Co., Ltd., China
Jing Wang	CCCC Fourth Harbor Engineering Institute Co., Ltd., China

Shuhui Lyu	CCCC Fourth Harbor Engineering Institute Co., Ltd., China
Zongjie An	Shenzhen Modern City Architectural Design Co. Ltd., China
Xudong Cao	MCC Capital Engineering & Research Incorporation Limited, China
Liang Chen	CSCEC4 Civil Engineering Co. Ltd., China
Xingzhou Chen	Xi'an University of Science and Technology, China
Chuang Feng	Nanjing Tech University, China
Peiyuan Lin	Sun Yat-sen University, China
Chao Liu	Guangzhou University, China
Hai Liu	Guangzhou University, China
Zhao Lu	The Hong Kong University of Science and Technology, China
Congshuang Luo	Henan University of Urban Construction, China
Qingzi Luo	Guangdong University of Technology, China
Feifan Ren	Tongji University, China
Jie Song	MCC Capital Engineering & Research Incorporation Limited, China
Wanying Wang	Guangdong University of Technology, China
Yonghong Wang	Qingdao University of Technology, China
Zhongwei Wang	Shenzhen Modern City Architectural Design Co. Ltd., China
Yong Wen	Zhongkai University of Agriculture and Engineering, China
Zhijun Xu	Henan University of Technology, China
Yang Wu	Guangzhou University, China
Houbing Xing	CSCEC4 Civil Engineering Co. Ltd., China
Xueqiang Yang	Guangdong University of Technology, China
Junhong Yuan	Inner Mongolia University, China
Shuang Zheng	CSCEC4 Civil Engineering Co. Ltd., China

Publication Chairs

Bingxiang Yuan	Guangdong University of Technology, China
Hüseyin Bilgin	EPOKA University, Albania
Qingzi Luo	Guangdong University of Technology, China
Zejun Han	Guangdong University of Technology, China

Contents

Structural Safety and Innovative Design of Bridges

Research on Load Bearing Performance of Pile Foundation for Superstructure Reconstruction of Bridge in Service

Kai-ren Li(✉), Ni-gui Qian, and Lei Xiong

Guangdong Jiaoke Detection Co., Ltd., Guangzhou 510000, Guangdong, China
1587441573@qq.com

Abstract. It is an important content of scientific evaluation of rationality, safety and economy of the bridge reconstruction plan to carry out the research on load-carrying properties of pile foundation for superstructure reconstruction of bridge in service. In order to understand the variation of load-bearing performance of pile foundation in the process of superstructure reconstruction of serving Bridges, based on the superstructure reconstruction project of Taiping Bridge and FEM software ABAQUS, the load-bearing performance of old piles, the difference in load bearing characteristics between old and new piles, and the load bearing performance under different pile lengths and diameters of new piles were studied respectively, providing references for the quality control in the process of superstructure reconstruction of serving Bridges. The results show that the vertical load bearing performance of the old pile foundation increases because the soil around the pile foundation is gradually consolidated. Vertical load-displacement change law of old and new piles are consistent, and the vertical differential settlement of pile top is small, which verifies the rationality of the new pile foundation design. When the length of new pile foundation grows by 9.1%, the settlement of pile top diminution by 1.14%. When the length of new pile decreased by 9.1%, the settlement of pile top grows by 1.5%. When the diameter of new pile foundation grows by 0.3m, the growth of pile carrying capacity is not enough to offset the influence of pile weight, and the vertical settlement of pile top increases. The differential settlement of old and new piles can be better controlled by increasing the pile length reasonably when designing the carrying capacity of super-long piles in superstructure reconstruction projects.

Keywords: Serving bridges · superstructure reconstruction · pile foundation · bearing performance

1 Introduction

With the development of our transportation business, the Bridges built in the early stage of our country have been overloaded for a long time, and the superstructure has been damaged to different degrees, which can no longer meet the increasing traffic demand [1, 2]. In the bridge reconstruction project, in order to save the project cost, the program

B. Yuan et al. (Eds.): ICCASE 2024, SUCI, pp. 3–12, 2024.
https://doi.org/10.1007/978-3-031-76102-7_1

of "replacing superstructure and strengthening substructure" is usually adopted. How to take full advantage of the carrying capacity of substructure pile and reasonably control the quality of service bridge superstructure reconstruction project is a topic in front of scholars. Relevant scholars have studied the load-bearing performance of pile of bridge reconstruction and expansion. For example, Song Wen-bin [3] studied the vertical load-bearing performance of existing bridge pile foundation and grouting reinforcement technology through ABAQUS finite element software. Feng Zhong-ju et al. [4] proposed the technical parameters of uneven displacement control of pile widened by existing bridge for highway reconstruction and expansion through centrifugal model test. Zhao Zhi-gang [5] analyzed the displacement control technology of the widened pile of the ancient bridge; Li Jingpei et al. [6, 7] analyzed the factors affecting the carrying capacity of old pile that have been subjected to long-term loads, studied the Load-displacement relation between new and ancient piles in natural saturated clay, and proposed that the characteristics of high carrying capacity and low settlement of old piles should be fully taken into account in the design of the mixed foundation of new and ancient piles. The research results of NG K W et al. [8–10] show that the load bearing capacity of pile foundation has a time effect, and the old pile foundation under long-term load usually has a higher load bearing capacity than the new pile.

At present, relevant scholars mainly carry out relevant technical studies on the splicing parts and ancient of new and ancient bridge pile foundation in the reconstruction and expansion of expressway, and few studies have been conducted on the load-bearing performance of the rebuilt bridge pile foundation in service. However, the bearing performance difference of the old and new bridge pile foundation is likely to cause internal forces in the bridge superstructure, leading to diseases, and affecting the construction safety and driving safety during the construction and operation period. In order to understand the variation of load bearing performance of pile foundation in the process of overframe reconstruction of bridge in service, this paper relies on the overframe reconstruction project of Taiping Bridge and uses FEM software ABAQUS to study the load bearing performance of old piles, the difference in load bearing characteristics between old and new piles, and the load bearing performance of new piles with different pile lengths and pile diameters. The research achievement can provide reasonable suggestions for the design and construction of overstructure pile foundation for Bridges in service, and have certain engineering application value.

2 Project Overview

Taiping Bridge is located in Guanfo Expressway, Guangdong Province, the upper structure is a four-span prestressed concrete continuous rigid frame bridge, the lower structure of the main pier is reinforced concrete I-section double-wall pier, the foundation is cast-in-situ bored pile, the total length of the main bridge is 330 m, opened to traffic in 1997. The reconstruction project adopts the plan of "superstructure replacement, substructure reinforcement", adding two piles between the two caps of the bridge cross bridge and the main pier foundation. The old piles are 64 m long and 1.8 m diameter. The new pile is 68 m long and 1.8 m in diameter.

3 Establishment of Finite Element Model

FEM software ABAQUS includes a wealth of material models, element types, loads and boundary conditions, can better deal with the nonlinear constitutive relations of rock and soil, load and boundary conditions of the complexity of the problem. Using FEM software ABAQUS, a 2D axisymmetric model of a single pile under vertical load is built. The soil length of the model is 2 times of the pile length and the width is 20 times of the pile diameter. The soil constitutive model is Mohr-Coulomb model, and the elastic model is used for pile concrete material. According to the design and geological exploration data, pile and soil parameters are set as shown in Table 1. The size of pile foundation model is shown in Table 2. The 2D axisymmetric pile-soil model under vertical load is shown in Fig. 1.

Table 1. Material parameters of pile soil

Name of the material	H/m	$\gamma/(\mathrm{g.cm}^{-3})$	e	Es/MPa	C/kPa	$\varphi(^{\circ})$	Constitutive model
Silty silt	0–8	17.1	200000	9	29.8	0.3	Mohr-Coulomb
Medium sand	8–16	19.4	200000	10	31.7	0.3	Mohr-Coulomb
Sandy clay	16–24	22	200000	14	34.5	0.3	Mohr-Coulomb
Silty clay	24–34	21	200000	8	27.8	0.3	Mohr-Coulomb
Heavily weathered granite	34–80	19.7	200000	22	25.8	0.3	Mohr-Coulomb
Concrete	/	24	3×10^7	——	——	0.25	elasticity

Table 2. Size of pile foundation model

Finite element model	pile length /m	Pile diameter /m
Old pile	64	2.1
New pile	66	2.1
New piles change pile length	60, 66, 72	2.1
New pile changed pile diameter	66	1.8, 2.1, 2.4

(a) Pile foundation model (b) Soil model around pile

Fig. 1. 2D axisymmetric pile-soil model under vertical load

4 Analysis of Model Calculation Results

4.1 Bearing Performance Analysis of Old Piles

In order to analyze the load bearing performance of old pile in the reconstruction of superstructure, three construction conditions were set in the model test through construction load estimation. Construction condition 1 was the load of the old superstructure of the bridge, and the vertical load of a single pile was 10000 kN. Construction condition 2 is that the old superstructure of the bridge is dismantled, the foundation is unloaded, and the vertical load capacity of a single pile is 0 kN; Construction condition 3 is the load of the new superstructure of the bridge, and the vertical load of a single pile is 4000 kN.

4.1.1 Vertical Load and Displacement Curve of Pile Top

The vertical load and displacement curve of old pile top is shown in Fig. 2, and the specific data are shown in Table 3.

Table 3. Vertical load-displacement of old pile

construction state	vertical load/kN	displacement of pile tip/mm	
1	Old superstructure load	10000	−6.626
2	The old superstructure was demolished	0	−0.386
3	New superstructure load	4000	−2.895

It can be seen from the chart that the displacement of old pile foundation top is −6.626 mm under the old superstructure load. After the removal of the old structure, the pile foundation is unloaded, the displacement of old pile foundation top is −0.386 mm, and the old pile foundation top floats 6.24 mm. When the new superstructure load acts, the displacement of old pile foundation top is −2.895 mm, and the settlement of pile top is 2.509 mm compared with the construction condition 2. Combined with the field monitoring, the old pile foundation top displacement in the simulation results is too large, indicating that under long-term load, due to the gradual consolidation of soil around the pile, the bearing capacity of the pile foundation of the bridge in service will change and its bearing capacity will be improved to a certain extent.

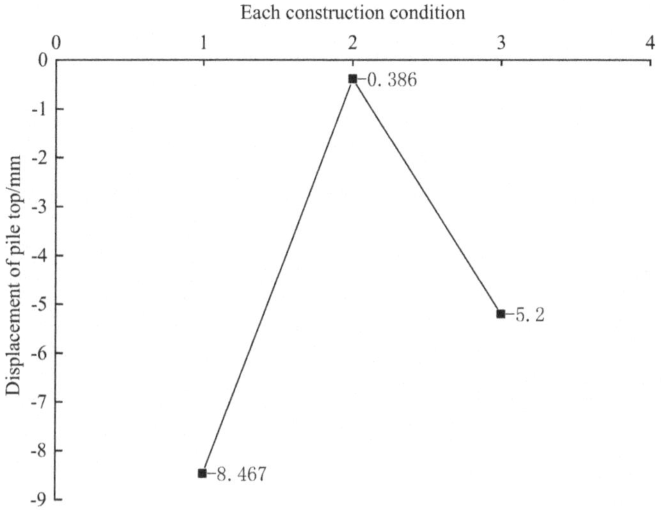

Fig. 2. Vertical load-displacement curve of old pile

4.1.2 Change Rule of Pile Side Friction Resistance

Figure 3 shows the distribution curve of pile side friction resistance under different construction conditions. It can be seen from the figure that under different construction conditions, the pile skin friction grows first, then decreases and then grows gradually with the increase of pile depth, and the maximum pile side friction appears in the middle and upper part of the pile foundation.

4.1.3 Load Sharing Between Pile End Resistance and Pile Side Friction Resistance

Considering the self-weight of pile foundation, the load sharing of pile end resistance and pile side friction resistance under different construction conditions is shown in Table 4.

It can be seen from the table that the load pile foundation presents obvious characteristics of friction pile, and the vertical bearing capacity of pile foundation is mainly borne by the side friction resistance of pile. Under the old superstructure load, the pile end resistance accounted for 18.2% of the total load, and the pile side resistance accounted for 81.8% of the total load. Under the new structure load, the pile end resistance accounts

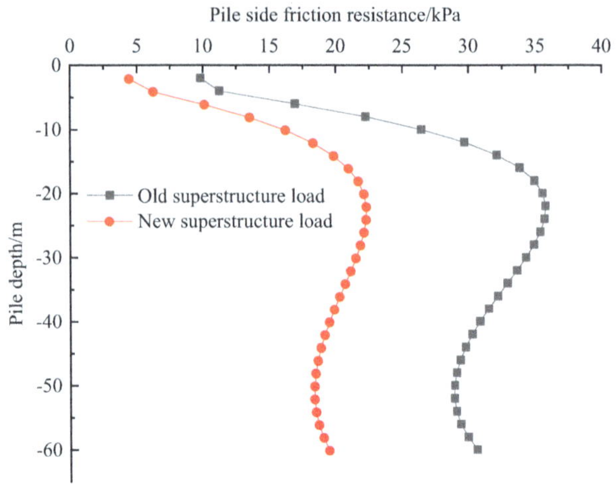

Fig. 3. Distribution curves of pile side friction under different construction conditions

Table 4. Load sharing of pile foundation under different construction conditions

analysis item	Construction conditions 1	Construction conditions 3
Pile foundation weight /kN	2215.6	2215.6
Upper load /kN	4000	10000
Total load /kN	6215.6	12215.6
Pile side resistance ratio /%	72.9	81.8
Pile end resistance ratio /%	27.1	18.2

for 27.1% of the total load, and the pile side resistance accounts for 72.9% of the total load. The analysis shows that the ratio of pile side friction to total load increases with the increase of pile top vertical displacement within the design load range. In this loading – unloading – loading construction condition, the soil around the pile is basically in the elastic stage, and the bearing capacity of the pile foundation does not decay, and the pile foundation can still bear the upper load normally.

4.2 Comparative Advantages of Load Bearing Performance of New and Old Piles

Under the action of new superstructure load, the new pile simulation analysis model is established, and the load-bearing performance of the new and old piles is compared and analyzed. Draw vertical load-displacement curves of new and old piles as shown in Fig. 4. As can be seen from the figure, when the new superstructure load is directly loaded, the vertical load-displacement changes of the new pile and the old pile are consistent, the maximum vertical displacement of the new pile and the old pile are −2.877 mm and −2.89 mm respectively, and the maximum vertical displacement of the new pile is reduced by 0.023 mm. The result of research shows that when the length of the new pile increases by 2 m, the vertical carrying capacity of the new pile is slightly increased, and the vertical differential settlement between the top of the new pile and the old pile is very small, which can reduce the influence of the internal stress changes caused by the differential settlement. At the same time, it shows that the design of the new pile is reasonable.

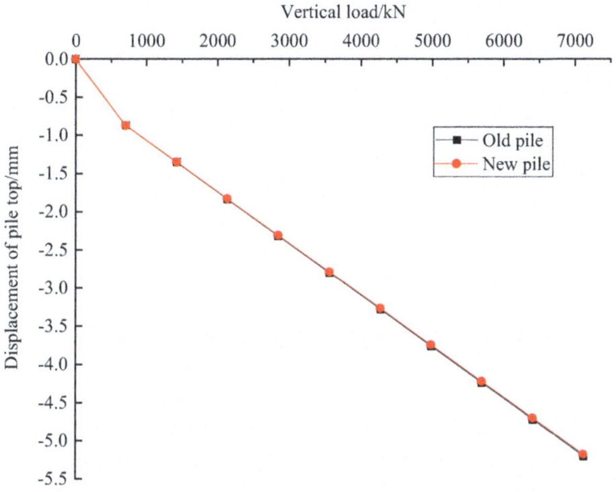

Fig. 4. Vertical load displacement curve of new and old piles

4.3 Influencing Factors Analysis of New Pile Load Bearing Performance

Under the same soil and load parameters, the pile diameter and pile length of the new pile are changed, and the simulation model is established to analyze the influence of different pile diameters and pile lengths on the bearing performance of the new pile.

4.3.1 Influence of Pile Length on Bearing Performance of New Pile

The vertical load-displacement curves of different pile lengths are shown in the Fig. 5. When the pile diameter is the same, the vertical load-displacement curves of different pile lengths vary according to the same law, and the vertical displacement of pile top decreases with the increase of pile length. Taking a new pile with a pile length of 66 m as the reference object, when the pile length increases by 6 m, that is, the pile length increases by 9.1%, the pile top displacement decreases by 1.14%, and when the pile length decreases by 9.1%, the pile top displacement increases by 1.5%. The results show that when the pile length is changed, the displacement of pile top is relatively small, and the vertical bearing capacity of pile foundation is basically unchanged, which also indicates that the design of the new pile length is economical and reasonable.

4.3.2 Influence of Pile Diameter on Bearing Performance of New Pile

The vertical load-displacement curves of different pile diameter are shown in the Fig. 6. When the pile length is the same, the vertical load-displacement curves of different pile

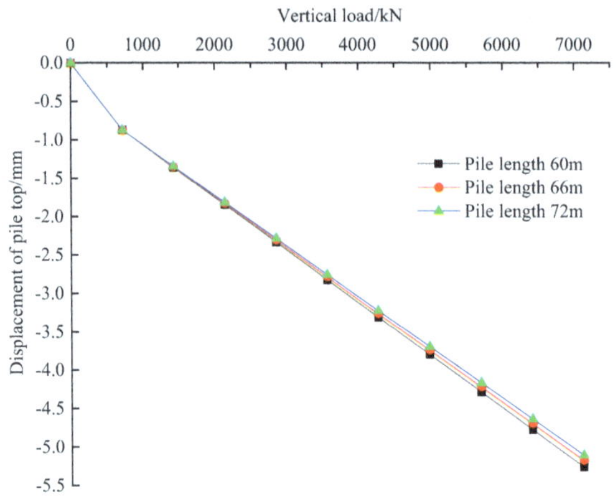

Fig. 5. Vertical load displacement curve of different pile lengths

diameter have the same change law, and the vertical displacement of pile top increases with the increase of pile diameter. When the pile diameter changes from 1.8 m to 2.4 m, the pile body weight increases by 1322.5 kN and 1526 kN each time the pile diameter increases by 0.3 m, accounting for 33.1% and 38.2% of the total load respectively, and the pile body weight accounts for a larger proportion of the total load. The results show that when the pile diameter increases, the pile side and bottom resistance will also increase correspondingly, but the increase of pile foundation carrying capacity is not enough to offset the increase of pile body weight, resulting in greater vertical settlement of pile

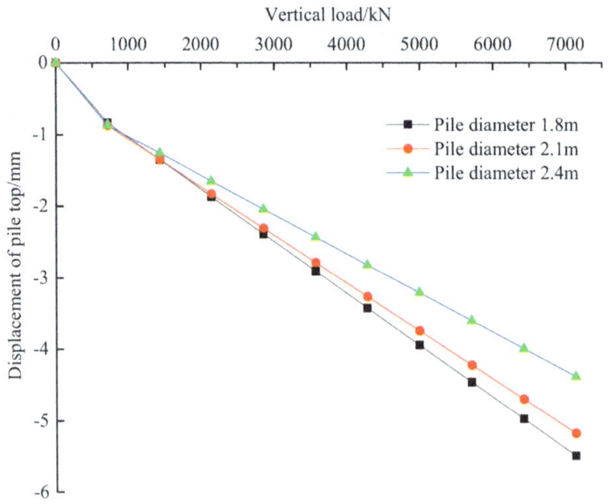

Fig. 6. Vertical load displacement curve of different pile diameter

top. The influence of pile body weight should be fully considered when analyzing the vertical carrying capacity of super-long piles.

5 Conclusions

Because of the gradual consolidation of soil around pile, the vertical bearing capacity of the old pile is increased to a certain extent; The old pile shows obvious characteristics of friction pile, and the vertical carrying capacity of pile foundation is mainly borne by the pile side friction resistance. In this loading – unloading – loading construction condition, the soil around the pile is basically in the elastic stage, and the bearing capacity of pile foundation does not decay, and the old pile can still bear the upper load normally.

The changes of vertical load-displacement of old and new piles are consistent, and the vertical differential settlement of pile top is small, which verifies the rationality of the new pile design; When the length of new pile increases by 9.1%, the displacement of pile top decreases by 1.14%. When the length of new pile is shortened by 9.1%, the displacement of pile top increases by 1.5%. When the diameter of new pile increases by 0.3 m, the increase of pile bearing capacity is not enough to offset the influence of pile body weight, and the vertical settlement of pile top increases. The differential settlement of old and new piles can be better controlled by increasing the pile length reasonably when designing the carrying capacity of super-long piles in superstructure reconstruction projects.

In the simulation analysis, based on some simplified treatment of pier soil materials and pier – soil contact surface, the established model can reflect the actual situation of the research object, but there are still some improvements in parameter value and boundary condition setting.

Refcrences

1. Feng, Z., Chen, J., Chen, J., Shao, J.: Settlement Difference Control Technology and Engineering Application of Widened Pile Foundation of Girder Bridge. China Architecture and Construction Press, Beijing (2014)
2. Wang, M.: Research on Settlement Difference Control Technology of Widened Pile Foundation of Girder Bridge in Expressway Reconstruction and Expansion Project. Chang'an University (2019)
3. Song, W.: Research on Load Bearing Performance and Grouting reinforcement Technology of Existing Bridge pile Foundation. Hebei University of Technology (2016)
4. Feng, Z., Chen, J., Liu, X., et al.: Centrifugal model test on differential settlement control of widened pile foundation of highway reconstruction and expansion. J. Chang'an Univ. (Nat. Sci. Edition) **42**(02), 1–9 (2012)
5. Zhi-gang, Z.: Construction control technology of settlement of widened pile foundation of old bridge. TranspoWorld **35**, 173–175 (2022)
6. Li, J., Zhang, L., Li, L., et al.: Study on load-settlement relationship between new and old piles in natural saturated clay. Chinese J. Rock Mech. Eng. **35**(09), 1906–1913 (2016)
7. Jing-pei, L.I., Zi-han, X.U.: Forecast of ultimate carrying capacity of pile under long-term load. J. Shanghai Jiao Tong Univ. **55**(04), 380–386 (2021)
8. Ng, K.W., Sritharan, S.: A procedure for incorporating setup into load and resistance factor design of driven piles. Acta Geotech. **11**(2), 347–358 (2016)

9. Kristinof, R., Thorn, M.: Setup of driven pile capacity with time in soft marine sediments. Australian Geomech. J. **52**(3), 1–9 (2017)
10. Doherty, P., Gavin, K.: Pile aging in cohesive soils. J. Geotech. Geoenviron. Eng. **139**(9), 1620–1624 (2013)

Analysis of the Influence of Bridge Pile Construction on Oil and Gas Pipelines

Jiaping Li(✉)

China Petroleum Pipeline Engineering CO., Ltd., Langfang 065000, China
454984630@qq.com

Abstract. With the large-scale construction of infrastructure such as highways and railways, the construction of infrastructure will inevitably have adverse influences on the normal use and safe operation of adjacent buried pipelines. Once an oil and gas pipeline leakage accident occurs, it will lead to huge economic losses and even casualties. This paper uses FLAC3D finite element software establish the bridge pile-pipeline soil mass model, and simulates the influence of bridge pile construction on the stress and deformation of nearby oil and gas pipelines. The research results indicate that after the construction and operation of bridge piles, their influence on the stress and displacement of the soil around the pipeline is relatively small, and they will not cause major safety hazards.

Keywords: bridge pile construction · oil and gas pipelines · numerical simulation · stress state · pipeline safety

1 Introduction

Pipeline transportation is the main way of transportation for oil and natural gas in China. According to the 14th Five Year Plan, the total length of newly built natural gas pipelines nationwide will reach 210000 km [1]. At the same time, a large number of highways and railways are also being constructed in China, and such new construction projects inevitably have parallel or intersecting situations with existing buried pipelines. The West East Gas Pipeline West Second Line is approximately 18 km along the same corridor as the Shenzhen Outer Ring Expressway, and the pipeline was constructed before the outer ring expressway. The pipeline runs parallel and crosses the proposed outer ring expressway multiple times, resulting in the inability to meet the safety distance specified in the Natural Gas and Oil Pipeline Protection Law [2] and the Highway Law of the People's Republic of China [3] during the construction of the outer ring expressway. Therefore, during the normal construction of bridge piles on the outer ring expressway in the future, pipeline damage and excessive displacement may occur. Once oil and gas pipelines are damaged, it may cause economic losses, serious secondary disasters, environmental pollution, and even casualties [4–7]. Therefore, the analysis of the influence of bridge pile construction on pipelines is of great significance.

At present, some scholars have carried out researches on the influence of tunnel construction on pipelines by using different methods such as theoretical analysis, model

B. Yuan et al. (Eds.): ICCASE 2024, SUCI, pp. 13–20, 2024.
https://doi.org/10.1007/978-3-031-76102-7_2

experiment and numerical simulation, but there are few researches on the influence of bridge pile construction on pipelines. Regarding the above problems, this paper uses FLAC3D software to simulate the influence of bridge pile construction on the stress and deformation of nearby pipelines. The research results can provide reference for the design and construction of the engineering.

2 Establishment of Finite Element Model and Parameter Selection

2.1 Establishment of Finite Element Model

During numerical simulation, a 50 m long pipeline located near the bridge pile was selected for analysis, with a diameter of 914 mm and a trench bottom width of 2.414 m. The elevated bridge is simulated with single pier and double piles, with a pile diameter of 1 m and a spacing of 5 m. The minimum distance between the piles and the pipeline is taken as 15 m. The three-dimensional finite element model takes a length of 50 m along the longitudinal direction, a width direction of 56 m, and a height direction of 30 m below the ground. In the model, eight node hexahedral elements are used. In order to optimize the mesh and meet the requirements of computational accuracy, the elements of the trench and its surrounding parts are densely distributed [8]. At the left and right boundaries of the model, fix the displacement in the X direction, at the front and back boundaries of the model, fix the displacement in the Y direction, and at the bottom of the model, fix the displacement in the Z direction. The three-dimensional solid finite element model is shown in Fig. 1.

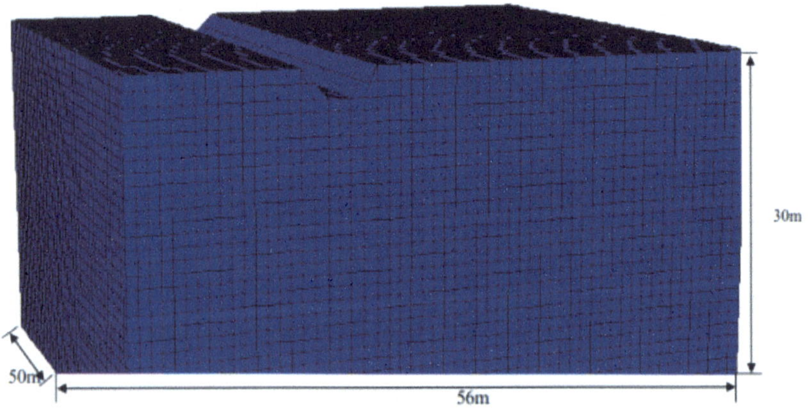

Fig. 1. Schematic diagram of the overall finite element model

2.2 Constitutive Relationship

Based on on-site sampling and rock mechanics test results, the Mohr-Coulomb yield criterion is used in the calculation to determine the failure of the rock mass, and its expression is:

$$f_s = \sigma_1 - \sigma_3 \frac{1 + \sin\varphi}{1 - \sin\varphi} - 2c\sqrt{\frac{1 + \sin\varphi}{1 - \sin\varphi}} \tag{1}$$

In the formula, σ_1 and σ_3 are the maximum and minimum principal stresses, respectively. c and φ are adhesive force and friction angle, respectively. When $f_s > 0$, the material would undergo shear failure. Under normal stress conditions, the tensile strength of rock mass is very low, so the tensile strength criterion $\sigma_3 > \sigma_T$ can be used to determine whether the rock mass has undergone tensile failure.

2.3 Mechanical Parameters

The calculation parameters were determined based on the geotechnical engineering survey report of the Guangzhou-Shenzhen branch of the West East Gas Pipeline Project (Qiuyuling-Dachan Island section). These parameters are reduced by referring other similar soil and rock parameters. Mechanical parameters are showed in Table 1.

Table 1. Mechanical parameters used in numerical simulation

Name	Density (kg/m^3)	Elastic modulus (MPa)	Poisson's ratio	Cohesive force (kPa)	Internal friction angle (°)
banket	1820	8.7	0.2	34.3	3.6
sandy soil	1880	4.74	0.25	26.4	19.1
clay	1900	15	0.24	18.2	16.2
granite	1900	34.3	0.18	37.8	24.3

3 Influence of Bridge Pile Construction on Pipeline Safety

The mechanical behavior of geotechnical engineering is closely related to the excavation history and process of the project. This simulation calculation mainly considers that the pipeline laying work has been completed, and focuses on analyzing the influence of bridge foundation construction outside the 15 m range on the pipeline. Due to the slope excavation method for pipeline trenches and the continuous installation of reinforced concrete cover plates above the pipelines, the influence of bridge foundation construction on the pipelines is mainly on the trench.

3.1 Stress and Displacement Analysis in the Bridge Pile Construction Process

The stress and strain state of the pipe trench before bridge pile construction will be superimposed with the stress generated by bridge pile construction, resulting in new displacement and affecting the safety of the pipe trench. From Fig. 2, it can be observed that before the construction of bridge piles, the stress distribution is mainly due to the layered distribution gradually increasing from top to bottom caused by the gravity of the rock and soil. The maximum principal stress in the stress field where the pipe trench is located is less than 0.1 MPa.

This paper numerically simulates the bridge pile as a drilled pile. During the drilling process, a compression effect is formed on the soil around the pile body, and the compression force is replaced by the stress along the pile hole towards the surrounding area. Figure 3 shows the distribution of maximum principal stress during the construction process. Compared with before construction, the trend of maximum principal stress distribution has not changed and still shows a layered and strip distribution, gradually increasing from top to bottom. However, in the soil around the bridge piles, the maximum principal stress significantly increases. Due to the influence of bridge pile construction, there is a slight change in the stress value of the soil around the pipe trench, but the change is small and will not cause damage to the soil.

In terms of displacement, numerical simulation assumes that the geological conditions and stress state along the pipeline laying direction of the trench are the same, so the model can be simplified as a plane strain model. Therefore, when analyzing the stability of the trench soil, only the displacement in the vertical and horizontal directions is considered. Figure 4 shows the vertical displacement diagram during the construction process. It can be seen that due to the influence of bridge pile construction, there is partial displacement of the soil around the bridge piles, manifested as settlement, with a maximum settlement of 3.3 mm. However, due to the distance from the bridge piles, there is no significant vertical displacement of the soil around the pipe trench. In terms of horizontal displacement, due to the squeezing effect of bridge pile construction, the soil on the side of the pipe trench near the bridge pile has shifted towards the inside of the pipe trench. However, the displacement is relatively small, with a maximum of about 2 mm, which will not have a significant impact on the safety of the pipe trench. Appropriate amount of monitoring points should be used during construction to monitor the settlement and displacement of pipeline.

After the completion of bridge pile construction, the upper bridge and vehicle loads are applied to the bridge piles through the bridge piers, thereby forming a new compression effect on the soil around the bridge piles. Figure 5 shows the distribution of the first principal stress of the model after construction is completed. Due to the distance between the simulated pipe trench and the bridge pile which is 15 m, the influence of the bridge pile on the pipe trench is relatively small. The stress changes in the soil around the pipe trench are not significant, and the maximum stress value is only 0.1 MPa, which will not cause significant safety effects.

For the displacement, the main consideration is the vertical and horizontal displacement of the soil around the vertical pipe trench. Numerical analysis results show that no significant vertical displacement was found in the soil around the pipe trench, with a maximum value less than 1 mm; In terms of horizontal displacement, due to the influence of bridge and upper vehicle loads, the maximum horizontal displacement of the soil around the pipe trench increases to 6 mm, which occurs in the upper soil near the bridge pile side of the pipe trench. The influence on the safety of the pipe trench is not very significant, but appropriate measures should be taken to avoid pipeline damage. At the same time, in the selection and construction method of bridge piles, the safety of pipelines should be considered and fully demonstrated.

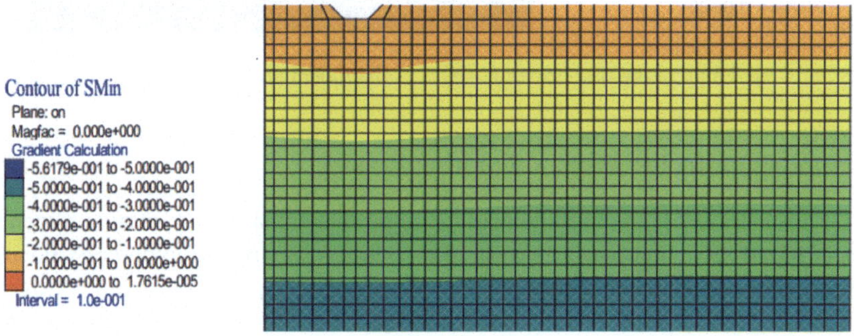

Fig. 2. First principal stress cloud map before bridge pile construction

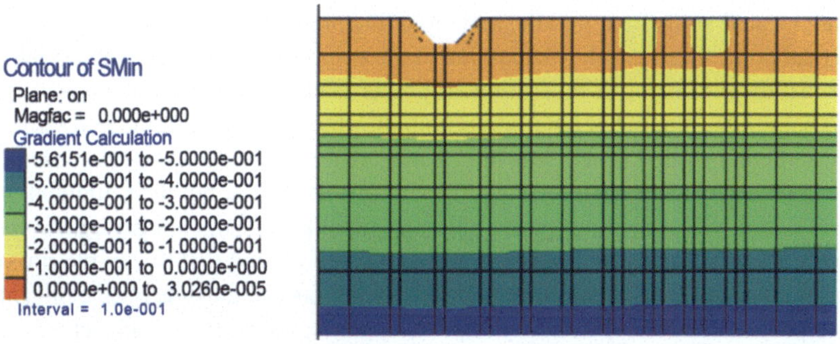

Fig. 3. Cloud diagram of the first principal stress during the construction process of bridge piles

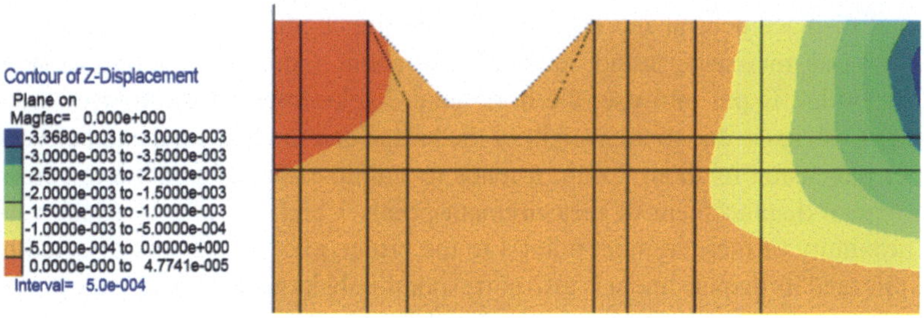

Fig. 4. Vertical displacement cloud map during bridge pile construction process

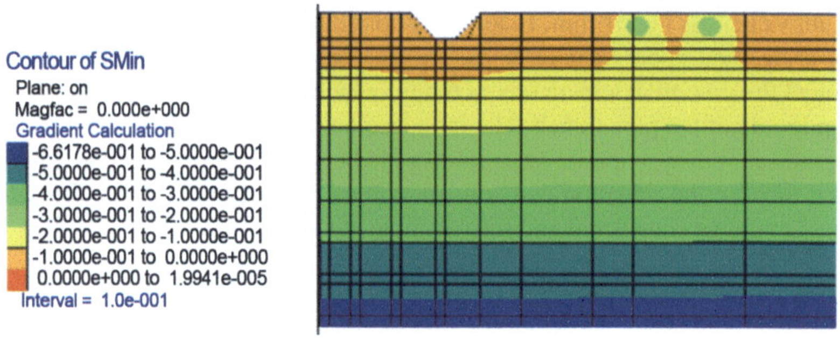

Fig. 5. Cloud map of the first principal stress of the model after construction completion

3.2 Analysis of the Key Point Monitoring Results

In order to better analyze the influence of bridge pile construction on pipe trenches, a series of monitoring points were set up in the numerical simulation process. Focus on monitoring the horizontal and vertical displacement of key points during the bridge pile construction process. In the setting of monitoring points, the main consideration is to set up measuring points at the vertical position of bridge piles and pipelines, as well as at a distance of 5 m from the vertical point. Measurement points 4, 5, and 6 are located at the foot, middle, and top of the pipe trench at the vertical point, and measurement points 1, 2, and 3 are located at the foot, middle, and top of the pipe trench at a distance of 5 m from the vertical point. Figure 6 shows that due to the compression effect of the bridge piles, the vertical displacement of measurement points 1 and 4 in the initial stage shows an upward bulge. The amount of bulge is very small, which then transforms into settlement, and the settlement value is also less than 1 mm. After the bridge is put into use, due to changes in the surrounding stress state, the soil further settles, but the settlement is still less than 1 mm, which will not pose a threat to pipeline safety. Comparing the vertical displacement of measurement points 1 and 4, it can be found that due to the proximity of measurement point 4 to the bridge pile, it is greatly affected by the bridge pile, and its displacement is also correspondingly large. Comparing the vertical displacement curves of measuring points 2 and 5, as well as measuring points 3 and 6, a similar trend can be observed.

From Fig. 7, it can be seen that due to the compression effect of bridge piles, the horizontal displacement at the foot of the pipe trench slope during the construction stage shows an increasing trend, increasing to about 1 mm and tending to stabilize. After the bridge is put into use, the horizontal displacement at the foot of the slope continues to increase due to the weight of the bridge and the load on the upper vehicles, and after increasing to about 2 mm, it tends to stabilize again. In addition, comparing the horizontal displacement of measurement points 1 and 4, it can be found that due to the proximity of measurement point 4 to the bridge pile, it is greatly affected by the bridge pile, and its displacement is also correspondingly large. Comparing the horizontal displacement curves of measuring points 2 and 5, as well as measuring points 3 and 6, a similar trend can be observed. The maximum displacement occurs at measuring point 6, which is the upper part of the vertical point of the bridge pile and pipeline trench, with a maximum displacement of 4 mm. A displacement of 4 mm does not pose a significant threat to the safety of the pipeline trench, but it is necessary to prevent sudden increase in displacement at geological conditions.

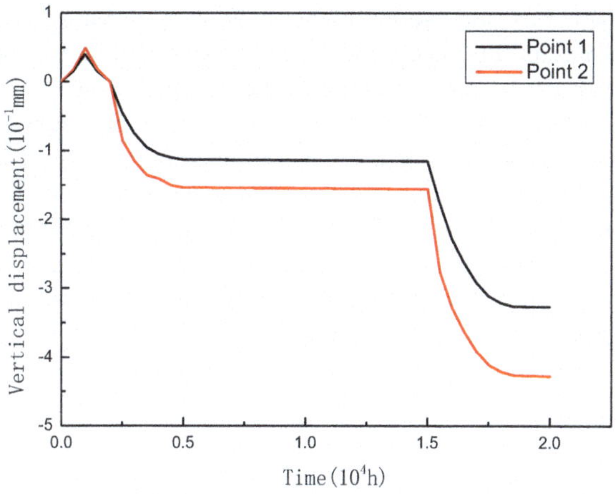

Fig. 6. Vertical displacement curves of measurement points 1 and 4

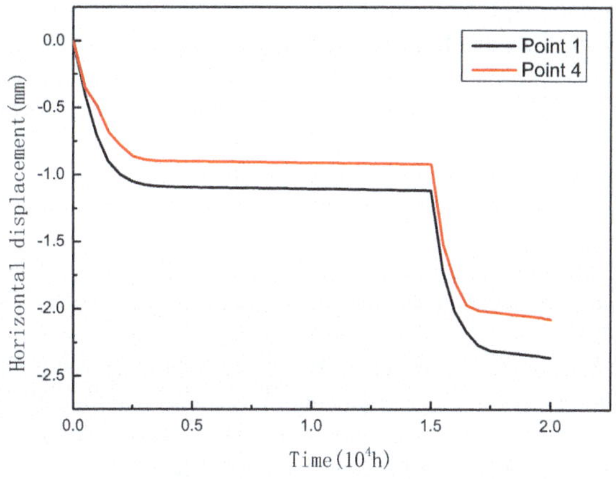

Fig. 7. Horizontal displacement curves of measurement points 1 and 4

4 Conclusion

This paper uses the simulation software FLAC3D to conduct three-dimensional nonlinear numerical calculation analysis on the stress and displacement status of the pipeline of the second-line West-East gas pipeline project before and after bridge pile construction. The stress and displacement trends of key parts and processes were recorded throughout the construction process, revealing the redistribution of stress in the soil during the construction process. The stability of the soil around the pipe trench during excavation was analyzed from two aspects: strength control and deformation control. Based on the calculation results, the following conclusions can be drawn:

(1) After the construction and use of bridge piles, there is little change in the stress inside the soil around the pipe trench, with a maximum stress value of 0.1 MPa, which does not pose a significant threat to the stability of the soil.
(2) The vertical displacement of the soil around the pipe trench shows settlement, with a maximum settlement of less than 1 mm. The maximum horizontal displacement is 6 mm, which occurs above the nearest position of the bridge pile in the pipe trench. It has a certain impact on the pipe trench, but will not cause major safety hazards.
(3) Detailed design should determine the bridge pile type, safety spacing between bridge pile and pipeline. Monitoring points near the pipeline should be used during construction process to monitoring pipeline displacement.

References

1. Gao, Z.Y., Zhang, H.Y., Gao, P.: New progress in China's oil and gas pipeline construction in 2022. Int. Pet. Econ. **31**(3), 16–23 (2023)
2. Law of the People's Republic of China on the Protection of Petroleum and Natural Gas Pipelines. Law Publishing, Beijing (2010)
3. Highway Law of the People's Republic of China. China Legal Publishing House, Beijing (2011)
4. Alamri, A.H.: Localized corrosion and mitigation approach of steel materials used in oil and gas pipelines–An overview. Eng. Fail. Anal. **116**, 104735 (2020)
5. Chen, C., Li, C., Reniers, G., Yang, F.: Safety and security of oil and gas pipeline transportation: a systematic analysis of research trends and future needs using WoS. J. Clean. Prod. **279**, 123583 (2021)
6. Soomro, A.A., Mokhtar, A.A., Kurnia, J.C., Lashari, N., Lu, H., Sambo, C.: Integrity assessment of corroded oil and gas pipelines using machine learning: a systematic review. Eng. Fail. Anal. **131**, 105810 (2022)
7. Peng, S., Zhang, Z., Liu, E., Liu, W., Qiao, W.: A new hybrid algorithm model for prediction of internal corrosion rate of multiphase pipeline. J. Nat. Gas Sci. Eng. **85**, 103716 (2021)
8. FLAC3D: Fast Langragian Analysis of Continua in 3 Dimensions, Version 4.00, User's Manual, Itasca Consulting Group, Minneapolis, Minnesota, USA (2009)

Research on Seismic Performance of Prefabricated Cantilever Beam Steel Structure Nodes

Qianwen Han[1], Ziming Fan[1(✉)], Jiajun Li[1], and Feng Wang[2]

[1] Key Laboratory of Building Structure of Anhui Higher Education Institutes, Anhui Xinhua University, Hefei 230088, China
253760722@qq.com

[2] College of Urban Construction, Anhui Xinhua University, Hefei 230088, China

Abstract. In this paper, the seismic performance of a cantilever beam steel structure with beam-column nodes is investigated.Using ABAQUS software, three types of nodes are modeled and analyzed with and without soft steel dampers as well as ordinary steel structures, and the seismic performance of the three types of nodes is quantitatively analyzed in terms of their hysteresis curves, skeleton curves, stiffness degradation, and other seismic performance. The results show that the nodes with dampers have the strongest mechanical properties in each, the nodes of ordinary steel structure are second, and the nodes without dampers are the worst. It can significantly consume the seismic energy of the building structure and reduce the seismic response in a controlled situation. Therefore, this node is conducive to the popularization and use of assembly structures in high-intensity areas under controlled circumstances, which can effectively improve the seismic performance of assembly structures and provide a reference for assembly building construction in practical application.

Keywords: Soft steel dampers · Cantilever beam steel structure · ABAQUS numerical simulation

1 Introduction

Rigid steel frames are widely used because of their good seismic performance, and beam-column connection nodes play a pivotal role in steel frames. Steel beams in frame structures are connected to columns using cantilevered beam segments, a connection form recommended by the code, and soft steel dampers, as energy-absorbing and vibration-reducing devices, are widely used in high-rise buildings and other fields. The novel combination of an "assembled steel structure + damper" can improve the level of earthquake resistance in the house. At the same time, the soft steel damper is easy to replace and can be adjusted by replacing the damper in case of damage.

Han et al. [1] proposed a static analysis of embedded reinforced nodes with cantilevered beam segments. The number of bolts in the nodes, the thickness of the cantilevered flange, the width of the flange, and the length of the cantilevered beam were

B. Yuan et al. (Eds.): ICCASE 2024, SUCI, pp. 21–30, 2024.
https://doi.org/10.1007/978-3-031-76102-7_3

changed. It is concluded that the number of flange bolts has less influence on the mechanical properties of the node. The thickness of the cantilevered beam section's outreach flange and the width of the cantilevered beam section's flange have more influence on the mechanical properties of the node, and the length of the cantilevered beam section also has a certain influence on the mechanical properties of the node. Appropriately increasing the length of the cantilever beam section increases the energy dissipation capacity, load-carrying capacity, and stiffness of the node.

Huang et al. [2] investigated the seismic performance of shaped nodes with cantilevered beam segments connected to columns with weak axes. They used ABAQUS software to carry out a proposed static analysis of the shaped nodes, a study of damage patterns, hysteresis properties, stiffness degradation, load carrying capacity, and ductility properties of shaped nodes with cantilevered beam segments, and compared with standard shaped nodes. The results show that the initial stiffness and load-carrying capacity of the shaped nodes with cantilever beam segments are reduced compared to the standard-shaped nodes, but the energy dissipation capacity and ductile properties are significantly improved.

Watakabe et al. [3] proposed a tubular viscoelastic damper; it is proved through the test that this kind of damper has the function of displacement amplification, and even if its time-course response acceleration is high, its structural deflection is small, so the deformation through the action of this damper is very small. It shows that tubular viscoelastic dampers acting in assembled buildings have good seismic performance through displacement amplification, and at the same time, the control effect on the seismic response of assembled buildings is obvious.

Baharmast et al. [4] numerically analyzed the seismic performance of four types of strengthened and weakened beam-column splice nodes with cantilevered beam sections. The results show that strengthened nodes increase the load-carrying capacity and bending capacity and improve the hysteretic properties as well as the energy dissipation capacity. In contrast, circular weakened nodes have lower load-carrying capacity compared to specimens with the same geometrical characteristics due to the reduced cross-section.

The above domestic and international studies are the seismic performance of nodes and dampers alone, and the combination of the two is less studied, so this paper will be the combination of steel structure and soft steel dampers for research. According to the seismic performance of cantilever beam steel nodes with and without soft steel dampers and ordinary steel nodes, the strength, stiffness, and equivalent viscous damping coefficient of the three types of nodes are compared to provide a reference for the design of the node and building construction.

2 Design of Assembled Beam-Column Nodes

In this paper, the node adopts a prefabricated steel structure, the beam height of the specimen is taken as 300 mm, the thickness of the web plate is taken as 8 mm, the effective length of the beam is taken as 1100 mm, the thickness of the flange of the specimen is 12 mm, and the thickness of the flange splicing plate is 8 mm, and all the specimens are made of 10.9-grade M20 high-strength bolts, the holes are drilled into holes at $d = 21.5$ mm, and the other steels are made of Q235B. The specimen constructional form and sectional The specimen construction form and cross-section size are shown in Fig. 1. The parameters of the steel structure components are shown in Table 1 [5].

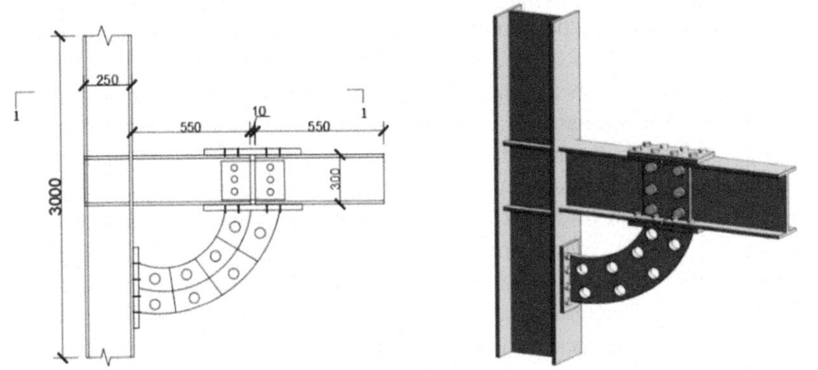

(a) Three-dimensional drawing of assembled steel structure nodes.

(b) Front view of assembled steel structure nodes.

Fig. 1. Specimen size and reinforcement diagram.

Table 1. Mechanical properties of steel components.

Form	Bolt diameter/plate thickness (mm)	Yield strength (MPa)	Maximum intensity (MPa)	Modulus of elasticity (GPa)	Poisson's ratio
Q235 steel plate	8	235	530	200	0.3
10.9 grade high Tensile bolts	21.5	940	1040	200	0.3

3 Finite Element Modeling

3.1 Cell Types and Meshing

In this paper, in the finite element modeling, the beams and columns are structurally meshed with C3D8R and C3D8I solid cells, respectively. The bolts are meshed with C3D8I sweep mesh, and the bolt load is applied on the bolt rods to simulate the pre-tension of the high-strength bolts. Considering the friction between the splice plate and the beam flange and web, contact pairs are set up between the flange splice plate and the beam flange, and the web splice plate and the web, and the friction property is defined, while the friction between the spigot and the hole wall is ignored [6].

3.2 Boundary Conditions and Loading Regime

(1) The degrees of freedom in X, Y, and Z directions are fixed at the top and bottom of the column of the assembled steel structure, with the center reference point of the top section of the column set as RP1 and coupled with the top surface of the column, The corresponding vertical load is applied in the negative direction along the Y axis, and the center reference point of the bottom section of the column is set as RP2 and coupled with the bottom surface of the column. Similarly, the lateral direction of the assembled rectangular beam end is also fixed, and the center reference point of the beam end section is set as RP3, which is coupled with the beam surface.
(2) The high-strength bolt screw and hole wall constraint is defined as binding, and the corresponding bolt preload is applied at different analysis steps, assigned to the bolt cross-section, and the direction is set. At the same time, the normal behavior contact is defined as a hard contact, the constraint enforcement method is defaulted, and separation after contact is allowed; the tangential behavior contact is defined as a penal contact, the friction directionality is isotropic in each direction, the friction coefficient is 0.3, and there is no restriction on the shear stress boundaries.
(3) Using the controlled displacement loading method, the beam-column node displacement loading curves are shown in Fig. 2.

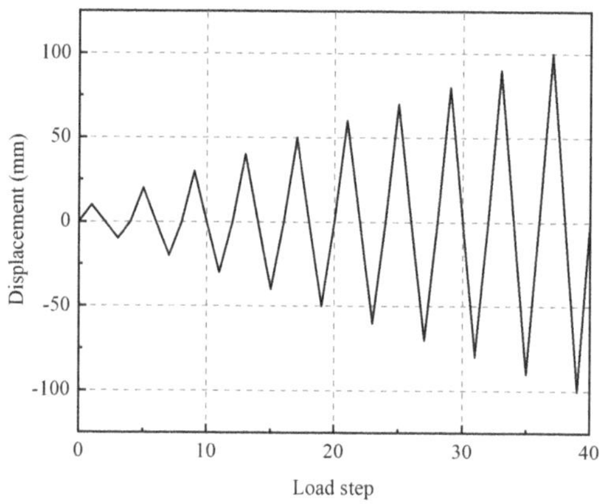

Fig. 2. Displacement loading curve.

4 Analysis of Finite Element Simulation Results

According to the simulation of ABAQUS finite element software, the nodes of cantilever beams of assembled steel structure installed with soft steel dampers are defined as controlled, and the nodes without soft steel dampers are defined as uncontrolled. Since the numerical simulation is an analysis in an ideal state, in the actual test process, there will be factors affecting the test results, such as friction between bolts and concrete holes, caulking between beams and columns, and so on. For this reason, tests will be conducted and comparatively analyzed in the subsequent study to verify the correctness of the theoretical calculations. The ordinary steel structure nodes are defined as ordinary nodes. The relationships between the hysteresis curve, skeleton curve, stiffness degradation curve, and equivalent viscous damping coefficient of the three types of nodes are shown below.

4.1 Hysteresis Curve

The hysteresis curve, i.e., the load-displacement curve, can well reflect the relationship between the bearing capacity and deformation of the structural members under the action of load displacement and can comprehensively reflect the seismic characteristics of the nodes [7]. For practical applications in building construction, the shape and characteristics of the hysteresis curve can be used to assess the seismic performance of a structure. The larger area of the hysteresis curve indicates that the structure has strong seismic performance, and the hysteresis curves of the three types of nodes are shown in Fig. 3.

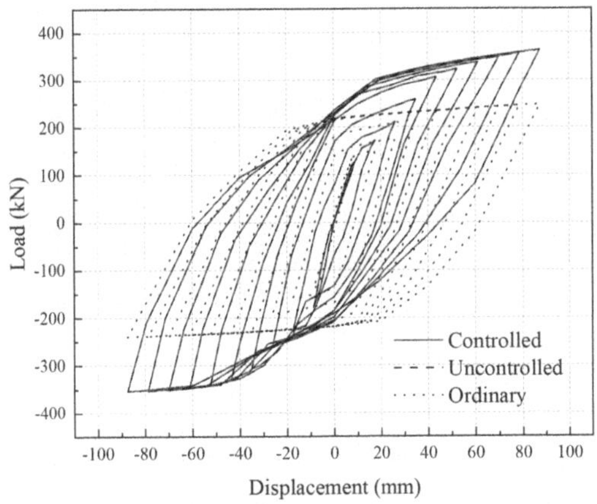

Fig. 3. Hysteresis curve.

The following pattern can be obtained from Fig. 3.

(1) The hysteresis curves of the three kinds of nodes are similar, all of which are shuttle-shaped. The hysteresis curves are fuller, the controlled load intensity is the largest, and the uncontrolled load intensity is the smallest. With the increase in load intensity, all three nodes begin to enter the yielding stage, at which time the stiffness of the structure starts to degrade.

(2) The hysteresis curve area of ordinary steel structure nodes is larger than that of uncontrolled nodes because the cantilever beam bolted steel structure adopts bolted connection, and the bolts will play a certain degree of elastic deformation and slip when the structure is under stress so that they can absorb part of the energy. The ordinary steel structure usually adopts a welded connection. The welded joints have high rigidity during stress, and the energy is difficult to absorb, which leads to a relatively large hysteresis curve area.

4.2 Skeleton Curve

Skeleton curves whose slopes indicate stiffness can better reflect the relationship between loads and displacements under various stages of the repeated action process of the member [8]. Among them, the controlled and uncontrolled skeleton curves are shown in Fig. 4.

The following pattern can be obtained from the skeleton curve of Fig. 4:

(1) In the preloading stage, all three nodes are located in the proportional stage, also called the elastic stage, where the load is proportional to the displacement. As the cyclic action increases step by step, the slopes of the skeleton curves of the three kinds of nodes begin to decrease, but the slopes under control are still larger than the slopes under uncontrolled.
(2) Through the comparison of the skeleton curves of the three nodes, it is found that the positive and negative trends of the controlled, uncontrolled, and ordinary steel nodes are approximately the same. The skeleton curves of the three nodes obtained by finite element simulation do not have a descending section, which is due to the fact that the finite element software simulation ignores the influence of the actual weld defects. For the controlled and uncontrolled nodes, the skeleton curve has no falling section because there is a 10 mm gap in the middle of the cantilever beam steel structure. The gap between the bolt and the hole wall and the gap between the steel plate and the beam column make the members not fully involved in the load during the loading period; for the skeleton curve, the ordinary steel structure has no falling section because the ordinary steel structure is usually used in the welded connection mode, and there is a weld seam. The decrease in the load capacity of the beam at the end of the actual test is caused by the cracking of the weld, which indicates that there is a certain error in the numerical simulation and the test. For this reason, tests and comparative analyses will be carried out in future research in order to verify the correctness of the theory.

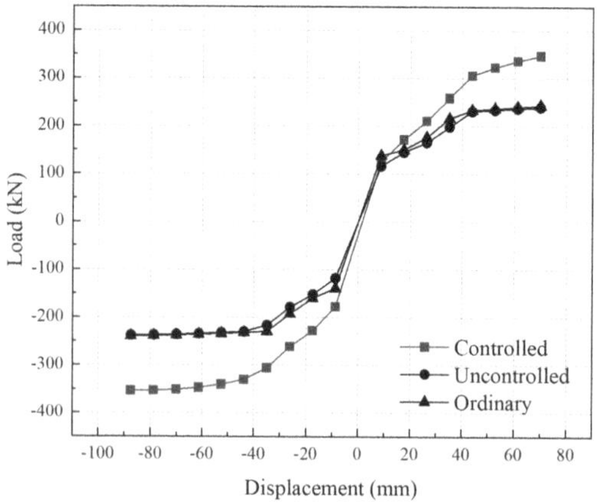

Fig. 4. Skeleton curve.

4.3 Stiffness Degradation Analysis

When the peak load remains consistent, the corresponding displacement increases with the increase in the number of cycles, which is, in essence, the cumulative damage of the structure under cyclic repetitive loading [9]. The stiffness of the specimen can be expressed by the cut line stiffness. In the practical application of building construction, stiffness degradation curves can be used to assess the durability of a structure. The flatter the stiffness degradation curve, the better the durability of the structure. The stiffness degradation curve of each specimen is shown in Fig. 5, and the stiffness degradation coefficient of the specimen can be calculated according to the following formula:

$$K\mathrm{i} = \frac{|+Fi| + |-Fi|}{|+X\mathrm{i}| + |-X\mathrm{i}|} \quad (1)$$

where, $+F\mathrm{i}$, $-F\mathrm{i}$—-value of the load at the i-th forward and reverse peak point;
$+X\mathrm{i}$, $-X\mathrm{i}$—displacement value of the i-th forward and reverse peak point.

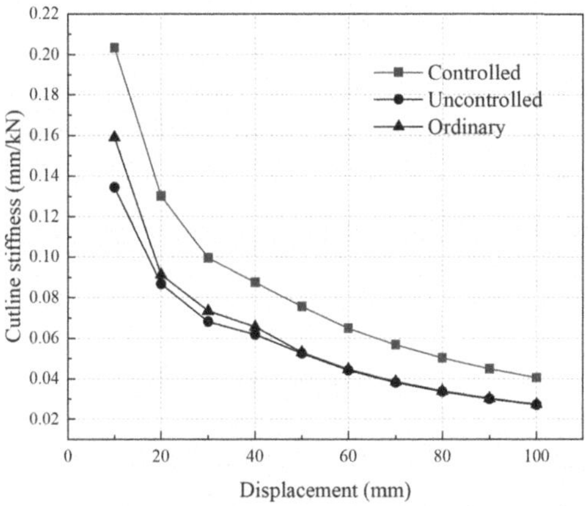

Fig. 5. Stiffness degradation curve.

The stiffness degradation curve in Fig. 5 leads to the following pattern:

(1) In the preloading stage, located in the elastic phase, the three kinds of node stiffness degradation curves are faster, the slope is larger, and the degradation curve is basically the same. However, the stiffness of the controlled is still consistently greater than that of the uncontrolled and ordinary steel structure.
(2) The reason that the hysteresis curve area of ordinary steel structure nodes is larger than that of uncontrolled nodes is that ordinary steel structures usually use welded connections, with welded joints that are fixed and rigid, and nodes with high stiffness. In contrast, in the steel structure with cantilever beam splicing, the cantilever beams and supports are connected flexibly, which enables the structure to undergo relative displacement and deformation under seismic loading, thus dispersing the seismic energy, resulting in a smaller stiffness.

4.4 Equivalent Viscous Damping Factor

The energy dissipation is measured in terms of the area enclosed by the hysteresis curve and is usually expressed in terms of the equivalent viscous damping coefficient, which is calculated according to the method in [10], and the calculation sketch is shown in Fig. 6.

$$\zeta\text{eq} = \frac{1}{2\pi} \cdot \frac{S_{\text{curved line AECF}}}{S\Delta ABO + S\Delta CDO} \tag{2}$$

where: $S_{\text{curved line AECF}}$ is the area enclosed by the hysteresis loop, and $S_{\Delta ABO}$ and $S_{\Delta CDO}$ is the area of the envelope triangle, where the controlled and uncontrolled variation with displacement is shown in Fig. 7.

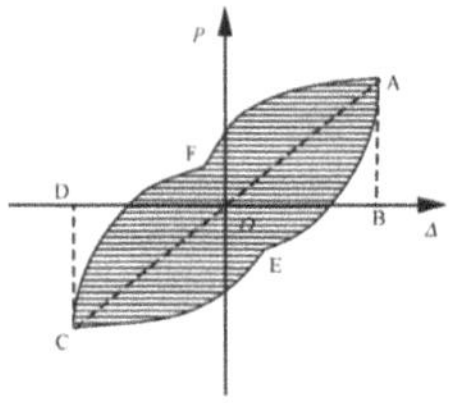

Fig. 6. Specimen energy consumption index.

The following pattern can be obtained by comparing Fig. 7:

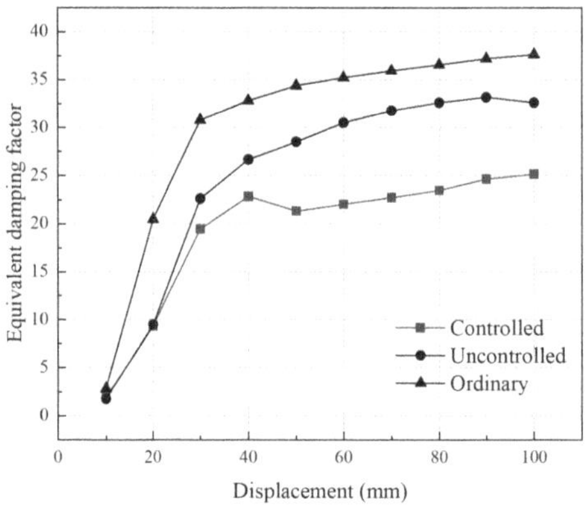

Fig. 7. Equivalent damping factor.

(1) The equivalent viscous damping coefficients of the three kinds of nodes all show the phenomenon of increasing in the early stage and stabilizing in the later stage. The equivalent viscous damping coefficient of the ordinary steel structure is the largest, and the equivalent viscous damping coefficient of the controlled one is the smallest.

(2) The reason that the equivalent viscous damping coefficient of ordinary steel nodes is higher than that of uncontrolled nodes is that the bearing capacity of ordinary nodes decreases too fast with the increase of loading displacement, which makes the triangular area of the denominator in the formula too large. Due to the poor ductility of ordinary steel structures, it enters into the plastic phase when loaded with smaller displacements, which increases energy consumption.
(3) The reason why the equivalent viscous damping coefficient of the controlled node is the lowest is that the equivalent viscous damping coefficient is also related to the stiffness of the structure, and the greater the stiffness, the smaller the equivalent viscous damping coefficient of the structure. Because a structure with a large stiffness produces a larger inertial force under seismic action, it requires a larger damping force to counteract it.

5 Conclusion

In this paper, the node mainly used is the cantilever beam steel structure through the steel plate and soft steel damper welding. Choosing the same geometry, materials, loads, etc., according to the presence or absence of soft steel damper as well as ordinary steel structure of the three forms of the node using ABAQUS on the model to compare and analyze, can be drawn the following conclusions:

(1) The use of an open-hole fan-shaped soft steel damper. This soft steel damper is simple in form, can avoid stress concentration and thus damage [11], is easy to replace, and has good energy dissipation capacity. It can be widely used in assembled buildings.
(2) Through Abaqus software simulation, it is found that the hysteresis curves of the three kinds of nodes are pike-shaped with certain energy dissipation characteristics; the skeleton curves and stiffness degradation capacity are all controlled maximum and uncontrolled minimum. The energy-consuming capacity displacement is the largest in ordinary steel structures before 70 mm, and the displacement is controlled maximum after 70 mm.

The node in this paper improves the seismic performance of the overall member through the connection of a soft steel damper. At the same time, the use of finite element analysis also proves that a soft steel damper has a very good seismic performance and can reduce the seismic response, so this node in the presence of control is conducive to the popularization of the use of the assembly structure in the high-intensity areas, which can effectively improve the safety of the assembled structure, and provide a reference for the construction of the building in the practical application.

Acknowledgments. This work was supported by the Key Laboratory of Building Structures in Anhui Ordinary Colleges and Universities Laboratory (Anhui Xinhua College) under Grant (KLBSZD202204); the Natural Science Research Program of Higher Education Institutions in Anhui Province under Grant (2022AH051869).

References

1. Han, M., Shi, J., Shi, Z., Wang, Y.: Performance analysis of assembled embedded reinforced nodes with cantilever beam segments. Steel Struct. **38**(01), 1–12 (2023). (in Chinese and English)
2. Huang, P., Lu, L., Zhang, S.: Finite element analysis of the seismic performance of shaped node weak-axis connections with cantilever beam segment splices. Construct. Technol. **48**(04), 371–374 (2017). (in Chinese)
3. Watakabe, T., Fujita, S., Omi, T., Kurabayashi, H., Ogata, K.: Research and development of vibration control device suitable for detached houses by using displacement amplification mechanism (fundamental characteristic and vibration control performance of pipe-type-damper): (fundamental characteristic and vibration control performance of pipe-type-damper). J. Syst. Des. Dyn. **3**(3), 368–379 (2009)
4. Baharmast, A., Razzaghi, J., Kamouneh, S.: Modification of pre-kobe column-tree moment resisting connection. Asian J. Civ. Eng. **19**(2), 223–237 (2018)
5. Ding, K., Ye, Y., Ma, W.: Seismic performance of precast concrete beam-column joint based on the bolt connection. Eng. Struct. **232**, 111884 (2021). (in Chinese)
6. Ding, K., Wang, K., Liu, Y.: Effects of hygrothermal environment on tensile capacity of adhesive-injected anchor bolt of ALC Panel and ETICS. KSCE J. Civ. Eng. **27**(3), 1107–1117 (2023). (in Chinese)
7. Ye, L.-P., Fang, E.-H.: A review of research on the force performance of steel-bone concrete members. J. Civ. Eng. **33**(5), 1–12 (2000). (in Chinese)
8. Miao, J., Cheng, W.-P., Song, Y.: Finite element analysis of prefabricated partially assembled steel and bone concrete frame beamolumn nodes. J. Dalian Univ. Tech. **54**(05), 568–574 (2014). (in Chinese)
9. Yu, J., Guo, Z., Guo, X.: Seismic performance of new assembled concrete frame beam-column nodes. Eng. Sci. Technol. **50**(03), 209–215 (2018). (in Chinese)
10. JGJ/T 101—2015, Building seismic test procedures. (in Chinese)
11. Li, G., Li, H.-N.: Research on vibration damping performance of new soft steel dampers. Vibr. Shock **03**, 66–72 (2006). (in Chinese)

Numerical Study of the Effect of Tsunami Waves on Offshore Box Beam Bridges

Yunpeng Shen, Minglei Zhang, Zhu Huang(✉), Xinyan Liao, Min Hu, Yuanfan Dai, Qian Li, and Yufei Feng

Power China Chengdu Engineering Corporation Limited, Chengdu 611830, China
Shen_yunpeng.chn@foxmail.com, {p2017250,p2015166,p2020432, 2022181}@chidi.com.cn, ricardo-huang@foxmail.com

Abstract. Coastal bridges are threatened by extreme loads such as typhoons, storm surges, and tsunamis due to their unique location. Ensuring the safety of coastal bridges during extreme weather events is crucial for minimizing loss of life, economic impacts, and expediting post-disaster recovery and reconstruction. A two-dimensional numerical water tank is established in OpenFOAM to simulate tsunami waves using the dam-break wave model. The Reynolds-averaged Navier-Stokes (RANS) equations and k-ε turbulence model are employed. The Volume of Fluid (VOF) method is utilized to capture the water surface. A downsized model at a scale of 1:20 was used to analyze the wave forces acting on the structure in terms of various tsunami like wave and bridge clearance scenarios. The results indicate that the effect of the tsunami like wave on the model can be broken down into three phases: initial impingement, decay and stabilization. The peak force acting on the model varies directly with wave height and inversely with bridge clearance.

Keywords: OpenFOAM · Box girder bridge · tsunami force · dam break wave

1 Introduction

With the steady advancement of the national strategy to become a maritime power and the implementation of the "Belt and Road" initiative, coupled with the rapid economic development, the construction of marine bridges has seen rapid growth, exemplified by structures such as the Zhoushan Cross-Sea Bridge, the Jiaozhou Bay Cross-Sea Bridge, and the Xinglin Bridge. Box girder bridges, known for their high torsional rigidity, light weight, and fast construction speed, are used extensively in the construction of bridges across the sea.

However, compared to traditional inland bridges that primarily bear conventional loads such as vehicles, trains, and pedestrians, marine and coastal bridges not only endure these typical loads but also face complex wave loads and the threats of extreme loads such as typhoons, storm surges, and tsunamis.

Tsunamis, among the most destructive natural disasters in the ocean, exert tremendous impact forces on bridge structures upon occurrence, leading to structural damage

B. Yuan et al. (Eds.): ICCASE 2024, SUCI, pp. 31–39, 2024.
https://doi.org/10.1007/978-3-031-76102-7_4

and significantly affecting transportation in coastal areas, as evidenced by events such as the 2020 Tonga Tsunami, Indian Ocean tsunami of 2004 [1] and the Japan tsunami 2011 [2]. Furthermore, damage to bridge structures hampers post-disaster evacuation and logistics operations. Therefore, studying the mechanisms of tsunami forces on near-sea box girder bridges is of significant importance for mitigating the impact of tsunami disasters and implementing post-disaster rescue and evacuation efforts.

Scholars both domestically and internationally have conducted wide-ranging investigation on the impact of tsunami like waves on sea-crossed bridges. In the studies of experimental, Nakao et al. [3] investigated the influence of different tsunami waves and bridge types on tsunami impact forces. Xu et al. [4] examined the regularity of the tsunami wave impact forces on box girders, while Farvizi et al. [5] examined the features of impact force of tsunami like waves on an inclined box girder. In numerical simulation studies, Xu et al. [6] makes a comparison of the forces caused by Stokes-5th order waves, freak waves and solitary waves. The hydrodynamic effects between a T-deck and a box deck under strange waves are investigated, while Motley et al. [7] researched the characteristics of tsunami wave impact forces on a inclined bridges. Wu et al. [8] used Smoothed Particle Hydrodynamics (SPH) to analyze bridge geometries and their effects on solitary wave forces, and performed a vulnerability assessment that resulted in recommended geometries for box beam bridges used to mitigate coastal hazards. In theoretical calculations, Ze Shi et al. [9] used the potential flow theory to establish an analytical model for the full second-order wave loading of a T-pontoon bridge, investigated the interaction between waves and the T-pontoon bridge, and found that the cantilever plate has a significant protective effect on the inner web girders. The above research mainly studied the T-girder bridges subjected to extreme wave action, and did not carry out detailed research on the superstructure stresses of box bridges widely used in China under extreme wave action.

In numerical simulations, all forces on the structure are the integral of the pressure applied to the bridge deck, thus the shape of the structure affects the magnitude of the force. The results of previous studies with T-girders may not be applicable to box girders, so it is necessary to study box girder superstructures subjected to extreme wave action. This study utilized OpenFOAM software to establish a two-dimensional numerical wave tank, employing the dam-break wave model to investigate the effects of different wave heights and bridge clearances on tsunami impact forces on coastal box girder bridges.

2 The Numerical Model Setup

2.1 Governing Equations and Boundary Conditions

Modeling of Reynolds Average Stress and the standard k-ε turbulent model are adopted in the numerical model, with the assumption of an incompressible fluid. The governing equations can be expressed as:

$$\frac{\partial \overline{u}_i}{\partial x_i} = 0 \tag{1}$$

$$\rho \frac{\partial \overline{u}_i}{\partial t} + \rho \overline{u}_j \frac{\partial \overline{u}_i}{\partial x_j} = \frac{\partial \overline{p}}{\partial x_i} + \mu \frac{\partial^2 \overline{u}_i}{\partial x_j \partial x_j} - \frac{\partial \rho \overline{u_i' u_j'}}{\partial x_j} \tag{2}$$

Among these, in the direction of i, $\overline{u}_i$ represents the average velocity and u_i' denotes the fluctuating component of velocity, while $\overline{p}$ stands for the mean pressure. In the i direction, if u_i is the velocity component, then $u_i = \overline{u}_i + u_i'$. In Eq. (2), the term for the Reynolds stress can be expressed as

$$-\rho\overline{u_i'u_j'} = \upsilon_t\rho\left[\frac{\partial\overline{u}_i}{\partial x_j} + \frac{\partial\overline{u}_j}{\partial x_i}\right] - \frac{2}{3}k\rho\delta_{ij} \quad (3)$$

In which k is the turbulent kinetic energy. The turbulent kinetic energy k and dissipation ε were solved by the standard $k-\varepsilon$ model.

$$\frac{\partial k}{\partial t} + \frac{\partial k\overline{u}_j}{\partial x_j} = \frac{\partial}{\partial x_j}\left[\left(\upsilon + \frac{\upsilon_t}{\sigma_k}\right)\frac{\partial k}{\partial x_j}\right] + \frac{P_k}{\rho} - \varepsilon \quad (4)$$

$$\frac{\partial \varepsilon}{\partial t} + \frac{\partial \varepsilon\overline{u}_j}{\partial x_j} = \frac{\partial}{\partial x_j}\left[\left(\upsilon + \frac{\upsilon_t}{\sigma_k}\right)\frac{\partial \varepsilon}{\partial x_j}\right] + C_{\varepsilon 1}\frac{\varepsilon}{k\rho}P_k - C_{\varepsilon 2}\frac{\varepsilon^2}{k} \quad (5)$$

where

$$P_k = -\rho\overline{u_i'u_j'}\left(\frac{\partial\overline{u}_j}{\partial x_i}\right)$$

The turbulent eddy viscosity υ_t is calculated using the formula:

$$\upsilon_t = C_\mu\frac{k^2}{\varepsilon}$$

In this paper, the aforementioned coefficients are $\sigma_k = 1.0$, $\sigma_\varepsilon = 1.3$, $C_{\varepsilon 1} = 1.44$, $C_{\varepsilon 2} = 1.92$, $C_\mu = 0.09$. To ensure stability during numerical computations, the computed time step δt is adjusted by the solver and the maximum Courant number is taken as 0.5. Table 1 shows the boundary conditions.

Table 1. Setting of model boundary conditions.

Boundary	α_1	epsilon	k	p_{rgh}	U
Flume Bottom	zeroGradient	epsilonWallFunction	kqRWallFunction	fixedFluxPressure	fixedValue
Left wall	zeroGradient	epsilonWallFunction	kqRWallFunction	fixedFluxPressure	fixedValue
Right wall	zeroGradient	epsilonWallFunction	kqRWallFunction	fixedFluxPressure	fixedValue
Atmosphere	inletOutlet	zeroGradient	zeroGradient	totalPressure	pressureInletOutletVelocity
Bridge model	zeroGradient	epsilonWallFunction	kqRWallFunction	fixedFluxPressure	fixedValue
Front and back wall	empty	empty	empty	empty	empty

2.2 The Layout of Numerical Models

In this study, that a two-dimensional (2D) numerical model is developed to investigate the effect of waves on bridge decks. The use of a 2D numerical model to predict wave forces is cost-effective in engineering terms and meets accuracy requirements. The layout of the numerical flume is shown in Fig. 1. For the Cartesian coordinate system, the origin of the coordinate system is positioned in the bottom left corner of the numerical flume. The total length of the flume is 30 m, of which the reservoir area is 12 m. The bridge model is located at 19.6 m. The specific parameters of the bridge model are shown in Fig. 2.

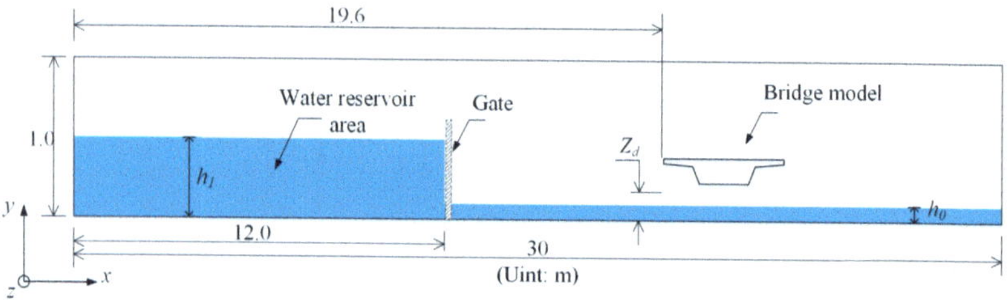

Fig. 1. Numerical flume arrangement

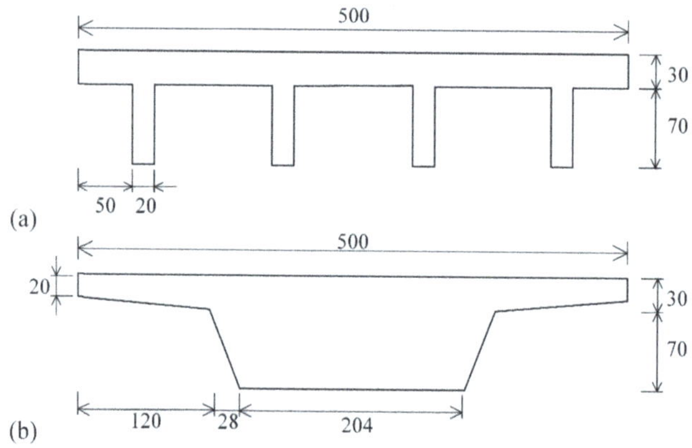

Fig. 2. Bridge model configuration (a) T-beam; (b) Box girder

2.3 Computational Mesh Division and Validation

As shown in Fig. 3, the computational grid is generated using a hierarchical refinement strategy, comprising background grid, primary computational zone grid, and near-model grid. The mesh in the 5.0 m long and 0.8 m high main computing zone was redefined to obtain detailed information about the flow field during wave impact. The cell lengths of the background and local refinement zones were 5×10^{-2} m and 6.25×10^{-3} m, respectively. The computer is configured with two Xeon Gold 5218R processor CPUs and 128 G of RAM. The cell lengths, total number of mesh and final computation time for different meshes are shown in Table 2.

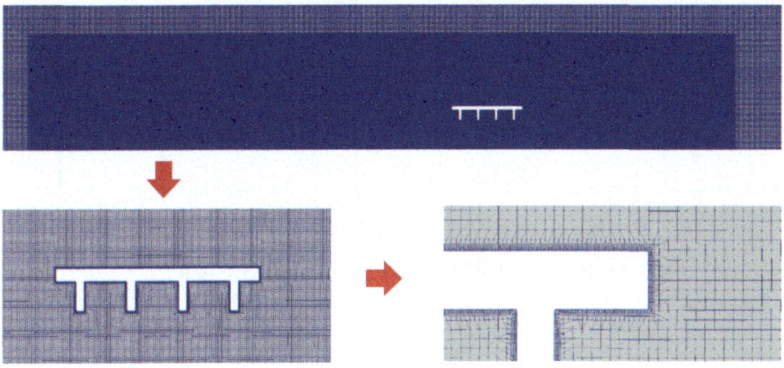

Fig. 3. Calculation mesh

Table 2. Grid Parameters

Mesh size	Main computing area mesh (m)	Total number of grids	Calculation time (h)
Coarse mesh	0.01	312482	4.98
Medium mesh	0.005	431324	7.37
Fine mesh	0.0025	909764	30.69

A comparison of the tsunami impact forces on the T-beam bridge (Fig. 2 (a)) using coarse, medium, and fine grid sets are as shown in Fig. 4. Nevertheless, there was no significant improvement in the computational results with further mesh refinement. A balance between accuracy and computational consumption was achieved by choosing a grid size of 0.0125 m. Figure 5 presents a comparison between the results obtained using the medium grid and the experimental results from Nakao et al.[3] and Motley's[7] numerical results. The comparison demonstrates the good accuracy of the computational results obtained using the proposed model.

3 Results

The analysis is conducted using a box girder model. Figure 6 depicts the time history of tsunami impact forces for $h_1 = 0.65$ m and $h_0 = 0.1$ m. It can be seen that the force exerted by the tsunami like wave on the deck can be broken down into three parts: the initial impact, decay, and stabilization. In the initial impact phase, the tsunami impact force is maximal, gradually decreasing thereafter to reach a stable range.

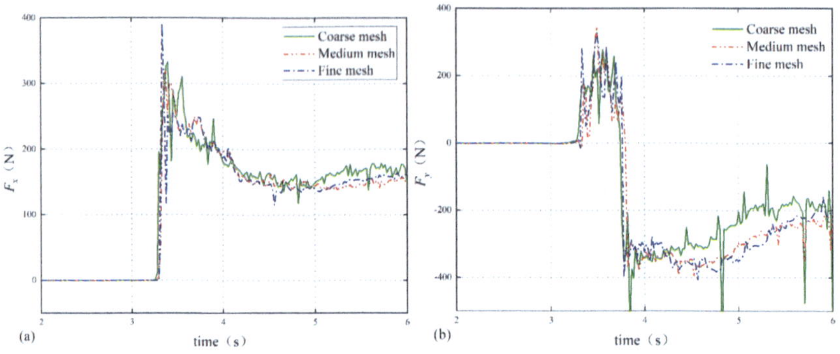

Fig. 4. Verification of mesh convergence

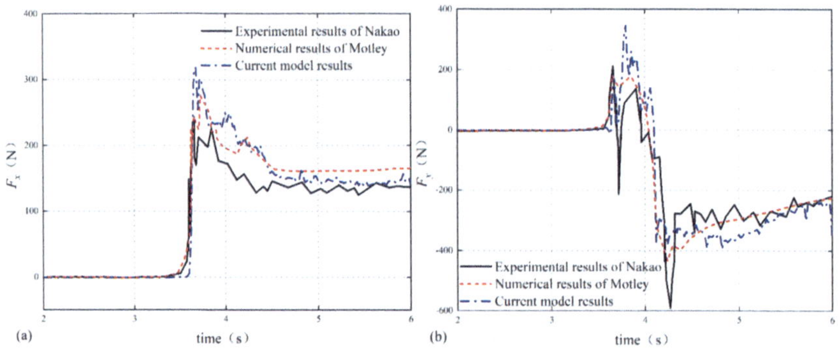

Fig. 5. Validation of calculation results

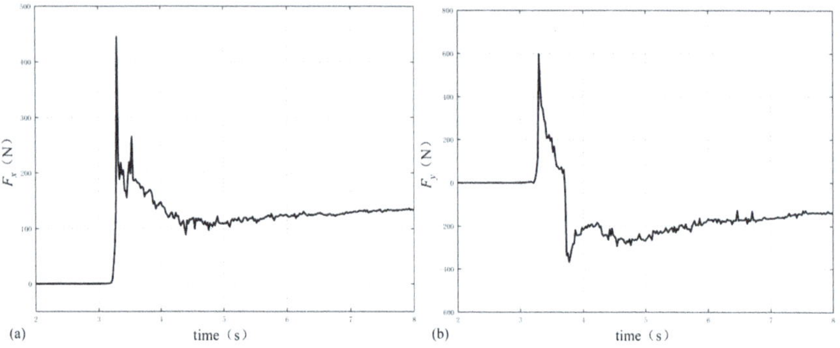

Fig. 6. Time course of the tsunami like wave impact force on box girder (a) horizontal; (b) vertical

3.1 Influence of Water Depth H1 in the Upstream Reservoir Area

Figure 7 illustrates the variation in peak force on the bridge deck with changes in the upstream reservoir depth h_1. It can be observed that the magnitude of the peak force increases with growing h_1. Furthermore, it is evident that, under the same conditions, the maximum value of the vertical wave force is larger than the maximum value of the horizontal wave force. This is attributed to the upstream reservoir serving as the energy storage area for the tsunami wave, with a larger h_1 implying a greater gravitational potential energy. Consequently, the increase in h_1 results in higher wave height and velocity of the dam-break wave, leading to greater impact force upon contact with the box girder.

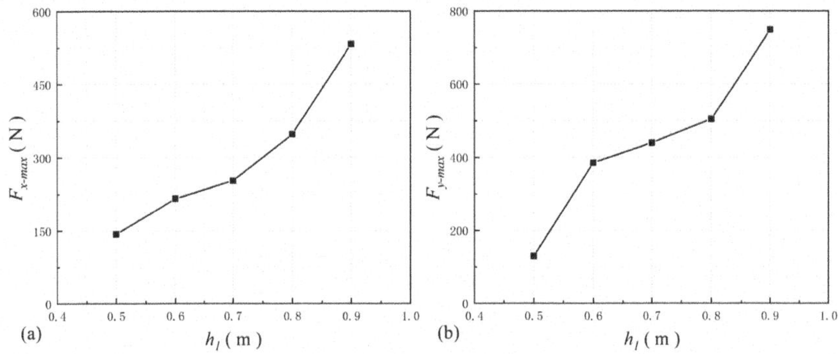

Fig. 7. (a) Horizontal force; (b) Variation of vertical force with water depth in the upstream storage area.

3.2 Influence of Girder Clearance

Figure 8 illustrates the variation of tsunami like wave impact force peaks on the bridge deck with changes in bridge clearance. It can be seen that the tsunami like wave impact force decreases with increasing bridge clearance. This is because as the clearance increases, the area of the girder subjected to the tsunami wave decreases, which leads to a diminution of the impact force. The horizontal peak force exhibits a fluctuating decreasing trend, attributed to the complex water-air interaction during impact. Conversely, the vertical peak force demonstrates a linear decreasing trend. Additionally, the influence of bridge clearance on vertical peak force exceeds that on horizontal peak force.

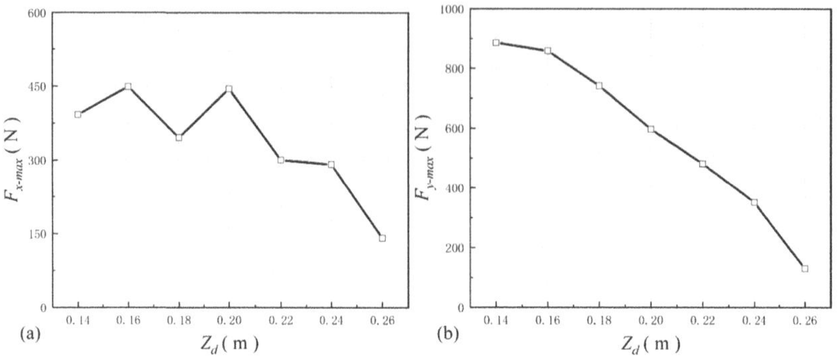

Fig. 8. (a) Horizontal force; (b) Variation of vertical force with bridge clearance.

4 Conclusion

In this paper, the wave force law of tsunami wave in box girder bridge superstructure is investigated, based on OpenFOAM, a 2D numerical wave model of box girder superstructure is established, a dam break wave is used to simulate the tsunami like wave, numerical calculations of the interaction between the tsunami like wave and the box girder superstructure are carried out, and parametric studies of the wave force in the different tsunami like wave heights and the clear height of the box girder is carried out, and the main conclusions are as follows:

(1) Tsunami impact forces on nearshore box girder bridges are be broken down into three stages: initial impact, decay, and stabilization, with the tsunami force being maximal during the initial impact stage, coastal bridges are mostly damaged during this stage.
(2) The peak tsunami impact force grows with the depth of the upstream reservoir, and under the same conditions, the vertical peak force exceeds the horizontal peak force. Higher vertical uplift forces make the bridge superstructure more susceptible to displacement damage, thereby destroying the coastal bridge structure.
(3) The peak tsunami impact force decreases with increasing bridge clearance. While the horizontal peak force exhibits a fluctuating decreasing trend, the vertical peak force shows a linear decreasing trend. In addition, bridge clearance has a much greater affect on the vertical peak force than what it does on the horizontal peak force. Coastal bridges can therefore minimize damage caused by extreme waves by increasing bridge clearances.

References

1. Ghobarah, A., Saatcioglu, M., Nistor, I.: The impact of the 26 December 2004 earthquake and tsunami on structures and infrastructure. Eng. Struct. **28**(2), 312–326 (2006). https://doi.org/10.1016/j.engstruct.2005.09.028
2. Unjoh, S.: Tsunami damage to bridge structures in Rikuzen-Takada City and the emergency road network recovery actions. In Proceedings of the International Symposium on Engineering

Lessons Learned from the 2011 Great East Japan Earthquake, pp. 1429–1438 (2012). https://www.jaee.gr.jp/event/seminar2012/eqsympo/pdf/papers/157.pdf
3. Nakao, H., Zhang, G., Sumimura, T., Hoshikuma, J.: Effect of cross-sectional configuration of superstructure on hydrodynamic behavior of bridges subjected to tsunami-induced loading. J. Japan Soc. Civil. Engineers, Ser. A1 **69**(4), I_42-I_54 (2013). https://doi.org/10.2208/jscejseee.69.I_42
4. Xu, Z., Melville, B., Nandasena, N.A.K., Whittaker, C., Shamseldin, A., Farvizi, F.: Tsunami loads on slab bridges. Coast. Eng. **165**, 103853 (2021). https://doi.org/10.1016/j.coastaleng.2021.103853
5. Farvizi, F., Melville, B.W., Shamseldin, A.Y., Shafiei, S., Hendi, E.: Experimental investigation of tsunami bore-induced forces and pressures on skewed box section bridges. Ocean Eng. **224**, 108730 (2021). https://doi.org/10.1016/j.oceaneng.2021.108730
6. Xu, G., Xie, W., Xu, Y., Xue, S., Chen, X., Ma, G.: Comparative study on hydrodynamic characteristics of T-deck and box-deck bridges under freak waves. Ocean Eng. **299**, 117301 (2024). https://doi.org/10.1016/j.oceaneng.2024.117301
7. Motley, M.R., Wong, H.K., Qin, X., Winter, A.O., Eberhard, M.O.: Tsunami-induced forces on skewed bridges. J. Waterw. Port Coast. Ocean Eng. **142**(3), 04015025 (2016). https://doi.org/10.1061/(ASCE)WW.1943-5460.0000328
8. Wu, G., Garlock, M.: Investigating the effects of box girder bridge geometry on solitary wave force using SPH modeling. Coast. Eng. **187**, 104430 (2024). https://doi.org/10.1016/j.coastaleng.2023.104430
9. Shi, Z., Cong, P., Teng, B.: Second-order wave loads on a floating bridge of T-type cross section. Appl. Ocean Res. **142**, 103851 (2024). https://doi.org/10.1016/j.apor.2023.103851

Load Limit Analysis Method for Highway Bridge Structures After Static Load Test Verification

Mengdie Chen, Ning An, and Songhui Li(✉)

Shandong University of Science and Technology, Qingdao, China
songhui.l@163.com

Abstract. In China, load testing is one of the important means to evaluate the safety and applicability of in-service bridges. However, the current evaluation regulations do not address the issue of bridge load limit analysis after static load test verification. To solve the problem, a highway bridge load limit analysis model considering the left truncated distribution characteristics of resistance is proposed for the verified short- to medium-span simply-supported RC bridges. Firstly, the live load to dead load ratios of RC beam bridges are introduced as the basic parameter in truck weight limit analysis, and the left-truncated distributions of resistances are constructed based on the left-truncated values of resistances relative to various test load to design load ratios in static load testing of highway bridges. Given that the resistance and dead load effect are independent random variables, an analytical model for back-calculating truck weight limit factors of highway bridges under different load levels is proposed in accordance with simplified truck weight limit model as used before. On the basis of this, the effect of test load to design load ratios on truck weight limit factors is discussed for bridges designed according to three bridge design codes of 85, 04 and 18 editions. The calculation results indicate that for the same bridge, the higher the load efficiency of the static load test used in the test, the larger the corresponding load limiting coefficient. The research results can provide scientific basis for formulating load limiting strategies for highway bridge structures after experimental verification.

Keywords: Highway bridge structures · Bridge load limit · Left truncated distribution · Reliability theory · Analysis model

1 Introduction

With the continuous development of the transportation industry, the number of cargo trucks has been increasing year by year [1]. In recent years, overloading transportation has become one of the main factors causing bridge collapse accidents. To solve the problem of bridge load limitation, scholars at home and abroad have conducted in-depth research work. Domestic scholars have studied the limit load values of bridges from the perspective of reliability [2–6], proposing load limits for highway bridges under different technical conditions [7], and some scholars have verified the adaptability of bridge load

B. Yuan et al. (Eds.): ICCASE 2024, SUCI, pp. 40–46, 2024.
https://doi.org/10.1007/978-3-031-76102-7_5

limit values [8, 9]. Foreign scholars James et al. [10] proposed a new bridge load limit formula (TTI formula) using the same overstress criterion as the bridge load limit formula (Formula B). Ghosn et al. [11] applied reliability theory to calibrate Formula B and provided a bridge load limit formula with a target safety level. However, in existing research, the issue of load limit values for bridges after static load test verification has not been addressed. The analysis of load limit values for such bridges has brought great difficulties and has become a technical problem that has troubled the maintenance and testing departments of highway bridges for many years [12].

Based on existing research, this article mainly solves the problem of load limit values for reinforced concrete beam bridges that have no significant degradation in resistance and have been verified through static load tests. Firstly, a simplified analysis model [4] for the limit load values of existing highway bridges is applied. By introducing the left truncated distribution of resistance, a bridge limit load analysis model considering the efficiency of static load testing is proposed. On this basis, with the static load test efficiency as the basic control parameter, the limit load values of bridges designed according to the "Design Specification for Highway Reinforced Concrete and Pre stressed Concrete Bridges and Culverts" (JTJ 023-85, JTG D62-2004, and JTG 3362-2018) (hereinafter referred to as the 85, 04, and 18 specifications) [13–15] are calculated, providing reliable theoretical basis for relevant departments to determine the limit load values of bridge structures after static load test verification.

2 Left Truncated Distribution Model of Resistance

After verification by static load tests, the resistance of a bridge must be a random variable greater than a certain value, which can be described by a left truncated distribution. If the resistance is known to be greater than a certain value, using a left truncated distribution for research and analysis is more realistic.

2.1 Model Construction

If the initial distribution of structural resistance R is known to be a lognormal distribution, its probability density function and probability distribution function are $f_R(r)$ and $F_R(r)$. If the lower limit value of the component resistance calculated from the static load test is r_L, taking this value as the left truncation point and normalizing it, the left truncation distribution model of the resistance random variable can be obtained, and the corresponding probability density function is

$$f(r|R \geq r_L) = \begin{cases} 0 & r < r_L \\ \frac{f_R(r)}{1-F_R(r_L)} & r \geq r_L \end{cases} \quad (1)$$

In the formula, $F_R(\gamma_L)$ is the distribution function value of the original distribution of resistance R at the left truncation point. Obviously, if the mean and coefficient of variation of the resistance R are known, the probability density function of the left truncated distribution of resistance R corresponding to any r can be inferred. This function is an important prerequisite for constructing a bridge load limit analysis model.

2.2 Calculation of Lower Limit Value of Resistance

If the load effects generated by the test load and the design vehicle load on the bridge loading control section are S_{LT} and S_{Qk} (including impact force), the corresponding static load test load efficiency $\eta_q = S_{LT}/S_{Qk}$. Furthermore, if the dead load standard value effect used in bridge design or actual use is S_{Gk}, referring to the American Bridge Evaluation Guidelines [16], the lower limit value r_L of the bridge's resistance after static load test verification is

$$r_{\mathrm{L}} = S_{\mathrm{Gk}} + S_{\mathrm{LT}} = S_{\mathrm{Gk}} + \eta_{\mathrm{q}} S_{\mathrm{Qk}} \quad (2)$$

Among them, r_L is called verification load or left truncation point of resistance. Only bridges that have passed the test load inspection and are still intact and undamaged can be described for their resistance using the left truncated distribution model.

Referring to existing literature [4], the design live and dead load ratio ($\rho = S_{Qk}/S_{Gk}$) is introduced, and the lower limit value of resistance can be expressed as

$$r_{\mathrm{L}} = S_{\mathrm{Gk}} + \eta_{\mathrm{q}} \rho S_{\mathrm{Gk}} \quad (3)$$

If η_q, ρ, and S_{Gk} are known, r_L can be calculated.

2.3 Model Analysis

If the random variables of the resistance, dead load, and vehicle load effects of the bridge structure are known to be R, S_G, and S_Q, respectively, and the resistance follows a left truncated distribution.

According to the simplified analysis model of bridge load limitation [15], the corresponding functional functions can be simplified as

$$Z = R_T - S_G - S_Q = R_T - S_G - S_{Q\mathrm{th}} \quad (4)$$

Among them, R_T is the random variable obtained after left truncation of R. $S_{Q\mathrm{th}}$ is the standard value effect generated by the load limit of the vehicle on the bridge.

If R_T and S_G are independent of each other and the corresponding probability density functions are $f(r|R \geq r_L)$ and $f_{SG}(s_G)$, the failure probability of the structure can be expressed as:

$$P_f = P(Z < 0) = \iint_{Z<0} f(r|R \geq r_{\mathrm{L}}) f_{\mathrm{SG}}(s_{\mathrm{G}}) dr ds_{\mathrm{G}} \quad (5)$$

In actual calculation, it can be reduced to the form of quadratic integration, that is:

$$P_f = \int_{r_L - s_{Q\mathrm{th}}}^{+\infty} \int_{r_L}^{s_G + s_{Q\mathrm{th}}} f(r|R \geq r_L) f(s_G) dr ds_G \quad (6)$$

The above equation is the bridge load limit analysis model that takes into account the left truncated distribution characteristics of resistance.

According to Eq. (6), if the allowable failure probability P_f of the structure is given, the corresponding limit value $S_{Q\text{th}}$ of the automotive load effect can be calculated in reverse. For the convenience of subsequent analysis, reference can be made to existing literature [4] to introduce a load limiting coefficient ζ_q and set it to

$$S_{Q\text{th}} = \zeta_q S_{Qk} \tag{7}$$

$S_{Q\text{k}}$ is the standard value effect of vehicle load adopted in the design.

After the above substitution, the problem of load limit analysis of the bridge after static load testing is transformed into the problem of calculating the load limit coefficient under a given failure probability.

ρ can be used as the basic parameter, develop an inverse program for the load limiting coefficient of the bridge after static load test verification using numerical integration method according to Eq. (6). For a given ρ, P_f and η_q, a unique corresponding ζ_q value can be obtained.

3 Calculation of Load Limiting Factor

According to the Unified Standard for Reliability Design of Highway Engineering Structures (GB/T50283-1999), the statistical parameters of the dead load effect S_G and resistance R are determined, and the limit load analysis program is applied to calculate the limit load coefficient of bridges designed according to different period specifications. The following takes reinforced concrete beam bridges as an example to calculate the ζ_q value corresponding to different static load test load efficiency η_q. To increase comparability, both 04 and 18 specifications are calculated based on the first level of load safety level.

According to the test procedure [17], η_q should not exceed 1.05. To systematically analyze the influence of different η_q on the load limit values of bridges after static load test verification, η_q is taken as 0.85–1.15.

In order to more intuitively analyze the impact of static load test efficiency on the load limit coefficient, the relationship curves between n and z are plotted for the calculation results, as shown in Figs. 1 and 2.

From Fig. 1, it can be seen that the load limiting coefficient of a bridge designed according to the 85 standard has the following characteristics:

(1) For the same live and dead load ratio ρ, the load limiting coefficient ζ_q increases with the increase of η_q value and shows an approximate linear relationship. As ρ increases, the $\eta_q - \zeta_q$ curve exhibits slight nonlinear characteristics.
(2) Under the same static load test efficiency, the larger the design live to constant load ratio ρ, the greater the corresponding ζ_q. If $\rho > 0.5$ and $\eta_q > 1.0$, the growth rate of ζ_q slightly increases with the increase of η_q.

Comparing Figs. 1 and 2, the variation pattern of the load limiting coefficient of the 04 and 18 standard bridges is very similar to that of the 85 standard bridges. From the two figures, it can be seen that if the same η_q is used, the static load test can more effectively improve the load limit calculation value of larger span bridges.

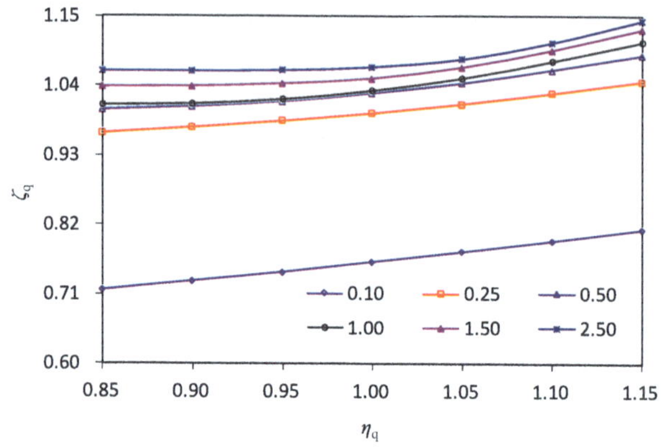

Fig. 1. Relations between η_q and ζ_q (code of 85 edition)

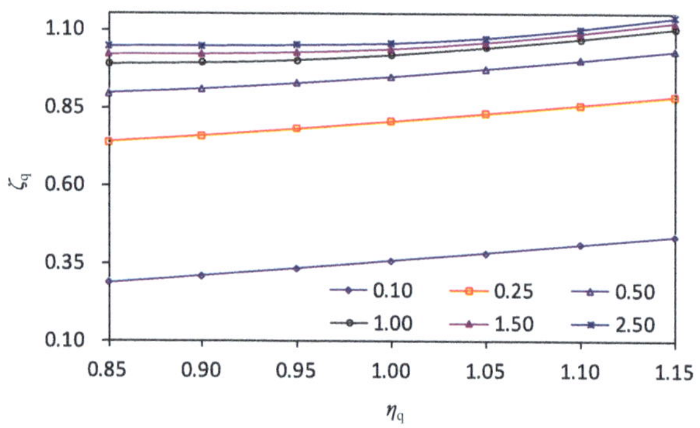

Fig. 2. Relations between η_q and ζ_q (codes of 04 and 18 editions)

In summary, after verification by load tests, if the influence of the left truncated distribution of resistance is taken into account, the load limiting coefficient of the bridge has been increased to varying degrees. The degree of improvement is related to the live and dead load ratio used in the bridge design, the load efficiency of static load tests, and the resistance sub coefficient.

4 Conclusion

Aiming at the probability distribution characteristics of bridge resistance verified by static load tests, a load limit analysis model for public bridges considering the left truncated distribution characteristics of resistance is proposed. Based on this, the load limit coefficients of reinforced concrete medium and small span bridges corresponding to different static load test efficiencies are calculated, and the necessity of introducing left truncated distribution of resistance is discussed. For reinforced concrete medium and small span bridges verified by static load tests, the main conclusions include:

(1) After considering the load efficiency of the static load test, the load limiting coefficient is larger than the calculation results without considering the verification factors of the static load test, and the increase is related to the load efficiency of the static load test, the design live and dead load ratio, and the resistance sub coefficient.
(2) Under the same conditions, the greater the load efficiency of the static load test, the greater the increase in the load limiting coefficient. Obviously, by reasonably improving the efficiency of static load tests, some uncertainties in resistance can be eliminated, which is beneficial for improving the efficiency of bridge structures.
(3) If the load efficiency of the test is less than 1.0, the impact of the static load test on the load limiting coefficient is extremely insignificant. In this case, the results of the static load test can only be used as a basis for judging whether it meets the design requirements and cannot be used as an adjustment basis for the bridge load limiting value.
(4) This article only provides the calculated limit load coefficients according to the standard design expression. In actual engineering, the verified limit load values of the bridge can also be determined by interpolation or extrapolation from Figs. 1 and 2 based on the design live and dead load ratio, static load test efficiency, and resistance sub coefficients used in the bridge structure. The article has certain scientific significance for the healthy and safe use of bridges and bridge risk management.
(5) In addition to the above conclusions, there are still some shortcomings in this article, that is, for actual bridges, it is necessary to solve the load limit problem of bridges with different resistance levels and the calculation problem of the total weight limit of typical limited load vehicles.

References

1. Ministry of Transport of the People's Republic of China, 2022. Statistical Bulletin on the Development of the Transportation Industry in 2021. https://xxgk.mot.gov.cn/2020/jigou/zhghs/202205/t20220524_3656659.html
2. Huang, C.J., Fang, S.S.: A study on the limited load tonnage of existing bridges based on structural reliability indicators. Highway **64**, 141–144 (2019). https://kns.cnki.net/kcms/detail/11.1668.U.20190715.1019.060.html
3. Zhu, R.: Research on Vehicle Load Limit Based on Bridge Structure Safety and Reliability. Urban Roads Bridges Flood Control **2021**(6), 253–257+281+27 (2021). https://doi.org/10.16799/j.cnki.csdqyfh.2021.06.066
4. Li, S.H.: Reliability-based analytical model for determining the truck weight limits on highway bridges. Chin. Civil Eng. J. **46**(9), 83–90 (2013). https://doi.org/10.15951/j.tmgcxb.2013.09.003
5. Ni, P., Li, J., Hao, H., Zhou, H.: Reliability based design optimization of bridges considering bridge-vehicle interaction by kriging surrogate mode. Eng. Struct. **246**, 112989 (2021). https://doi.org/10.1016/j.engstruct.2021.112989
6. Yin, Z.K.: Reliability analysis and remaining life study of prestressed concrete continuous box girder bridges. Central South Univ. Forestry Technol. (2020). https://doi.org/10.27662/d.cnki.gznlc.2020.000596
7. Li, S.H., Li, S., Nie, R.F., et al.: Load limits for highway bridges under different levels of technical conditions. J. Transp. Eng. **20**(6), 126–134 (2020). https://doi.org/10.19818/j.cnki.1671-1637.2020.06.011

8. Li, G.Y.: Adaptability analysis of bridge load limit values under normal use limit state. Eng. Technol. Res. **8**(17), 14–16 (2023). https://doi.org/10.19537/j.cnki.2096-2789.2023.17.005
9. Xu, T.: Adaptability analysis of bridge load limit values under normal use limit state. Shandong Univ. Sci. Technol. (2020). https://doi.org/10.27275/d.cnki.gsdku.2020.001443
10. James, R.W., Noel, J.S., Furr, H.L., et al.: Proposed new truck weight limit formula. J. Struct. Eng. **112**(7), 1589–1604 (1986). https://doi.org/10.1061/(ASCE)0733-9445(1986)112:7(1589)
11. Ghosn, M., Moses, F.: Effect of changing truck weight regulations on U.S. bridge network. J. Bridge Eng. **5**(4), 304–310 (2000). https://doi.org/10.1061/(ASCE)1084-0702(2000)5:4(304)
12. Li, S.H., Li, C., Nie, R.F., et al.: Load limit values for RC simply supported beam bridges considering resistance correction coefficient and static load test efficiency. J. Civ. Eng. **53**(10), 99–105 (2020). https://doi.org/10.15951/j.tmgcxb.2020.10.10.10
13. China Communications Press: JTJ 023-85 Code for design of highway reinforced concrete and prestressed concrete bridges and culvers (1985). https://www.mot.gov.cn/
14. China Communications Press: JTG D62-2004 Code for design of highway reinforced concrete and prestressed concrete bridges and culverts (2004). https://www.mot.gov.cn/
15. China Communications Press: JTG 3362-2018 Code for design of highway reinforced concrete and prestressed concrete bridges and culverts (2018). https://www.mot.gov.cn/
16. American Association of State Highway and Transportation Officials. MBE-3 The Manual for Bridge Evaluation (2018). https://trid.trb.org/view/887610
17. China Communications Press: JTG/T J21-01-2015 Load test methods for highway bridge (2015). https://www.mot.gov.cn/

Prediction of Bridge Monitoring Data and Time-Variant Reliability Assessment Based on Kalman Filtering

Zi-yi Zhang[1], Qian-hui Pu[1(✉)], Yu Hong[1], and Xu-guang Wen[2]

[1] School of Civil Engineering, Southwest Jiaotong University, Chengdu 610031, China
zyzhang@my.swjtu.edu.cn, qhpu@vip.163.com, hongyu@swjtu.edu.cn
[2] Guangxi Key Laboratory of International Join for China-ASEAN Comprehensive Transportation, Nanning University, Nanning 530000, China

Abstract. Effectively using the massive data of bridge health monitoring and controlling the safety status of bridge structure in real time is the research focus in the field of bridge health monitoring. In order to process the monitoring data online and evaluate the time-varying reliability of the bridge structure in real time, this paper proposes a dynamic prediction model based on Kalman filter to predict the daily maximum stress monitoring data during the service of the bridge by using an engineering example of a long-span cable-stayed bridge. Finally, the time-varying reliability index of the structure is established by using the prediction data, and the failure probability of the structure is obtained, and the online real-time dynamic prediction of the structural reliability is realized. The results show that the predicted value obtained by the dynamic prediction model based on Kalman filter has a high degree of fitting with the original data, and can accurately reflect the change trend and range of the original data. The convergence speed of prediction error variance is fast, and finally tends to 0. With the increase of the number of iterations, the absolute value of the relative error tends to decrease gradually, and the prediction accuracy of the model is higher. The time-varying reliability index and failure probability of the structure calculated by the predicted value based on the prediction model are highly fitted with the reliability index and failure probability based on the original data, which can reflect the change trend and range. In summary, the dynamic prediction model based on Kalman filter can still perform well in the case of less initial data, and realize real-time dynamic and long-term prediction of monitoring data. The time-invariant reliability of the structure can only be evaluated by offline data. In this paper, the dynamic prediction model based on Kalman filter is used to evaluate the time-varying reliability of the structure, which can realize the online real-time evaluation of the structural reliability and control the state of the structure or component in real time. It has important engineering significance.

Keywords: Bridge engineering · data prediction · time-dependent reliability · Kalman filtering · maximum stress · health monitoring

B. Yuan et al. (Eds.): ICCASE 2024, SUCI, pp. 47–56, 2024.
https://doi.org/10.1007/978-3-031-76102-7_6

1 Introduction

The current bridge health monitoring system has matured in the development and application of sensors, data transmission, and data acquisition [1]. However, a significant challenge in the field of health monitoring is the rational utilization of vast amounts of monitoring data. Further investigation is needed on how to effectively utilize monitoring information to establish dynamic prediction models for monitoring response data. This will enable the prediction of dynamic response behavior and time-varying reliability of in-service bridges.

Currently, experts and scholars both domestically and internationally are conducting research on three main approaches to monitoring data prediction: traditional prediction methods, intelligent algorithms, and dynamic system prediction methods. Traditional time series models include Autoregressive (AR) models [2], Moving Average (MA) models, Autoregressive Moving Average (ARMA) models [3] and Autoregressive Integrated Moving Average (ARIMA) models [4]. These time series prediction models have the characteristic of a single structural form, with model parameters remaining largely unchanged during the prediction process. They are primarily used for static predictions and effectively handle linear relationships in data. In comparison to time series prediction models, intelligent algorithm models have the ability to learn and extract features from input training data, enabling the handling of highly non-linear and non-stationary data. Intelligent algorithms currently used for monitoring data prediction include Artificial Neural Networks [5], Fuzzy Neural Networks [6], and Support Vector Machines [7]. These intelligent algorithms exhibit biomimetic or evolutionary functionality, continuously improving and enhancing themselves during learning. When predicting information, they quickly react based on processes that closely resemble the given information, resulting in more accurate predictions. However, the parameters of intelligent algorithm prediction models typically remain unchanged during testing and application after training, making it challenging to capture the changing patterns in monitoring data. Additionally, these models require sufficient learning time and a large number of training samples to improve prediction accuracy. Beyond the aforementioned time series and intelligent algorithm prediction models, there is a category of probability-based prediction models, including Grey Prediction Models [8], Bayesian Dynamic Models [9], and Kalman Filter Dynamic Models [10]. These models fully consider the uncertainty of monitoring data. During the prediction process, they automatically update model parameters based on changes in monitoring data to ensure accuracy. Grey Prediction Models and Bayesian Dynamic Models typically have certain requirements for the quantity of historical data to ensure the accuracy of data predictions. The Bayesian algorithm performs exceptionally well in addressing nonlinear system issues but demands a higher level of prior knowledge. For linear systems, the Kalman Filter algorithm is more flexible and efficient than the Bayesian algorithm, especially in real-world engineering applications where rapid response is needed. The Kalman Filter algorithm is more suitable, providing quick and real-time state estimates.

For existing structures, reliability is, in fact, a dynamic process that changes continuously over time. Using the off-line monitoring information of bridge structure to evaluate the reliability of bridge structure, a lot of research has been carried out at home and abroad [11, 12]. However, research on how to use real-time monitoring data for online dynamic reliability assessment is still in its early stages[13]. Currently, studies on time-varying reliability are focused on combining Bayesian theory with reliability assessment. Liu et al.[9] used actual monitoring data combined with Bayesian theory and a Gaussian Copula model to achieve dynamic reliability prediction of a steel box girder structure. However, employing Bayesian theory to realize dynamic reliability prediction of structures requires a significant amount of historical data to determine the system's initial state, demanding a high level of prior knowledge. This approach is less suitable for real-world engineering applications that require swift responses. In contrast, the Kalman filtering algorithm can swiftly correct the prior state to a more accurate state estimate by computing the Kalman gain, with lower requirements for the quantity of historical data. Therefore, the Kalman filtering algorithm exhibits superior performance in real-time dynamic reliability assessment, particularly excelling in real-world engineering applications that demand rapid responses.

Based on the above research status, this paper proposes a method for monitoring data prediction and time-varying reliability assessment of bridge structures based on a Kalman filter dynamic model. Dynamic prediction models based on Kalman filtering are established using real-time monitoring data of the dynamic response of bridge structures. Subsequently, the prediction models are combined to accomplish the prediction and assessment of time-varying reliability. The accuracy and engineering practicality of the proposed method are validated through engineering examples.

2 Dynamic Prediction Model for Monitoring Data Based on Kalman Filtering

The Kalman filter algorithm, proposed by R. E. Kalman in 1960, was initially designed to address the issue of inaccurate estimation of state information such as position and velocity due to noise interference in measurement sensor data in the aerospace field [14]. The Kalman filter algorithm is a Bayesian probability inference-based algorithm that combines prior information and measurement data to extract the true part from a signal contaminated by noise, thereby estimating the true state of the system. Currently, the Kalman filter algorithm has been widely applied in civil engineering fields such as traffic flow prediction, foundation settlement prediction, and structural deformation prediction. The Kalman filter dynamic prediction model effectively combines the system state with monitoring data, directly updating the state estimate when new data is introduced, without the need to recalculate all past data. This enables rapid, real-time, and dynamic prediction of monitoring data.

The Kalman filter dynamic prediction model consists of a state equation and an observation equation:

The state equation is:

$$x_t = F_t x_{t-1} + Q_t \tag{1}$$

The observation equation is:

$$z_t = H_t x_t + R_t \tag{2}$$

where x_t is the system state vector; zt is the observation vector; Ft is the state transition matrix; H_t is the observation matrix; Q_t is the system noise matrix, following a standard normal distribution; Rt is the observation noise matrix, following a standard normal distribution.

The Kalman filter dynamic prediction model assumes knowledge of x_0 (the estimated initial state of the system) and P_0 (the error covariance matrix). Additionally, it assumes that the initial state and noise at each time step are mutually independent. The Kalman filter's dynamic prediction model consists of two recursive processes: prediction and correction. In the prediction phase, the Kalman filter uses the estimate of the state from the previous time step to predict the state at the current time step. In the correction phase, the Kalman filter incorporates the observed values at the current time step to adjust the predicted values obtained in the prediction phase, yielding a new estimate closer to the true state, i.e., the posterior state. The detailed recursive process of the Kalman filter dynamic prediction model is as follows:

Step1. Predict the current state of the system at the current time step based on the optimal estimate of the system's state at the previous time step:

$$\hat{x}_t^- = F_t\hat{x}_{t-1} \tag{3}$$

where $\hat{x}_t^-$ is the predicted estimate of the system state vector at the current time step, and $\hat{x}_{t-1}$ is the estimate of the system state vector at the previous time step.
Step2. Calculate the deviation of the predicted system state vector at the current time step in the first step:

$$P_t^- = F_tP_{t-1}F_t^T + Q_t \tag{4}$$

where P_t^- represents the predicted covariance matrix of the estimation error for the current time step, and P_{t-1} represents the covariance matrix of the estimation error from the previous time step.
Step3. Calculate the Kalman Gain for the current time step:

$$K_t = P_t^-H_t^T(H_tP_t^-H_t^T + R_t)^{-1} \tag{5}$$

where K_t represents the Kalman Gain at the current time step, which is utilized to balance the correction of the predicted estimate based on both the prediction and observation of the system state.
Step 4. Incorporate the current observation z_t to correct the predicted estimate of the current system state vector obtained in step 1, yielding the optimal estimate at the current time step:

$$\hat{x}_t = \hat{x}_t^- + K_t(z_t - H_t\hat{x}_t^-) \tag{6}$$

where $\hat{x}_t$ denotes the corrected estimate of the system state vector at the current time step.
Step5. Compute the deviation of the optimal estimate of the system state vector at the current time step in step 4:

$$P_t = (I - K_tH_t)P_t^- \tag{7}$$

where P_t is the corrected value of the covariance matrix of the system state estimation error at the current time step, and I is the identity matrix.

3 Time-Varying Reliability Assessment of Bridge Structures

This paper employs the First-Order Second-Moment (FOSM) method [15] to establish the reliability index of a bridge structure or component, simultaneously considering the mean and standard deviation of monitoring response information. Assuming the structural performance function is:

$$Z = g(R, S) = R - S \tag{8}$$

where R is the random variable for structural resistance, and S is the random variable for load effects.

In the performance function of Eq. (8), if both R and S follow normal distributions with means and standard deviations denoted as μ_R, μ_S, σ_R and σ_S respectively, then the performance function Z also follows a normal distribution with mean μ_Z and standard deviation σ_Z given by:

$$\mu_Z = \mu_R - \mu_S \tag{9}$$

$$\sigma_Z = \sqrt{\sigma_R^2 + \sigma_S^2} \tag{10}$$

The expression for the reliability index is:

$$\beta = \frac{\mu_Z}{\sigma_Z} = \frac{\mu_R - \mu_S}{\sqrt{\sigma_R^2 + \sigma_S^2}} \tag{11}$$

This paper uses a long-span cable-stayed bridge as an engineering example for the analysis of time-dependent reliability prediction. The daily extreme stress at the mid-span of the bridge deck is monitored. The FOSM method is employed based on six months of monitoring data to predict the time-dependent reliability, considering the mean and variance of the monitored extreme stress. For practical engineering applications, this method provides sufficient calculation accuracy.

The limit state equation for the mid-span girder is:

$$g = R - \gamma_M M \tag{12}$$

where R is the yield strength of the steel plate, α is the correction coefficient for the sensor, and M is the extreme stress predicted based on the Kalman filter.

Based on Eqs. (8), (9), (10), (11) and (12), the formula for predicting the time-varying reliability of the bridge component is obtained as follows:

$$\beta = \frac{\mu_R - \gamma_M \mu_M}{\sqrt{\sigma_R^2 + (\gamma_M \sigma_M)^2}} \tag{13}$$

where μ_M and σ_M represent the predicted mean and standard deviation of daily extreme stresses obtained based on Kalman filtering, while μ_R and σ_R represent the mean and standard deviation of steel resistance calculated according to the specifications, respectively.

4 Verification of Bridge Structure

The monitoring data for this bridge health assessment are derived from a large-span cable-stayed bridge on a highway, with a total length of 1641 m. The main span is 860 m, arranged with spans of (215 + 430 + 215) m. The total height of the main bridge tower is 137.5 m, featuring 68 pairs of inclined cables anchored to steel anchor beams on the main tower longitudinal beam. The bridge deck is 38 m wide, comprising six lanes in both directions. The main beam is designed as a steel box girder and the steel grade is Q345. The health monitoring system for this bridge primarily observes environmental temperature, structural temperature, strain, cable force, deflection, and vibration acceleration. A total of 208 sensors of 10 different types are installed on the bridge. Due to the abundance of monitoring points in this project, we will take the example of stress and strain monitoring points on the beam top at the mid-span, selecting data for daily extreme stresses from a period of six months for data prediction based on Kalman filtering and time-dependent reliability analysis. The span arrangement of the bridge and the positions of the selected sensors are shown in Fig. 1. Firstly, we establish a dynamic prediction model based on Kalman filtering and use this model to predict six months of monitoring data. Subsequently, we employ the First Order Second Moment (FOSM) method, combining the predicted data to analyze the time-dependent reliability of the structure.

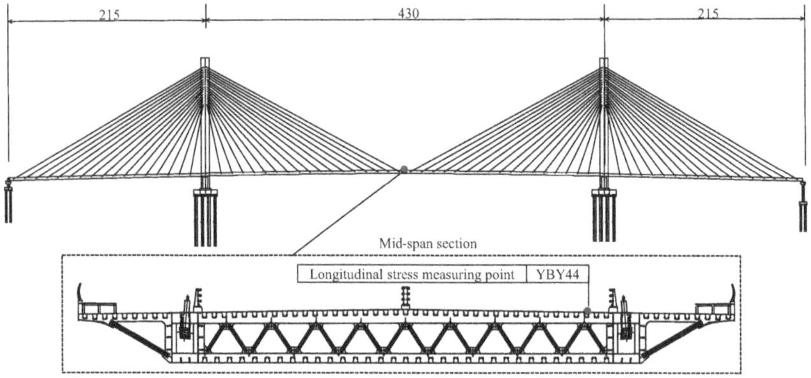

Fig. 1. Bridge layout (unit: m)

4.1 Dynamic Prediction Model Based on Kalman Filtering

Select monitoring data of daily extreme stress on the mid-span cross-section beam top from June 16, 2022, to December 15, 2022, for the validation of the prediction algorithm and the prediction analysis of time-dependent reliability.

The average value of daily extreme stress from May 15th to June 15th is used as the initial state estimate, with the observation variance of the daily extreme stress as the initial state covariance matrix, i.e., $x_0 = 26.986$, $P_0 = 1.164$. The observation noise covariance matrix is typically determined based on the precision of the observation instrument and measurement methods, and according to the monitoring instrument model, the variance of the observation noise is set as $R_t = 2.387$. The system noise covariance matrix is usually determined empirically based on the trial calculation of the filtering error sequence. In this study, Q_t is set to 0.1 [10]. Combining the Kalman filter recursion formula and the initial state estimate, the predicted results of daily extreme stress for six months are shown in Fig. 2.

The accuracy of the prediction can be expressed by the relative error (RE). The relative error is the absolute error between the predicted value and the true value divided by the true value, multiplied by 100%, and presented as a percentage. The calculation formula is shown in Eq. (14), and the results of the RE are illustrated in Fig. 3.

$$\mathrm{RE} = \frac{\left|\hat{x}_t^- - z_t\right|}{z_t} \times 100\% \tag{14}$$

where $\hat{x}_t^-$ and z_t have the same meanings as previously defined, representing the predicted estimate of the current system state vector and the observed value at the current time, respectively.

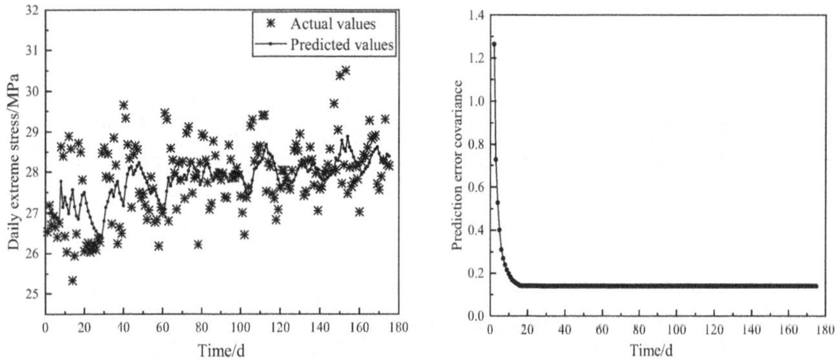

a) Prediction results b) Prediction error covariance

Fig. 2. Daily extreme stress prediction results of prediction model

From the results in Figs. 2 and 3, it can be observed that the predicted estimates closely match the actual monitoring data. The predicted values accurately reflect the true trends and magnitudes of the monitoring data. The convergence speed of the predicted error covariance is fast, gradually approaching 0. The relative error between the actual monitoring data and the predicted values decreases with an increasing number of iterations. This indicates that as the model continuously adjusts to the system state, the prediction accuracy improves, confirming the rationality of the initial state selection.

4.2 Time-Varying Reliability Assessment of Bridge Structures

According to the "Code for Design of Steel Structures" (GB 50017–2017) [16], the design parameters for the steel of grade Q345 used are as follows: $\mu_R = 345\mathrm{MPa}$, $\sigma_R = 33.8\mathrm{MPa}$. The correction coefficient for the sensors γ_M is 1.15 [9].

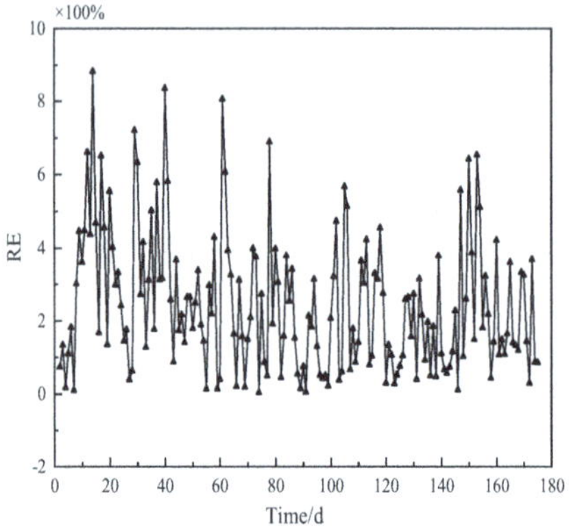

Fig. 3. RE of prediction

Substituting the above parameters into Eq. (13), the time-varying reliability index of the bridge component can be obtained:

$$\beta = \frac{345 - 1.15\mu_M}{\sqrt{33.8^2 + (1.15\sigma_M)^2}} \tag{15}$$

The time-varying reliability index obtained from the calculation in the above equation for the mid-span top stress monitoring point is shown in Fig. 4, and the corresponding failure probability is shown in Fig. 5. As observed from Figs. 4 and 5, the time-varying reliability index and failure probability obtained from the Kalman filter dynamic prediction model reflect the changing trends and ranges of the reliability index and failure probability obtained from actual monitoring data. The time-varying reliability index obtained based on the prediction model considers the uncertainty and randomness of the daily peak stress, reflecting the dynamic nature of the reliability index. This is advantageous for real-time monitoring of the structural state during service.

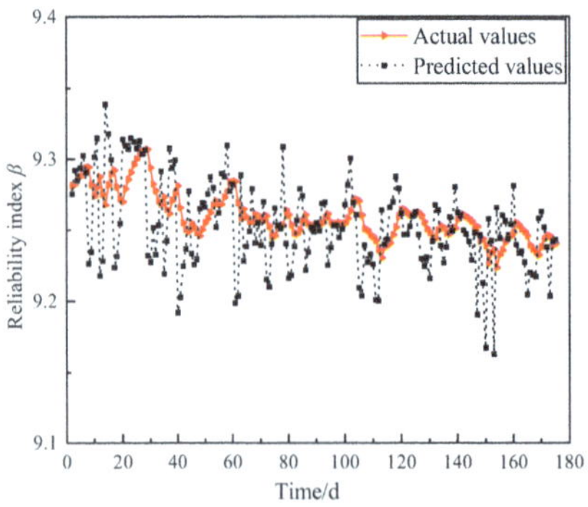

Fig. 4. Time-dependent reliability index

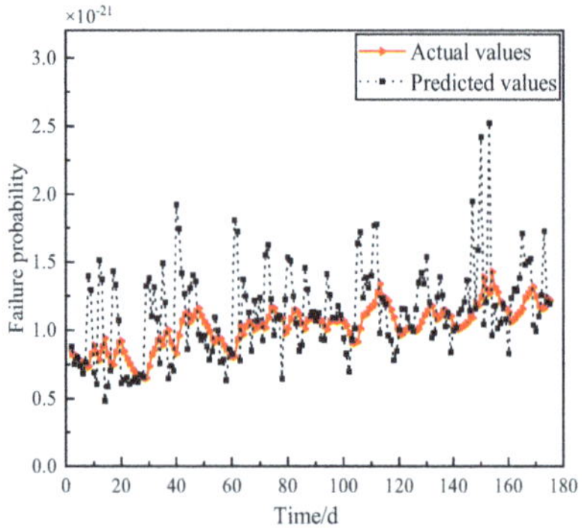

Fig. 5. Failure probability

5 Conclusions

In conjunction with the engineering case, this paper proposes and validates a dynamic prediction model based on the Kalman filter and a method for evaluating the time-varying reliability of structures. The following conclusions have been drawn:

1. The dynamic prediction model based on the Kalman filter achieves a high level of fit with the original data, accurately reflecting the changing trends and ranges of the original data. Additionally, the convergence speed of the prediction error variance is rapid, ultimately approaching 0. With an increase in the number of iterations, there is a gradual reduction in the absolute value of the relative error. In each iteration, the system state estimate is adjusted based on newly added monitoring data, gradually bringing the system state estimate closer to the true value and continually improving the predictive accuracy of the model.
2. Through analysis, it is observed that the structural time-dependent reliability indices and failure probabilities calculated based on the predictive model can accurately reflect the changing trends and ranges of reliability indices and failure probabilities obtained from the original data.
3. The time-invariant reliability of the structure can only be assessed based on offline data. This study employs a dynamic prediction model based on the Kalman filter to assess the time-varying reliability of structures, enabling real-time online evaluation and monitoring of structural or component conditions. This research holds significant engineering significance as it facilitates the real-time prediction of structural reliability, allowing for timely monitoring of structural health status. By promptly identifying potential issues and implementing necessary measures for repair and maintenance, the safety and stability of structures can be ensured.
4. In future research, stress data from multiple cross-sections should be included in the evaluation to comprehensively assess the time-varying reliability of bridge structures. The plan involves analyzing the correlation between stress variations in different cross-sections and investigating their impact on the overall reliability of bridge structures. This will contribute to a more comprehensive understanding of the dynamic behavior and reliability of bridge structures, providing valuable insights for optimizing design and maintenance strategies.

Acknowledgments. The research described in this paper was supported by Science and Technology Project of Guangxi (Guangxi Science AA21077011), which is greatly appreciated.

References

1. Ju, H.W., Deng, Y., Li, A.Q.: Correlation model of deflection-temperature-vehicle load monitoring data for brige structures. J. Vibr. Shock **42**(06), 79–89 (2023). https://doi.org/10.13465/j.cnki.jvs.2023.06.010
2. Yao, X.J., Yang, X.: Automated modal identification method in frequency domain improved by AR model power spectrum. J. Chongqing Jiaotong Univ. (Nat. Sci.) **42**(02), 1–7 (2023). https://doi.org/10.3969/j.issn.1674-0696.2023.02.01

3. Li, P., Zhu, H.Z., Luo, G.J., et al.: Simulation of stochastic wind field of offshore wind turbines based on ARMA model. Eng. J. Wuhan Univ. **57**(01), 112–120 (2024). https://doi.org/10.14188/j.1671-8844.2024-01-014
4. Liang, N.X., Yan, J., Yang, W.C., et al.: A combined prediction model for highway traffic safety based on ARIMA-LSTM. J. Chongqing Jiaotong Univ. (Nat. Sci.) **42**(04), 131–138 (2023). https://doi.org/10.3969/j.issn.1674-0696.2023.04.17
5. Moon, H.S., Ok, S., Chun, P., Lim, Y.M.: Artificial neural network for vertical displacement prediction of a bridge from strains (Part 1): girder bridge under moving vehicles. Appl. Sci. **9**(14), 2881 (2019). https://doi.org/10.3390/app9142881
6. An, J., Wang, M.L.: Hybrid fuzzy neural network based on error distribution analysis for time-series prediction model. J. East China Univ. Sci. Technol. **49**(04), 589–597 (2023). https://doi.org/10.14135/j.cnki.1006-3080.20220212001
7. Zhang, X.S., He, B., Sabri, M.M.S., et al.: Soil liquefaction prediction based on bayesian optimization and support vector machines. Sustainability **14**(19), 11944 (2022). https://doi.org/10.3390/SU141911944
8. Yousuf, M.U., Al-Bahadly, I., Avci, E.: A modified GM(1,1) model to accurately predict wind speed. Sustain. Energy Technol. Assessments **43**, 100905 (2021). https://doi.org/10.1016/j.seta.2020.100905
9. Liu, Y.F., Fan, X.P.: Dynamic reliability prediction for the steel box girder based on multivariate Bayesian dynamic Gaussian copula model and SHM extreme stress data. Struct. Control Health Monit. **27**(6), e2531 (2020). https://doi.org/10.1002/stc.2531
10. Zhang, H., Zhang, L.S.: Research on prediction of deformations in long-span cable-stayed bridges based on Kalman filter. J. Zhejiang Univ. Technol. **49**(06), 650–655 (2021). https://doi.org/10.3969/j.issn.1006-4303.2021.06.009
11. Mao, J.-X., Wang, H., Li, J.: Fatigue reliability assessment of a long-span cable-stayed bridge based on one-year monitoring strain data. J. Bridge Eng. **24**(1), 05018015 (2019). https://doi.org/10.1061/(ASCE)BE.1943-5592.0001337
12. Hou, N., Sun, L., Chen, L.: Cable reliability assessments for cable-stayed bridges using identified tension forces and monitored loads. J. Bridge Eng. **25**(7), 05020003 (2020). https://doi.org/10.1061/(ASCE)BE.1943-5592.0001573
13. Fan, X.P., Zhou, H., Liu, Y.F.: Time-variant reliability analysis of bridge members based on Bayesian dynamic models and copula theory. J. Jilin Univ. (Eng. Technol. Edition) **25**, e2185 (2023). http://kns.cnki.net/kcms/detail/22.1341.T.20220803.1555.007.html
14. Kalman, R.E.: A new approach to linear filtering and prediction problems. J. Basic Eng. **82**(1), 35–45 (1960)
15. Liu, Y.F., Fan, X.P.: Use of monitoring extreme strain data for failure probability analysis of concrete bridge girder: an optimal vine gaussian-copula model. Structures **44**, 274–283 (2022). https://doi.org/10.1016/J.ISTRUC.2022.08.016
16. GB 50017-2017, Standard for Design of Steel Structures

Revit Secondary Development of Excavation Support Design Software for Fabricated Steel Support System of Prestressed Fish-Bellied Beams

Peiyao Zhang[1], Kai Ma[2(✉)], Sen Pang[3], Ziqing Zhou[2], and Hongguang Zhu[2]

[1] China Nuclear Power Engineering Co., Ltd, Beijing 100048, China
zhang.py@buaa.edu.cn

[2] China University of Mining and Technology, Beijing 100083, China
makai056600@163.com, zhg@cumtb.edu.cn

[3] Beijing Building Research Institute Co., Ltd., CSCES, Beijing 100073, China

Abstract. Fabricated steel support system of prestressed fish-bellied beams has the advantages of effective displacement control and large mechanical operation space, and has been widely used. The proposed layout method of fabricated steel support system of strip fish-bellied beams has promoted its development. In order to further promote the application of fabricated steel support system of prestressed fish-bellied beams, the simulation module of excavation support design for the internal support system of prestressed fish-bellied beam was developed by using Revit secondary development technology and plane layout method. The module has been successfully practiced and applied in some construction units and teaching units, helping the promotion of new fabricated steel support system of prestressed fish-bellied beams.

Keywords: fish-bellied beams · plane layout · Revit secondary development

1 Introduction

Prestressed fish-bellied beam steel support can provide open excavation space for the foundation pit, which is convenient for earth excavation and transportation; secondly, the fish-bellied beam system can adjust the vertical and horizontal displacement of the surrounding soil layer for deformation control through the application of prestressing. Especially in soft soil areas or when there are sensitive pipelines and other key protection objects around the foundation pit, the system has good deformation control ability and control effect [1, 2]. In order to promote the practical application of the new support structure of fabricated steel support system of prestressed fish-bellied beams, the Chinese Society of Civil Engineering published the "Technical specification for fabricated steel support system consisting of prestressed fish-bellied beams in excavation" [3], which suggested the selection and design process of prestressed fish-bellied beam. In order to promote the application of pre-stressed fish-bellied beam support system of

B. Yuan et al. (Eds.): ICCASE 2024, SUCI, pp. 57–62, 2024.
https://doi.org/10.1007/978-3-031-76102-7_7

prefabricated steel structure, the design and simulation module of excavation support system of prestressed fish-bellied beam of prefabricated steel structure is developed by using Revit parameterization technology and secondary development of Revit software [4–6]. It solves the problems that the plane layout design is difficult and the fish-bellied beam design is too dependent on the experience of the designer. so as to facilitate the practical application of construction units and teaching units, and help the promotion of fabricated steel support system of new prestressed fish-bellied beams.

2 Revit Secondary Development of Simulation Module of Excavation Support Design for Fabricated Steel Support System of Prestressed Fish-Bellied Beams

The professional term of Revit is "building information model", which is also understood as "virtual architecture". It can be regarded as an interactive platform that can integrate various professional information models. The application of Revit technology has brought great convenience to the current construction industry [7–9].

2.1 Establishment of Fish-Bellied Excavation Component Family Library

In the process of establishing the standard component family library of fish-bellied excavation, there can be multiple component sizes for the same type of component family, so it is necessary to parameterize the component family, so as to adapt to the needs of different excavation projects. For the fish-bellied excavation, the standard family members that need to be established can be divided into two categories: vertical members and horizontal supporting members. The standard family of fish-bellied beam is shown in Fig. 1.

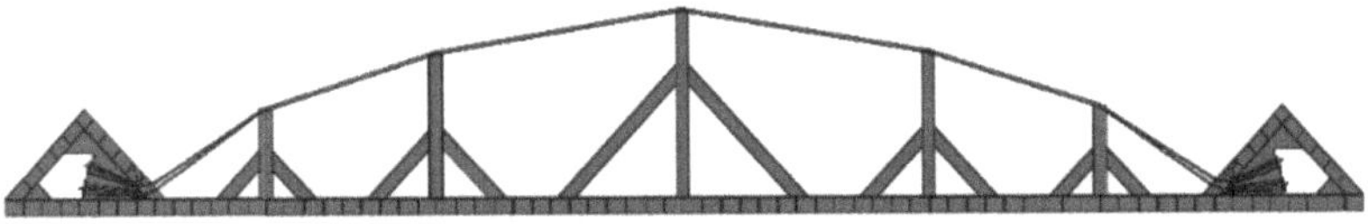

Fig. 1. Standard family model of prestressed fish-bellied beam

2.2 Establishment of Fish-Bellied Excavation Internal Support Model

In the actual project, the shape of the excavation is very different. However, in the process of this secondary development, the realization of various functions for the most common long pit is carried out.

In order to realize the functions of the secondary development of the fish-bellied excavation, it is necessary to determine the shape of the excavation and the elevation of each support. In order to obtain this information, it is necessary to realize information interaction with users. Therefore, you need to create a windows form in Visual Studio.

A drop-down menu is set at the number of supports to allow the user to complete the selection, and a window appears after the selection is complete, allowing the user to enter the elevation of each support.

After obtaining the basic information of the excavation, it is necessary to store the information in the field attributes for later modeling.

In Revit, the obtained parameters are used to establish the corresponding elevation, shaft network and basic model of excavation. Then you need to create a windows form again for the user to enter the basic selection of fish-bellied internal support members. The automatic layout of the excavation can be carried out after the given excavation shape and the selection of the fish-bellied inner supporting members. Parameter modification and loading of the same class of standard parameterized components can be programmed as a method, setting the parameters that need to be passed in. After the method is written, it can be repeated in the main function, eliminating the need to write repeated code. Taking the loading of the fish-bellied beam standard family as an example, the description of the method preparation (Fig. 2):

```
private void FishBeam(string s1, string s2, int n, Variable va, Document doc, double
d)
  { if (d != 0)
    { FamilySymbol getSymbol = null;
     FamilyInstance fishBeam = null;
     if (doc.LoadFamilySymbol(s1, s2, out getSymbol))
        {fishBeam = doc.Create.NewFamilyInstance(new XYZ(va.Width / 3 +
4000 * yingchi, va.PileDiameter / 2 + 400 * n * yingchi, 0), getSymbol, Structural-
Type. NonStructural);}                                    //Insert a family of
fish-bellied beam members into the model
          foreach (Parameter param in fishBeam.Symbol.Parameters)
       { if (param.Definition.Name == "Fish-bellied beam span")  //The span fam-
  ily parameters of fish-bellied beam were modified
          {param.Set ( va.FishSpan ); }
    if (param.Definition.Name == "Vector height")          //Modify the height
  family parameters of the fish-bellied beam vector
          {param.Set ( va.FishRise ); }}
   ElementTransformUtils.Move Element(doc, fishBeam.Id, new XYZ(0, 0, -d));
  //Move the family to the corresponding elevation
```

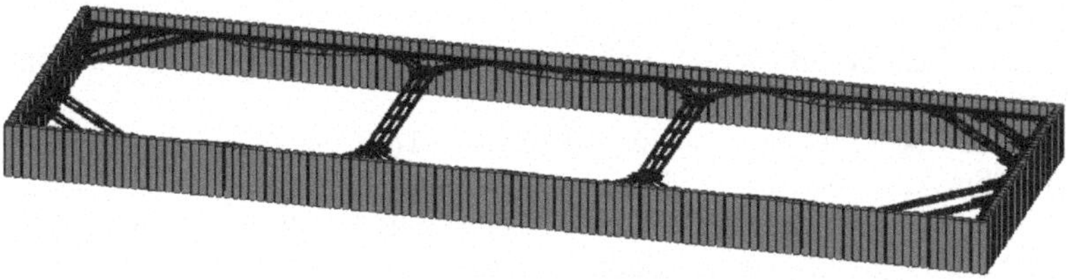

Fig. 2. Three-dimensional model map

2.3 The Realization of Secondary Development Programming

First, a prompt window is given to enable the user to choose whether to perform structural calculation or not. Then in the structure calculation, we need to obtain enough calculation information. There are two aspects of information to be obtained in the calculation process. One is the internal force design value obtained by the finite element simulation of the integral internal support, which is used to check the calculation results. On the other hand, the ratio of the prestress value of the component to the design value of the axial force of the component needs to be confirmed by the designer. Therefore, it is necessary to build an interactive windows form for data acquisition. After obtaining and storing the corresponding data, the calculation formula is programmed. For each calculation formula written as a method, easy to call in the main function. Take the calculation of prestress of steel strand as an example to illustrate:

```
private void GangJiaoXianPrestress(double q ,double ln ,double s, Variable va)
{double a=Math.Sin(s*3.1415926535/180);
  va.NGangJiaoXian = 1.1 * q * ln* 0.5 / a;
  va.NNGangJiaoXian = Convert.ToInt32((va.NGangJiaoXian * 1000 / 0.8 /
1320 / 140) * 1.05); }
```

After all formulas are written and calculated, a calculation book needs to be formed and exported. The generation of the calculation book needs to use word secondary development function, the basic idea of secondary development is: write the template file of the calculation book, the format is.docx, edit the part of the calculation book that needs to be rewritten as a bookmark, and then obtain the prepared calculation book template document in VS, identify the bookmark in the document, and then rewrite the content of the bookmark.

In the secondary development process, you should first add Microsoft office references about word.

The ability to rewrite the content of the bookmark is then written as a method to facilitate a loop call in the main function. The code of the method is:

```
private string GetBookContent( string name,Variable v)
    {string value = null;
     switch (name)
     {case "excavation length": value = Convert . ToString ( Convert .ToInt32 (
v . Length) ); break;
      case "excavation width": value = Convert . ToString ( Convert .ToInt32 (v
. Width)); break;
      case " excavation depth": value = Convert . ToString ( Convert .ToInt32 (v
.JKDepth)); break;
      case " soil pressure ": value = Convert . ToString ( Convert .ToInt32 (v .
Ptuyali)); break;
         case "steel strand Axial force design value ": value = Convert . ToString
( Convert .ToInt32 (v.NGangJiaoXian)); break;
      ......
    }return value; }
```

The code for calling a method in the main function is:

```
The code for calling a method in the main function is:
Microsoft.Office.Interop.Word.Application app = new MSWord.Application();
string version = app.Version;
Microsoft.Office.Interop.Word.Document docu = app.Documents.Add(@"C:\
Desktop\family\ calculation book.docx"); //Get Calculator Template Document
docu.ActiveWindow.Visible = true;
foreach (Bookmark bk in docu.Bookmarks) //Recognize bookmarks in the doc-
ument and call methods to complete the rewriting of the bookmark content
{bk.Range.Text = GetBookContent(bk.Name, va); }
```

After the calculation book is generated, a prompt window is generated.

2.4 Statistics of the Number of Fish-Bellied Excavation Components

Realizing the statistics of the number of components can facilitate the designer to calculate the engineering quantity. The statistics of the number of components include the following parts: the types of components, the models and specifications of components, the number and the total weight. The data extraction comes from two aspects: one is the component selection information input by users in the early stage of modeling, and the other is the process of model building load family. Take the statistics of the components surrounding purlin bracket as an example to illustrate. Different models correspond to different bracket specifications. The statistical number of brackets can be obtained by cyclic statement assignment when loading bracket family.

3 Summary

In this paper, a standard parametric family library for fish-bellied excavation is established by applying the above plane layout method to the fish-bellied excavation with the help of Revit secondary development. Moreover, the interactive interface of programming design in Revit is completed. The internal force calculation, pre-stress value calculation, component strength and stability checking calculation are realized by programming. Finally, the statistics of the number of fish-bellied excavation components are realized by programming. Through the above secondary development, the rapid modeling of the fish-bellied steel support can be realized.

This simulation design module has been applied in the engineering project practice of China Construction Bureau and the teaching of our school, and has achieved good results, and is rated as Beijing level virtual simulation experiment project. It is believed that in the following practice, it can effectively boost the application of fabricated steel support system of prestressed fish-bellied beams and achieve better development.

References

1. Zhu, X., Zhao, Y., Fang, R.: Analysis of advantages of prestressed fish-bellied steel structure support in deep excavation system. Building Mater. Dev. Guidance **15**(20), 83–85 (2017). https://doi.org/10.16673/j.cnki.jcfzdx.2017.0113

2. Kyung, K.N., Sik, P.J., Joon, J.H., et al.: Performance of innovative prestressed support earth retention system in urban excavation. J. Korean Geotech. Soc. **21**(2), 27–36 (2005)
3. Technical specification for fabricated steel support system consisting of prestressed fish-bellied beams in excavation: T/CCES 3-2017. China Architecture and Construction Press. Beijing (2017). ISBN: 15112·30171
4. Jx, W.: Research on Design Optimization of Fish-bellied Excavation Support System and BIM Software Development. China University of Mining and Technology (Beijing), Beijing (2019)
5. Pang, S., Zhou, Z., Ma, K., et al.: Analysis of displacement influence of span of prestressed fish-belly beam on foundation pit plane. In: The 10th International Conference on Civil Engineering. Nanchang (2023)
6. Pang, S., Ma, K., Zhou, Z., et al.: Study on Influence of Tension Height of Steel Strand on Displacement of Excavation Supported by Fish-bellied beams. In: The 10th International Conference on Civil Engineering. Nanchang (2023)
7. Xu, Z., Zhang, Y.l., et al.: Revit secondary development manual creation of bridge component parametric model. J. Jilin Jianzhu Univ. (06), 28–32 (2023).
8. Li, Y., Cheng, H., et al.: Secondary development of c#-based prefabricated component library with parametric modeling. J. Inform. Technol. Civil Eng. Architect. (06), 101–105 (2023). https://doi.org/10.16670/j.cnki.cn11-5823/tu.2023.06.17
9. Saeed, H.R.U., Aamer, M.R., Rehan, M., et al.: A multi-facet BIM based approach for Green Building design of new multi-family residential building using LEED system. Int. J. Constr. Manag. **23**(12), 2024–2038 (2023)

Key Technology of Intelligent Construction of Cast-In-Place Bridge Tower Based on Factory-Like Working Conditions

Yang Li[1,3](✉), Bazhou Li[1,3], and Wei Tian[2,3]

[1] CCCC Wuhan Harbor Engineering Design and Research Institute Co., Ltd., Wuhan 430040, China
896083327@qq.com
[2] CCCC Second Harbor Engineering Company Ltd., Wuhan 430014, China
[3] Hubei Key Laboratory of Advanced Materials and Reinforcement Technology Research for Marine Environment Structures, Wuhan 430040, China

Abstract. To cater to the need for safe construction practices within China's transportation infrastructure, this paper introduces key technologies for the intelligent construction of cast-in-place bridge towers under factory-like conditions, not only enhances the quality and efficiency of construction projects, but also marks a preliminary step towards the intelligent construction of cast-in-place bridge towers.

Keywords: super-high bridge towers · factory-like conditions · tower building machine · reinforcement components

1 Introduction

As the foundation of economic and social development, the construction of large-scale transportation infrastructure has strong demand for long-span bridge construction in China and even in the world, which brings problems such as harsh working environment at altitude, high construction risk and difficult to guarantee project quality. Due to its high stiffness, good corrosion resistance and suitable cost, concrete bridge tower is the most widely used in China, accounting for more than 90%, and has always been the focus and difficulty of long-span bridge construction, accounting for about 1/3 of the construction period of long-span bridge [1, 2].

2 Current Status of Cast-In-Situ Bridge Tower Construction Domestically and Internationally

At present, the construction of ultra-large span cable-supported bridges in China is thriving. With the advancement of bridge construction technologies and building materials, the spans and heights of bridges are increasing. According to statistics, there are 59

B. Yuan et al. (Eds.): ICCASE 2024, SUCI, pp. 63–69, 2024.
https://doi.org/10.1007/978-3-031-76102-7_8

bridges in China with main spans equal to or greater than 400 m, and the tallest tower column in the world has reached a height of 332 m [3]. Concrete tower columns, due to their high rigidity, superior mechanical properties, and low maintenance costs, are most widely applied in ultra-large span bridges [9].

The current construction process for cast-in-situ concrete tower columns in China is characterized by numerous operational procedures, high labor intensity, and a low level of integration, marking it as a typical high-energy consumption, high-risk, labor-intensive, and extensive industry.

Improving the quality and efficiency of construction for super-high concrete bridge towers and reducing safety risks in high-altitude operations in China urgently require an upgrade to the current construction mode to change the backward situation of cast-in-situ scenarios.

3 Intelligent Construction System for Cast-In-Situ Bridge Towers Based on the Concept of Factory-Like Operations

3.1 Inspiration from Intelligent Manufacturing

The future of industrialized construction is moving in two directions: one is the factory production of prefabricated structures with assembly on-site, and the other is the construction of a system for the intelligent construction of cast-in-situ structures, which remains irreplaceable for large bridges and super-tall buildings, centered around the collaborative operation of equipment groups [7].

The construction industry, including bridge building, has been slow to upgrade due to the difficulty in standardizing products and numerous constraints in the construction process. In recent years, scholars and experts have begun to explore the development of intelligent construction. For instance, Academician Ding has pointed out that intelligent construction is not just a transformative innovation in construction technology but will also reshape the construction industry from aspects such as product form, construction methods, business philosophy, market structure, and industry management [5]. Academician Zhang believes that the core of intelligent construction should include intelligent materials, structures, construction processes, and maintenance [8].

3.2 Intelligent Construction of Cast-In-Situ Bridge Towers Based on Factory-Like Concepts

The research related to the intelligent construction system for cast-in-situ structures is still in its initial stages. For instance, equipment development often focuses on localized upgrades with less emphasis on integrated R&D efforts; there is insufficient digital integration of manpower, machines, and materials on-site, and coordination and control over equipment collaboration need strengthening. Construction robots are more suited to standardized, relatively simple cast-in-situ components, with less involvement in complex concrete components.

Realizing intelligent construction during the bridge engineering construction phase (specifically, from design drawings to the project delivery phase) requires a core focus on digital construction. A comprehensive and systematic technical system must be established, enabling precise perception, simulation, analysis, and control throughout the construction process to progressively achieve the construction objectives (Fig. 1).

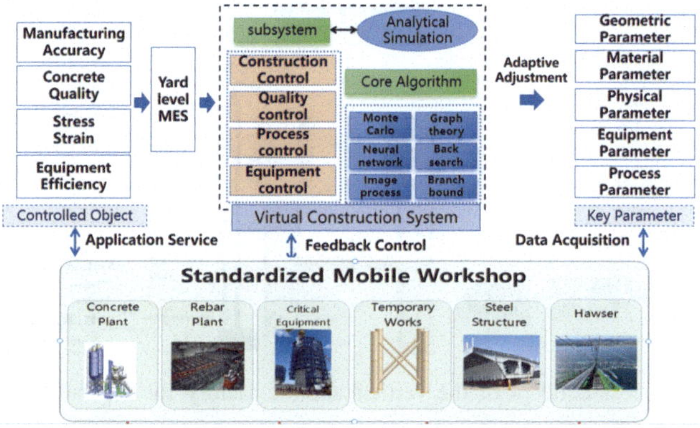

Fig. 1. Bridge intelligent construction system based on the mobile factory concept.

This paper proposes transforming the on-site construction of concrete structures in complex or special construction environments into mobile factories, shifting some cast-in-situ operations to backend assembly line construction, and completing the necessary front-end operations at factory-like fixed workstations established on-site. Analogous to the industrial production stages of prefabricated components, this involves creating standardized mobile workshops and conducting research into integrated construction equipment, assembly of reinforcement components, automation in concrete construction, and digital quality control technologies.

4 Key Technologies for Intelligent Construction of Cast-In-Situ Bridge Towers Based on Factory-Like Concepts

In constructing super-high concrete bridge towers, drawing from industrial manufacturing concepts involves transferring most steel reinforcement and concrete operations to backstage standard workshops and establishing connections with the supply systems for concrete and steel, the primary raw materials. In these workshops, raw materials (steel, aggregate, cement, etc.) are processed into semi-finished products (reinforcement components, concrete), which are then fabricated into final products (bridge towers) in the construction forefront, thus transforming the site construction of concrete structures under complex or special environments into a mobile factory. This creates a "vertical mobile factory" for the industrial construction process of concrete bridge towers. Research needs to be conducted in three key areas: integrated intelligent tower building technologies to provide a factory-like environment on-site; rapid assembly technologies for bridge tower reinforcement components to significantly reduce the time spent on steel reinforcement projects on-site; and quality feedback control technologies for cast-in-situ concrete, utilizing automation to enhance the construction efficiency and quality of cast-in-situ concrete (Fig. 2).

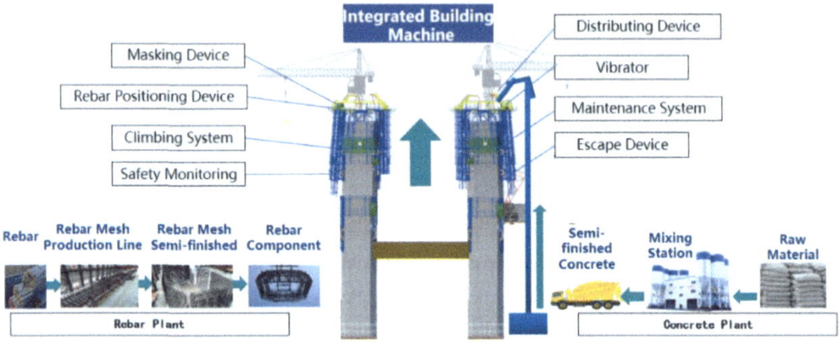

Fig. 2. Overall process concept of the vertical mobile factory for super-high concrete bridge towers.

4.1 Integrated Intelligent Tower Building Technologies

Hydraulic climbing formworks, serving as the primary construction equipment for most bridge towers and merely functioning as platforms for personnel operations, fail to meet the requirements for factory-like production. Hence, it is necessary to establish an intelligent factory-like tower-building machine and its supporting equipment on concrete bridge towers to conduct "integrated" construction in an assembly line manner.

Fig. 3. Bridge Intelligent Construction System Based on the Mobile Factory Concept.

This equipment, referred to as the tower-building machine, covers four segments of the bridge tower in height. According to the construction characteristics of concrete bridge towers, the tower building machine's framework is spatially divided into functional zones for steel binding, concrete pouring, concrete curing, and surface finishing, designated as the pouring work layer, curing work layer, finishing work layer, and elevator entrance layer, respectively (Fig. 3).

4.2 Rapid Assembly Technologies for Bridge Tower Reinforcement Components

The rapid assembly construction of bridge tower reinforcement components transfers most of the tower's steel reinforcement from site construction to a ground-level factory for industrial production of steel mesh. Using an assembly method for segment assembly of tower columns enhances the construction quality of tower columns and reduces safety risks. Inside the steel processing factory, steel mesh is produced almost entirely through mechanization. Then, the steel mesh is transported to the bridge site, where the entire segment of the bridge tower's reinforcement components is assembled within a specialized frame, followed by the overall lifting and rapid high-precision docking of segment reinforcement components using tower cranes and rigging, achieving assembly construction of large bridge tower steel reinforcement, significantly shortening the duration of steelwork operations, reducing the number of construction personnel on high towers, and lowering safety risks.

4.3 Quality Feedback Control Technology for Cast-In-Situ Concrete

The production of concrete involves the collaboration of multiple trades, with its quality defects intricately linked to the process control of production, transportation, pouring, compaction, and curing of concrete. Through research on quality feedback control technology for cast-in-situ concrete, it's possible to achieve real-time quantification of changes in concrete properties during the processes of pouring, compaction, and curing, and adjust control measures accordingly. This enables a more scientific and rational guarantee of quality control throughout the entire cycle of concrete in cast-in-situ scenarios, reducing the occurrence of concrete quality defects and thereby crafting quality concrete projects [4].

In terms of automation in pouring, addressing the issues of over-compaction and under-compaction in bridge concrete construction, due to narrow spaces and dense steel reinforcement, a coiled concrete variable frequency compaction device and linked servo control system have been developed. This not only ensures the quality of concrete compaction but also doubles the on-site construction efficiency. To tackle the challenge of controlling the appearance quality of cast-in-situ concrete in real time, an integrated smart template featuring industrial cameras and intelligent visual recognition technology has been devised. Utilizing the YOLO-v5 deep neural network model for systematic detection and intelligent judgment of concrete appearance, in conjunction with an intelligent compaction system, it achieves real-time smart identification and targeted elimination of surface air voids in concrete, ensuring the quality of concrete appearance with surface air void area ratios less than 0.2%.

5 Engineering Application Demonstration

The Shenzhen-Zhongshan Corridor project, located at the mouth of the Pearl River, the bridge faces challenging construction conditions, including open waters, frequent strong winds, and a harsh environment characterized by high temperatures, humidity, and salinity [6]. The project applied key technologies for the intelligent construction of cast-in-situ bridge towers based on factory-like operating conditions. The project realized seven days of enclosed curing, enhancing concrete quality, achieving construction progress of 1.2 m per day per tower column segment, requiring only 8–10 workers for high-altitude steel reinforcement tasks—reducing labor by 60%, achieving main reinforcement placement accuracy within 3 mm, stirrup placement accuracy within 2 mm, and a 100% qualification rate for steel reinforcement protective layer thickness. Compared with other similar technologies, this technology has the following advanced nature (Table 1).

Table 1. Comparison of technical advantages

Category	Similar Method	Integrated Tower Builder
Concrete Placement	Hand Casting	Mechanical Castingc
Concrete Vibration	Manual Vibration	Mechanical Vibration
Concrete Curing	Non-maintenance	7 Days Maintenance Time
Personnel Input	18	9
Construction Efficiency	0.6m/d	1.2m/d

6 Conclusion

To meet the need for safe construction for China's transportation infrastructure, the author proposes key technologies for the intelligent construction of cast-in-situ bridge towers based on factory-like operating conditions, centering on generalized factory-like intelligent equipment and employing digital twin platforms. By modularizing and partitioning cast-in-place components and adopting an industrial assembly line production method for cast-in-place concrete components, an intelligent tower-building machine suited to factory-like operating conditions was developed. This innovation enhanced project quality and construction efficiency, achieving the essential safety requirements of mechanization reduction in labor, automation substitution for labor, and intelligent minimization of labor.

References

1. Qiuhong, X., Xiangyang, L.: The application of fully intelligent controlled hydraulic steel formwork in the prefabrication of box girders for the Shenzhen-Zhongshan corridor. World Bridges **47**(6), 36–40 (2019)

2. Wen, S., Zhilong, W., Guohua, J., et al.: Integrated modular factory assembly of hydraulic climbing formwork and rapid dismantling construction technology. Construct. Technol. **48**(20), 43–45 (2019)
3. Dongzhi, W., Hang, Y., Wen, Z.: Integrated intelligent construction technology and experimental study for super high cable towers. Highways **5**(5), 122–126 (2021)
4. Hanbin, Z.: Key construction technologies of the Izmit bay bridge in Turkey. Construct. Technol. **49**(3), 29–32 (2020)
5. Ding, L.: Reflections on cultivating innovative engineering and technological talents in intelligent construction. Higher Eng. Educ. Res. (05), 1–4+29 (2019)
6. Weile, C.: Intelligent construction of the Shenzhen-Zhongshan corridor. China Highways **17**, 52–54 (2019)
7. Anonymous: ACI excellence in concrete construction awards announce a new bridges category. Concrete Int. **46**(4), 30–30 (2024).
8. Reitsema, A.D., Luković, M., Grünewald, S., Hordijk, D.A.: Future infrastructural replacement through the smart bridge concept. Materials **13**(2), 405 (2020)
9. Hossain, R.M., Waliullah, M.S.: Identification of necessity behind the construction of the third Shitalakhya bridge based on level of service (LOS) measurement parameters applicable for developing countries. J. Urban Manage. **13**(1), 88–106 (2024)

Research on the Mechanical Performance of Asymmetric Continuous Beam Bridge Ball Joint Based on Horizontal Rotation Construction

Kun Guo[1(✉)], Xiangheng Niu[2], He Feng[1], Kui Huang[1], Hualong Li[1], Yinjun Jing[1], and Zhengfei Zhao[1]

[1] China Construction Third Engineering Bureau Group Northwest Co., Ltd., Xi'an, China
gk_changan@163.com

[2] CCCC First Highway Consultants Co., Ltd., Xi'an, China

Abstract. For large-span concrete continuous beam bridges constructed by the horizontal rotation method, the force on the concrete ball joint during the horizontal rotation process is complex, and simplified algorithms are often used in practice. However, this simplified calculation generally does not consider the uneven distribution of load and changes in friction coefficient, resulting in large errors and a large calculation amount. This paper conducted a specific study on the elevated project of the North Ring Road for cross freight transportation on Eurasia Avenue in Xi'an International Port Area. A refined finite element model of the ball joint was established using ANSYS, and the actual stress state of the ball joint was simulated; And considering the complex stress state and deformation of the ball joint under different load distribution forms and friction coefficients, the stress change curves under different parameter settings were drawn based on the yield strength theory. The results indicate that the research method presented in this article is effective and feasible. The established model can effectively simulate the situation of ball joints during construction. In response to changes in load and friction coefficient, adjustments can be made in the model in a timely manner, effectively solving the problem of traditional simplified calculations.

Keywords: Rotating construction · Adjustable height ball joint · Large span continuous beam · Finite element analysis

1 Introduction

The application of swivel bridges is becoming increasingly widespread, and the key component is the ball joint that bears the gravity of swivel construction [1]. The force characteristics of the flat rotating ball joint are complex, and the theoretical calculation method of the ball joint widely uses a static state calculation model, which has a large error. The current standard adopts a simplified calculation method to simplify the contact surface of the ball joint into a plane contact calculation model, and simplifies the stress of the ball joint into a plane uniformly distributed force [2].

B. Yuan et al. (Eds.): ICCASE 2024, SUCI, pp. 70–78, 2024.
https://doi.org/10.1007/978-3-031-76102-7_9

In practice, the upper part of the ball joint transfers the load to the lower turntable through the concrete upper turntable, and the size, shape, and diffusion angle of the upper turntable all affect the final force transmission path [3]. In terms of structure, due to the imbalance of the upper structure of the bridge, the load distribution of the upper turntable is initially uneven [4].

On the basis of the calculation method for the static state of the ball joint [5, 6], this article analyzes the actual stress state of the ball joint contact surface according to the yield strength theory of material mechanics. By comparing the friction coefficient and different load distribution forms of the ball joint, and combining with the finite element analysis software ANSYS, a finite element model is established. Through practical engineering cases, the force and transmission mechanism of the ball joint suitable for different load distribution states are provided.

2 Engineering Survey

The entire bridge of this project is located on a straight line, and the main bridge is set up with 2 × 60 m prestressed concrete continuous beams in sections, with a length of 120 m. The width of the left bridge is 24.3 m, and the width of the right bridge is 20.1 m. The main beam is designed with equal width and follows the road line through markings. The minimum distance between the bottom of the bridge and the rail surface is 12.97 m for the left bridge and 11.79 m for the right bridge; The main bridge adopts the planar pivot construction method to cross the existing railway line, and adopts a pivot system with ball joint center support as the main support and ring road support as the auxiliary.

The swivel structure is set at the top of the bridge pier, consisting of a turntable, ball joint (including adjustable height mechanism), support feet, circular slide, traction system, and other parts. The weight of the swivel is 14500t on the left and 11000t on the right. The lower turntable serves as the foundation for supporting the entire weight of the rotating structure, and is equipped with a lower ball joint and a circular downward slide of the rotating structure system. The upper turntable is an important structure of rotation, forming a multi-directional and three-dimensional force state throughout the entire rotation process. The diameter of the turntable is 11.0 m for the left bridge and 9.5 m for the right bridge, with a height of 1.25 m. The turntable is the part where the ball joint, support foot, and upper turntable are connected, and it is also the part where the traction force of the turntable is directly applied. The rendering of this project is shown in Fig. 1.

3 Theoretical Calculation Method for Ball Joint Strength

The theoretical calculation method for ball joint strength is based on the model established under the actual stress situation during the rotation of the ball joint [7]. This calculation method is based on the derivation of contact stress calculation theory for solving spatial problems in elasticity, considering the influence of frictional force generated by the traction torque of the ball joint. The contact part of the translational ball joint is locally abstracted as a half space body under uniformly distributed stress, And calculate the three-dimensional principal stress of the complex stress state of the half

Fig. 1. 3D Realistic View of the Project

space body, and obtain the strength condition of the normal stress on the ball joint based on the material yield strength theory [8].

3.1 Solving Contact Surface Stress in Half Space by Contact Stress Method

The calculation model for the contact stress on the contact surface of a half space body is shown in Fig. 2.

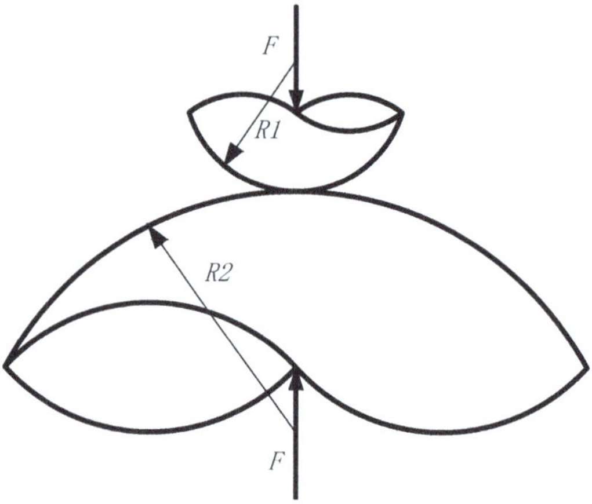

Fig. 2. Simplified diagram of contact stress calculation for half space bodies

According to the knowledge of elastic mechanics in solving the contact problem of a half space body [9], when the half space body contacts at the boundary and the Poisson's ratio is 0.3, the expression for the maximum contact stress is

$$q = 0.388\left[\frac{FE^2(R_1 + R_2)^2}{R_1^2 R_2^2}\right]^{\frac{1}{3}} \tag{1}$$

E: The elastic modulus of the contact material;

F: The gravity of the upper structure of the ball joint;

R_1 and R_2: The radii of the lower and upper spherical joints, respectively. For the structural characteristics of the spherical joint of the rotating construction bridge, R_1 takes a negative value and R_2 takes a positive value.

3.2 Establishing a Uniformly Distributed Load Model for a Half Space Body

Enlarge the contact boundary of the two ball joints locally, examine the structural characteristics and stress state of the maximum contact stress contact part [10]. Due to the curved surface of the ball joint contact surface and the uniformly distributed load on the ball joint, calculate the stress state of the contact surface according to the model of uniformly distributed load on the half space body. The force state is shown in Fig. 3.

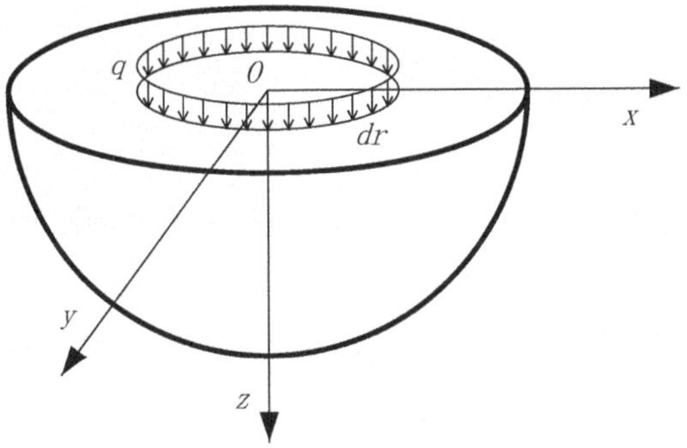

Fig. 3. Uniform load on half space body

According to the method of solving spatial problems in elasticity, the derivation formula under uniformly distributed loads is as follows [11]:

$$\sigma_z = -q\left[1 - \frac{z^3}{\left(z^2 + a^2\right)^{\frac{3}{2}}}\right] \tag{2}$$

$$\sigma_x = \sigma_y = -\frac{q}{2}\left[(1 + 2\mu) + \frac{z^3}{\left(z^2 + a^2\right)^{\frac{3}{2}}} - \frac{2(1 + \mu)z}{\left(z^2 + a^2\right)^{\frac{1}{2}}}\right] \tag{3}$$

According to the formula, it is necessary to find the extremum of the z-function. By calculation, it is obtained that when $a = 0$, i.e. directly below the half space body, σ_z takes the extreme value $-q$; According to engineering practice, Poisson's ratio $\mu = 0.3$, when $z = 2.55a$, σ_x, σ_y Take the extreme value $-0.0068q$.

4 Structural Modeling and Analysis

4.1 Finite Element Model

This chapter calculates and analyzes the stress situation of ball joint structures under different parameters, mainly studying two aspects: The influence of unbalanced torque of the upper structure on the ball joint structure, and the influence of curvature radius of the sliding surface on the ball joint.

Taking the left span bridge of this project as the research object, the upper part of the ball joint bears a rotating weight of 14,500 tons, and the friction coefficient of the sliding surface is 0.06. The finite element model is simulated using ANSYS, and the sliding surface between the upper and lower ball plates is simulated using contact elements. The finite element model is shown in Fig. 4.

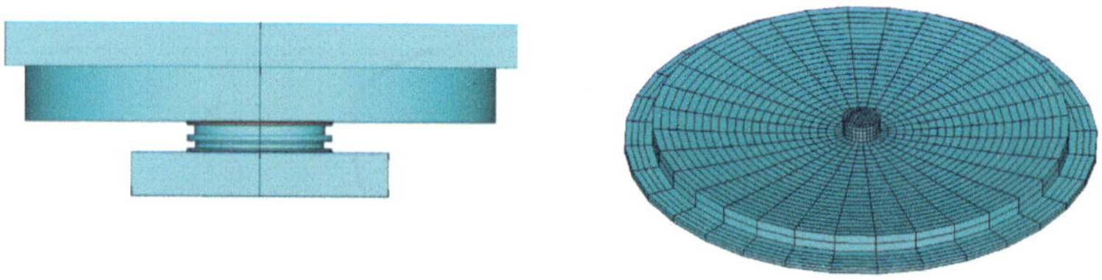

Fig. 4. Finite element model of ball joint structure

4.2 Analysis of the Influence of Unbalanced Moments on Ball Joint Structures

Seven sets of analysis data are provided in the article. While keeping the total load of the upper rotating body constant, the eccentricity values of 0 m, 0.05 m, 0.1 m, 0.15 m, 0.2 m, 0.25 m, and 0.3 m are taken for analysis. The stress state of the ball joint structure is obtained as follows. The calculation results are shown in Figs. 5 and 6 (Table 1).

Table 1 Unbalanced bending moment values corresponding to different eccentricities

Eccentricity (m)	0	0.05	0.1	0.15	0.2	0.25	0.3
Unbalanced moment (kN·m)	0	7250	14,500	21,750	29,000	36,250	43,500

From the above analysis results, it can be seen that when the eccentricity is 0.1 m, compared with the working condition without eccentricity, the maximum Von Mises stress of the upper ball plate increases by 2.1%; The maximum Von Mises stress on the lower ball plate increases by 5.3%; The contact stress on the contact surface increases by 3.5%. When the eccentricity is 0.2 m, compared with the working condition without eccentricity, the maximum Von Mises stress of the upper ball plate increases by 5.8%;

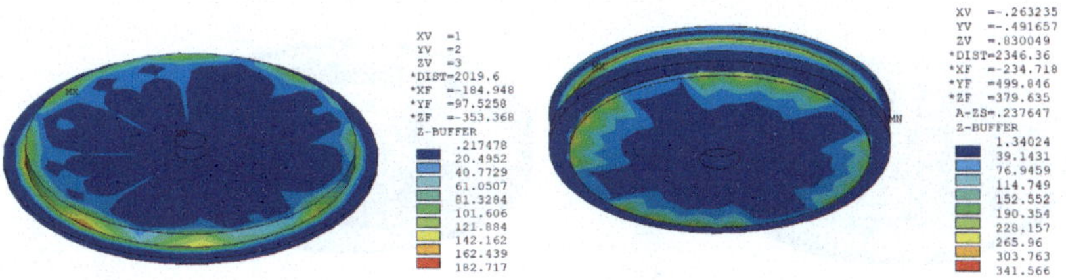

Fig. 5. Von Mises stress on the lower and upper ball plate at an eccentricity of 0.3 m

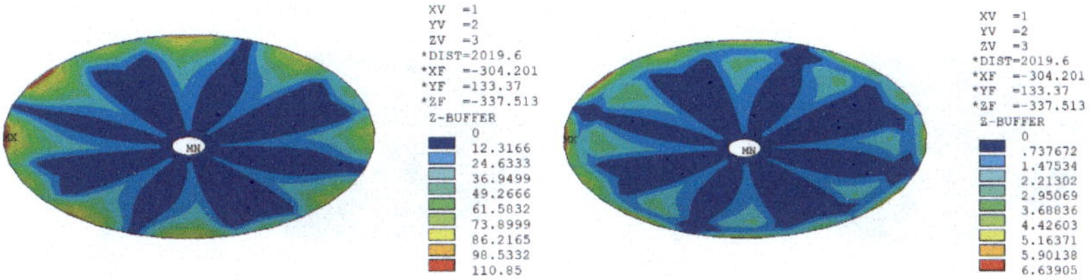

Fig. 6. The total stress and friction force of the contact surface at an eccentricity of 0.3 m

The maximum Von Mises stress on the lower ball plate increases by 10.1%; The contact stress on the contact surface increases by 8.1%. When the eccentricity is 0.3 m, compared with the working condition without eccentricity, the maximum Von Mises stress of the upper ball plate increases by 10.3%; The maximum Von Mises stress on the lower ball plate increases by 14.3%; The contact stress on the contact surface increases by 12.3%.

After calculation, when the eccentricity of this bridge is greater than 0.38 m, the local Von Mises stress of the upper ball plate will exceed its yield strength of 355 MPa, and yield phenomenon will occur here. At this time, the corresponding unbalanced load on the left and right sides is 9180 kN. Therefore, when pouring the main beam with cantilever, it is necessary to avoid an unbalanced load greater than 9180 kN on both sides.

4.3 Analysis of the Influence of Curvature Radius on Sliding Surfaces

The curvature radii R of the sliding surface of the ball joint are taken as 9 m, 13 m, 17 m, 25 m, 50 m, and infinity for analysis, and the influence of different curvature radii on the ball joint structure is studied. The calculated location is shown in Fig. 7. The calculation results are shown in Figs. 8 and 9.

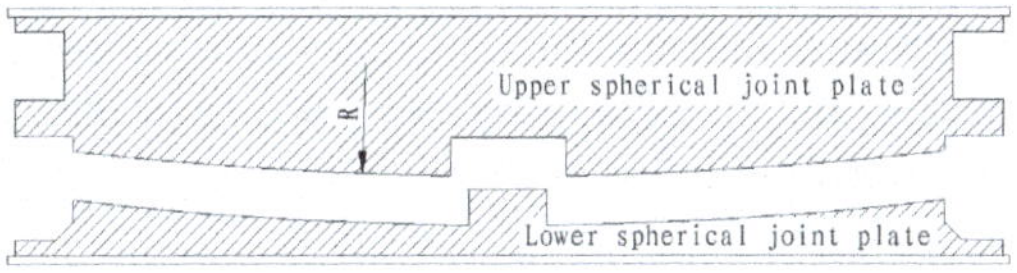

Fig. 7. Schematic diagram of curvature radius of ball joint

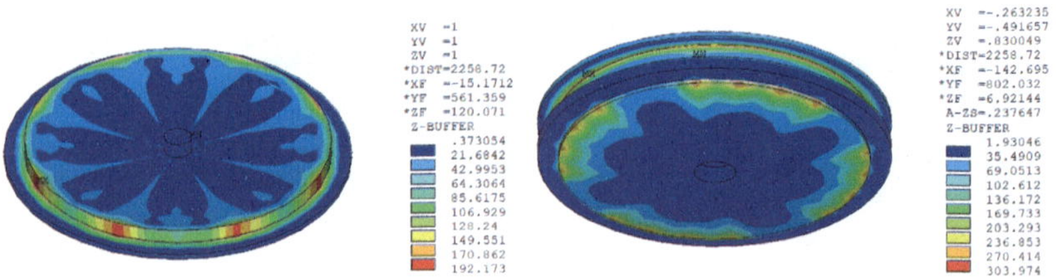

Fig. 8. Von Mises stress in the lower and upper spherical plate with a curvature radius of 9 m

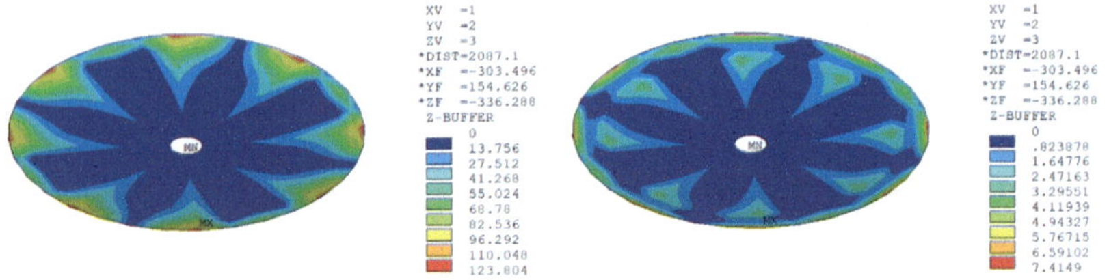

Fig. 9. The total stress and friction stress of the contact surface at a curvature radius of 9 m

From the above analysis, it can be seen that the curvature radius has a significant impact on the ball joint structure. As the curvature radius increases, the Von Mises stress on the upper and lower ball plates decreases, but the trend becomes smoother. From this perspective, when the curvature radius of the ball joint increases to a certain value and then continues to increase, the effect of stress reduction is not significant.

The curvature radius used for the ball joint in this project is about 13 m. In comparison, when the curvature radius R = 9 m, the maximum Von Mises stress on the upper ball plate increases by 14.3%; The maximum Von Mises stress on the lower ball plate increases by 3.0%; The contact stress on the contact surface increases by 20.9%; When the curvature radius R = 17 m, compared with the working condition with a curvature radius of 13 m, the maximum Von Mises stress of the upper ball plate is reduced by 11.2%; The maximum Von Mises stress on the lower ball plate is reduced by 1.8%; The contact stress on the contact surface is reduced by 14.4%. When the curvature radius R is 25 m, compared with the working condition with a curvature radius of 13 m, the maximum Von Mises stress on the upper ball plate is reduced by 29.3%; The maximum Von Mises stress of the lower ball plate is reduced by 3.9%; The contact stress on the contact surface decreased by 31.4%. When the curvature radius is further increased, the decrease in Von Mises stress on the upper and lower ball plates tends to be gradual. After the curvature radius R > 50 m, the stress on the upper and lower ball plates remains almost unchanged. The contact stress of the ball joint shows a trend of first decreasing and then increasing with the increase of curvature radius. When the curvature radius is 25 m, the contact stress reaches its minimum and then gradually increases. After the curvature radius R > 50 m, the contact stress remains almost unchanged.

5 Conclusions and Suggestions

This article takes the under construction railway section bridge in the northern part of Xi'an, China as the engineering background, and based on the structural characteristics of the swivel bridge, conducts in-depth research on the stress performance and structural parameters of the ball joint during the swivel construction process. Taking the equivalent stress of the ball joint structure as the control objective and considering the contact stress of the contact surface, sensitivity analysis was conducted on the two main design parameters of curvature radius and eccentricity. The stress performance and distribution law of each part in the rotating system were studied, providing theoretical basis and technical support for the structural design and construction of the ball joint. The main research conclusions of this article are as follows:

- By using finite element software, a local finite element analysis was conducted on the rotating system of the rotating bridge, comparing and verifying the theoretical values and finite element results of the stress on the ball joint contact surface, and analyzing the safety of the rotating system. During the process of rotating construction, there may be adverse working conditions such as unbalanced torque, which can affect the contact state and stress distribution of the concrete ball joint. After calculation and analysis, it is proposed that the maximum limit of the eccentric load that this bridge can withstand is 9180 kN, which provides guidance for the subsequent rotation construction of this bridge.
- For the ball joint structure of this bridge, the contact stress of the ball joint shows a trend of first decreasing and then increasing with the increase of curvature radius. At a curvature radius of 25 m, the contact stress reaches its minimum, and as the curvature radius increases, the contact stress begins to increase and tends to flatten. The curvature radius of the ball joint used for the cross railway bridge in the research background of this article is 22.5 m, which is relatively close to the optimal curvature radius studied in this article. The results of this study can provide reference and guidance for the construction of similar large tonnage swivel bridges in the future.
- The ball joint, as the core component of the rotating construction bridge, is a crucial part of the rotating construction. However, there is currently limited theoretical analysis on the design parameters of the ball joint, and many design parameters are determined based on engineering practice summaries. In view of this, this article systematically studies the force behavior of the ball joint through local analysis based on the characteristics of rotational construction, providing theoretical support for further optimizing the rotational device and ensuring construction safety.

Acknowledgements. This paper is sponsored by Shaanxi Provincial Construction Science and Technology Plan Project (Program No.2023-K87).

References

1. Zhang, S.: Research on Key Structures and Stress Performance of Asymmetric Single Tower Cable Stayed Bridge during Rotation Construction. Chongqing Jiaotong University (2017)

2. Yu, P.: Research on Construction Control and Spherical Joint Parameter Analysis of Large Tonnage Swiveled Cable Stayed Bridges. Chongqing Jiaotong University (2022)
3. Liu, T.: Research on Key Issues of Concrete Ball Joint in Large Span Bridge Rotation Construction. Wuhan University of Technology (2018)
4. Xu, C., Xu, X., Huang, J., et al.: Study on the mechanical properties of self weight unbalanced swivel bridge ball joints. J. Railw. Sci. Eng. **19**(10), 3063–3069 (2022)
5. Zhao, H.: Study on the mechanical properties of steel ball joints in large-span asymmetric rotating cable-stayed bridges. Industrial Architecture **51**(10), 116–121 (2021)
6. Chen, X.: Mechanical Performance Analysis of Asymmetric Swivel Bridge Rotating System. Lanzhou Jiaotong University (2023)
7. Liu, X., Wang, H.: Structural design of a railway crossing swivel bridge. Urban Architecture **20**(23), 198–201 (2023)
8. Zuo, M., Jiang, K.: Research on stress calculation method for the rotation process of the spherical hinge of a rotating bridge. Railway Standard Design **59**(12), 36–39 (2015)
9. Campos, L.T., Oden, J.T., Kikuchi, N.: A numerical analysis of a class of contact problems with friction in elastostatics. Methods in Applied Mechanics & Engineering **34**(1), 821–845 (1982)
10. Ohte, S.: Finite element analysis of elastic contact problems. JSME International Journal **16**(313), 2210–2216 (1973)
11. Parsons, B., Wilson, E.A.: A method for determining the surface contact stresses resulting from interference fits. J. Eng. Indu., 208–218 (1970)

Seismic Response Analysis and Seismic Isolation Design of Large-Span Through Arch Bridges

Xiaowei Deng[1(✉)], Zisong Gong[1], and Haiqiang Zhong[2]

[1] Tongji Architectural Design (Group) Co., Ltd., Shanghai 200092, China
sz22dxw@tjad.cn

[2] State Key Laboratory of Disaster Reduction in Civil Engineering, Tongji University, Shanghai 200092, China

Abstract. Large-span through arch bridges have complex structures and certain spatial coupling effect. Based on a 216 m long span through arch bridge with steel-box ribs and steel-box girders, a three-dimensional spatial finite element model is established to study the dynamic characteristics and seismic response through response spectrum analysis. The fixed bearings and the fixed piers are vulnerable parts for controlling seismic design. And then cable-sliding friction aseismic bearings (CSFAB) are adopted for seismic isolation design. The analysis results indicate that CSFAB can effectively reduce the seismic response of the substructure, and also prevent excessive displacement of beams. It is a cost-effective means of seismic isolation.

Keywords: Large-span through arch bridges · dynamic characteristics · seismic response · seismic isolation design · cable-sliding friction aseismic bearings

1 Introduction

The large-span through arch bridge are desirable for their economic feasibility and esthetic pleasure [1]. Formed by beams, arches, and suspenders, they have complex composite structures and certain spatial coupling effect in dynamic effects [2]. So the seismic behavior and seismic response analysis is relatively complex [3]. The design specifications also provide very little guidance and explanation for it. In view of this, in order to further study the the seismic performance and seismic isolation measures, a 216 m long span through arch bridge is taken as the background project. Three-dimensional spatial finite element model is established to study the dynamic characteristics and seismic response through response spectrum analysis, identifying the vulnerable parts of the bridge [4], and using nonlinear time-history analysis methods to study the seismic isolation effect of cable-sliding friction aseismic bearings (CSFAB) [5]. It can provide reference for seismic design of similar bridges.

B. Yuan et al. (Eds.): ICCASE 2024, SUCI, pp. 79–88, 2024.
https://doi.org/10.1007/978-3-031-76102-7_10

2 Project Overview

The project is located in Shenzhen, Guangdong Province. A new bridge will be built to cross over the Jihe Expressway, which is a double-layer transportation system, including 8 lanes on the ground and 8 lanes in the underground tunnels. The main bridge adopts a through arch bridge with a span of 216 m and a width of 28.5 m. The facade layout is shown in Fig. 1:

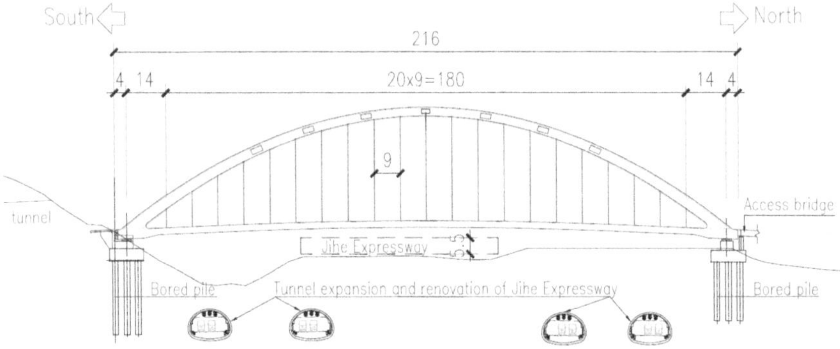

Fig. 1. The facade layout of the main bridge. (Unit: m)

The main arch adopts a basket shaped arch with a rectangular steel box section. The two arch ribs are inclined inward by 10°, with a calculated span of 208 m and a rise to span ratio of 1/5. The tie beam adopts double main beams, with the section of steel box girder, connected by several crossbeams. The main bridge contains a total of 21 pairs of suspenders, which are located at the aforementioned crossbeams with the longitudinal spacing of 9 m.

The north side is an embedded bridge abutment, while the south side is a double-column pier. The column adopts a rectangular cross-section, with a longitudinal dimension of 4.5 m and a transverse dimension of 5.0 m. The foundation below the bearing platform adopts D200 cm drilled cast-in-place piles, and the bearing layer at the pile end is slightly weathered sandstone.

3 Analysis of Structural Dynamic Characteristics

3.1 Finite Element Model

The structural analysis adopts the finite element software Midas/Civil 2021 to establish a three-dimensional finite element model. The beams and arches are made of Q345qD steel, while the abutment and the pier are made of C40 concrete, as well as the suspenders are made of wire 1670. The beams, the arches, the abutment and the pier are simulated using spatial beam elements, while the decks are simulated using plate elements, as well as the suspenders are simulated using cable elements. Secondary dead load is considered as additional mass. In order to consider the pile-soil interaction, pile foundations are simulated by equivalent soil springs. The model is shown in Fig. 2:

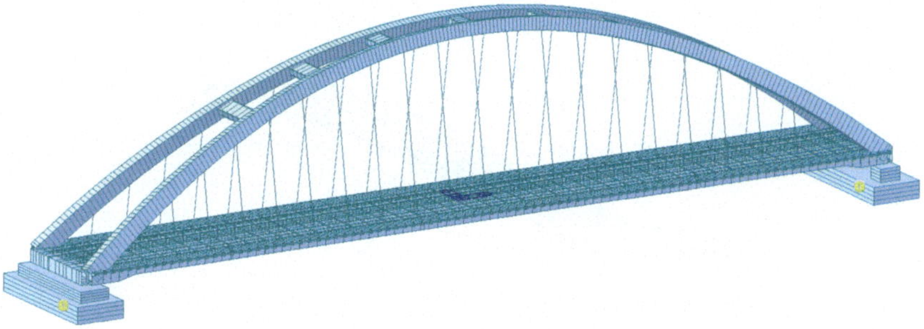

Fig. 2. Finite element model in Midas/Civil.

The main bridge has four bearings, including one fixed bearing (GD), two unidirectional movable bearings (DX) and one bidirectional movable bearings (DX). Each of them, with the vertical bearing capacity of 45000 kN, is located below the intersection of the main arch and the tie beam. The spherical bearings are initially adopted, simulated by elastic connections in the model, and the layout plan diagram is shown in Fig. 3:

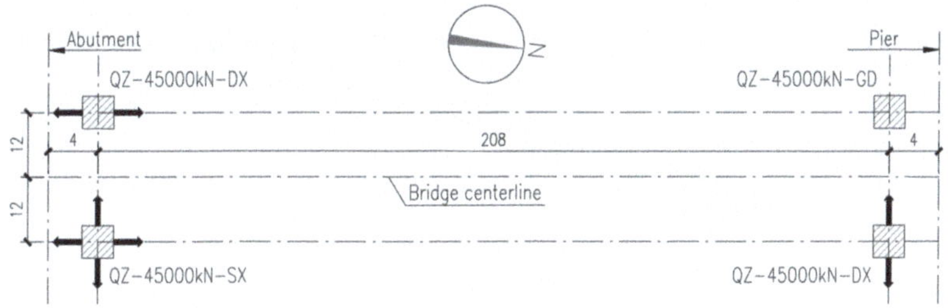

Fig. 3. The layout plan diagram of bearings. (Unit: m)

3.2 Dynamic Characteristic Analysis

Based on the analysis of dynamic characteristics, the first four vibration mode diagrams are shown in Fig. 4:

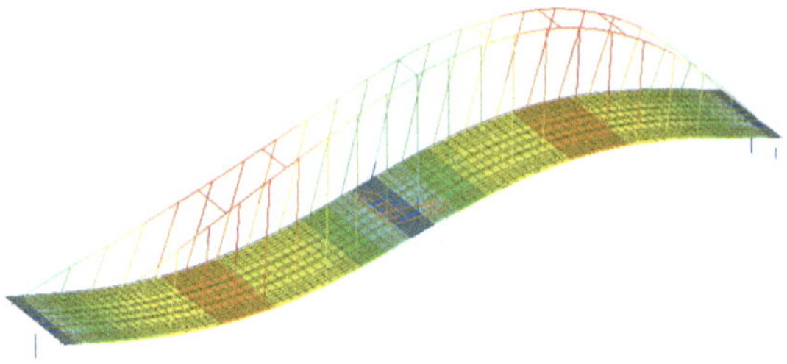

(a) First vibration mode

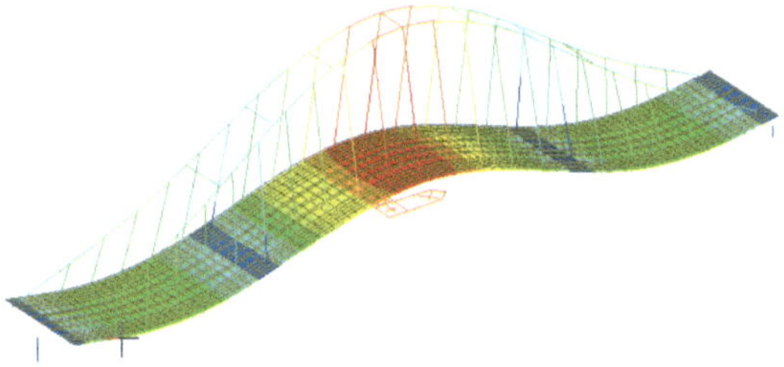

(b) Third vibration mode

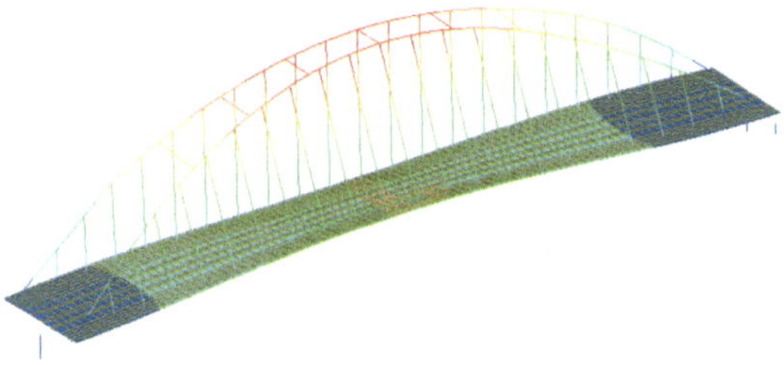

(c) Second vibration mode

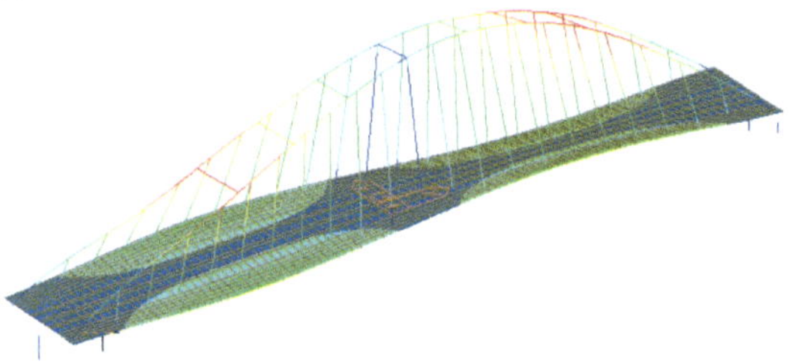

(d) Fourth vibration mode

Fig. 4. The first four vibration mode diagrams.

The vibration modes of large-span through tied arch bridges are complex, mainly including vertical vibration, lateral vibration, and torsional vibration of arch ribs and tie beams. The period and mode characteristics of the first 10 vibration modes are listed in Table 1.

Table 1. The period and mode characteristics of the first 10 vibration modes.

Mode order	Period (s)	Mode characteristics
1	2.30	The first order antisymmetric vertical bending of main arch and tie beam
2	1.54	The first order symmetric transverse bending of main arch
3	1.11	The first order symmetric vertical bending of main arch and tie beam
4	0.99	The first order antisymmetric transverse bending of main arch, and the first order antisymmetric torsion of tie beam
5	0.75	The first order symmetric transverse bending of tie beam
6	0.66	The second order symmetric vertical bending of main arch and tie beam
7	0.60	The first order symmetric transverse bending of main arch, and the first order symmetric torsion of tie beam
8	0.57	The second order antisymmetric vertical bending of main arch and tie beam
9	0.42	The first order asymmetric torsion of tie beam
10	0.41	The second order asymmetric torsion of tie beam

4 Seismic Response Analysis

4.1 Seismic Fortification Goal

The main bridge of this project is a large-span arch bridge with a span exceeding 150 m. According to the design specification [6], the seismic fortification is classified as A-Class and adopts two levels of seismic fortification. Under the action of E1 and E2 earthquakes, the seismic fortification goals are shown in Table 2.

Table 2. Seismic fortification goals of the main bridge.

Classification	E1 earthquake (10% exceedance probability in 50 years)		E2 earthquake (2% exceedance probability in 50 years)	
	Post-earthquake service requirements	Damage state	Post-earthquake service requirements	Damage state
A-class	Continue to use immediately	The overall response of the structure is in the elastic range basically with no damage	Return to use after simple repair	May cause minor local damage

4.2 Earthquake Action and Ground Motion Parameter

The earthquake action is determined based on the seismic safety evaluation of the engineering site. The basic seismic intensity is seven degrees (0.15 g). The peak ground motion acceleration of the bedrock with 10% exceedance probability in 50 years is 0.8187 cm/s^2, while the peak ground motion acceleration of the bedrock with 2% exceedance probability in 50 years is 1.5557 m/s^2.

For this steel structure arch bridge, the damping ratio is taken as 0.03. The horizontal seismic acceleration response spectrum curves of E1 and E2 earthquakes are shown in Fig. 5.

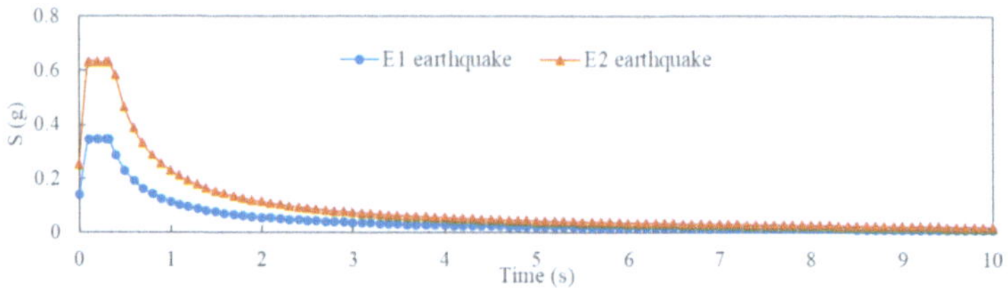

Fig. 5. The horizontal seismic acceleration response spectrum

Considering the randomness of earthquake acceleration time-history, 3 sets of time-history curves are selected for analysis, and the absolute value of the correlation coefficient between any two sets of time-history curves is less than 0.1. The calculation result takes the maximum of the 3 sets of time-history analysis. 3 sets of E2 earthquake horizontal acceleration time-history curves are shown in Fig. 6:

The vertical seismic action is obtained by multiplying the horizontal one by 2/3.

4.3 Seismic Response Spectrum Analysis

4.3.1 Superstructure Response

All the superstructures, including main arch, tie beam, crossbeams, and suspension rods, remain elastic under the earthquakes of both El and E2 levels.

4.3.2 The Reaction Force of Bearings

Earthquakes in the X, Y, and Z directions will generate certain reaction force in various directions, which may exceed the performance of conventional bearings and cannot be ignored. The horizontal reaction force of the fixed bearing reaches 26% and 51% of the vertical bearing capacity of the bearing, respectively under E1 and E2 earthquakes. The reaction force results are shown in Table 3.

4.3.3 Substructure Response

Under E1 earthquakes, the bridge piers and pile foundations remain elastic while under E2 earthquakes, the pier and pile foundation yield, which does not meet the seismic fortification goals in Table 2.

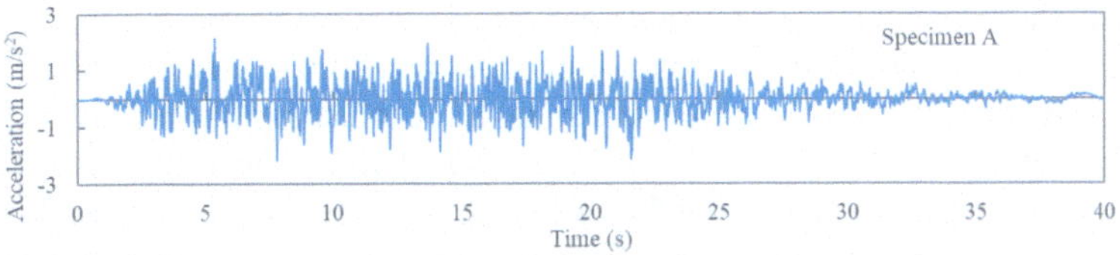

(a) Specimen A

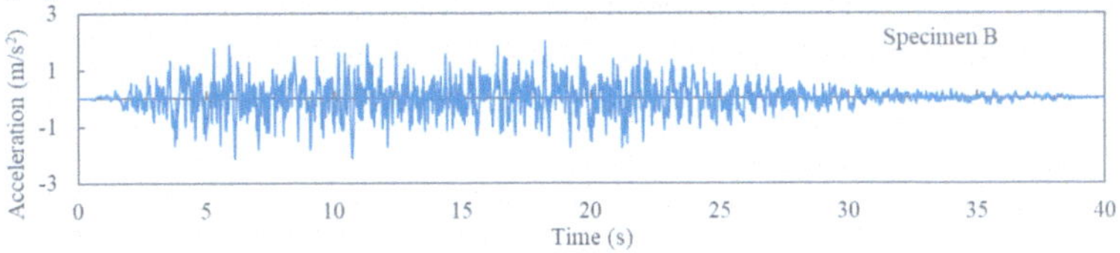

(b) Specimen B

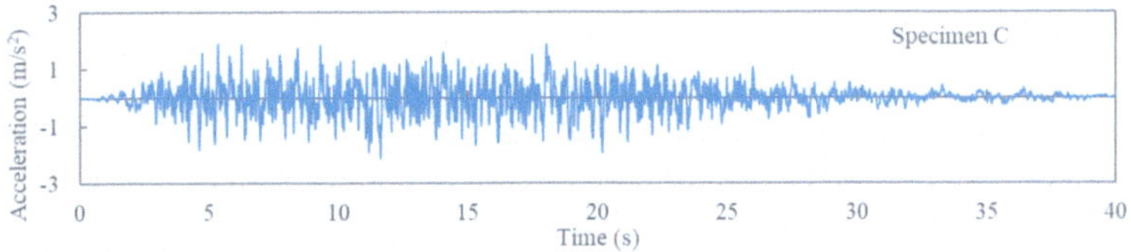

(c) Specimen C

Fig. 6. The horizontal seismic acceleration time-history curves of E2 earthquake

Table 3. The reaction force of fixed bearing under earthquakes based on response spectrum analysis.

Earthquakes	The reaction force (kN)			Earthquakes	The reaction force (kN)		
	FX	FY	FZ		FX	FY	FZ
E1-X	10265	176	2454	E2-X	20361	321	4870
E1-Y	5057	3986	1831	E2-Y	10296	7723	3721
E1-Z	898	188	1857	E2-Z	1767	342	3735
E1-XYZ	11479	3994	3581	E2-XYZ	22885	7737	7178

According to calculations, for large-span through arch bridges, the horizontal reaction force of the fixed bearings and the internal force of the corresponding piers are easily exceeded, becoming vulnerable parts for controlling seismic design. Therefore, seismic isolation design is considered, as follows.

5 Seismic Isolation Design

5.1 Cable-Sliding Friction Aseismic Bearings (CSFAB)

Shown in Fig. 7, Cable-sliding friction aseismic bearings are new type of cost-effective seismic isolation bearings [7], which are formed by adding cable restrainers on the basis of very mature spherical bearings. Under the action of normal use loads or E1 earthquake, the performance is the same as that of spherical bearings. Fixed bearings undertake the vast majority of horizontal forces, while sliding bearings slide to isolate the upper and lower structures, reduce horizontal force transmission, and utilize friction to dissipate energy. Under the action of E2 earthquake, the shear connectors are cut off and fixed supports are transformed into sliding bearings, thereby releasing seismic energy, effectively reducing the seismic response of the structure, and limiting the relative displacement of piers and beams through cable restrainers to prevent beams falling [8]. The restoring-force model of cable-sliding friction aseismic bearings is Shown in Fig. 8.

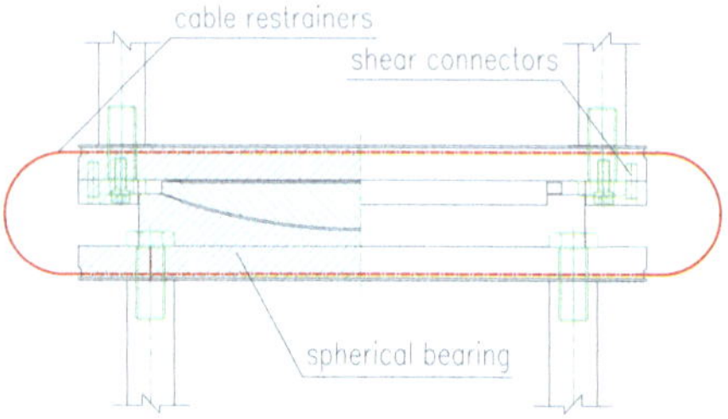

Fig. 7. Structure diagram of CSFAB

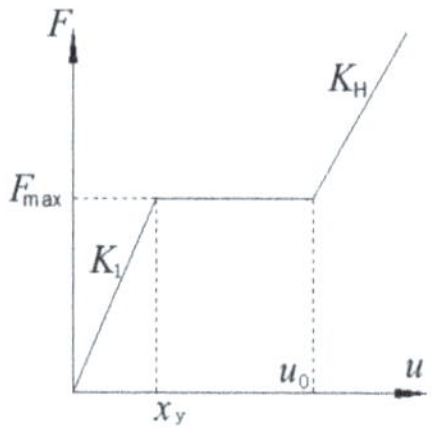

Fig. 8. Restoring-force model of CSFAB [9]

Shown in Fig. 8, $F_{\max}$ means the critical sliding friction force, x_y means the bearing displacement of critical sliding, u_0 means the bearing displacement when cable restrainers begin to function. $K_1 = F_{\max}/x_y$, which means the initial horizontal stiffness of sliding bearings. K_H means the horizontal stiffness of cable restrainers.

For the bearings of this bridge, according to the seismic fortification goals, the shear capacity is set at 30% of the vertical bearing capacity, i.e. 13500 kN. Taking into account the displacement requirements caused the earthquakes, the temperature effect and the braking force, u_0 is proposed to be 0.15 m.

5.2 Nonlinear Time-History Analysis

Under the action of E2 earthquake, the shear connectors are cut off, all four bearings become sliding bearings, and the displacement is limited by the cables. They are simulated by nonlinear connections in the finite element model Based on the nonlinear time-history analysis, the maximum displacement of the bearings in the longitudinal direction is 0.106 m, as shown in Fig. 9, and the maximum displacement in the transverse direction is 0.10 m.

Superimposing half of the temperature effect, i.e. 0.038 m, the total displacement of the bearing along in the longitudinal direction is 0.144 m, and the cable restrainers have not been activated. Therefore, the proposed value of u_0 as 0.15 m is appropriate.

After seismic isolation design, the maximum horizontal force of the fixed bearing reduces to 13500 kN, which is the shear capacity of the shear connectors. The seismic mitigation ratio of E2 earthquakes reaches 41%, and the pier and pile foundation are still in the elastic range, which can meet he seismic fortification goals.

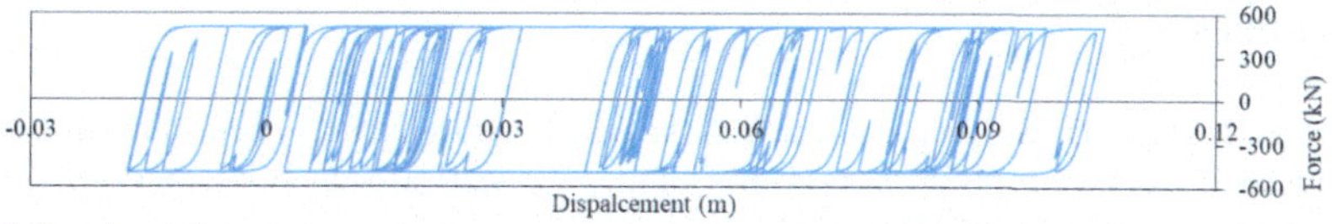

(a) Specimen A

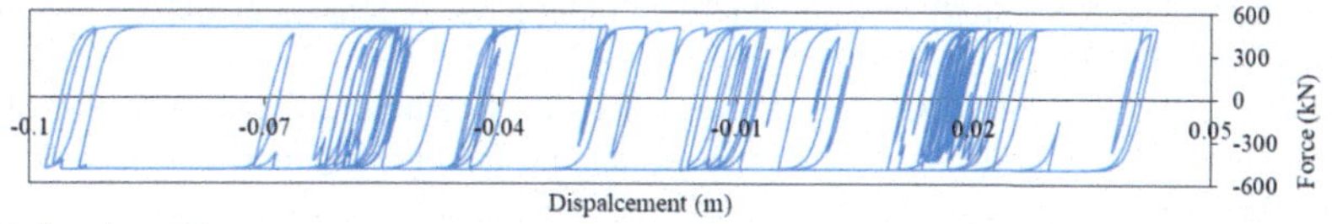

(b) Specimen B

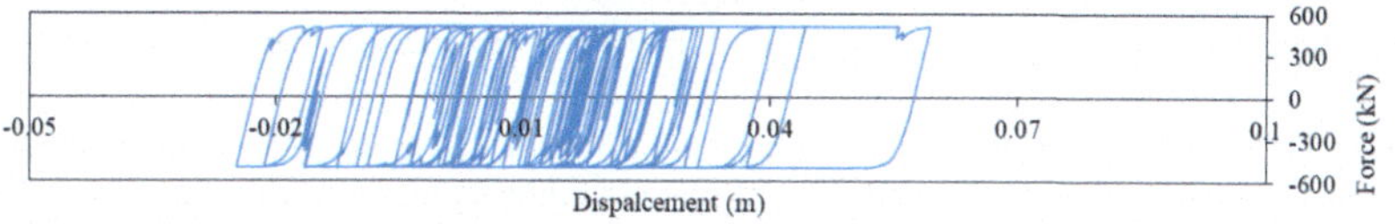

(c) Specimen C

Fig. 9. The hysteretic curve of bearings in the longitudinal direction

6 Conclusions

This paper studies the dynamic characteristics and seismic response of large-span through arch bridges, and conducts seismic isolation design, forming the following conclusions:

(1) The vibration modes are complex, mainly including vertical vibration, lateral vibration, and torsional vibration of arch ribs and tie beams. The influence of higher-order vibration modes must be considered in seismic analysis.
(2) Earthquakes in the X, Y, and Z directions will generate certain reaction force in various directions. The coupling effect of earthquakes in three directions must be considered.
(3) Under the earthquakes of both El and E2 levels, all the superstructures remain elastic while the horizontal reaction force of the fixed bearings and fixed piers become vulnerable parts for controlling seismic design. Special attention should be paid to fixed piers in seismic design.
(4) CSFAB can effectively reduce the seismic response of the substructure, and also prevent excessive displacement of beams. It is a cost-effective means of seismic isolation.
(5) There may be some residual displacement after E2 earthquakes, as CSFAB do not have self-centering capacity after the shear connectors are cut off. This is the point that needs improvement.

Acknowledgments. This work is sponsored by Shanghai Rising-Star Program (Grant No. 21QB1406000).

References

1. Salar, F.T., Majid, B.: Seismic evaluation of the bridge with a hybrid system of cable and arch: Simultaneous effect of seismic hazard probabilities and vertical excitations. Mech. Based Des. Struct. Mach. **52**(4), 2136–2152 (2024)
2. Chen, X.C., Xia, X.S.: Study on ground motion input mode of long-span arch bridges. World Earthq. Eng. **34**(02), 26–32 (2018)
3. Mohseni, I., Lashkariani, H.A., Kang, J., Kang, T.H.-K.: Dynamic response evaluation of long-span reinforced arch bridges subjected to near- and far-field ground motions. Appl. Sci. **8**(8), 1243 (2018)
4. Yu, H.: Study on seismic performance and seismic isolation schemes of large-span simple-supported bowstring arch bridge. Urban Roads Bridges & Flood Control **08**(8), 280–283+28–29 (2020)
5. Zhong, H.Q., Guo, J.J., Yuan, W.C., Wang, Z.Q., Dang, X.Z., Deng, X.W.: Numerical study on seismic performance of a novel multi-level aseismic highway bridge with cable-sliding friction aseismic bearings and dissipative controlled rocking piers. Eng. Struct. **275**, 115253 (2023)
6. CJJ 166-2011, Code for Seismic Design of Urban Bridges
7. Yuan, W.C., Wei, Z.H., Cao, X.J., Rong, Z.J.: Cable–sliding friction aseismic bearing and its application to bridge seismic. Engineering Mechanics **28**(S2), 204–209 (2011)
8. Chen, W.W., Yuan, W.C., Yang, H.L.: Application of the cable–sliding friction aseismic bearing in a tied-arch bridge. Structural Engineers **29**(5), 84–90 (2013)
9. DB 51/T2241-2016, Manual & application guidelines of cable-sliding friction aseismic bearings

Internal Force Correction Method of Solid Element Finite Element Model of Continuous Beam Arch Bridge Based on Stress-Free State Control Method

Tao Han, Xiaozhong Li(✉), and Hao Yang

School of Civil Engineering, Lanzhou Jiaotong University, Lanzhou 730070, China
69530763@qq.com

Abstract. Different construction methods and bridge completion time have great influence on the internal force of the main beam of concrete bridge. The traditional solid finite element modeling scheme is difficult to consider the influence of construction methods and bridge completion time on the internal force of the main beam of concrete bridge, and the finite element accuracy of the rod system is difficult to meet some engineering and research requirements. Taking a three-span continuous beam arch bridge as an example, based on the theory of stress-free state control, this paper proposes a method to first calculate the difference between the stress-free state quantity of the main beam with and without considering the construction method and the bridge completion time, and then convert the stress-free state quantity into the temperature field of a section beam through the temperature linear expansion coefficient to correct the internal force of the solid finite element model. It is found that the corrected results are good and the accuracy can meet the needs of practical engineering or bridge research.

Keywords: stress-free state control method · Full bridge solid element finite element model · Bridge construction method · Fine model · Internal force correction

1 Introduction

According to structural mechanics theory and concrete creep theory, different construction methods and bridge completion time have a greater impact on the bending moment of the main beam of a concrete bridge. Taking three-span continuous beams as an example, the maximum positive and negative bending moments of the main beam are close to each other when the bridge construction method is adopted, as shown in Fig. 1 (a). When the cantilever method is used, the main beam is dominated by negative bending moments, as shown in Fig. 1 (b). When the simple supported continuous construction method is adopted, the bending moment of the main beam is dominated by the positive bending moment, and its bending moment diagram is similar to that of the three-span simple supported beam. At the same time, shrinkage and creep of concrete Bridges

B. Yuan et al. (Eds.): ICCASE 2024, SUCI, pp. 89–99, 2024.
https://doi.org/10.1007/978-3-031-76102-7_11

under long-term loading will further change the bending moment of the main beam [1, 2], which is the key factor affecting the normal stress and principal stress of the bridge structure. Although in the past, we have some methods to take the influence of construction methods, shrinkage and creep and other factors on the stress of the bridge structure into account when we conduct the finite element modeling of the whole bridge solid element [3]. For example, the life-and-death unit technology is used to simulate the phased construction of Bridges [4, 5], but there is a problem of poor model convergence. The simulation of concrete shrinkage and creep is realized through the secondary development of finite element software [6], but the operation is too complicated and the iterative calculation cost is high. Therefore, it is necessary to study a simple finite element modeling method for the whole bridge solid element, which can consider the influence of construction method and concrete shrinkage and creep on the internal force of the bridge structure.

Stress-free state control method, as a control method to solve the phased construction of Bridges, has been widely used in practical projects. Yujie Yu [7] and Pengfei Li [8] et al. proposed a cable-force calculation method based on stress-free state control method in order to solve the defects of inaccurate calculation of cable force by traditional methods and repeated iterative calculation. Wenming Wang et al. [9] proposed a closing scheme of main beam considering closing temperature based on stress-free state control method, and proved that this scheme is superior to the traditional closing scheme. Haijun Wu et al. [10] used stress-free state control method to control the construction of concreted-filled steel tube arch, and successfully achieved the arch state to reach the predetermined goal. It can be seen that stress-free state control method can solve the role of improving the internal force and linear shape of the structure in practical engineering, but no one has

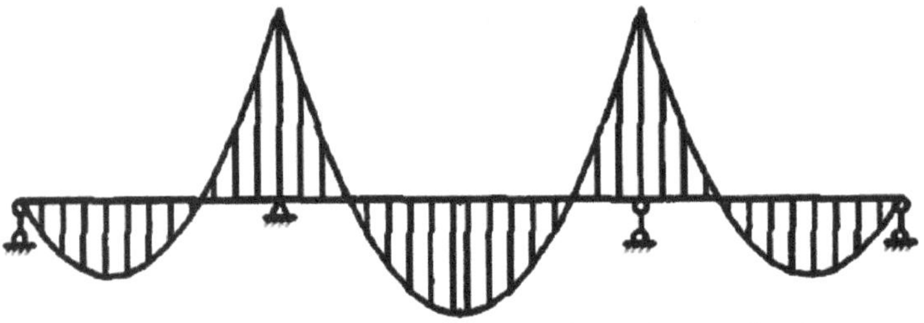

(a) Bending moment diagram of a three-span continuous beam under the construction method of a single-pass bridge

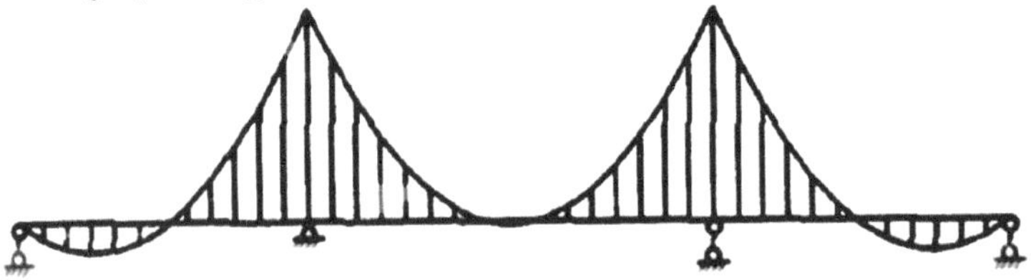

(b) Bending moment diagram of three-span continuous beam under cantilever construction method

Fig. 1. Bending moment diagram of three-span continuous beam under different construction methods applied it to correct the internal force of the finite element model. Therefore, this paper puts forward a calculation method based on stress-free state control method to solve the influence of construction method on the finite element model of the whole bridge solid element.

2 Stress-Free State Control Method

The stress-free state control method is a theoretical method to solve the secondary internal forces of statically indeterminate structures. When the external load of the structure, the structural system, the supporting boundary condition, the stress-free length of the element and the stress-free curvature are certain, the corresponding internal force and displacement of the structure are unique and have nothing to do with the formation process of the structure, which is precisely difficult to realize in the finite element simulation of solid elements. This method can well avoid the influence of the formation process of the structure on the secondary internal force of the bridge. When the external load on the structural unit is relieved, the statically indeterminate structure is disassembled into several statically indeterminate structures, and the structure is in a free state. At this time, the structure has two stress-free state quantities: stress-free length and stress-free curvature [11].

For any structure, according to the energy balance equation $\delta\Pi = \delta U + \delta W$, the balance equation of the phased construction structure can be derived:

$$[K]\{\sigma\} = \{P\} + \{L_0\} \tag{1}$$

where: L_0 is the stress-free state load matrix generated by the subsequent structure on the existing deformed structure during the phased construction.

The structural balance equation of the construction method of a single bridge is as follows:

$$[K]\{\sigma\} = \{P\} \tag{2}$$

It can be found that compared with formula (2), there is more generalized load term L_0 on the right side of formula (1). For the construction method of a one-off bridge, it can be considered that $L_0 = 0$. At this time, formula (1) is completely equal to formula (2). The internal force of the structure is determined and unique, and has nothing to do with the formation process of the structure.

3 Project Overview and Solid Element Finite Element Model Parameters

3.1 Project Overview

In this paper, a railway continuous beam arch bridge is selected for research. The upper part adopts (90 + 180 + 90) m span prestressed concrete continuous beam, and the main beam section adopts single box and double chamber variable height box section. The beam height at the middle span and side fulcrum is 4.0 m, and the beam height at the middle fulcrum is 9.5 m. The arch structure is 180 m concrete filled steel tube suspension arch (as shown in Fig. 2), and the construction method of cantilever construction is beam first and arch later.

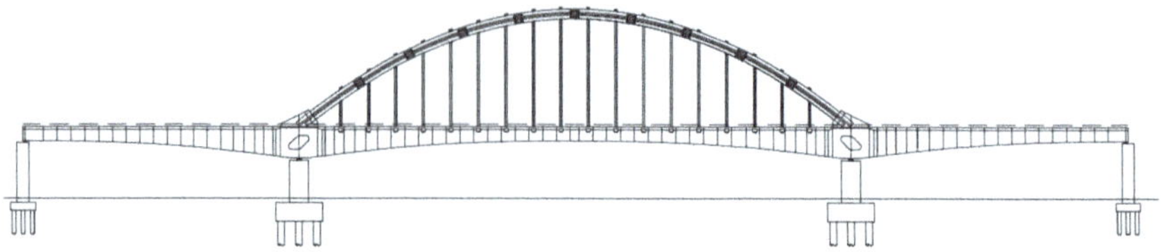

Fig. 2. Bridge elevation structure diagram

3.2 Solid Element Finite Element Modeling Parameters

In this paper, Midas civil software and large general finite element software abaqus (as shown in Figs. 3 and 4) are used to establish the beam element finite element model using cantilever construction mode and the solid element finite element model based on stress-free state control method. The beam element model established by Midas civil is the contrast model of abaqus solid element model, and its model parameters will not be described here. This paper focuses on the parameters of abaqus solid element finite element model. In abaqus model, solid elements are used for main beams and arches, and hexahedral mesh is used for all solid elements to ensure simulation accuracy. Beam element is used after the section of arch rib concrete filled steel tube is converted to equivalent steel section. Truss element is used for longitudinal prestressed steel beam and arch suspension rod of the whole bridge, and the prestress is applied by cooling method. The point and plane coupling constraints are adopted between the main beam and the boom, between the arch seat and the arch rib, and the common joint constraints are adopted between the arch rib and the boom. In the finite element model, the upper support plate is simplified as a rigid body plane, and coupled with the RP point of the rotation center, and the spring connected with the earth is set in the rotation center.

Fig. 3. Midas beam element finite element model

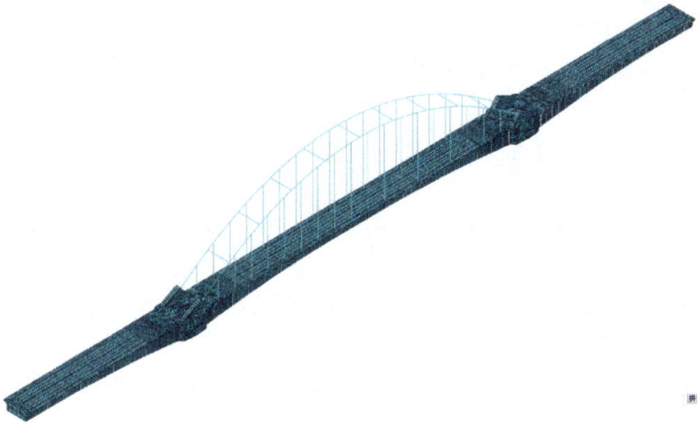

Fig. 4. Abaqus solid element finite element model

4 Bending Moment Correction of Main Beam Based on Stress-Free State Control Method

It can be seen from the above that after the stress-free state quantity is determined, the internal force of the structure is only determined, and the stress-free state quantity that affects the bending moment of the bridge structure is stress-free curvature. In the solid finite element model, the stress-free state quantity of the structure can be changed by setting the linear expansion coefficient of the element temperature and changing the element temperature. In order to simulate the secondary internal force caused by cantileving construction and concrete shrinkage and creep [12], the stress-free curvature of the beam body is modified to be consistent with Midas model by applying temperature gradient to the middle span main beam. Due to the limited length of the mid-span segment, the distance from the key segment such as the beam arch joint segment is much higher than its beam height, and the influence of temperature load applied to the mid-span segment on the bearing maintenance system is very limited according to the Saint-Venant principle. And the self-stress is small, so it is feasible to modify the curvature of the state without stress by means of temperature equivalence method in the middle span.

In order to reduce interference variables, the model without arch is first modified: By superimposing the constant load bending moment of the structure extracted by Midas model and the 10-year contracting and creep sub-bending moment (as shown in Fig. 5 (a)), the beam internal force of the bridge ten years after the bridge was completed under constant load was obtained, and compared with the dead weight bending moment of the beam extracted by abaqus model, the difference between the two was obtained (as shown in Fig. 5 (b)), and the curvature difference L1 0 of the main beam in the non-stress state was calculated. The temperature gradient required to change the curvature L1 0 in the non-stress state of the mid-span mid-segment is inversely calculated, and the bending moment of the main beam under the corrected constant load and concrete shrinkage and creep is shown in Fig. 5 (c).

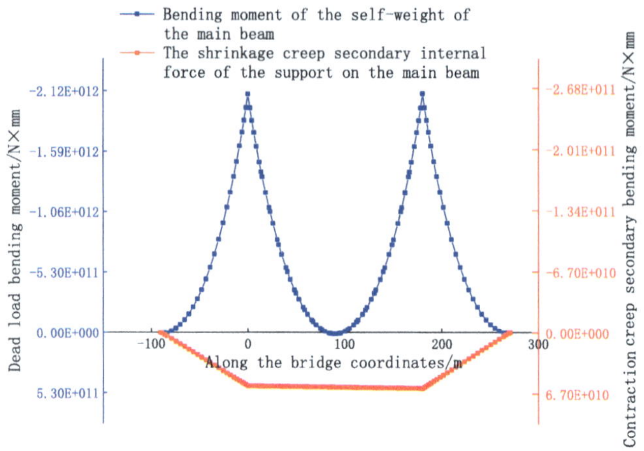

(a) Midas model bending moment at constant load and shrinkage creep

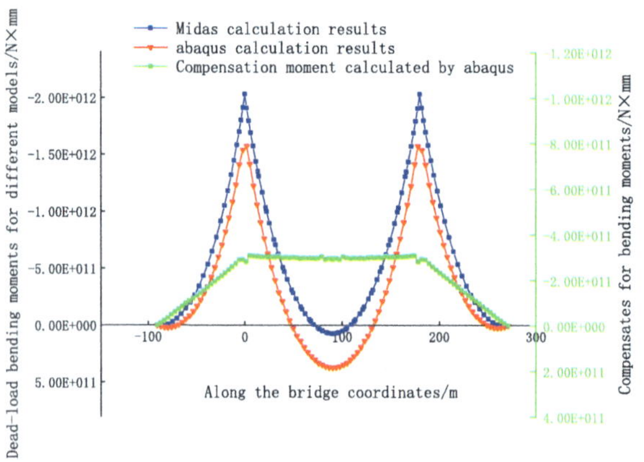

(b) Modify the bending moment comparison diagram of the main beam of the first two models

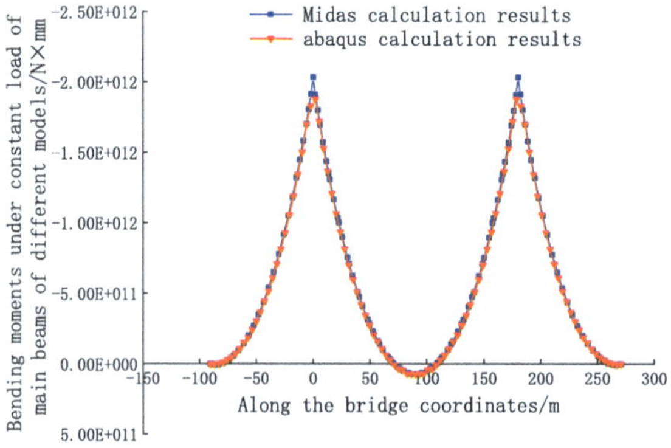

(c) Comparison of bending moments of the main beams of the two models after modification

Fig. 5. Bending moment of the main beam under constant load and concrete shrinkage and creep

At the same time, the construction method also affects the bending moment generated by prestressing on the main beam, and abaqus model does not take into account the prestressing loss of the steel beam, resulting in deviations between the prestressing of the steel beam and the precise calculation results. Therefore, based on the stress-free state control method and Midas calculation results, the initial premoment and secondary premoment of the prestressing on the concrete section of the main beam are modified respectively [13]. The specific methods are as follows:

(1) Input the prestressed tendons into the model after correcting the curvature L1 0 of the non-stress state, and apply the prestressed tendons according to the design drawings. Compare the initial premoment extracted from the Midas model with the bending moment of the main beam under the prestress action extracted from abaqus, obtain the bending moment amplitude ratio Δ of the two, and then fold it into the abaqus prestress to reduce the prestress loss, where:

$$\Delta = \frac{M_1^b - M_1^z}{M_2^b - M_2^z} \tag{3}$$

where: Mb 1, Mz 1:are the bending moments of mid-span section and mid-pier section extracted by Midas model, respectively.

Mb 2, Mz 2: the bending moments of mid-span and mid-pier sections extracted from abaqus model, respectively.

(2) Superposition the initial premoment and secondary premoment of Midas model (as shown in Fig. 6 (a)), and compare them with the bending moment of the main beam extracted after reducing the prestress of abaqus model, and calculate the difference between the two (as shown in Fig. 6 (b)), so as to calculate the curvature difference of the main beam in the non-stress state L1 0 + L2 0. The temperature gradient required to change the stress-free state curvature L2 0 in the middle span span is inversely calculated, and corrected by applying temperature gradient compensation to the middle span span span segment (Fig. 6 (c)).

Figure 7 is the comparison between the bending moment of the revised abaqus main beam model and the bending moment of the Midas model main beam under constant load and prestressed load. On this basis, arch rib suspensions and other arch structures are added to the abaqus model, and the suspensions' tension is applied to the design state through multiple iterations by cooling method. At this time, the extracted bending moment of the main beam of abaqus model is the bending moment of the main beam in the completed bridge state. Figure 8 is the comparison diagram of the bending moment of the main beam in the abaqus completed bridge state model and the Midas completed bridge state model.

It can be seen from the figure that the maximum negative bending moment of the main beam at the middle pier position is -5.53×10^{11} in abaqus model and -5.16×10^{11} in Midas model. The simulated value of abaqus model is about 6.9% of the total Midas amplitude compared with that of Midas model. The maximum bending moment of the mid-span main beam in abaqus model is -4.93×10^{10}, and the maximum negative bending moment in Midas model is -5.59×10^9. The simulated value of abaqus model is about 8.1% of the total Midas amplitude compared with that of Midas model. It can

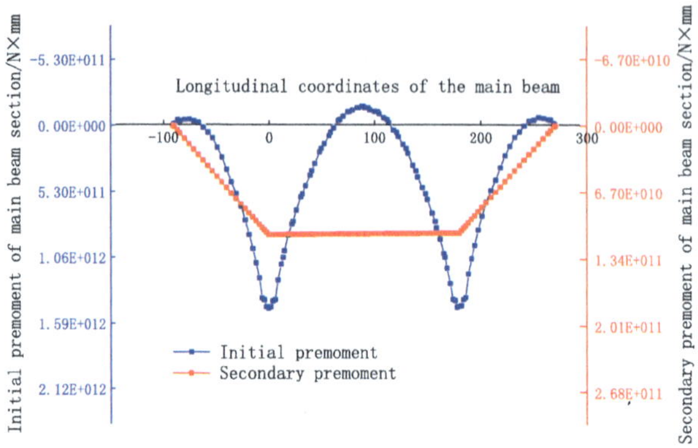

(a) Bending moment diagram of the initial premoment and secondary premoment of the Midas model steel bundle

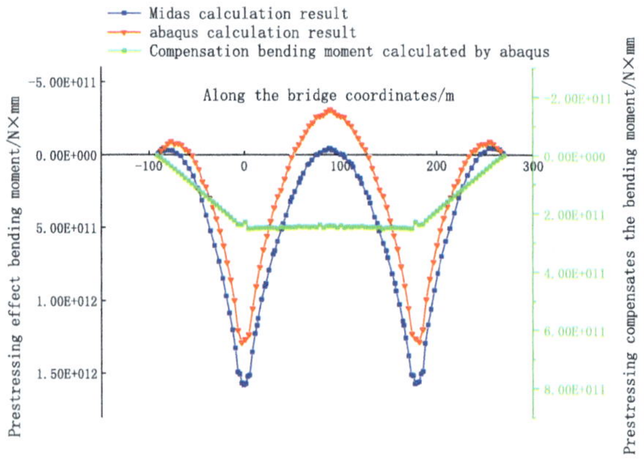

(b) Modify the bending moment comparison diagram of the main beam of the first two models

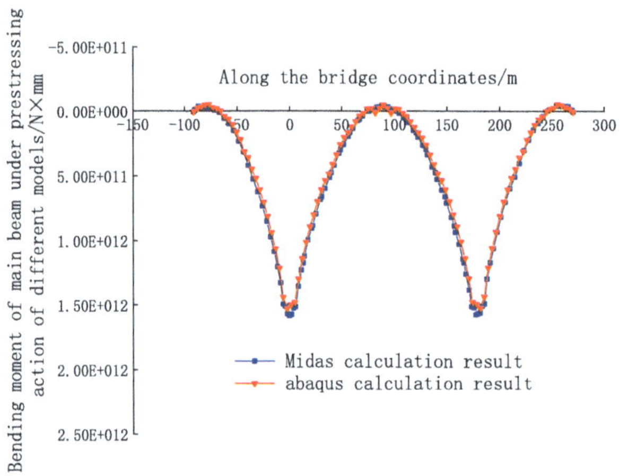

(c) Comparison of bending moments of the main beams of the two models after modification

Fig. 6. Internal diagram of the main beam under the action of prestress

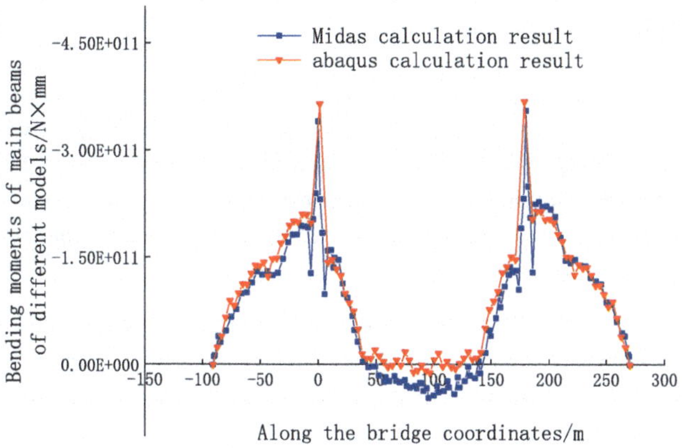

Fig. 7. Does not include the bending moment diagram of the main beam of the arch model

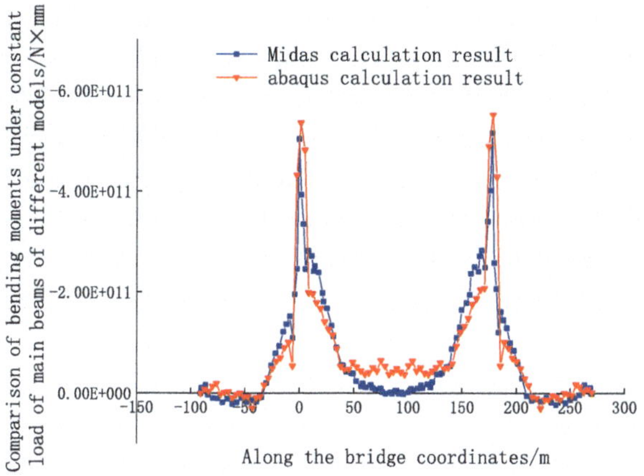

Fig. 8. Bending moment of the main beam of the bridge after ten years

be seen that Midas is basically consistent with the bending moment diagram of the main beam of abaqus model, and the solid finite element model of abaqus basically meets the accuracy required by various engineering and research.

However, the finite element simulation with this method still has some shortcomings. Because the bending moment of the whole bridge is corrected by correcting the stress-free curvature in the middle span of the main beam in this paper, the mechanical behavior of the middle span main beam in the bridge state cannot be accurately simulated. This method corrects the overall stress-free curvature of the main beam by applying temperature to individual segments. Although the overall stress-free curvature reaches the target state, it cannot achieve that the stress-free curvature of any local beam segment is consistent with the target state, thus causing the bridge alignment to fail to reach the target state. Therefore, this correction method can only be used to correct the internal force state.

5 Conclusion

Based on the stress-free state control theory, this paper proposes to modify the internal force of the solid finite element model by calculating the difference between the stress-free state quantity of the main beam with and without considering the construction method and the bridge completion time, and converting the stress-free state quantity into the beam body temperature field through the temperature linear expansion coefficient. The main conclusions are as follows:

(1) The stress-free state method has a good effect on the internal force correction of the solid finite element model, and the maximum bending moment deviation of the solid element finite element model is about 8.1%, which basically meets the accuracy required by various engineering and research.
(2) At present, the stress-free state method can only correct the internal force of the solid finite element model, and it cannot meet the linear requirements of the beam body for the time being, which needs to be further studied.
(3) Compared with other methods, the solid finite element model modified by the stress-free state method has a simpler operation method and lower calculation cost, so it is suitable for the foundation bridge model establishment in the field of engineering and bridge research except bridge construction and shrinkage creep research itself.

In view of the shortcomings of this method, this paper gives some suggestions for future research: the key to make the beam body alignment meet the requirements is how to solve the fine temperature field, so that the stress-free curvature of any local beam section is consistent with the target state, which can be further studied from this aspect in the future.

Acknowledgments. This paper is financially supported by the National Natural Science Foundation of China (52162043).

References

1. Marino, J., Neda, B., Dragan, Ć., Alen, H.: Strain analysis on cast-in-place balanced cantilever prestressed concrete bridges during construction. Structural Engineering International
2. Yoon, Y.-S., Lee, S.-K., Lee, M.-S., Kim, J.-K., Kwon, K.-Y.: Rheological concrete creep prediction model for prestressed concrete bridges constructed by the free cantilever method. Magazine of Concrete Research **63**(9), 645–653 (2011)
3. Hoffman, I.S., et al.: Finite element numerical simulation of a cable-stayed bridge construction through the progressive cantilever method. Structural Concrete **23**(2), 632–651 (2022)
4. Jie, W.: Force analysis of steel-mixed composite beam bridge under shear nail cluster and uniform arrangement. Wuhan University of Technology (2015)
5. Li, Y.: High piers large span continuous rigid frame bridge construction control and the key problems in research. Lanzhou Jiaotong University (2022). https://doi.org/10.27205/dcnki.Gltec.2021.000053
6. Weng, Y.: Analysis of creep effect of PC box girder based on Secondary development of ANSYS. Central South University (2012)

7. Yu, Y., Luo, Y., Zhou, W., et al.: Determination method of cable force for cable-stayed suspension construction based on zero displacement and stress-free state control. Eng. Sci. Technol. 1–8 [2024-03-14]. https://doi.org/10.15961/j.jsuese.202300186
8. Li, P., Wang, S., Wei, S., et al.: Cable tension control of large space shaped steel tower cable-stayed bridge. J. Highw. Transport. Sci. Technol. **39**(10), 49–58 (2022)
9. Wang, W., Chai, S., Nie, N., et al.: Closing scheme of main beams of prestressed concrete cable-stayed bridges based on stress-free state method. Railway Construction **62**(06), 93–98 (2022)
10. Haijun, W., He, L., Wang, S., et al.: Linear control method of arch rib of long-span concrete-filled steel tube arch Bridge based on stress-free state method. Bridge Construction **50**(06), 20–26 (2019)
11. Qin, S.: Theory and Practice of Bridge Construction control + stress-free state Method. People's Communications Press, Beijing (2006)
12. Yin, Z., et al.: Arch Calculation and Control Method of 500 m class Steel pipe arch Bridge. China J. Highw. Trans. **35**(5), 60–72 (2022). The 10.19721 / j.carol carroll nki. 1001-7372.2022.05.006
13. Ma, K., Guo, Y.: Mechanical analysis of joint joints of flexible arch bridges with continuous beams in high-speed railway. J. Railw. Eng. **37**(05), 23–29 (2020)

Research on Shear Lag Effect of Curved Box Girder with Cantilever Plate Variable Width Section

Jinbo Yang(✉) and Nanhong Ding

School of Civil Engineering, Lanzhou Jiaotong University, Lanzhou 730070, China
421161255@qq.com

Abstract. In order to meet various terrains and special conditions, variable-width curved box girder bridges have been widely used in urban transportation road construction. In this study, the energy variational principle is used to establish the control differential equations and boundary conditions for the shear hysteresis effect of a box girder with cantilevered slabs with variable width curves, and the formula for the positive stresses in the cross section is derived by the difference method. Combined with examples, the analytical solution of this paper and the numerical solution of finite element under concentrated load and uniform load are in good agreement. In comparison with the equal cross-section curved box girder, the positive stresses in the cross-section of the cantilever plate widened curved box girder change by a greater magnitude and have a higher peak value.

Keywords: curved box girder · cantilever plate variable width section · shear lag effect · energy variational principle · difference method

1 Introduction

At present, the study of shear hysteresis effect on box girders with variable width curves is limited to finite element numerical analysis, and the analytical method is insufficient. Luo Q.Z and Zhang Y.H investigated the shear hysteresis effect, constrained torsion and distortion of curved box girders of equal cross-section using the energy variational method [1, 2]. Lu M.F established the control differential equation of shear hysteresis effect of linear box girder with variable width cross section using the principle of energy variational and investigated the effect of width variation [3]. Marcello A, Li Z and Luo X analyzed the shear hysteresis effect of equal cross section curved box girder using the finite element method and analyzed the key factors such as the radius of curvature and the warping displacement mode [4–6].

The cubic parabola is used as the shear lag warping displacement function [7]. The effect of linear variation of the width of cantilever plate of curved box girder on the strain energy is considered, and the equation of positive stress in the cross section of curved beam is derived according to the energy variational and differential methods. And compared with the equal cross-section curved beam, The effect of cantilever plate width variation on the shear hysteresis effect of curved box girders is analyzed.

B. Yuan et al. (Eds.): ICCASE 2024, SUCI, pp. 100–106, 2024.
https://doi.org/10.1007/978-3-031-76102-7_12

2 Objects of Study and Underlying Assumptions

The plan and cross-section of cantilever plate widening curve box girder are shown in Figs. 1 and 2. z-axis is the direction of beam axis.

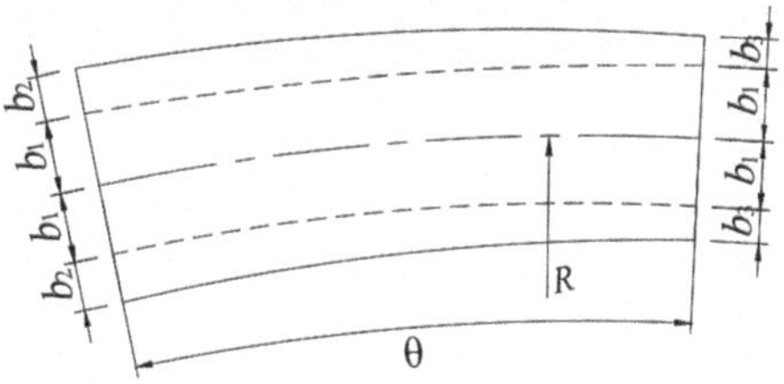

Fig. 1. Plane diagram of curved box girder with variable width

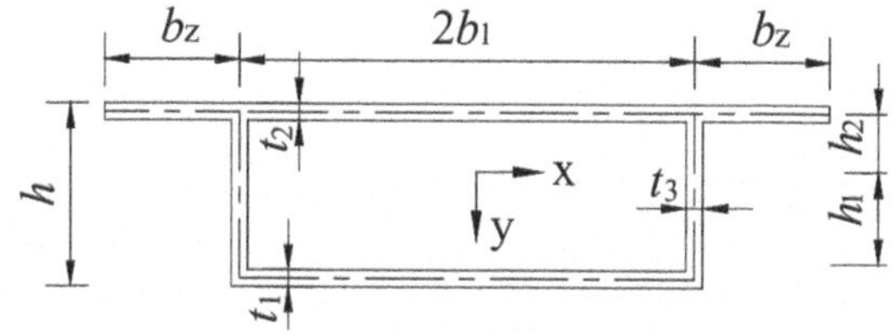

Fig. 2. Cross section of curved box girder with variable width

The basic assumptions are as follows

(1) Small deformation assumption and rigid perimeter assumption;
(2) The transverse bending strain energy of the web plate is neglected;
(3) The flange plate is not extruded vertically.

It is assumed that the shear-lag warpage displacement pattern at any point in the cross-section of the airfoil is

$$u_i(x, y, z) = -h_i\left(1 - \frac{z^3}{b_i^3}\right)U(z) \tag{1}$$

where U(z) is the shear-lag generalized displacement.

$$b_z = b_1 - \frac{b_1 - b_2}{L}z = b_1 - \alpha z \tag{2}$$

3 Control Differential Equations for Curved Box Girders with Variable Width Cross Section

3.1 Total Potential Energy Expression

According to the assumption, the strains on the wing plate of the box girder with variable width curve have positive strain ε_{z1} generated by shear lag warping displacement, positive strain ε_{z2} generated by obeying the flat cross-section assuming the uniform displacement of the rigid cross-section, and tangential strain γ_i.

The strain energy of each plate of the box girder in vertical deflection is given by

$$\overline{V}_i = \frac{1}{2}\iint t_i\left[E(\varepsilon_{z1}+\varepsilon_{z2})^2 + G\gamma_i^2\right]\mathrm{d}x\mathrm{d}z \tag{3}$$

where: E is the modulus of elasticity; G is the shear modulus.

The total potential energy of the box girder with cantilever plate widening curve is deduced to be

$$\begin{aligned}\Pi &= \overline{V}_1+\overline{V}_2+\overline{V}_3+\overline{V}_4+\overline{V}_J+\overline{V}_w+\overline{V}_m+W\\ &= \frac{1}{2}\int\{EI\left[(v'')^2+\frac{\varphi^2}{R^2}-2\frac{\varphi}{R}v''\right]+EI_s[-\frac{3}{2}\frac{\varphi}{R}U'+\frac{3}{2}v''U'+\frac{9}{14}(U')^2\Big]\\ &+EI_u\frac{9G}{5Eb_1^2}U^2+EI_3(-\frac{\alpha}{2b_z}\frac{\varphi}{R}U+\frac{\alpha}{2b_z}v''U+\frac{5\alpha}{14b_z}U'U+\frac{3\alpha^2}{35b_z^2}U^2+\frac{9G}{5b_z^2}U^2\Big)\\ &+GJ\left(\varphi'-\frac{v'}{R}\right)^2+EI_w\left(\varphi''-\frac{v''}{R}\right)^2-(qv+m\varphi)\}\mathrm{d}z\end{aligned} \tag{4}$$

where: I_1, I_2, I_3, I_f are the moments of inertia of the base plate, cantilever plate, cantilever plate and web plate, respectively; J is the torsional moment of inertia; I_w is the fan moment of inertia. $I = I_1 + I_2 + I_3 + I_f, I_s = I_1 + I_2 + I_3, I_u = I_1 + I_2$.

3.2 Establishment of Elastic Control Differential Equations

According to the principle of minimum potential energy, the variation of the total potential energy of the system is zero.

By dividing the total potential energy, the bending-torsion-shear hysteretic elasticity control differential equation for a box girder with a variable width cross-section curve can be obtained as

$$\left(EI+\frac{EI_w}{R^2}\right)v'''' - \frac{GJ}{R^2}v'' + \frac{EI_w}{R}\varphi'''' - \frac{EI+GJ}{R}\varphi'' + \frac{3}{4}EI_sU' + \frac{\alpha}{4b_z}EI_3U'' = q \tag{5}$$

$$\frac{EI_w}{R}v'''' - \frac{EI+GJ}{R}v'' + EI_w\varphi'''' - GJ\varphi'' + \frac{EI}{R^2}\varphi - \frac{3}{4}\frac{EI_s}{R}U' - EI_3\frac{\alpha}{4b_z}\frac{U}{R} = m \tag{6}$$

$$\begin{aligned}&\frac{3}{4}EI_sv' + \frac{\alpha}{4b_z}EI_3v'' - \frac{3}{4}\frac{EI_s}{R}\varphi' + \frac{\alpha}{4b_z}EI_3\frac{\varphi}{R} + \frac{9}{14}EI_sU'' -\\ &(EI_u\frac{9G}{5Eb_1^2} + \frac{3\alpha^2}{35b_z^2}EI_3 + \frac{9G}{5Eb_z^2}EI_3)U = 0\end{aligned} \tag{7}$$

boundary condition

$$\left[\left(EI+\frac{EI_w}{R^2}\right)v' - \frac{GJ}{R^2}v' + \frac{EI_w}{R}\varphi' - \frac{EI+GJ}{R}\varphi' + \frac{3}{4}EI_sU'' + EI_3\frac{\alpha}{4b_z}U'\right]\delta v\Big|_0^l = 0 \tag{8}$$

$$\left[\left(EI+\frac{EI_w}{R}\right)v''+\frac{EI_w}{R}\varphi''-\frac{EI}{R}\varphi+\frac{3}{4}EI_sU'+\frac{\alpha}{4b_z}EI_3\,U\right]\delta v'\Big|_0^l=0 \tag{9}$$

$$EI_w(\varphi''+\frac{v''}{R})\delta\varphi'\Big|_0^l=0 \tag{10}$$

$$(-\frac{EI_w}{R}v'+\frac{GJ}{R}v'-EI_w\varphi'+GJ\,\varphi')\delta\varphi\Big|_0^l=0 \tag{11}$$

$$[EI_s\left(\frac{3}{4}v''+\frac{9}{14}U'-\frac{3}{4}\frac{1}{R}\varphi\right)+\frac{5\alpha}{28b_z}EI_3\,U]\delta U\Big|_0^l=0 \tag{12}$$

where b1 = b2, α = 0, At this point, the above control differential equation for the variable width curved beam degenerates into the control differential equation for the shear-lag effect of the curved box girder with equal cross-section. It can be seen that the radius of curvature R is infinite, which can be changed into the control differential equation of the straight box girder, the same as the literature [1]. So the control differential equation established in this paper is more general.

The solution domain is divided into a differential grid, and the points on the beam are listed as a system of differential equations, which are substituted into the corresponding boundary conditions, the MATLAB is utilized to solve for the positive stresses in the cross-section of the box girder with variable width curve..

4 Calculus Analysis

It is known that in practical engineering, a certain cantilever plate width varies linearly along the axis of the variable width section simply supported curved steel box girder. Curvature radius R = 120 m, span L = 30 m. The material properties $E = 2.06◊10^5$ MPa, $G = 0.79◊10^5$ MPa, $\mu = 0.31$, $\alpha = 0.0278$.(1) On the top surface of the web of the bridge span, symmetrically acting uniform load $q = 50$ kN/m; (2) Symmetrically acting a pair of centralized load $2P = 2 \times 50$ kN on the top surface of the web in the span.

The ANSYS finite element analysis model of the variable width section simply supported curved steel box girder is established by using shell181 shell unit.

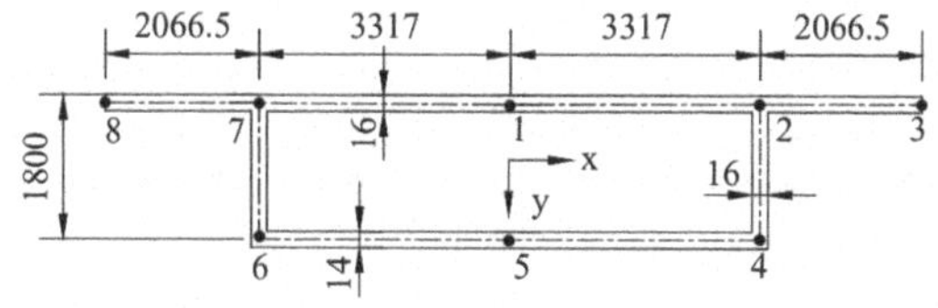

Fig. 3. Cross section of the widening curved box girder (mm)

According to the calculation method of this paper and ANSYS to calculate the two kinds of loads under the action of the $L/2$ section stress, the calculation of the cross-section and the location of the stress calculation point is shown in Fig. 3, the calculation results are shown in Table 1.

Table 1. Section stress of curved beam with variable (MPa)

Position	Uniform load		Concentrated load	
	Variational solution	Finite element solution	Variational solution	Finite element solution
1	−25.5529	−26.2497	−1.3376	−1.4152
2	−27.1791	−28.0568	−2.9253	−3.0847
3	−24.4168	−25.7371	−1.6145	−1.7925
4	40.5299	42.8207	3.0159	3.2612
5	37.8661	38.5917	2.0651	2.1802
6	54.5998	56.8036	4.3087	4.6431
7	−36.8465	−38.4953	−3.2949	−3.5192
8	−33.0173	−34.9864	−1.9473	−2.1325

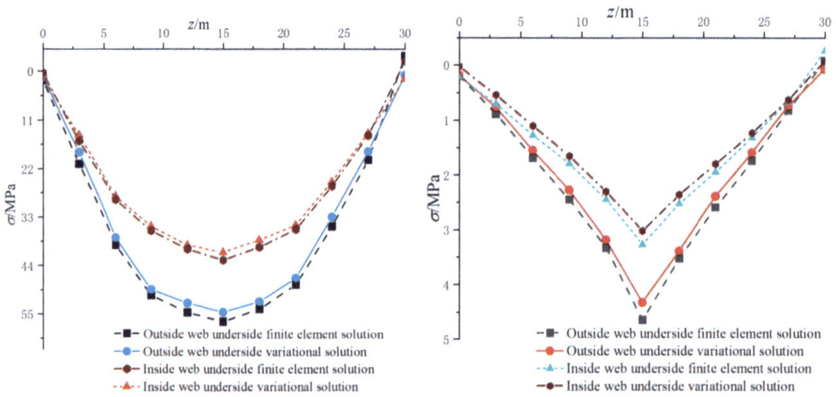

Fig. 4. Variational solution and finite element solution(MPa)

The mean distribution load and centralized load are applied, and the stress on the bottom surface of the web of the box girder with cantilever plate widening curve along the beam axis is taken for longitudinal comparison, which is shown in Fig. 4. As can be seen from the charts, the results calculated according to the method of this paper and the results of ANSYS calculations coincide with each other well on the whole, which indicates that the method proposed in this paper is feasible.

Under the action of the same loading conditions, the top surface stresses of the web plate on the outside of the curve and the positive stresses of the cantilever plate in L/2 cross-section of cantilever plate are compared between the cantilever plate widened curve box girder and the equal cross-section curved box girder to analyze the difference in the distribution of shear hysteresis between the two. The L/2 cross-section of the box girder with variable width curve is selected as the cross-section size of the box girder with equal cross-section curve, and the cross-section stresses of the box girder with equal cross-section are calculated by the method in the literature [1]. The distribution of positive stresses in the cantilever plate of the mid-span section of the cantilever plate of the widened curved box girder and the equal section curved box girder are shown in Fig. 5.

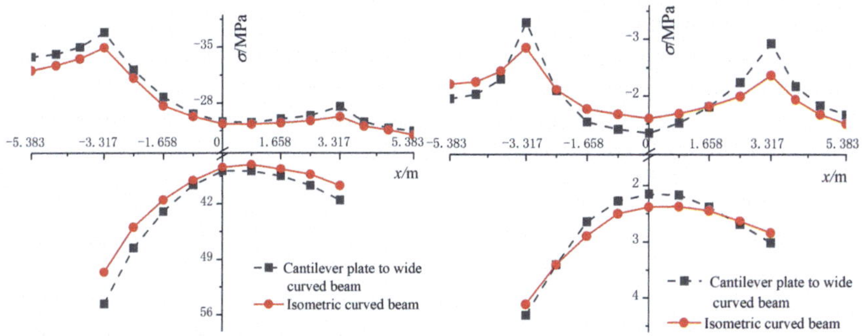

Fig. 5. The stress of the mid-span section(MPa)

From Fig. 5, it can be seen that the stress distribution patterns in the mid-span section of the widened curved girder and the equal cross-section curved girder are more consistent. However, compared with the equal-section curved box girder, the cantilever plate stress of the variable-width curved box girder has a larger variation and a higher peak value under uniform and concentrated loads, and the top surface of the web outside the curve has the largest difference, which is about 18.87%, under concentrated loads. It shows that the change of width has some influence on the shear hysteresis of the box girder with variable width curve.

5 Conclusions

1. The differential equation for controlling the shear lag effect of box girder with variable width curve established in this study can be degraded to the differential equation for controlling the shear lag effect of box girder with equal cross-section when the cantilever plate of box girder remains unchanged along the span length; and it can be degraded to the differential equation for controlling the shear lag effect of straight box girder bridge if the radius of curvature is infinitely large. Therefore, the equations constructed in this paper are more general.
2. With uniform and concentrated loads, the cross-section stresses of box girder with variable width curves calculated in this paper are in good agreement with the ANSYS results, which indicates that the method proposed in this paper is feasible.
3. With the same load, the cross-section distribution pattern in the span of variable width curved box girder and equal cross-section curved girder is more consistent, but with the uniform load and centralized load, the cross-section stress of variable width curved box girder varies more with a higher peak, which indicates that the change of cantilever plate width has a certain effect on the shear hysteresis of variable width curved box girder.

References

1. Luo, Q.Z.: Analysis of shear lag effect in thin walled curved box girder bridges. J. China Railw. Soc. **05**, 88–93 (1999). https://doi.org/10.3321/j.issn:1001-8360.1999.05.021
2. Zhang, Y.H., LI, Q.: Flexural-torsional analysis of curved box-girders with the consideration of the shear lag effect. Eng. Mech. **26**(10), 123–129 (2009). JournalArticle/5af45799c095d718d818a65a
3. Lu, M.F., Ding, N.H.: Research on shear lag effect of box girder with variable width section. Chinese J. Computat. Mech. **39**(06), 818–823 (2022). https://doi.org/10.7511/jslx2021050201
4. Marcello, A., Fabio, M.G., Giuseppe, L.: Symplectic analysis of thin-walled curved box girders with torsion, distortion and shear lag warping effects. Thin-Walled Structures, 175 (2022). https://doi.org/10.1016/j.tws.2022.109244
5. Li, Z., Ray, S.K., Ming-Jie, L.: Finite beam element with 26 DOFs for curved composite box girders considering constrained torsion, distortion, shear lag and biaxial slip. Eng. Strut. 232 (2021). https://doi.org/10.1016/j.engstruct.2020.111797
6. Luo, X., Ni, J.M., Cheng, Z.Q.: Research on influence parameters of shear lag effect of PC curved box girders with corrugated steel webs. China Measurement & Test **49**(03), 166–171 (2023). https://doi.org/10.11857/j.issn.1674-5124.2021060154
7. Huang, M.H., Zhang, Y.H.: Influence of different warping displacement functions on shear lag effect of thin-walled box girders. J. Lanzhou Jiaotong Univ. **41**(03), 1–7 (2022). https://doi.org/10.3969/j.issn.1001-4373.2022.03.001

Optimization of Cable Force for Hybrid Girder Cable-Stayed Bridge Based on Particle Swarm Optimization Algorithm

Wenhao Xiao(✉)

School of Civil Engineering, Changsha University of Science & Technology, Changsha, China
694655956@qq.com

Abstract. In order to make the hybrid girder cable-stayed bridge force reasonable, in line with the reasonable bridge formation state, the adaptive particle swarm optimization algorithm is used to optimize the cable-stayed bridge of hybrid girder cable-stayed bridge, the adaptive particle swarm optimization algorithm is written by using MATLAB software, and the finite element model is established by using Midas software for the analysis and calculation, and the objective function is constructed by using the influence matrix, and the value of the fitness is calculated, and the adaptive particle swarm optimization algorithm is used to find the global optimal solution. The results show that the vertical displacement of the main beam and the lateral displacement of the main tower are obviously reduced after optimization, and the bending moment of the main tower is obviously improved.

Keywords: Particle swarm optimization algorithm · Hybrid girder cable-stayed bridge · Cable-stayed force optimization · Influence matrix

1 Introduction

Hybrid girder cable-stayed bridges are widely used in the design of large-span bridges because the structure consists of a combination of steel and concrete, which improves its mechanical properties and increases economy. Hybrid girder cable-stayed bridge means that the main girder of a cable-stayed bridge consists of two different materials along the length of the girder, with the girder of the main span being a steel girder, and the girder of the side spans being a concrete girder. Hybrid girder cable-stayed bridges have the advantage of large spanning capacity due to the use of steel girders in the main spans, while concrete girders are used in the side spans, thus providing good anchorage and reducing the cost of bridge construction [1].

In cable-stayed bridges with hybrid girder structures, the main challenge lies in achieving an optimal distribution of forces between the cables and the concrete girders. Due to the differing material properties such as elastic modulus and thermal expansion coefficients between steel cables and concrete girders, there is a risk of uneven force distribution. This non-uniform distribution can lead to instability in the structure and potential damage at the cable-girder connections. Therefore, the critical task is to

B. Yuan et al. (Eds.): ICCASE 2024, SUCI, pp. 107–113, 2024.
https://doi.org/10.1007/978-3-031-76102-7_13

adjust the force distribution through careful design and construction measures to ensure harmonious mechanical performance between the cables and the concrete girders.

Another crucial challenge is the design of connections between the cables and the concrete girders. These connections play a pivotal role in transmitting and distributing cable forces effectively while maintaining structural integrity. However, achieving reliable and durable connections poses significant challenges due to the differences in material properties and the dynamic nature of cable forces. Inadequate connection designs may result in stress concentration, cracking, or loosening at the cable-girder interfaces, compromising the overall stability and safety of the bridge structure. Therefore, it is imperative to employ appropriate connection forms and construction techniques, along with reinforcing measures, to ensure the safety and reliability of cable-girder connections in cable-stayed bridges with hybrid girder structures.

In the design and construction process of cable-stayed bridges, the determination of the cable force is a very critical content, cable-stayed bridges as a high degree of superstatic structure, the force of the structure is affected by the change of the cable force, so how to determine the reasonable cable force is critical for the state of the cable-stayed bridge. Traditional cable-stayed bridge into the bridge cable force optimization methods include: the optimization method of the specified structural state, bending energy minimization method, the influence matrix method, mathematical optimization methods [2], and so on. However, with the rise of intelligent optimization algorithms in recent years, a new theoretical basis and practical methods are proposed for the cable-stay optimization problem.

The adaptive particle swarm optimization algorithm proposed in this paper is based on the particle swarm optimization algorithm [3], which changes the inertia weights of the algorithm in real time according to the evolutionary state of the particle swarm and adds the perturbation processing when the algorithm falls into the local optimum. This method can improve the convergence speed of the algorithm, and at the same time enhance its ability to jump out of the local optimum and the accuracy of finding the global optimal solution.

APSO considers both local and global search simultaneously by simulating particles flying through the search space. This allows APSO to more easily find the global optimal solution compared to traditional methods, which may get stuck in local optima. APSO has a straightforward principle and is easy to implement, requiring no complex mathematical models or sophisticated algorithms. This makes APSO convenient and flexible for practical applications in cable force optimization of cable-stayed bridges.

2 Principles of Adaptive Particle Swarm Optimization Algorithm

Particle swarm optimization algorithm is an evolutionary computation technique based on swarm intelligence method proposed by James Kennedy and Russell Eberhart in 1995 [4]. The updating formula for particle velocity and position is:

$$\begin{aligned} v_{id} &= w \times v_{id} + c_1 \times rand \times (p_{id} - x_{id}) + c_2 \times rand \times (p_{gd} - x_{id}) \\ x_{id} &= x_{id} + v_{id} \end{aligned} \tag{1}$$

where, w is the inertia weight, c_1, c_2 is the acceleration factor, and rand is a random number between [0,1].

The inertia weight w of the above formula is proposed to balance the global search and local search, when w is small, the algorithm's local search ability is stronger, and when w is large, the algorithm's global search ability is stronger [5], which can be either a normal number or a linear or nonlinear positive time function. The adaptive particle swarm optimization algorithm, based on the classical particle swarm optimization algorithm, determines the evolutionary state of the particle swarm by calculating the average Euclidean distance between the particles, so as to realize the dynamic adjustment of the inertia weights. The average Euclidean distance between particles is calculated by the formula [6] as:

$$d_i = \frac{1}{N-1} \sum_{j=1, j \neq i}^{N} \sqrt{\sum_{k=1}^{D} (x_i^k - x_j^k)^2} \tag{2}$$

where N, D are the population size and dimension. An evolutionary factor f [6] is introduced, which is calculated as:

$$f = \frac{d_g - d_{\min}}{d_{\max} - d_{\min}} \in [0, 1] \tag{3}$$

where d_g is the average distance of the globally optimal particle.

The evolution factor f reacts to the evolution state of the particle swarm through the relationship between the average distance between the global optimal particle and other particles. Mapping the inertia weight w with the evolution factor f is more accurate to adjust the inertia weight w than its mapping with the number of iterations, which allows the algorithm to better balance between the global search and the local search, and enhances the performance of the algorithm. The inertia weight w is mapped to the evolutionary factor f using the tanh function with the expression:

$$w(f) = \frac{1.8e^f - e^{-f}}{e^f + e^{0.837f}} \in [0.4, 0.9] \tag{4}$$

The inertia weight w takes values in the range [5, 7] [0.4,0.9].

When $f \in [0, 0.2]$, the particle swarm is defined as being in a state of convergence, at which point a perturbation process is performed, in which the globally optimal particle p_{gd} is processed as follows:

$$p_{gd} = p_{gd} + (x_{\max} - x_{\min}) \cdot Gaussian(\mu, \sigma^2) \tag{5}$$

where x_{max}, x_{min} are the limiting values of the particle swarm perturbation changes; $Gaussian(\mu, \sigma^2)$ is a standard normally distributed random number. If the result after the perturbation is better than the original particle, it is set as the new globally optimal particle, and vice versa, it is left unchanged. This increases the ability of the particle swarm to jump out of the local optimum. The fitness function of the algorithm is the bending energy function.

APSO has a more powerful global search capability and the ability to jump out of the local optimum compared to PSO. It judges the evolutionary state of the particle swarm by calculating the average distance between particles, and maps the inertia weights to the evolutionary factors with a functional relationship, which can realize the dynamic adjustment of the inertia weights. When the particle swarm is in the convergence state, APSO can add perturbations to the particles to reduce the possibility of the particles falling into the local optimum.

3 Optimization Model

3.1 Constraints

The constraints, as required by the specification, are as follows:

$$\begin{cases} \sigma_{i\min} \le \sigma_i \le \sigma_{i\max} \\ T_{i\min} \le T_i \le T_{i\max} \\ |T_{i+1} - T_i| \le \Delta \end{cases} \tag{6}$$

Among them, the extreme values of critical section stresses $\sigma_{i\min}$, $\sigma_{i\max}$ are determined according to the code, and the extreme values of the cable force $T_{i\min}$, $T_{i\max}$ are determined by combining the actual situation of the model calculations with the requirements of the code being satisfied, and Δ is the floating interval of the cable force.

3.2 Optimization Objectives

According to the principle of influence matrix method [8], the influence matrix of cable force on bending energy is constructed. Taking the bending energy as the objective function, the expression for the bending energy of the structure is:

$$U = \int_S \frac{M^2(S)}{2EI} dS \tag{7}$$

The structure can be expressed after discretizing it as:

$$U = \sum_{i=1}^{m} \frac{L_i}{4E_i I_i} (M_{L_i}^2 + M_{R_i}^2) \tag{8}$$

where, m is the total number of structural units; L_i, E_i, I_i denote the length, elastic modulus and sectional moment of inertia of unit No. i; and M_{L_i}, M_{R_i} denote the bending moments at the left and right ends of the unit. Taking the minimum bending energy of the cable-stayed bridge as the target, while controlling the bending moment, it also has a better control effect on the deformation.

4 Project Example

4.1 Project Overview

A cable-stayed bridge with a total length of 760 m double-tower and double-surface hybrid combination girder cable-stayed bridge, the cable-stayed cables are parallel to the cable face, fan-shaped arrangement, a total of 144 cable-stayed cables throughout the bridge, the longest length of about 234 m, the cable basic spacing of 3 m. The structure is a semi-floating system, the side span concrete girders are Π-type section, the girder height is 3 m, and the width is 23.75 m. The bridge layout is shown below as Fig. 1:

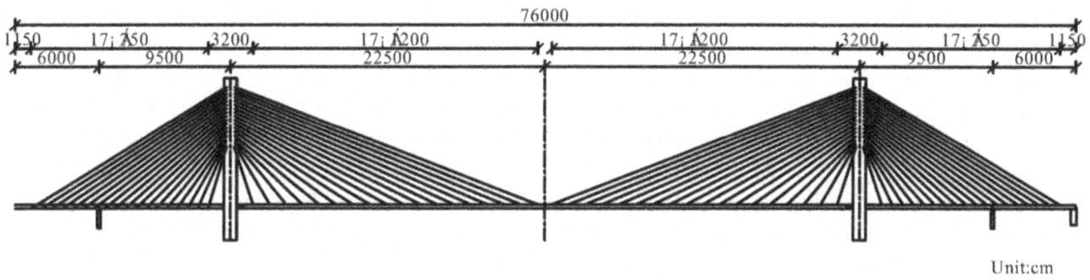

Fig. 1. Bridge Layout

4.2 Parameter Setting

Considering the symmetry of the bridge structure, one side of the cables is taken as the design variable, the initial value of the inertia weight w is set to 0.9 [3], the learning factor c1, c1 are set to 1.496189 [3], the number of populations is set to 36, the number of iterations is set to 200, the position limit is set to 800, and the velocity limit is set to 0.1 times of the position limit. The range of the cable force is obtained by calculating the range of the cable force when the bridge is in the constant load condition. So the ranges of the mid-span cable force are [1500,5000], and the ranges of side The range of mid-span force is [1500,5000], and the range of side-span force is [2000,6500].

4.3 Optimization Results

After the particle swarm algorithm for intelligent optimization results compared with the results under the design cable force, the displacement of the main girder was significantly improved, mainly in the middle of the mid-span steel girder, the maximum displacement was reduced from −122.92 mm to −33.28 mm, and the maximum bending moment was reduced from −183492.1 kN·m to −168339.7 kN·m. The displacement of the main tower at the top of the girder was reduced from 43.49 mm to 6.09 mm. The displacement of the top of the main tower was reduced from 43.49 mm to 6.09 mm, and the improvement effect was remarkable. The bending moment distribution of the main tower is more reasonable, from the original bending moment distributed only on the right side to both left and right sides and the extreme value is reduced from 52490.91 kN·m to 30012.28 kN·m. The results are shown in Fig. 2.

This calculation example also has some limitations, for example, it does not consider the geometric nonlinear effects of cable-stayed bridges as well as the shrinkage and creep under the state of bridge formation, and it does not consider the effect of prestressing on the bending moment of the bridge. In addition, during the calculation process, I found that between the auxiliary pier and the bridge tower, the main girder produces a large displacement and bending moment. This is rare for hybrid girder cable-stayed bridges.

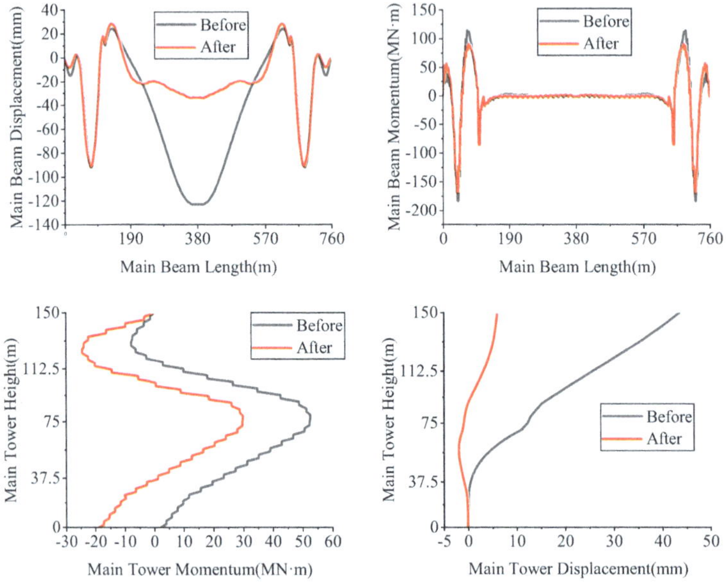

Fig. 2. Optimization Effect Comparison

5 Conclusion

The adaptive particle swarm optimization algorithm proposed in this paper determines the evolutionary state of the particle swarm based on the average distance between particles, and at the same time adjusts the inertia weight of the algorithm accordingly, which can effectively improve the algorithm's ability to find the optimum, and additionally carries out perturbation processing when the algorithm enters into the convergence stage, which can increase the ability of the algorithm to jump out of the local optimum. The optimization results show that the displacements and bending moments of the main beam and the main tower are improved to different degrees.

References

1. Song, C., Xiao, R., Wang, Z.: Cable force optimization of cable-stayed bridges: a surrogate model-assisted differential evolution method combined with B-Spline interpolation curves. Engineering Structures **283**, 2023 (2023). 115856, ISSN 0141-0296, https://doi.org/10.1016/j.engstruct.2023.115856

2. Alberto, M., Luís, S., João, N.: Optimization of under-deck cable-stayed concrete bridges. Computers & Structures **296**, 2024 (2024). 107323, ISSN 0045-7949, https://doi.org/10.1016/j.compstruc.2024.107323
3. Wen, C., Wang, L., Ren, X.: General particle swarm optimization algorithm. In: 2023 IEEE 2nd International Conference on Electrical Engineering, Big Data and Algorithms (EEBDA), pp. 1204–1208. Changchun, China (2023). https://doi.org/10.1109/EEBDA56825.2023.10090725
4. Tareq, S., Ayman, E., Mohammed, A., Qasem, A.: Particle swarm optimization: a comprehensive survey. IEEE Access **10**, 10031–10061 (2022). https://doi.org/10.1109/ACCESS.2022.3142859
5. Liu, W., Wang, Z., Yuan, Y.: A novel sigmoid-function-based adaptive weighted particle swarm optimizer. IEEE Transactions on Cybernetics **51**(2), 1085–1093 (2021). https://doi.org/10.1109/TCYB.2019.2925015
6. Hu, Y., Zhang, Y., Gong, D.: Multiobjective particle swarm optimization for feature selection with fuzzy cost. IEEE Transactions on Cybernetics **51**(2), 874–888 (2021). https://doi.org/10.1109/TCYB.2020.3015756
7. Wang, Z., Zhan, Z., Kwong, S., Jin, H.: Adaptive granularity learning distributed particle swarm optimization for large-scale optimization. IEEE Transactions on Cybernetics **51**(3), 1175–1188 (2021). https://doi.org/10.1109/TCYB.2020.2977956
8. Wang, Z., Zhang, N., Cheng, Q.: Multi-objective optimization-based reasonable finished state in long-span cable-stayed bridge considering counterweights. Structures **51**, 1497–1506 (2023). ISSN 2352-0124, https://doi.org/10.1016/j.istruc.2023.03.061

Sensitivity Analysis on Structural Parameters of Curved Extradosed Cable-Stayed Bridge

Lei Xu(✉), Haoliang Tian, and N. R. Obouaka

CCCC First Highway Consultants Co., Ltd., Xi'an, China
leixus@foxmail.com, onilfa@foxmail.com

Abstract. Extradosed cable-stayed bridge which is featured in beautiful shape, distinctive structural characteristics and the dual advantages of girder and cable-stayed bridge, is vigorously developed in the world in recent years. Despite the fact that there are extensive researches concerning on extradosed cable-stayed bridges in the world, but most of them almost focus on the linear extradosed cable-stayed bridge, while the research on Curved Extradosed Cable-Stayed bridge is rarely involved. This paper outlines the current research status and development of Curved Extradosed Cable-Stayed bridges, analyzes and discusses the proportion of load sharing of extradosed cable-stayed bridge, and proposes a formula for the load sharing proportion of curved extradosed cable-stayed bridge. Taking the Loong River Bridge as an engineering example, a finite element analysis model is established to analyze the influencing factors of the load sharing proportion of curved extradosed cable-stayed bridge. Five structural parameters are mainly selected for comparative analysis, including main girder stiffness, prestressing in the main girder, cross-sectional area of the stay cable, main tower stiffness, and main tower height. The paper compares and analyzes the effects of variations in these five structural parameters on the structural performance of Curved Extradosed Cable-Stayed bridges, and obtains the sensitivity laws of the main girder, main tower, stay cables, and main piers to variations in these five structural parameters. Therefore, the research on the structural behavior and structural design of Curved Extradosed Cable-Stayed bridge is not only favorable for expanding the scope of application of extradosed cable-stayed bridge, but also favorable for providing technical supports to the design and construction of curved extradosed cable-stayed bridge.

Keywords: Curved Extradosed Cable-Stayed bridge · structural performance · cables-girder load sharing proportion · structural parameters

1 Introduction

The extradosed cable-stayed bridge, also known as a partial cable-stayed bridge, represents a composite bridge system situated between cable-stayed bridge and continuous beam bridge. As a novel bridge form, it boasts graceful aesthetics and distinctive structural features, combining the advantages of both cable-stayed bridge and beam bridge. Over the past two decades, it has experienced vigorous development and widespread application both in China and internationally [1, 3, 4].

B. Yuan et al. (Eds.): ICCASE 2024, SUCI, pp. 114–125, 2024.
https://doi.org/10.1007/978-3-031-76102-7_14

A review of the current research status and development trends of extradosed cable-stayed bridges, both in China and abroad, reveals that although numerous such bridges have been constructed worldwide, the majority of these existing or under-construction bridges are linear in shape. Curved Extradosed Cable-Stayed bridges are exceptionally rare. A scant number of reports on Curved Extradosed Cable-Stayed bridges can be found in existing literature. Specific instances include: ① Pont de Saint-Rémy-de-Maurienne Bridge in France, with a radius of R = 500 m and spans of (52.5 + 48.5) m. ② Sunniburg Bridge in Switzerland, with a radius of R = 500 m and spans of (59 + 128 + 140 + 134 + 65) m. ③ Shin-Kanto Bridge in Japan, with a radius of R = 400 m and spans of (74.1 + 140 + 69.1) m. ④ Kack-Hwa First Bridge in South Korea, with a radius of R = 900 m and spans of (55 + 115 + 100) m. ⑤ Pyung-Yeo 2 Bridge in South Korea, with a radius of R = 820 m and spans of (65 + 120 + 65) m.

The Curved Extradosed Cable-Stayed bridge has advantages in structural performance compared to the curved continuous rigid frame bridge. It can effectively solve the problem of excessive mid-span deflection, improve the structural durability, and it has better torsion resistance. The research on Curved Extradosed Cable-Stayed bridge is a new method to solve the structural problems of long-span curved bridges in mountainous areas. Its starting point is to use extradosed cables to resist the torsional effect of the structure due to plane bending. Curved Extradosed Cable-Stayed bridges have broad application prospects in the construction of long-span curved bridges in mountainous areas. Currently, almost all research on extradosed cable-stayed bridges, both in China and abroad, focuses exclusively on linear extradosed cable-stayed bridges, with very few studies addressing Curved Extradosed Cable-Stayed bridges [4, 5]. Therefore, conducting research on the structural performance and design aspects of Curved Extradosed Cable-Stayed bridges holds significant importance. Such endeavors will expand the applicability of extradosed cable-stayed bridges and provide technical support for the design and construction of Curved Extradosed Cable-Stayed bridges.

2 Engineering Example and Finite Element Analysis Model

2.1 Engineering Example

This paper investigates the sensitivity of structural parameters in Curved Extradosed Cable-Stayed bridges using the Loong River Bridge as an engineering example. The Loong River Bridge spans a "V"-shaped valley between mountains, avoiding environmental damage caused by excavating the mountainside. The bridge layout consists of (6 × 30 m T-beams) + (86 + 160 + 86 m extradosed cable-stayed bridge) + (5 × 30 m T-beams). The main bridge is a three-span prestressed concrete Curved Extradosed Cable-Stayed bridge, with a span layout of (86 + 160 + 86)m. The abutments, piers, and beams are interconnected. The main bridge is situated on a curvature radius of R = 850 m on a horizontal curve, The cantilever casting construction method is used, the elevation view of this bridge is shown in Fig. 1. The bridge has a total width of 28 m and adopts a separated double-box design, with each box having a single-chamber section and a width of 12 m. The bridge towers and stay cables are arranged within a 4 m-wide space in the middle of the bridge, with crossbeams placed between the double boxes to anchor the stay cables. The height of the box beam at the tower location is

8.0 m, while the height at the mid-span and at the edge support points is 3.6 m, utilizing a three-way prestressed system. The main tower adopts a single-column tower with a dumbbell-shaped solid concrete cross-section, measuring 4.5 m longitudinally and 3.0 m transversely, with a height of 37.5 m and embedded cable saddles inside the tower column. On layer of stay cables is used and symmetrically placed in relation with the tower, with cable spacings of 4.0 m or 4.5 m on the box beam and 2.0 m on the tower. Low-relaxation high-strength steel strands are used for the stay cables, with each cable having a $31\text{-}\Phi^{j}15.24$ cross-section. There are 6 pairs of cables on each upper tower column, totaling 12 pairs of cables for the entire bridge.

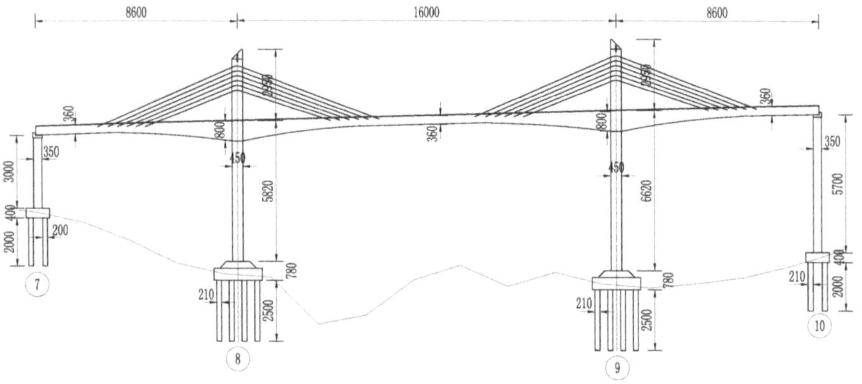

Fig. 1. Layout of the main bridge of the Loong River Bridge (unit: cm)

2.2 Finite Element Analysis Model

A numerical simulation analysis of the main bridge of the Loong River Bridge was conducted using finite element software. The structural analysis employed a spatial beam element model, with the main tower, box girder, and main pier represented by spatial beam elements, while the stay cables of the cable-stayed bridge were modeled using planar truss elements [2, 6]. Abutments were treated as rigid in the calculations. The box girder was divided into segments according to each construction section and considering the anchoring positions of the stay cables. Calculation parameters were based on actual bridge design values. The finite element analysis model is illustrated in Fig. 2.

The finite element analysis model has a total of 503 nodes and is divided into 437 elements, of which the main beam is divided into 244 elements, the main tower and main pier are divided into 61 elements, the cross beams are divided into 84 elements, and the stay cables are divided into 48 elements. The structural analysis introduced here is based on the linear elasticity assumption. Since the normal use stage of prestressed concrete structures is mainly in the linear elastic working stage of the material, this assumption is also true.

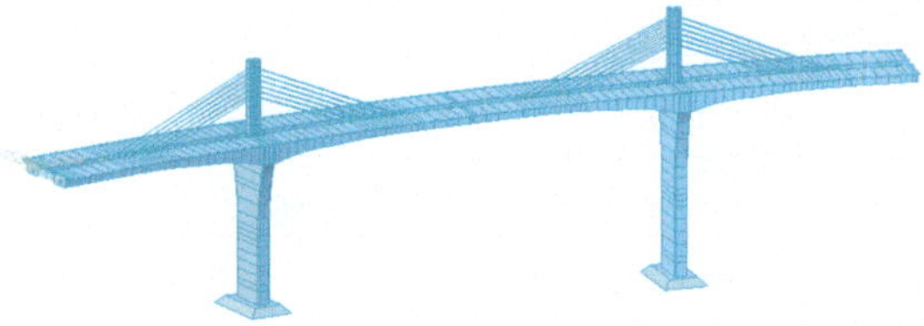

Fig. 2. Model of the Main Bridge of the Loong River Bridge

3 Proportion of Load Sharing Between Stay Cables and Girder in Curved Extradosed Cable-Stayed Bridges

The primary load-bearing components of a extradosed cable-stayed bridge under external loading are the box girder and the stay cables. The box girder plays a predominant role, while the stay cables serve an auxiliary adjustment function. The coordinated action of these two elements forms the structural characteristics of the extradosed cable-stayed bridge [3]. Regarding the proportion of load sharing between stay cables and girder, depending on the classification of loads, corresponding to the conditions of permanent loads, it can be referred to as the permanent load sharing ratio of stay cables and girder. Corresponding to live loads, it can be referred to as the live load sharing ratio of stay cables and girder [7, 9]. For a extradosed cable-stayed bridge, among the numerous permanent load forces, there must be a state of cable force that minimizes the cost of the superstructure, known as the economical cable force. The permanent load sharing ratio of stay cables and girder in this state is termed as the economical permanent load sharing ratio of stay cables and girder. This concept is significant for investigating the optimal distribution ratio of stay cable and girder prestress in extradosed Cable-Stayed Bridges [7, 9].

According to existing literature, research on the proportion of load sharing between stay cables and girder in extradosed Cable-Stayed Bridges, both domestically and internationally, has been primarily focused on straight extradosed Cable-Stayed Bridges. The stay cables of straight extradosed Cable-Stayed Bridges can be considered arranged in the same plane and analyzed accordingly in the planar arrangement. Formulas only consider the angle between the stay cables and the box girder in the vertical plane. However, the formulas for the proportion of load sharing between stay cables and girder in straight extradosed Cable-Stayed Bridges are not applicable to curved extradosed Cable-Stayed Bridges. This is because the stay cables of curved extradosed Cable-Stayed Bridges are not arranged in a planar manner but in a spatially curved configuration. For curved extradosed Cable-Stayed Bridges, consideration must be given not only to the vertical angle between the stay cables and the curved box girder but also to the angle φ_i between the projection of the spatial stay cables in the plane of the box girder and the tangent line at the anchor point of the main girder. Moreover, the box girder is curved, further adding to the complexity of the proportion of load sharing between stay cables and girder in curved extradosed Cable-Stayed Bridges. Based on the analysis of the structural characteristics of curved extradosed Cable-Stayed Bridges, the formula for the proportion of load sharing between stay cables and girder in curved extradosed Cable-Stayed Bridges can be derived as follows:

$$\psi = \frac{\sum_i^n (1/\delta_{ci})}{(1/\delta_{cg\max})} = \frac{\sum_i^n E_{ci}A_{ci}\sin^2\alpha_i\cos\phi_i/L_{ci}}{E_{cg}I_{cg}/L_{cg}^3}$$

In the equation:

δ_c represents the vertical component of the elongation of unit tension of cable i;

δ_{cgmax} represents the vertical displacement of the main girder at the location of the cable when subjected to unit vertical force;

E_{ci}, A_{ci}, and L_{ci} denote the elastic modulus, cross-sectional area, and length, respectively, of the ith stay cable;

α_i represents the vertical angle between the ith stay cable and the plane of the main girder;

φ_i represents the angle between the projection line of the ith stay cable in the plane of the main girder and the tangent line at the anchor point of the main girder;

E_{cg}, I_{cg}, and L_{cg} respectively denote the elastic modulus, moment of inertia, and calculated midspan length of the curved main girder;

4 Sensitivity Analysis of Structural Parameters in Curved Extradosed Cable-Stayed Bridges

There are numerous factors that influence the proportion of load sharing between stay cables and girder in curved extradosed Cable-Stayed Bridges. By analyzing the structural parameters that affect this proportion, and comparing the variations in various structural parameters to analyze the changes in the forces and deformations of curved extradosed Cable-Stayed Bridges, one can reflect on how changes in the load sharing ratio affect the structural performance of curved extradosed Cable-Stayed Bridges [8, 9]. Based on the structural characteristics of curved extradosed Cable-Stayed Bridges, the main structural parameters that affect the proportion of load sharing between stay cables and girder are as follows: ① main girder stiffness, ② prestressing within the main girder, ③ cross-sectional area of the stay cables, ④ main tower stiffness, and ⑤ main tower height.

Variations in structural parameters of curved extradosed Cable-Stayed Bridges directly affect the forces and deformations of the main structural components, thereby influencing the overall structural performance. However, different variations in structural parameters have varying degrees of influence on the forces and deformations of the main structural components of curved extradosed Cable-Stayed Bridges, i.e., different sensitivities. Therefore, it is necessary to analyze the sensitivity of curved extradosed Cable-Stayed Bridges to variations in structural parameters. Analyzing the sensitivity of curved extradosed Cable-Stayed Bridges to variations in structural parameters will enable more directed and targeted optimization of their design [8, 9]. Considering the structural characteristics of curved extradosed Cable-Stayed Bridges, the five main structural parameters mentioned above are selected as independent variables for sensitivity analysis of curved extradosed Cable-Stayed Bridges to variations in structural parameters. The sensitivity analysis employs a single-variable variation calculation mode.

In this study, the Loong River Bridge (86 m + 160 m + 86 m) is taken as an engineering example for analysis and computation. The load calculation includes permanent loads, traffic loads, and temperature loads. The recommended scheme of the Loong River Bridge is taken as the original model for computation analysis. The selected target structural parameters are decreased by 20%, decreased by 10%, increased by 10%, and increased by 20%, respectively, forming five calculation models for each structural parameter ξ as a variable [8]. Thus, a set of five calculation models is established with variations in structural parameters: 1.2ξ, 1.1ξ, ξ, 0.9ξ, and 0.8ξ. Finite element simulation analysis is performed on these five models under the variations in structural parameters, and the sensitivity laws of curved extradosed Cable-Stayed Bridges to variations in structural parameters are obtained based on the calculation results.

4.1 Sensitivity of the Main Girder to Variations in Structural Parameters

The main girder of the Loong River Bridge is of a separated design. Based on the relative positions of the main girder with respect to the bridge's centerline, it can be divided into outer and inner girders, connected by transverse beams. Based on the computed analysis data, the sensitivity of the outer and inner girders to variations in structural parameters exhibits the following characteristics:

(1) The mid-span bending moment of the outer and inner girders is most sensitive to variations in prestressing within the main girder, as shown in Fig. 3-a.
(2) The mid-span deflection of the outer and inner girders is most sensitive to variations in prestressing within the main girder and variations in the main tower height, as shown in Fig. 3-b.
(3) The bending moment of the main girder at the pier top of both the outer and inner girders is most sensitive to variations in the main girder stiffness, followed by variations in the main tower height, as shown in Fig. 3-c.

(4) The axial force in the main girder at the pier top of both the outer and inner girders is most sensitive to variations in prestressing within the main girder and variations in the main tower height, as shown in Fig. 3-d.
(5) The torsional moment in the outer girder at the pier top is most sensitive to variations in prestressing within the main girder, followed by variations in the main girder stiffness, as shown in Fig. 3-e. The torsional moment in the inner girder at the pier top is most sensitive to variations in the main girder stiffness.

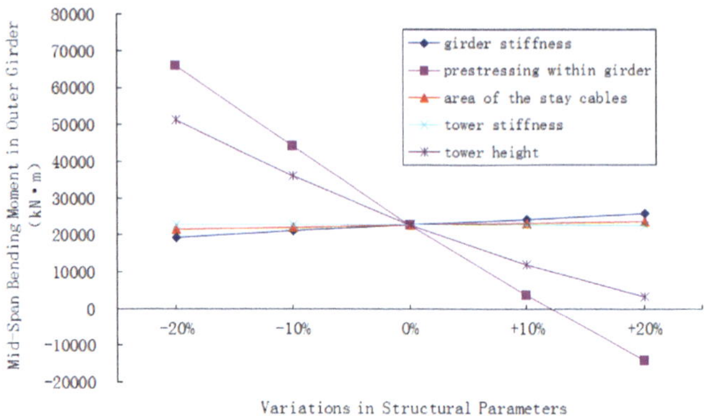

(a) Comparison of Mid-Span Bending Moment in Outer Girder

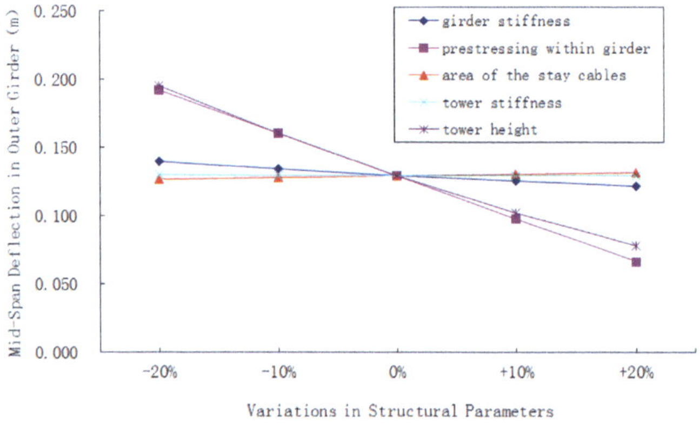

(b) Comparison of Mid-Span Deflection in Outer Girder

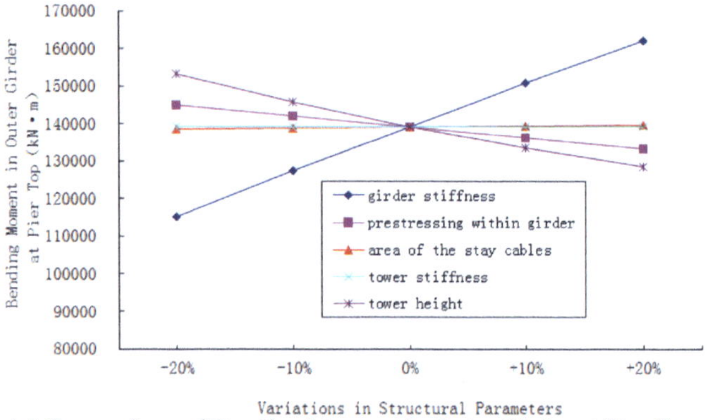

(c) Comparison of Bending Moment in Outer Girder at Pier Top

Fig. 3. Sensitivity Comparison of Outer Girder to Structural Parameter Variations

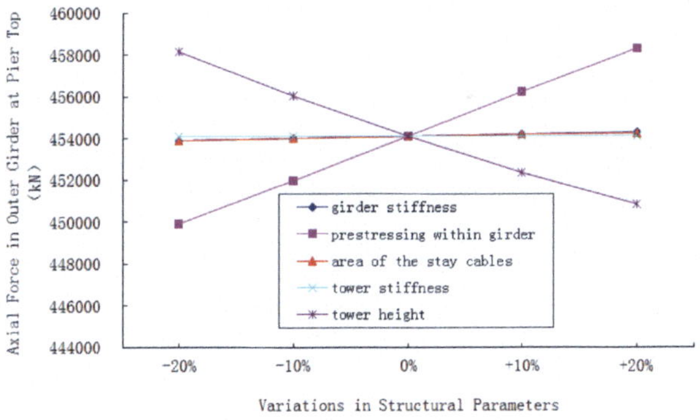

(d) Comparison of Axial Force in Outer Girder at Pier Top

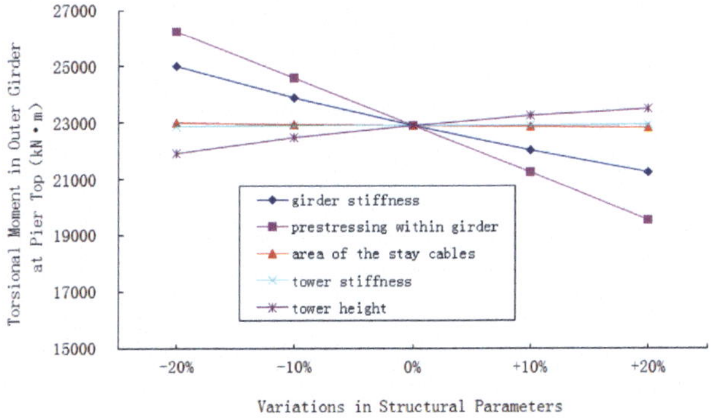

(e) Comparison of Torsional Moment in Outer Girder at Pier Top

Fig. 3. (*continued*)

4.2 Sensitivity of the Main Towers to Variations in Structural Parameters

The main towers of the Loong River Bridge are anchored to the main piers, situated between the two box chambers of the separated main girder and are not connected to the main girder. As the two main towers are symmetrically positioned about the midpoint of the bridge, the sensitivity comparison analysis is conducted on the main tower of Pier 8. Based on the computed analysis data, the sensitivity of the main towers to variations in structural parameters exhibits the following characteristics:

(1) The bending moment in the longitudinal direction at the base of tower is most sensitive to variations in prestressing within the main girder, followed by variations in the main tower height, as shown in Fig. 4-a. The bending moment in the transverse direction at the base of tower is most sensitive to variations in the main tower height.
(2) The displacement in the Y-axis direction on the top of tower is most sensitive to variations in the main tower height, followed by variations in main tower stiffness, as shown in Fig. 4-b. The displacement in the X-axis direction on the top of tower is less than the displacement in the Y-axis direction.

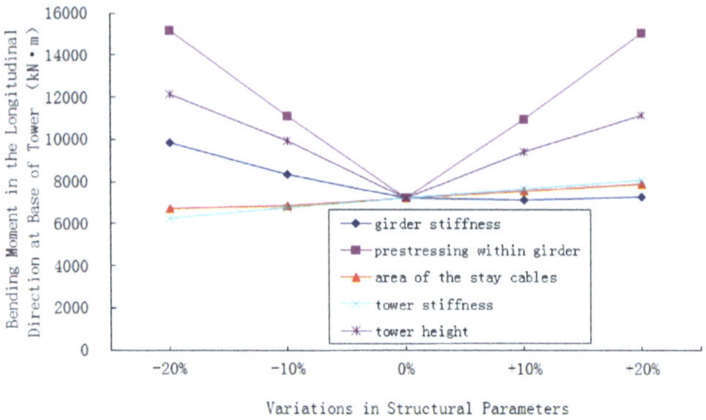

(a) Comparison of Bending Moment in the Longitudinal Direction

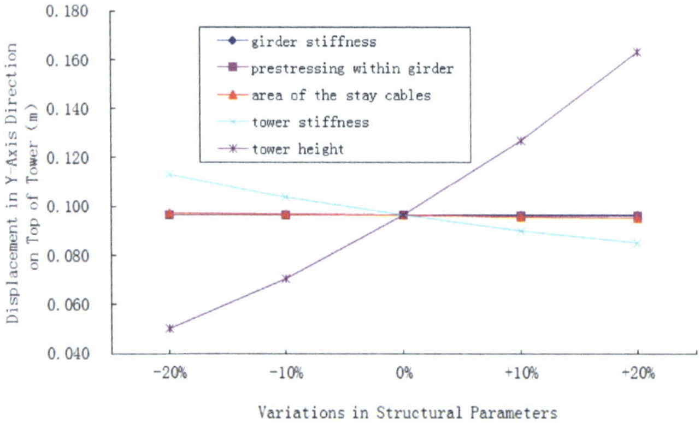

(b) Comparison of Displacement in Y-Axis Direction on Top of Tower

Fig. 4. Sensitivity Comparison of Main Towers to Structural Parameter Variations

4.3 Sensitivity of Stay Cables to Variations in Structural Parameters

The stay cables of the Loong River Bridge are symmetrically arranged, with the entire structure being symmetrical about the midpoint of the bridge and the stay cables being symmetrical about the main towers. The forces and stresses of the stay cables at Tower 8 are selected for sensitivity comparison analysis. Based on the computed analysis data, the sensitivity of stay cables to variations in structural parameters exhibits the following characteristics:

(1) The maximum cable forces is most sensitive to variations in prestressing within the main girder, followed by variations in main tower height, as shown in Fig. 5-a.
(2) The maximum cable stresses is most sensitive to variations in the cross-sectional area of the stay cables, followed by variations in prestressing within the main girder, as shown in Fig. 5-b.

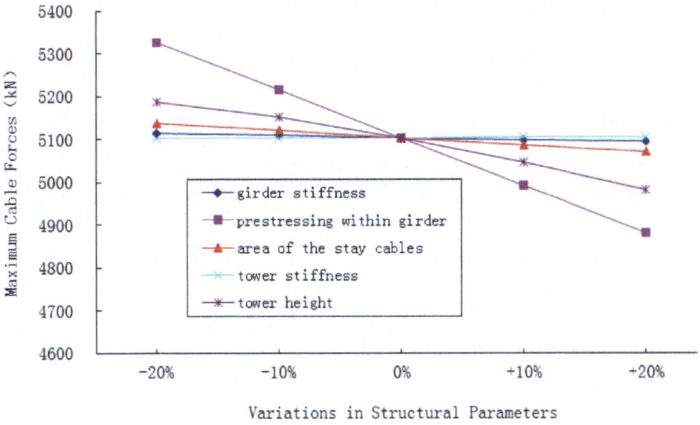

(a) Comparison of Maximum Cable Forces

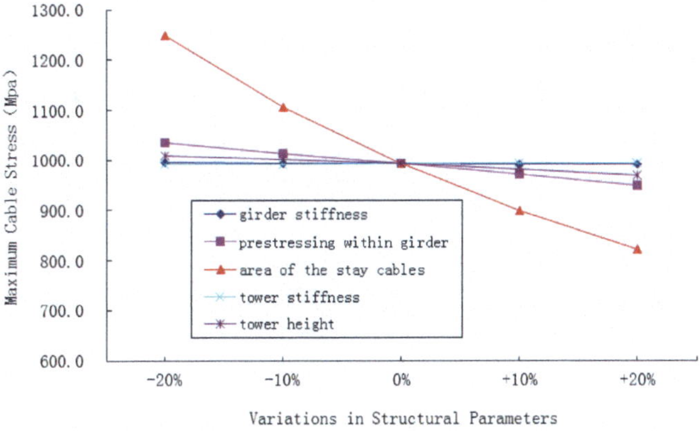

(b) Comparison of Maximum Cable Stress

Fig. 5. Sensitivity Comparison of Stay Cables to Structural Parameter Variations

4.4 Sensitivity of Piers to Variations in Structural Parameters

The piers of the Loong River Bridge are integrated with the separated main girders, resembling a continuous rigid-frame bridge structure. The pier cross-section varies. Pier 8 is selected for sensitivity analysis. Based on the calculated data analysis, the sensitivity of piers to structural parameter variations exhibits the following characteristics:

(1) The bending moment in the longitudinal and transverse directions at the bottom of pier are most sensitive to variations in the internal pre-stress of the main girder, followed by variations in the height of the main tower, as shown in Fig. 6-a.
(2) The displacement in the X-axis direction on the top of pier is most sensitive to variations in the internal prestress of the main girder, as shown in Fig. 6-b.

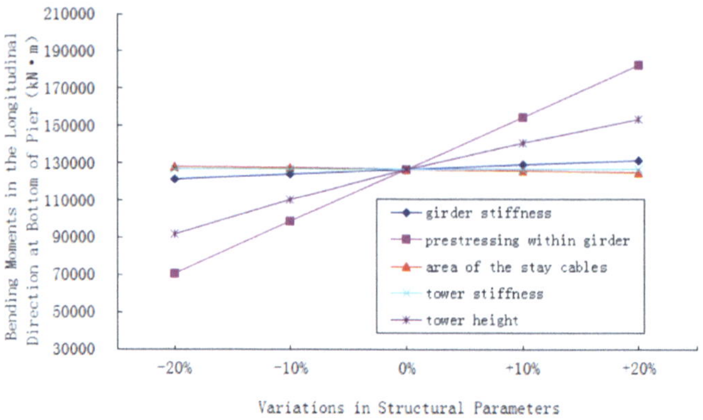

(a) Comparison of Bending Moments in the Longitudinal Direction

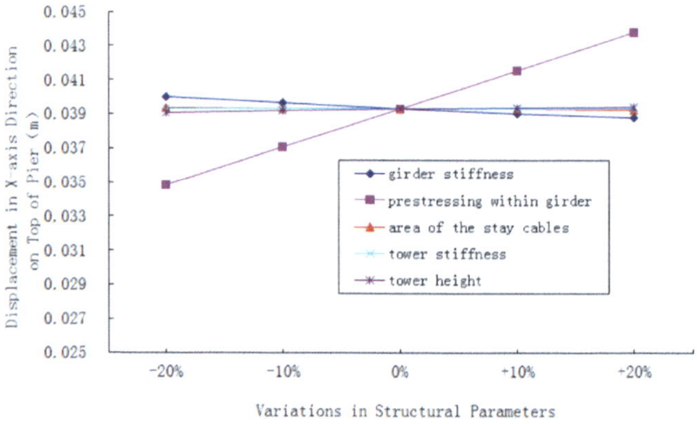

(b) Comparison of Displacement in X-axis Direction on Top of Pier

Fig. 6. Sensitivity Comparison of the Main Pier to Structural Parameter Variations

5 Conclusion

The analysis of the effects of variations in five structural parameters, namely main girder stiffness, internal prestress of the main girder, cross-sectional area of stay cables, main tower stiffness, and main tower height, on the structural performance of curved extradosed cable-stayed bridges yields the following main conclusions:

(1) Curved extradosed cable-stayed bridges constitute spatial structural systems, rendering planar structural systems inadequate. In determining the load-sharing ratio of stay cables in curved extradosed cable-stayed bridges, considerations extend beyond the parameters applicable to straight extradosed cable-stayed bridges to include the spatial curvature of stay cables. Therefore, an equation for the load-sharing ratio of stay cables in curved extradosed cable-stayed bridges should incorporate the angle φi between the projection line of the stay cable in the main girder plane and the tangent line at the main girder anchorage point.

(2) Variations in structural parameters of curved extradosed Cable-Stayed Bridges directly affect the forces and deformations of the main structural components, thereby influencing the overall structural performance. Different variations in structural parameters have varying degrees of influence on the forces and deformations of the main structural components of curved extradosed Cable-Stayed Bridges. Therefore, it is necessary to analyze the sensitivity of curved extradosed Cable-Stayed Bridges to variations in structural parameters. Analyzing the sensitivity of curved extradosed Cable-Stayed Bridges to variations in structural parameters will enable more directed and targeted optimization of their design. Such endeavors will expand the applicability of extradosed cable-stayed bridges and provide technical support for the design and construction of Curved Extradosed Cable-Stayed bridges. Relevant rules can be found in the previous analysis.

The paper also has some limitations or deficiencies, this article's research on the structural performance of curved extradosed cable-stayed bridges is mainly conducted on the static performance of the structure. However, the research on the dynamic performance of the curved extradosed cable-stayed bridges has not been involved. Further in-depth research is needed on the structural dynamic performance of curved extradosed cable-stayed bridges.

References

1. Yan, G.M.: On "Partial Cable-Stayed Bridges"—Japan Watanabe Nanobashi, Watanabe Kitabashi, and Odawara Port Bridges. Foreign Bridges **1**, 47–50 (1996)
2. Shen, M.W.: Research on Static Performance of Partial Cable-Stayed Bridges (Master's thesis). Tongji University (2002)
3. Chen, H.J., Wang, K., Li, C.G.: Discussion on Partial Cable-Stayed Bridges. Bridge Construction **1**, 44–47 (2002)
4. Collings, D., Gonzalez, A.S.:. Extradosed and cable-stayed bridges, exploring the boundaries. In: Proceedings of the Institution of Civil Engineers-Bridge Engineering, Vol. 166, No. 4, pp. 231–239. Thomas Telford Ltd. (2013)
5. Ruiz-Teran, A.M., Aparicio, A.C.: Two new types of bridges: under-deck cable-stayed bridges and combined cable-stayed bridges—the state of the art. Can. J. Civ. Eng. **34**(8), 1003–1015 (2007)
6. Liu, S.L., Wang, S.S.: Cable-Stayed Bridge Design. People's Communications Press (2006)
7. Chen, C.C.: Research on Core Issues of Extradosed Cable-Stayed Bridge Design Theory (Doctoral dissertation). Tongji University (2005)
8. Saad, F.: Structural optimization of extradosed bridges. In IABSE Symposium Report, Vol. 88, No. 6, pp. 212–223. International Association for Bridge and Structural Engineering (2004)
9. Sardesai, M.V., Desai, A.K.: Investigation into Cable-Structure Interaction For Extradosed Bridge. Int. J. Eng. Res. Appl. **3**, 1424–1429 (2013)

Elasticity Solutions for Continuous Laminated Beams Subjected to Arbitrarily Distributed Loads

Huixuan Han[1,2](✉), Xuxu Zhou[3], and Zhiyi Pei[4]

[1] School of Civil Engineering, Southeast University, Nanjing 211189, China
huixuanhan@just.edu.cn
[2] School of Civil Engineering and Architecture, Jiangsu University of Science and Technology, Zhenjiang 212100, China
[3] Zhenjiang Construction Engineering Quality Supervision Station, Zhenjiang 212000, China
[4] Huaxin Design Group Co., Ltd., Wuxi 214072, China

Abstract. Discontinuous high-speed vehicle loads may induce highly localized stresses that can lead to cracking and delamination failure of simply supported composite girder bridges. An elasticity analytical method is proposed to predict the displacement and stress distributions of multi-span laminated beams under arbitrary static distributed loads. The reaction forces of the intermediate supports are considered to be unknown vertical external forces acting on the bottom surface of the continuous beam. Based on the basic equations of plane stress problems in theory of elasticity, the general solutions of displacements and stresses for each single layer are derived by adopting the Fourier series expansion method. Then, the elasticity analytical solutions of the laminated beam can be obtained by the continuity conditions of the interfaces and the boundary conditions of the top and bottom surfaces. The convergence and comparison studies of the solution are conducted through two numerical examples. And the numerical results using the present method are in good agreement with those by using the finite element software, thus effectively verifying the accuracy and correctness of the present method.

Keywords: Continuous laminated beams · Intermediate supports · Fourier series expansion · Elasticity solutions

1 Introduction

The laminated composite structures are widely used in engineering field due to their ability to fully utilize the mechanical properties of materials, such as multi-layer steel bridge deck pavement systems, steel-concrete or wood-concrete composite members [1, 2]. With the rapid development of high-speed railways and highways, the simply-supported continuous girder bridges occupy a large proportion because of their excellent mechanical properties, easy standardization and ride comfort. Those structures may reveal complex behavior subjected to external loading owing to the mismatched material properties and various interface connections between the interlayers [3] such as

B. Yuan et al. (Eds.): ICCASE 2024, SUCI, pp. 126–134, 2024.
https://doi.org/10.1007/978-3-031-76102-7_15

continuum, delamination [4], containing rubber columns [5], and material infiltration of interfaces [6]. Therefore, refined mechanical analysis of multi-span laminated structures is of great significance.

There are mainly two categories of theories adopted to the mechanical analysis of beams, namely the analytical methods (Euler-Bernoulli beam theory, Timoshenko beam theory, elasticity theory) and the numerical methods using finite element method. Xi and Chen [7] proposed a first-order shear deformation theory for laminated beams based on the Mindlin assumption. Xu et al. [8] used the Fourier series expansion method to study the elasticity solution of clamped-simply supported laminated beams considering interface slips. Based on the Hellinger-Reissner variational principle, Zhang et al. [9] proposed a semi-analytical model for laminated beams that can deal with interface delamination. The solution based on Euler-Bernoulli beam theory neglects shear deformation and is therefore less accurate for thick beams. Timoshenko beam theory provides better results for moderate thickness beams. However, the displacement assumptions in both theories are independent of material properties leading to inaccurate prediction of deflections and stresses. Compared with the results from the above beam theories, the solution from the elasticity theory is more accurate because it abandons all assumptions for displacements. The accuracy and numerical stability of the finite element solution depend greatly on the parameter settings and the computational cost of numerical simulation is high. Therefore, this paper attempts to develop a simple and efficient elasticity analytical method to predict the displacement and stress distributions in multi-span laminated beams under arbitrary static distributed loads.

2 Theoretical Formulations

Consider a multi-span simply supported laminated beam with p layers, as shown in Fig. 1. It is assumed that each single layer of the beam is composed of homogeneous isotropic materials. The length and thickness of the laminated beam are L and H, respectively. A global coordinate system with the axes along the length and thickness directions represented by x and y respectively is established. The thickness, Young's modulus and Poisson's ratio of the ith layer are denoted as h_i, E_i and μ_i, respectively. A distributed load $Q(x)$ is applied on the top surface of the beam. There're K intermediate supports on the bottom surface of the laminated beam.

The ith layer of the multi-span laminated beam is analyzed according to the plane stress model of the theory of elasticity. The stress equilibrium differential equations expressed in terms of displacements can be obtained from the constitutive equation of elasticity:

$$\begin{cases} \frac{\partial^2 u^{(i)}}{\partial x^2} + \frac{1-\mu_i}{2}\frac{\partial^2 u^{(i)}}{\partial y_i^2} + \frac{1+\mu_i}{2}\frac{\partial^2 v^{(i)}}{\partial x \partial y_i} = 0 \\ \frac{\partial^2 v^{(i)}}{\partial y_i^2} + \frac{1-\mu_i}{2}\frac{\partial^2 v^{(i)}}{\partial x^2} + \frac{1+\mu_i}{2}\frac{\partial^2 u^{(i)}}{\partial x \partial y_i} = 0 \end{cases} \tag{1}$$

where $u^{(i)}$ and $v^{(i)}$ are the displacements of the ith layer of the laminated beam in the x axis and y axis directions, respectively.

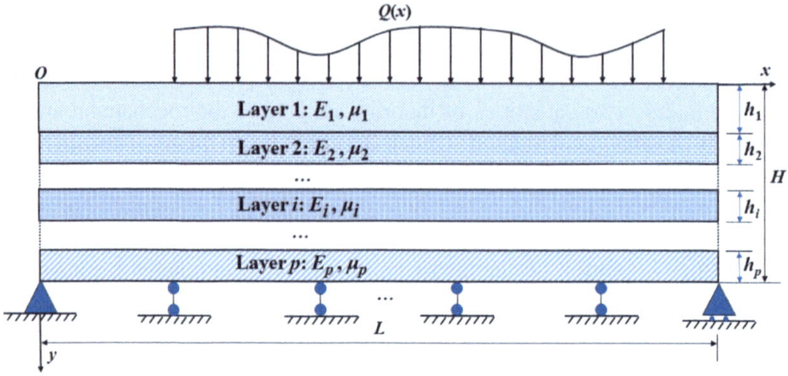

Fig. 1. Plane stress model of a multi-span laminated beam in the global coordinate system

For the multi-span laminated beam, the boundary conditions on the two edges ($x = 0, L$) satisfy the following equations:

$$\sigma_x(0, y_i) = \sigma_x(L, y_i) = 0, \quad v(0, y_i) = v(L, y_i) = 0 \tag{2}$$

The displacement functions can be expressed in the form of Fourier series as follows to satisfy the boundary conditions in Eq. (2).

$$u^{(i)} = \sum_{m=1}^{\infty} U_m^{(i)}(y_i)\cos(a_m x), \quad v^{(i)} = \sum_{m=1}^{\infty} V_m^{(i)}(y_i)\sin(a_m x) \tag{3}$$

Substituting Eq. (3) into Eq. (1) yields the general analytical solutions of the displacement and stress distributions of the ith layer:

$$\begin{aligned}
u^{(i)}(x, y_i) &= \sum_{m=1}^{\infty} \cos(a_m x)\{\cosh(a_m y_i)A_m^{(i)} + \sinh(a_m y_i)B_m^{(i)} + [\frac{3-\mu_i}{a_m(1+\mu_i)} \\
&\sinh(a_m y_i) + y_i\cosh(a_m y_i)]C_m^{(i)} + [\frac{3-\mu_i}{a_m(1+\mu_i)}\cosh(a_m y_i) + y_i\sinh(a_m y_i)]D_m^{(i)}\} \\
v^{(i)}(x, y_i) &= \sum_{m=1}^{\infty} \sin(a_m x)\{\sinh(a_m y_i)A_m^{(i)} + \cosh(a_m y_i)B_m^{(i)} + y_i\sinh(a_m y_i)C_m^{(i)} + y_i\cosh(a_m y_i)D_m^{(i)}\} \\
\sigma_x^{(i)}(x, y_i) &= -\frac{E_i}{1+\mu_i}\sum_{m=1}^{\infty} \sin(a_m x)\{a_m\cosh(a_m y_i)A_m^{(i)} + a_m\sinh(a_m y_i)B_m^{(i)} \\
&+[\frac{3+\mu_i}{1+\mu_i}\sinh(a_m y_i) + a_m y_i\cosh(a_m y_i)]C_m^{(i)} + [\frac{3+\mu_i}{1+\mu_i}\cosh(a_m y_i) + a_m y_i\sinh(a_m y_i)]D_m^{(i)}\} \\
\sigma_y^{(i)}(x, y_i) &= \frac{E_i}{1+\mu_i}\sum_{m=1}^{\infty} \sin(a_m x)\{a_m\cosh(a_m y_i)A_m^{(i)} + a_m\sinh(a_m y_i)B_m^{(i)} \\
&+[\frac{1-\mu_i}{1+\mu_i}\sinh(a_m y_i) + a_m y_i\cosh(a_m y_i)]C_m^{(i)} + [\frac{1-\mu_i}{1+\mu_i}\cosh(a_m y_i) + a_m y_i\sinh(a_m y_i)]D_m^{(i)}\} \\
\tau_{xy}^{(i)}(x, y_i) &= \frac{E_i}{1+\mu_i}\sum_{m=1}^{\infty} \cos(a_m x)\{a_m\sinh(a_m y_i)A_m^{(i)} + a_m\cosh(a_m y_i)B_m^{(i)} \\
&+[\frac{2}{1+\mu_i}\cosh(a_m y_i) + a_m y_i\sinh(a_m y_i)]C_m^{(i)} + [\frac{2}{1+\mu_i}\sinh(a_m y_i) + a_m y_i\cosh(a_m y_i)]D_m^{(i)}\}
\end{aligned} \tag{4}$$

in which,

$$a_m = \frac{m\pi}{L},\ y = \begin{cases} y_i & i = 1 \\ y_i + \sum\limits_{j=1}^{i-1} h_j & i = 2, 3 \cdots p \end{cases} \tag{5}$$

$A_m^{(i)}$, $B_m^{(i)}$, $C_m^{(i)}$, $D_m^{(i)}$ are the coefficients to be determined.

The distributed load Q(x) can be rewritten in the form of Fourier series expansion:

$$Q(x) = \sum_{m=1}^{\infty} q_m \sin(a_m x),\ q_m = \frac{2}{L}\int_0^L Q(x)\sin(a_m x)dx \tag{6}$$

The boundary conditions of the top surface of the beam is

$$\sigma_y^{(1)}(x, 0) = Q(x),\ \tau_{xy}^{(1)}(x, 0) = 0 \tag{7}$$

The reaction forces of all the rigid intermediate supports on the bottom surface of the laminated beam are regarded as unknown vertical external forces, as follows:

$$\sigma_y^{(p)}(x, \sum_{j=1}^{p} h_j) = \sum_{k=1}^{K} g_k \delta(x - x_k),\ v^{(p)}(x_k, \sum_{j=1}^{p} h_j) = 0, \tau_{xy}^{(p)}(x, \sum_{j=1}^{p} h_j) = 0 \tag{8}$$

in which,

$$g_k \delta(x - x_k) = \sum_{m=1}^{\infty} g_m \sin(a_m x),\ g_m = \frac{2}{L}\int_0^L g_k \delta(x - x_k)\sin(a_m x)dx,\ x_k \in (0, L), k = 1, 2 \cdots K \tag{9}$$

According to the continuity conditions of the interfaces, the displacements and stresses at the interlayer interfaces satisfy the following equations:

$$\begin{aligned} u^{(i)}(x, h_i) &= u^{(i+1)}(x, 0),\ v^{(i)}(x, h_i) = v^{(i+1)}(x, 0) \\ \sigma_y^{(i)}(x, h_i) &= \sigma_y^{(i+1)}(x, 0),\ \tau_y^{(i)}(x, h_i) = \tau_y^{(i+1)}(x, 0) \end{aligned} \tag{10}$$

Substituting Eqs. (6) ~ (10) into Eq. (4) and truncating the terms of the Fourier series up to N gives $4pN + K$ equations. The unknown coefficients $A_m^{(i)}$, $B_m^{(i)}$, $C_m^{(i)}$, $D_m^{(i)}$ can be obtained by solving these equations. Then, by bringing these coefficient values back to Eq. (4), the distributions of displacements and stresses at any point within the laminated beam under any distributed load can be obtained.

3 Numerical Examples

3.1 Example 1: Equal Span Continuous Beam

In order to illustrate the whole process of the present method, a three-span continuous laminated beam with 5 layers subjected to a locally distributed load, as shown in Fig. 2, is introduced as Example 1. The values of the thickness, Young's modulus, Poisson's ratio of each layer; the length of the beam; the magnitude, applied area and direction of the distributed load are listed in the Fig. 2.

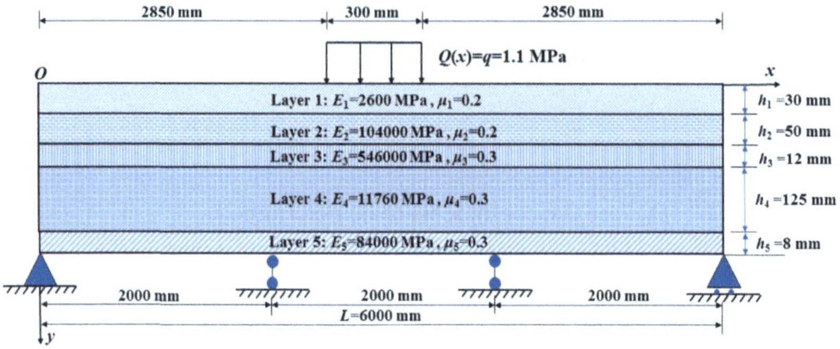

Fig. 2. Three-span laminated beam under a locally distributed load

Table 1 shows the normal stress at the point ($x = 2000$ mm, $y = 31$ mm) in layer 2 of left intermediate support and shear stress at the point ($x = 0$ mm, $y = 100$ mm) in layer 4 of left side support by truncating the Fourier series terms up to N. The results in Table 1 show the good convergence and numeric stability. It is shown that $N = 80$ can guarantee the satisfactory accuracy, therefore $N = 80$ is used in the following computations.

Table 1. The Convergence of stresses of Example 1

Truncated term number N	σ_x/MPa	τ_{xy}/MPa
20	4.2818	0.08
40	4.5135	0.07
60	4.5739	0.06
80	4.5749	0.06

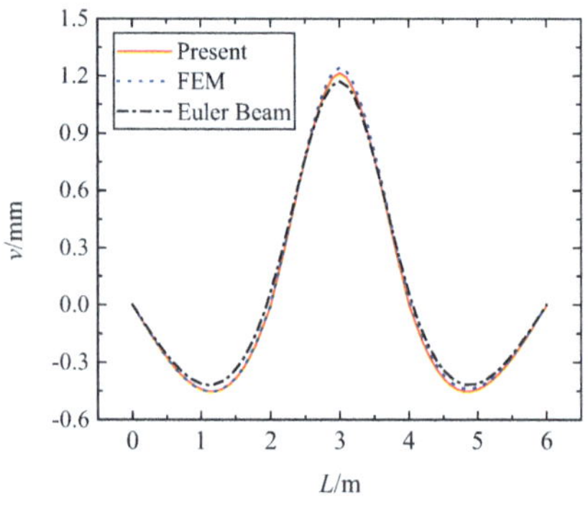

Fig. 3. Comparison of the deflection v along the length of the beam

To validate the accuracy of the present method, the computational solutions of the displacement and stress distributions are compared with those obtained by the finite element software ANSYS, where the Plane 13 element is chosen with a mesh size of 5 mm. Figure 3 presents the comparisons of the deflection v along the length of the beam among the present solution, the 1-D Euler beam solution and the Finite element solution. Figure 4 presents the comparison of the normal and shear stress distributions (x = 2000 mm, 0 mm) along the thickness direction of the beam by using the present method and the finite element method, respectively. It is noted from Figs. 3 and 4 that the present solutions agree well with the finite element solutions (FEM) using the ANSYS. Since the beam theory deflection solution does not take into account the deformation of the laminated beam itself, the corresponding results are slightly smaller than the analytical present and finite element solutions.

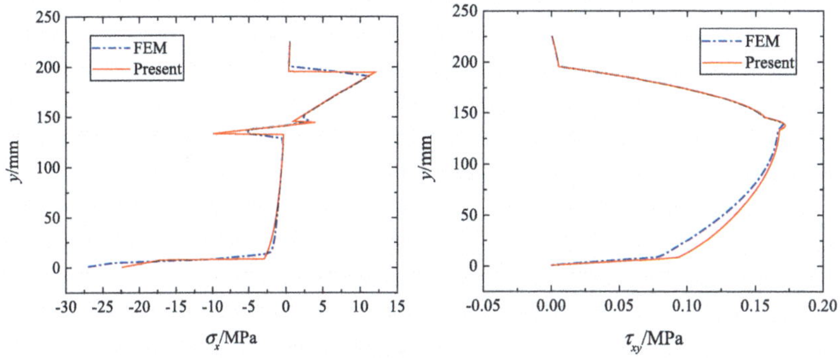

Fig. 4. The distribution of stress along the thickness direction of the beam in Example 1 by using the present method and the finite element method

3.2 Example 2: Unequal Span Continuous Beam

A five-span continuous laminated beam with 3 layers subjected to a fully distributed load, as shown in Fig. 5, is introduced as Example 2. The values of the thickness, Young's modulus, Poisson's ratio of each layer; the length of the beam; the magnitude, applied area and direction of the distributed load are listed in the figure. In the finite element analysis, the mesh size of 5 mm and the Plane element is used. Figures 6 and 7 present the deflection in vertical direction v along the length of the beam, the normal and shear stress distributions (x = 3000 mm, 0 mm) along the thickness direction of the beam by using the present method and the finite element method, respectively. It is found from the figures that the present solution and the finite element solution show a good consistency, which verifies the accuracy of the present method proposed in this paper.

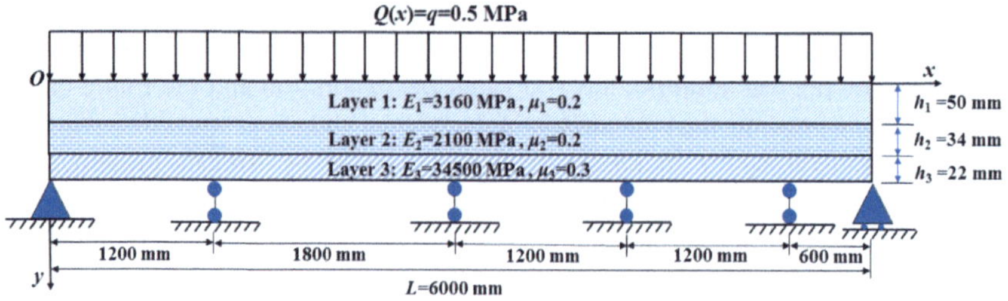

Fig. 5. Five-span laminated beam under a fully distributed load

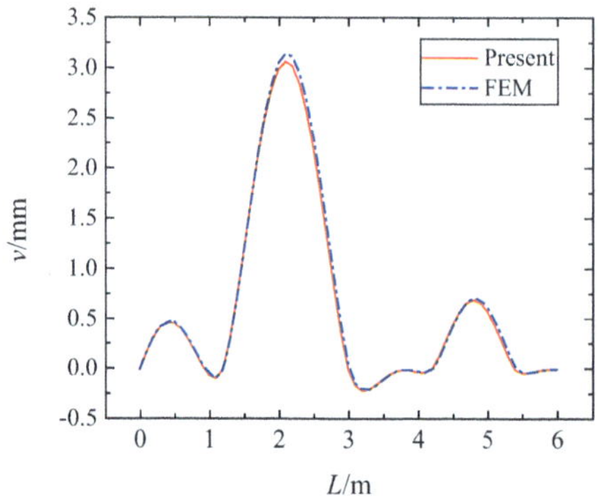

Fig. 6. Comparison of the deflection v along the length of the beam in Example 2 between the present solution and the Finite element solution

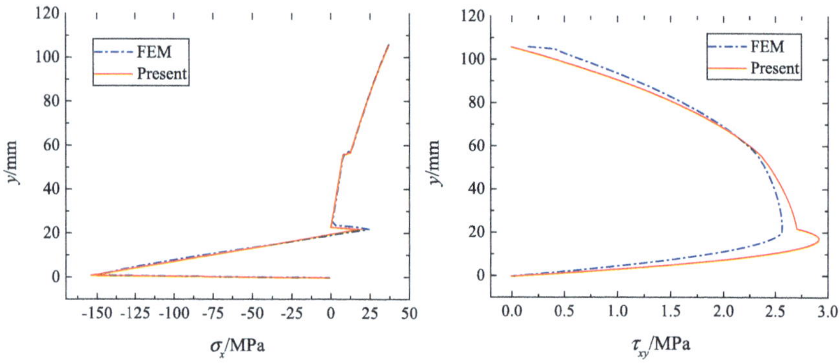

Fig. 7. The distribution of stress along the thickness direction of the beam in Example 2 by using the present method and the finite element method

4 Conclusions

In this paper, an exact analytical solution model based on elasticity theory for multi-span laminated beams under arbitrary distributed loads is established. The reaction forces of the rigid intermediate supports are considered as unknown external forces acting on the bottom surface of the laminated beam. Then, the analytical solutions of the laminated beam can be obtained by the continuity conditions of the interfaces and the boundary conditions of the top and bottom surfaces. The present solutions rigorously satisfy the boundary conditions and hence have high accuracy and efficiency which has been verified by convergence and comparison studies.

In practical engineering, the present analytical method can be applied to design and optimize continuous girder bridges for high-speed railways and highways under vehicle loads, as well as to predict unfavorable locations. It is applicable to the stress and displacement analysis of laminated beams under different layers and spans, as well as different interlayer interface conditions. In future work, the present model can be extended to explore the mechanical properties of steel bridge deck pavement under different mechanical loads. By calculating the stresses and displacements of each layer, the interface damage problems of steel bridge deck pavement can be effectively analyzed.

Acknowledgments. This work was supported by the National Natural Science Foundation of China (52108149), China Postdoctoral Science Foundation (2021M690623) and Fundamental Research Program for the Jiangsu Provincial Colleges and Universities (21KJB560004).

References

1. Serubibi, A., Hazell, P.J., Escobedo, J.P., Wang, H.X., et al.: Fibre-metal laminated structures: high-velocity impact, penetration, and blast loading—a review. Compos. Part A-Appl. S. **173**, 1076674-1–1076728 (2023). https://doi.org/10.1016/j.compositesa.2023.107674
2. Chen, L.L., Zhao, X.Y., Qian, Z.D., Li, J.Q.: A systematic review of steel bridge deck pavement in China. J. Road Bridge Eng. **3**, 1–15 (2023). https://doi.org/10.1016/j.jreng.2023.01.003
3. Liew, K.M., Pan, Z.Z., Zhang, L.W.: An overview of layerwise theories for composite laminated and structures: development, numerical implementation and application. Compos. Struct. **2019**, 240–259 (2019). https://doi.org/10.1016/j.compstruct.2019.02.074
4. Gu, Y.W., Nguyen, T., Donough, M.J., Prusty, B.G., Wang, J.: Mechanisms of pop-up delamination in laminated composites pierced by the initial pure waterjet in abrasive waterjet machining. Compos. Struct. **297**, 115968-1–116015 (2022). https://doi.org/10.1016/j.compstruct.2022.115968
5. Zhu, B., Zhou, D., Liu, W.Q.: Effect of resinic columns on bucking and vibration of a composite sandwich plate. J. Vib. Shock **30**(4), 8–11 (2011). https://doi.org/10.13465/j.cnki.jvs.2011.04.019
6. Hu, Z.J., Shen, P., Jiang, Q.C.: Developing high-performance laminated Cu/Tic composites through melt infiltration of Ni-doped freeze-cast preforms. Ceram. Int. **45**(9), 11686–11693 (2019). https://doi.org/10.1016/j.ceramint.2019.03.043
7. Xi, Z.C., Chen, H.R.: The theory for composite laminated beams. Acta Mech. Solida Sin. **11**(2), 1–6 (1994). https://doi.org/10.13801/j.cnki.fhclxb.1994.02.001

8. Xu, R.Q., Wu, Y.F.: Two-dimensional analysis solutions of simply supported composite beams with interlayer slips. Int. J. of Solids Struct. **44**(3), 165–175 (2007). https://doi.org/10.1016/j.ijsolstr.2006.04.027
9. Zhang, W.Z., Fu, M.H., Lin, J.H.: A precise method for delamination analysis on laminated composite beams. Eng. Mech. **29**(7), 35–41 (2012). https://doi.org/10.6052/j.issn.1000-4750.2010.10.0764

Research on the Suspender Tension Method of Beam-Arch Combined Bridge with Beam-First-and-Arch-Next

Xigen Zhang[1], Jincheng Gao[2], Hui Zeng[3], Shenglong Li[4], Zhenquan Jia[5], and Renjun Yan[1](✉)

[1] College of Marine and Energy Power Engineering, Wuhan University of Technology, Wuhan, Hubei, China
zhanggengen1989@foxmail.com
[2] No. 719 Research Institute of CSIC, Wuhan, Hubei, China
[3] Hubei Railway Transportation Vocational College, Wuhan, Hubei, China
[4] China Construction Third Engineering Bureau International, Beijing, China
[5] China Construction Third Bureau First Construction Engineering Limited Liability Company, Shanghai, China

Abstract. As the beam-arch combination system bridge gives full play to the structural characteristics of beam bending, arch compression and their combination, the structural stress is more complex. Therefore, the internal force and linearity of the bridge meet the design requirements, and the suspender tensioning control is very important. In order to avoid multiple tensioning, cumbersome error correction, too much choice of calculation methods, this paper uses finite element calculation software MIDAS CIVIL to establish the whole bridge construction stage model and arch-beam combination bridge force characteristics to determine the number of times of tensioning, tensioning sequence and the size of the tensile force, etc., to form a set of simple and practical suspender tensioning program. In combination with the actual tensioning of the suspender of Caojiaping Bridge, the frequency method is used to measure the cable force of the suspender, and the stress-strain and deformation of the main girder and arch ribs are verified and measured. Practice has shown that following this simple and practical hanger tensioning scheme can meet the requirements of the arch ribs, main girder forces, and linear shape. And after the bridge is completed, the single cable strength and design of the cable strength in the error is less than 10%, the sum of the suspender cable strength error is less than 1%. The practical tensioning program formed in this paper has high practical value and reference value in the actual construction process.

Keywords: beam-arch combined bridge · finite element simulation · orthotropic iteration method · suspender tension method

B. Yuan et al. (Eds.): ICCASE 2024, SUCI, pp. 135–142, 2024.
https://doi.org/10.1007/978-3-031-76102-7_16

1 Introduction

In recent years, with the development of production technology and the advantages of reasonable force, good wind and earthquake resistance, simple construction and various forms of beam-arch composite Bridges [1], they have been widely used. However, because the external structure is statically indeterminate and the internal structure is of higher order statically indeterminate, the tension control is particularly important because the tension process of the derrick will affect the internal force state of the whole bridge. Zhu Min et al. [2] formed the optimal bridge state through the optimal combination of the zero displacement method and the energy minimum method, established the relationship between the structural state variables and the bridge state objectives during the construction period, and thus obtained the reasonable derrick tension at different construction stages. Wang Li [3] studied the advantages and disadvantages of four different initial tensioning schemes starting from the quarter point by establishing a scale model. Zhou Xiaoyong et al. [4] obtained the optimized suspender tensioning sequence through different influences on the distribution and magnitude of structural internal forces caused by the tensioning sequence and cable force of the suspender. At present, the theoretical research on the suspender tensioning scheme is quite sufficient, but in the actual construction process, due to the limitation of the theoretical knowledge of the construction personnel, it is difficult for them to make complex and accurate calculation. Therefore, how to simplify the formulation of the suspender tensioning sequence and number, and realize that the bridge cable force and the design cable force are within the error range, so as to avoid the tedious adjustment of the internal force of the suspender? It has very important practical significance in construction [5]. In this paper, the tensioning sequence and number of boom are analyzed from the engineering practice, and the finite element model is established by MIDAS CIVIL to carry out the formal assembly iterative analysis of the tensioning force of each tensioning process. Then, a set of simple and reliable tensioning scheme of boom is summarized to guide the actual construction of boom tensioning.

2 Overview of the Project

Caojiaping Bridge is located in Yichang City, Wufeng New City, Caojiaping Landscape Avenue, for the sole arch rib steel tube concrete arch girder combination bridge, bridge length 97.2 m, width 24 m, calculated span is 90 m. Main arch rib height 140 cm, width 310 cm, left and right steel tube outer diameter are 140 cm, wall thickness is 25 mm, the upper and lower steel plates are 1700 mm (width) × 25 mm (thickness) steel plate; the arch axis for the secondary. The arch axis is quadratic parabolic, sagittal height is 18 m, sagittal span ratio is 1/5. The arch longitudinal girder adopts integral single-box and five-compartment box girder structure, with the girder height ranging from 2.2 to 2.44 m. The whole bridge is equipped with 16 pairs of suspenders, with a spacing of 5 m.

3 Theoretical Analysis for Determining the Suspender Tensioning Scheme

3.1 Selection of Suspender Tensioning Times

Commonly used suspender tensioning method has a tensioning method, initial tensioning and adjusting the tensioning of the two tensioning method as well as many times to adjust the tensioning method. In actual construction, due to the consideration of construction safety and construction costs and other factors, a tensioning method due to the tensioning process may appear individual suspender force is too large, or the main girder, the main arch and the foot of the arch of the internal force changes significantly, there is a large peak situation [5], so a tensioning method is not much used in the actual construction. It is also important to point out that, taking into account the impact of construction errors, structural nonlinearities, process and equipment accuracy and other factors, engineering experience has shown that too many adjustments to the suspender tensioning of the tie arch bridge, the final accuracy is not necessarily good [6]. So the secondary tensioning method [7] is a good choice in actual construction, no matter from the factors of structural safety or construction cost.

3.2 Selection of Suspender Tensioning Sequence

For the final result of the suspender force, and the order of the suspender tensioning is irrelevant [8], however, in the actual construction process, due to tensioning equipment, human factors and other constraints, it is not possible to tension all the suspender at the same time, so it is necessary to determine the order of tensioning. From economic considerations, the construction side will generally only provide a pair of symmetrical tensioning equipment, so that Caojiaping Bridge has a total of 40320 kinds of tensioning sequence, but according to the principle of trying to make the structural internal force and deflection changes in the magnitude of the small and the actual construction experience [8], the commonly used tensioning sequence is attributed to three, respectively, from the arch to the arch order, from the arch to the arch order, as well as from the four-point radial tensioning sequence. Also for safety construction considerations, which from the foot of the arch to the vault order in the tensioning of the first pair of suspenders, the arch of the upper edge of the stress increase is more obvious, there may be the actual stress is too large [5], the same way according to the order from the arch to the foot of the arch, the foot of the arch stress is also the same, so the comprehensive consideration of the four-point radial tensioning order is the most reasonable practical suspender adjustment order.

3.3 Practical Methods for Determining Suspender Tensioning Forces

In recent years, bridge workers have proposed several methods to determine the reasonable construction cable force, mainly including: inverted disassembly method, positive loading-inverted disassembly iteration method, positive loading iteration method, stress-free state control method, [9] Each of these methods has its own advantages and disadvantages, some are cumbersome to solve but with high accuracy, some are simple and with large errors. In this paper, the target cable force of boom is taken as the control target, and there are no additional constraints, and the formal installation iteration method is a simple and practical explicit calculation method to solve the cable force of boom. This method can be divided into the following processes to calculate the cable force of the derrick [10]:

(1) According to the actual construction stage of the bridge, establish the calculation model and input the various parameter cases needed;
(2) Assuming that the bridge-forming cable-stay is $\{F_0\}$, define the initial cable-stay entered for the i th iteration of the calculation as $\{F_i\}$, and the resultant cable-stay obtained from this calculation is denoted as $\{F\}$;
(3) Use the corresponding program in the finite element software to input the command:$\{F_i\}$=$\{F_0\}$ to carry out the first iteration of the calculation and obtain the suspender force $\{F\}$;
(4) Check whether $\frac{\{F_0\}-\{F\}}{\{F_0\}}$ is smaller than the preset permissible value. If it is less than, then the commanded tension force for each suspender is $\{F_i\}$, otherwise make $\{F_{i+1}\} = \{F_0\} + \{F_i\} - \{F\}$ and repeat steps 3 and 4 until $\frac{\{F_0\}-\{F\}}{\{F_0\}}$ is less than the permissible value, and the tension force for each suspender is the last iteration of the input $\{F_i\}$.

MIDAS CIVIL was used to establish the finite element model. The whole bridge structure is discretized into 118 units with a total of 236 nodes, and the structure is discretized as shown in Fig. 1.

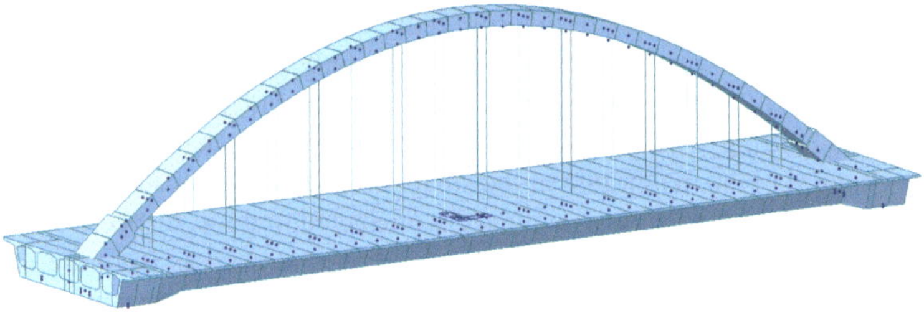

Fig. 1. Caojiaping Bridge Model

In accordance with the previously determined number of tensioning and tensioning sequence, Caojiaping Bridge suspender tensioning is divided into two tensioning, the first time for the initial tensioning, in order to form the "arch ribs - suspender - main girder" of the overall force system, so that the suspender to bear part of the main beam weight, to improve the main girder The first is the initial tensioning, in order to form "arch rib - suspender - main girder" overall stress system, make the suspender bear part of the main girder weight, improve the main girder's stress characteristics, and facilitate the removal of the main girder bracket, the purpose of which is to remove the frame, so the range of error can be larger, generally taken into the bridge rope force of 10% to 30%; after the second adjustment of tension as the final tension, the tension of each tension stage is calculated by the formal installation iteration method, so that the internal force state and linear shape of the structure after the second paving bridge can meet the design requirements. The numerical values of the formal installation iteration method are shown in Tables 1 and 2.

Table 1. Initial tensile suspender tension

Tensioning sequence	Suspender number	Single suspender force (KN)
1	5, 12	155.59
2	8, 9	155.59
3	2, 15	152.39
4	7, 10	155.59
5	4, 13	155.48
6	6, 11	155.59
7	3, 14	154.63
8	1, 16	77.23

Table 2. Adjustment of tensioning suspender tensioning force

Tensioning sequence	Suspender number	Single suspender force (KN)
1	5, 12	2027.3
2	8, 9	1684.04
3	2, 15	1820.61
4	7, 10	2270.96
5	4, 13	1792.61
6	6, 11	2139.53
7	3, 14	1507.34
8	1, 16	783.93

Then, after substituting the theoretically calculated values into the model checking, the main arch stress and deformation of the main girder in each tensioning stage are obtained, as shown in Figs. 2 and 3, and their values are all within the permissible range of engineering construction.

3.4 Suspender Tensioning Program Field Measurement

According to the previous theoretical analysis to get a set of simple and practical tensioning program, and used in Caojiaping Bridge suspender tensioning construction, as shown in Fig. 4, to get the measured adjustment of the cable force and the bridge cable force, as shown in the following Tables 3 and 4:

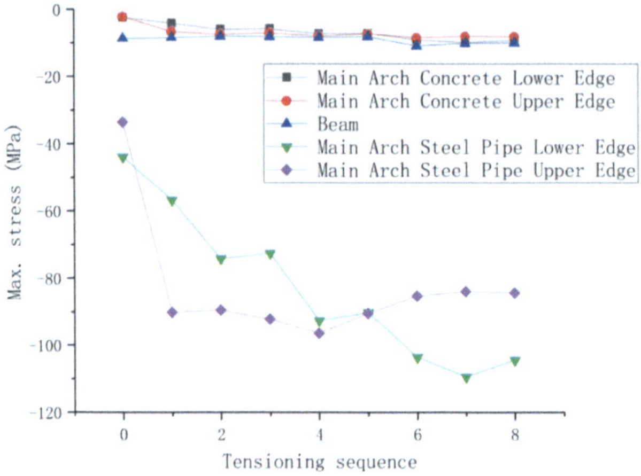

Fig. 2. Maximum stress diagram of main arch of main girder at each tensioning stage

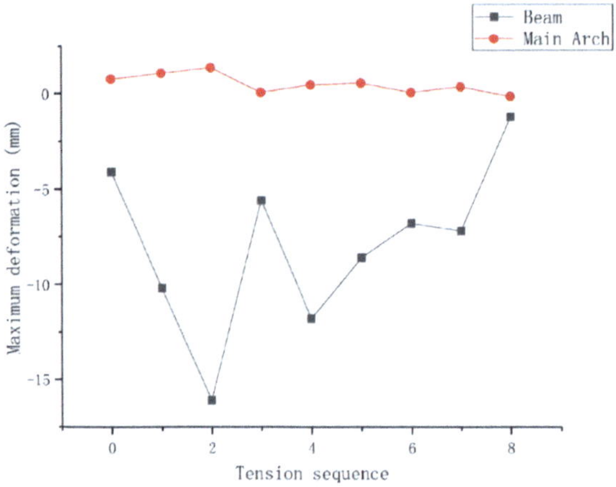

Fig. 3. Maximum deformation of main arch of main girder at each tensioning stage

Through the measured results, after the bridge into the suspender cable force and the design of the error between the cable force are less than 10%, and all the suspender force and the design of the sum of the cable force is less than 1%, are within the allowable error, of which the 16th suspender is the shortest length of the suspender, the stiffness is the largest, so the measured using the frequency method of measurement of the existence of a large error, the actual readings of the jack oil pump shall prevail. This can be obtained, the simple and practical tensioning program in Caojiaping Bridge suspender construction process using successful.

Fig. 4. Caojiaping Bridge

Table 3. Measurements after the completion of the second tensioning of the suspender

Suspender number	measured value (t)	Actual force (t)	Design value (t)	inaccuracies
U 01	80.5	76.7	71.25	8%
U 02	109.7	141.4	137.32	3%
U 03	99.8	147.3	132.24	11%
U 04	80.6	129.8	130.08	0%
U 05	87.7	145.9	131.99	11%
U 06	81.8	129.2	119.3	8%
U 07	88.8	151.5	142.95	6%
U 08	73.6	123.0	127.62	-4%

Table 4. Measured suspender force after completion of Phase II laying

Suspender number	Actual force (t)	Design value (t)	inaccuracies
U 01	79.7	85.3	–6.5%
U 02	152.3	156.2	–2.5%
U 03	162.4	153.3	5.9%
U 04	145.1	151.8	–4.4%
U 05	167.0	154.0	8.5%
U 06	147.6	141.3	4.4%
U 07	169.8	165.0	2.9%
U 08	142.7	149.6	–4.6%
Combined force of all suspenders	1166.6	1156.5	0.9%

4 Conclusion

In view of the cumbersome formulation of the current suspender tensioning scheme and the difficulty of engineering applicability, this paper, based on theoretical analysis and engineering practice, ensures the reliability of theoretical analysis as well as the operability and convenience of on-site construction, forms a set of simplified and practical tensioning scheme of the first beam and then arch tied pole arch bridge, and proposes the use of two tensioning times, radiating the tensioning sequence from four points. The use of formal assembly iterative method to calculate the tension of boom is more reasonable and simplified in the actual construction, and the practical scheme has been well applied in the actual construction of Caojiaping bridge, providing a feasible solution for the subsequent construction of this kind of bridge boom, in order to promote the steady development of the theoretical engineering of boom tension.

References

1. Lin, J.F.: Arc bridge construction process of derrick cable force analysis. J. Fujian Traff. Sci. Technol. **4**, 3 (2011). CNKI:SUN:FJJT.0.2011-04-016
2. Zhu, M., Sun, J.Y., Li, G.P., et al.: Boom times tensioning beam arch combination bridge analysis. J. East China Highway **5**, 4 (1999). CNKI:SUN:HDGL.0.1999-05-008
3. Wang, L.: Experimental study on the optimization of tensioning scheme of suspender in tie arch bridge. Railway Constr. **60**(1), 4 (2020). CNKI:SUN:TDJZ.0.2020-01-003
4. Zhou, X.Y., Lv, Z.M., Wang, J.H., et al.: Optimization of suspension rod tensioning scheme for braced cast-in-place tie arch bridges. Chin. Foreign Highway **002**, 042 (2022)
5. Xiao, R.C., Xiang, H.F.: Cable-stayed bridge cable force optimization and its engineering application. J. Comput. Mech. **15**(1), 9 (1998). CNKI:SUN:JSJG.0.1998-01-016
6. Zhang, T.F.: Permanent fix frame arc bridge construction monitoring key technology research. Central south university, 2010. https://doi.org/10.7666/d.y.1720808
7. Hai, T., Xiang, F.S.: Strongly subfeasible sequential quadratic programming method of cable tension optimization for cable-stayed bridges. Chin. J. Theor. Appl. Mech. **38**(3), 381–384 (2006). https://doi.org/10.3321/j.issn:0459-1879.2006.03.013
8. Zhao, G.M., Chen, K.C.: Timing of cable adjustment process for tie arch bridge. Zhongnan Highway Eng. **3**(1), 28–31 (1997)
9. Arai, H., Kasuga, A., Breen, J.E., et al.: Optimum cable-force adjustments in concrete cable-stayed bridges. J. Struct. Eng. **121**(4), 685–694 (1995)
10. Xu, Z.: Research on the control and optimization of suspension rod tensioning of steel-tube concrete tie arch bridge. Chang'an University (2014)

Research on the Mechanism and Countermeasures of Temperature Cracks in Large-Span Prestressed Concrete Continuous Beam Bridges

Jing Huang(✉)

Wuchang Shouyi University, Wuhan 430064, Hubei, China
290896438@qq.com

Abstract. With the continuous improvement of construction quality and operation maintenance requirements for highway bridges in China, crack control during bridge construction has become particularly important. Significant temperature difference stress will be generated inside the bridge, leading to cracks.

This article applies the principle of finite element analysis to analyze the temperature stress of the Wuhan Jingzhou Expressway Hanbei River No.2 Extra Large Bridge, and verifies that the temperature stress of the bridge meets the requirements of the specifications under the most unfavorable conditions. Further suggestions from the aspects of design and construction are proposed as follows: appropriately increase the net distance between pipelines and between pipelines and the lower edge, and prohibit the implementation of key construction pouring or key parts such as closure section construction when the temperature difference between day and night is large. This not only provides important guidance for the design and construction of large-span prestressed concrete continuous beam bridges, but also provides theoretical support and practical reference for crack prevention and control in similar bridge structures.

Keywords: finite element analysis · temperature cracks · stress control · solutions

1 Introduction

With the continuous increase of traffic volume, controlling temperature cracks has become one of the key goals to improve the quality of bridge construction. The industry generally adopts the method of strengthening construction monitoring and controlling hydration heat to avoid temperature cracks. However, due to the poor tensile performance of concrete materials, cracking problems are prone to occur during bridge construction and operation, seriously affecting the bearing performance and durability of bridge structures [1]. The physical and chemical properties of concrete raw materials are relatively unique, making them prone to temperature cracks. Construction personnel need to prevent temperature cracks in the selection of raw materials, concrete mixing, curing and other processes [2].

B. Yuan et al. (Eds.): ICCASE 2024, SUCI, pp. 143–150, 2024.
https://doi.org/10.1007/978-3-031-76102-7_17

2 Principle of Temperature Cracks

2.1 The Main Research Content of This Paper

(1) Derive and organize the calculation formulas for internal forces, temperature effects, and creep effects of existing continuous beam bridges.
(2) Structural analysis of Hanbei River No. 2 Grand Bridge. Summarize the successes and shortcomings of these technical measures through finite element calculations and the current engineering situation, and identify areas for further improvement in crack control.

2.2 Principle of Temperature Load

The finite element method discretizes the continuum region with elements, discretizes the temperature at each point of the element to nodes, and replaces the temperature field inside the element with node temperature. The temperature at any point within a unit can be expressed as:

$$T(x, y, z) = \sum_{i=1}^{20} N(\xi, \eta, \zeta) T_i \tag{1}$$

Take the discrete element as a sub region R for solving region R, and its temperature field satisfies the following conditions:

$$\frac{\partial^2 T}{\partial x^2} + \frac{\partial^2 T}{\partial y^2} + \frac{\partial^2 T}{\partial z^2} = 0 \tag{2}$$

According to the variational principle, the problem is equivalent to the extremum of the following functional:

$$I^e(T) = \frac{1}{2} \iiint\limits_R \left[\left(\frac{\partial T}{\partial x} \right)^2 + \left(\frac{\partial T}{\partial x} \right)^2 + \left(\frac{\partial T}{\partial x} \right)^2 \right] dxdydz \tag{3}$$

According to the extreme conditions, there are:

$$\begin{Bmatrix} \frac{\partial I^e}{\partial T_1} \\ \frac{\partial I^e}{\partial T_j} \\ \frac{\partial I^e}{\partial T_k} \\ \bullet \\ \bullet \\ \bullet \end{Bmatrix} = \frac{\partial I^e}{\partial (T)^e} = \left[h^e \right] \{T\}^e = 0 \tag{4}$$

$$h_{ij}^e = \frac{1}{2} \iiint\limits_{\Delta R} \left[\frac{\partial N_1}{\partial x} \frac{\partial N_1}{\partial x} + \frac{\partial N_1}{\partial y} \frac{\partial N_1}{\partial y} + \frac{\partial N_1}{\partial z} \frac{\partial N_1}{\partial z} \right] dxdydz \tag{5}$$

In the formula: $\left[h^e\right]$ is the unit conduction matrix Integrating all units within the solution area yields:

$$\frac{\partial I}{\partial (T)} = [H]\{T\} = 0 \quad (6)$$

In the formula: $\{T\} = \left\{T_1\ T_2 \bullet\bullet\bullet\ T_n\right\}^T$ is the temperature array of structural nodes. Using $[H]$ as the conduction matrix and introducing boundary conditions to solve 2–6 yields all node temperatures. Where T_1 is the assumed known temperature of a certain node.

The above method is also the principle of finite element analysis software simulating temperature loads by simulating temperature fields.

3 Application Instance

3.1 Overview of Hanbei River No.2 Extra Large Bridge Project

This paper is based on the Wuhan Jingzhou Expressway Hanbei River No. 2 Grand Bridge. The total length of Hanbei River No. 2 Grand Bridge is 1530 m, with a continuous T-beam of 30 m per span as the approach bridge and a three span variable cross-section continuous box beam of 55 + 100 + 55 m as the main bridge. MIDAS/CIVIL calculations were used and verified by a bridge doctor.

3.2 Establishment of Calculation Model

3.2.1 MIDAS/CIVIL and Hanbei River No. 2 Bridge Model

The calculation and analysis of this bridge adopts the spatial finite element analysis program (MIDAS/CIVLL). The finite element model of the cantilever method construction for the Hanbeihe No.2 super large continuous beam bridge is shown in Fig. 1.

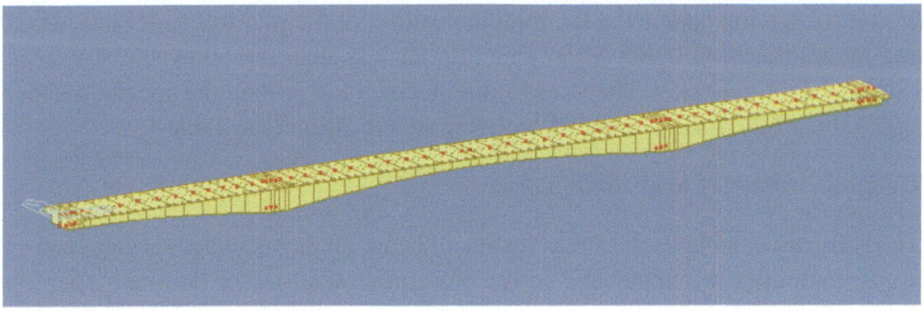

Fig. 1. Finite Element Model of Main Bridge Completion Status

3.2.2 Main Calculation Parameters

The closing temperature of the main bridge box girder is set at 22 °C, with an increase in temperature of 20 °C and a decrease in temperature of 20 °C. The temperature rise and fall modes are specified in the gradient temperature regulations of JTGD60-2004 in China [3], and are adjusted accordingly according to the actual situation.

3.3 Comparison and Analysis of Model Calculation Results

3.3.1 Main Calculation Loads and Usage Load Combinations

The custom load case combination according to relevant specifications is:

Combination 1: Constant load + Construction load + Steel tendon primary + Steel tendon secondary + Creep secondary + Shrinkage secondary.

Combination 2: Constant load + Construction load + Steel tendon primary + Steel tendon secondary + Creep secondary + Shrinkage secondary + Overall heating + (JTG2004) heating mode.

Combination 3: Constant load + Construction load + Steel tendon one time + Steel tendon two times + Creep two times + Shrinkage two times + Overall cooling + (JTG2004) cooling mode.

Combination 4: Constant Load + Construction Load + Steel Bundle Primary + Steel Bundle Secondary + Creep Secondary + Shrinkage Secondary + Overall Heating + (JTG2004) Heating Mode + Moving Load.

Combination 5: Constant load + Construction load + Steel tendon primary + Steel tendon secondary + Creep secondary + Shrinkage secondary + Overall cooling + (JTG2004) cooling mode + Moving load.

3.3.2 Analysis of Normal Stress Calculation Results for Box Girder Control Section

(1) (l) Under single load action

Under the single load action, the maximum normal compressive stress of the box girder control section appears at the middle position of the pier top under the JTG2004 heating mode, reaching 0.465 MPa, which is far less than the allowable stress value. This single factor generates The stress value is much lower than the specified concrete strength value of 24.4 MPa in the 04 highway specification (see Table 1).

(2) Under the action of load combination

The software calculates the stress results of beam elements under various load combinations, including the upper and lower box girder sections of the main span beam elements Comparison of edge stresses, the table shows the stress values of the control section of the main beam.

From Table 2, it can be seen that under the five load combinations, the maximum normal stress value of the box girder control section is 8.55 MPa, which only accounts for 35.04% of the specified allowable value of 24.4 MPa, and there is still a relatively large safety reserve of The distribution of stress can be seen in Figs. 2 and 3. According

Table 1. Stress values of the control section of the main beam under single load action (unit: MPa)

Load combination	Stress position	Side span 3/8	Side span 1/4	Middle of pier top	Main span ¼	Main span 3/8	Main span middle
Overall heating up	Upper margin	0.0473	–0.0805	–0.0310	–0.0183	0.0811	0.1010
	Lower margin	0.0585	–0.0920	–0.0329	0.0238	–0.118	–0.1510
Overall cooling	Upper margin	0.0473	0.0805	0.0310	–0.0200	–0.0811	–0.1010
	Lower margin	–0.0585	–0.0920	–0.0329	0.0273	0.1180	0.1510
JTG2004 heating mode	Upper margin	0.3600	0.2980	0.1610	0.3330	0.3760	0.4170
	Lower margin	0.2960	0.2790	0.1770	0.3950	0.4420	0.4650
JTG2004 cooling mode	Upper margin	–0.3600	–0.2980	–0.1610	–0.3330	–0.3760	–0.417
	Lower margin	–0.2960	–0.2790	–0.1770	–0.3950	–0.4420	–0.4650

Table 2. Stress values of the control section of the main beam under various load combinations (unit: MPa)

Load combination	Stress position	Side span 3/8	Side span 1/4	Middle of pier top	Main span 1/4	Main span 3/8	Main span middle
Combination 1	Upper margin	–5.47	–5.87	–4.09	–8.09	–8.03	–4.42
	Lower margin	–6.67	–7.25	–6.68	–5.40	–6.92	–8.33
Combination 2	Upper margin	–5.17	–5.67	–3.97	–7.73	–7.56	–3.88
	Lower margin	–6.30	–6.86	–6.46	–5.04	–6.63	–7.94
Combination 3	Upper margin	–5.77	–6.06	–4.21	–8.44	–8.50	–4.95
	Lower margin	–7.04	–7.64	–6.90	–5.76	–7.22	–8.55
Combination 4	Upper margin	–4.18	–4.72	–3.19	–7.39	–7.25	–3.59
	Lower margin	–5.46	–6.43	–6.36	–4.3	–4.83	–5.62
Combination 5	Upper margin	–4.78	–5.12	–3.43	–8.10	–8.20	–4.66
	Lower margin	–6.20	–7.22	–6.79	–4.75	–5.42	–6.18

to the verification results, the Hanbeihe No.2 Extra Large Bridge has been successful in controlling the stress at the upper and lower edges of the box girder section, and has achieved good results in use.

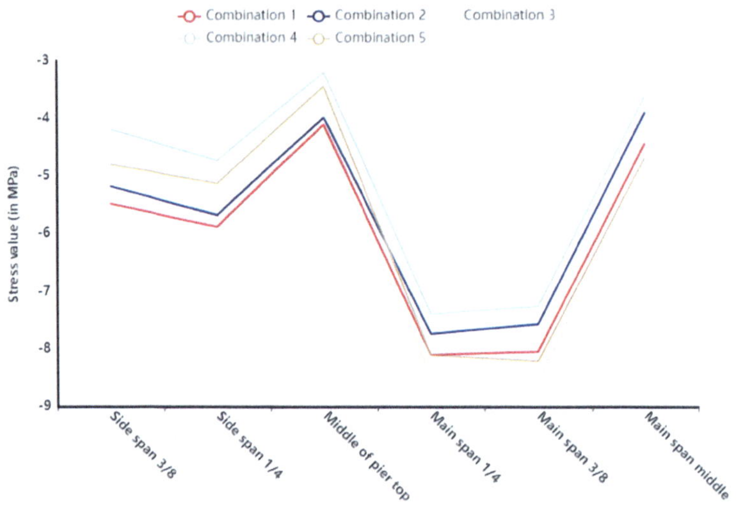

Fig. 2. Control section stress values at the upper edge of the beam under various load combinations

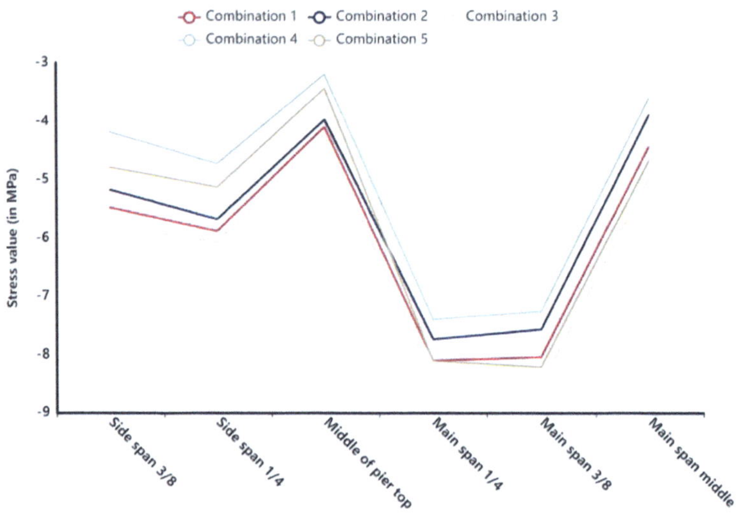

Fig. 3. Control section stress values at the lower edge of the beam under various load combinations

4 Conclusion

Based on the temperature load analysis results of the Hanbei River No. 2 Bridge, it is explained that the internal force caused by temperature changes is an important cause of box girder cracking, especially when the temperature changes are large, the comprehensive factors caused by temperature are easy to cause cracks in the beam. The contact type crack width observation device can be used to measure the crack width of concrete bridge decks, and crack images of concrete bridges can be obtained under noise and radiation environments, respectively [4]. The combination of the above methods can be widely applied to crack prevention and monitoring of similar bridge projects, which is of great significance for improving the structural integrity and durability of similar bridges.

4.1 Measures to Eliminate Temperature Effects

The existing mature measures to reduce the impact of temperature changes generally include [5]: firstly, to reduce the temperature rise of hydration heat, it is necessary to control the amount of cement used; secondly, to do a good job in insulation and maintenance, especially during winter construction. When the ambient temperature is low, it is prohibited to use the method of watering for curing, because the presence of temperature stress will lead to concrete shrinkage, and different degrees of cracks will appear in the concrete structure after forming [6].

4.2 Response Measures for Material Factors

In engineering, local shrinkage cracks caused by excessive local water cement ratio are more common, which often occur on the top surface of each poured concrete, such as the bridge deck pavement layer or the integrated layer, the top of beams and columns. This is the concentrated area of floating slurry after concrete vibration [7]. The maximum value (peak) of plastic shrinkage area of concrete corresponds to a water cement ratio of 0.5. In order to reduce shrinkage, the water cement ratio can be controlled between 0.35 and 0.45, and measures such as adding water reducing agents can be taken to minimize its value.

4.3 Suggestions and Measures for Design

(1) It is recommended to increase the clear distance between pipes and between pipes and the lower edge appropriately, which is beneficial for compacting the concrete and preventing uneven temperature distribution from causing internal cracks to develop.
(2) The combination of tooth plate anti collapse reinforcement and lower edge anti collapse reinforcement should use closed hoop reinforcement to strengthen the constraint of the bottom plate concrete and prevent the bottom plate from cracking due to temperature loads.
(3) The collapse failure of straight section steel bundles caused by pipeline positioning errors requires not only the calculation and setting of downward anti cracking steel bars, but also attention to the upward cracking trend and measures (there have been examples of damage to the upper edge of the bottom plate).

4.4 Outlook

The study of temperature cracks in bridges needs to be simulated using different types of analysis software to verify the crack propagation under various temperature change combinations and working conditions. Accumulate more data and provide rational suggestions for subsequent design and construction.

References

1. Yong, Z., Xiaolong, Z., Lina, D., et al.: Causes and preventive measures of concrete cracks in municipal bridge construction. Sichuan Building Mater. **49**(12), 171–173 (2023)
2. Hongwen, J.: Concrete temperature stress control and crack optimization measures in highway and bridge construction. Transp. Manager World **05**, 79–81 (2022)
3. JTG D60-2004. General Specification for Design of Highway Bridges and Culverts. Beijing, People's Transportation Press, P16–P37 (2004)
4. Xing, W.: Algorithm for detecting crack width in concrete bridges under complex environments. J. Shandong Jiaotong Univ. **31**(04), 127–132 (2023)
5. Luo, Q, Yu, J.: Lateral cracks in the wing plates of reinforced concrete continuous box girder bridges. Bridge Constr. (01), 43–47 (1997)
6. Leqi, D.: Exploring concrete crack prevention and control technology in road and bridge engineering. Sichuan Building Mater. **50**(01), 112–114 (2024)
7. Swamy, R.N., Hung, H.: Engineering properties of high volume fly ash concrete. In: Fly ash, Silica Fume, Slag&Natural Pozzolans in Concrete, Proceeding Sixth CANMET/ACI International Conference, Bangkok, Thailand, pp. 331–359, May 31-june 5,1998

Optimal Permutation for Nodes of Bridge Structure by Using Object Oriented Method

Yongyi Yang[1] and Zizhen Zhang[2](✉)

[1] XIHUA University, Chengdu 610031, Sichuan, China
[2] Kunming Atide Software Co., Ltd., Kunming 650106, Yunnan, China
zhangzz@atidesoft.com

Abstract. Enhancing the speed of finite element analysis in bridge structures not only improves the efficiency and quality of engineering design, but also plays a crucial role in bridge safety assessment, emergency management, and cost control. It is a key direction in the development of modern bridge engineering technology. The total stiffness half bandwidth of a bridge structure is a critical factor that affects the analysis speed, as its size directly influences the computational resources and time required to solve structural equations. This paper analyzes the characteristics of node arrangement when the total stiffness half bandwidth is minimized, and summarizes three principles. The AD algorithm is applied for node optimization ordering, and the algorithm is further optimized in light of the specific characteristics of bridge structures. An object-oriented bandwidth optimization program was developed using VC++. The correctness and reliability of the program were verified through comparative case studies, demonstrating that the method can optimize the half bandwidth value significantly, thereby reducing the computer memory usage and enhancing the computational analysis efficiency.

Keywords: bridge structure · optimization · object-oriented method · semi band width · FEM

1 Introduction

The finite element analysis of bridge structure ultimately boils down to solving a large linear equation system. The formal kernel characteristics of the bridge structure determine that the coefficient matrix of the equations is a sparse structure, usually the non-zero term is concentrated near the diagonal, and the position of the submatrix in the coefficient matrix is closely related to the node numbering order. In order to improve the efficiency and accuracy of solving equations and reduce the memory consumption of the machine, the node sorting must be optimized.

An automatic algorithm is presented for reducing the bandwidth of symmetric sparse connectivity matrices [1]. Akhras G. introduces the development and application of the finite element software NASAB, which can be used for linear, geometric nonlinear and material nonlinear analysis and nonlinear aerostatic stability analysis of cable-supported bridge structures. The software program consists of two main parts: the programming

B. Yuan et al. (Eds.): ICCASE 2024, SUCI, pp. 151–157, 2024.
https://doi.org/10.1007/978-3-031-76102-7_18

part and the computing part [2]. Grooms H R introduces the finite element method and bandwidth optimization algorithm in engineering science [3].

In summary, most research at home and abroad on bandwidth optimization focuses on solid or plate and shell elements. However, typical bridge structures are composed of beam and rod elements, and certain node arrangement characteristics exist when the structural half bandwidth is minimized. Optimizing the half bandwidth is crucial for improving the efficiency and quality of bridge engineering design, safety assessment, emergency management, and cost control. This paper addresses this situation by applying the AD algorithm [4] for node optimization ordering, discussing the data composition and algorithm optimization issues of bridge structures, and designing the corresponding program structure. Utilizing the powerful data structures and algorithms of STL, an object-oriented approach was employed to develop the corresponding software. Extensive numerical examples have verified the correctness and practicality of this program.

2 Grid Node Sorting Optimization Method

2.1 Sorting Guidelines

In the pre-processing of finite element simulation, it is usually the initial grid and The node serial number is carried out according to a certain rule. But generally speaking, choreography The half-band of the finite element total stiffness matrix formed by the resulting grid and nodes The width value is very large, and the grid nodes need to be reordered to reach half-band. The smallest purpose [5]. The AD optimization sorting method used in this article is one A method of induction and deduction. Through the simple network with the best arrangement of nodes The following rules can be summarized by the analysis of the grid diagram.

2.1.1 And Guidelines

In the case that the number of node correlations (adjacent nodes) is the same, with the increase of the node serial number, the sum of the relevant node serial numbers is arranged in an incremental state.

2.1.2 Average Criterion

With the increase of node serial numbers, the average and incremental states of the relevant nodes are arranged.

2.1.3 Spacing Criteria

As the number of the node increases, the sum of the maximum and minimum serial numbers of the relevant node is arranged in an incremental state.

2.2 Programming Ideas

The idea of compiling an optimization program to reduce the half bandwidth of the grid node matrix is to meet the above three conditions. In this case, the half-bandwidth value of node arrangement is the smallest. The program consists of two parts of alternating iterations [5–7].

2.2.1 Part I

Renumber the node according to the span and average criteria.

(1) Apply the span criterion, adjust the initial order, so that the span sequence is arranged in ascending order, and then calculate the bandwidth. If the bandwidth increases, maintain the order of the previous step.
(2) Apply the average criterion to adjust the initial order so that the node number and average sequence are arranged in ascending order.
(3) Discrimination: When the overall number of iterations is greater than the empirical value, or the algorithm transfers to the second part when the number of adjacent times in the bandwidth sequence is the same, otherwise it will be transferred to 2.2.1 (1).

2.2.2 Part II

Calculate and adjust according to the criteria and average criteria. Full number.

(1) Application and guidelines adjust the initial order to make the order of the parts and parts Columns are arranged in ascending order. If the bandwidth increases, the order of the previous step will be maintained.
(2) Apply the average criterion to adjust the initial order to make the average order Columns are arranged in ascending order. If the bandwidth increases, the order of the previous step will be maintained.
(3) Discrimination: terminate when the overall number of iterations is greater than the experience value. Or when the adjacent three times in the bandwidth sequence are the same, the algorithm will be transferred to the first part. Score, otherwise turn 2.2.2 (1).

According to the characteristic that the beam and pole units are composed of 2 nodes, the calculation The method is optimized to update only the nodes to be exchanged and the corresponding units. Node. In addition, a dynamic array storage is established during node update. The corresponding relationship between the new and old node numbers is updated before passing through the array updates the node coordinates to reduce the number of operations.

3 Object-Oriented Programming

Object-oriented programming is a new programming method. The basic idea of law is to use objects, classes, inheritance, encapsulation, aggregation and messages. The concepts of transmission and polymorphism are used to construct the system for programming.

Face The object method is simple to maintain, has good scalability and high code reuse. Other characteristics have attracted more and more attention and have become today's computers. Mainstream technology in the field [8].

The connection between various classes and management classes is established through a two-way linked list structure. The new unit can be inherited from the FemElement class. Derived, the unit management class uses polymorphism to achieve different types of Unit management. The sorting class is connected with each management class through the interface. System, so it is easy to sort and update nodes (as shown in Fig. 1) As shown).

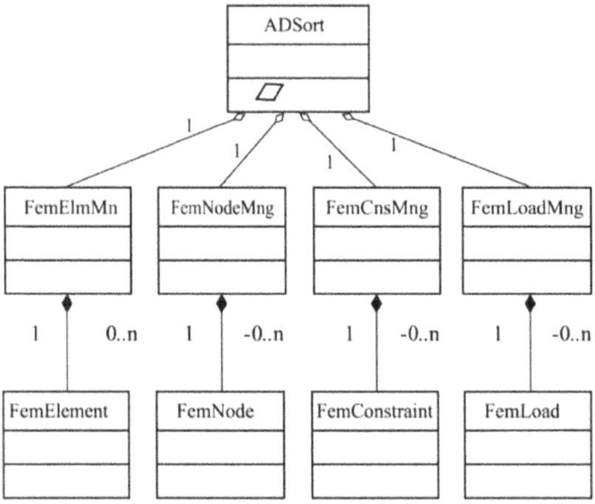

Fig. 1. Schematic diagram of optimization algorithm class aggregation relationship

Throughout the two parts of the algorithm, whether the span criterion and the average criterion are applied. Or the guidelines, both involve more time-consuming operations - sorting and comparison. Dynamic array of large storage space. In response to this situation, in programming The C + + container standard library and algorithm standard library (STL) are applied to Structures and algorithms such as volume, linked list, search, sorting, etc. manage data. Action.

4 Arithmetic Example

Example 1: Optimal sorting of fishbone beam nodes commonly used in bridge structures (Fig. 2).

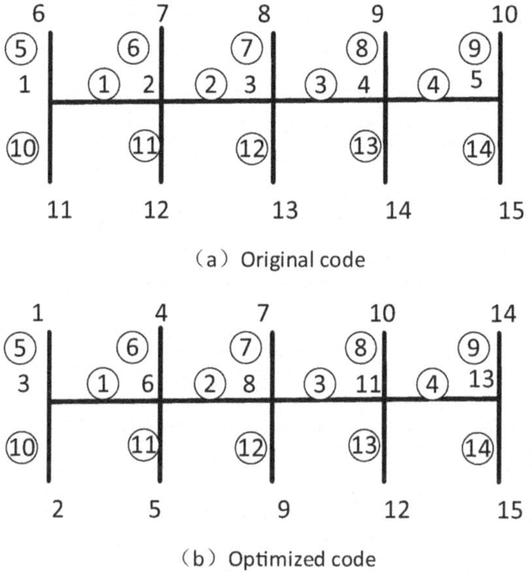

Fig. 2. Fishbone beam coding optimization

Table 1. Coded contact table (optimized results in parentheses)

Point order	Contact form	Sum	Average amount	Spacing
1	1 6 11(1 2 3)	18(6)	6(2)	12(4)
2	2 7 12(1 2 3)	21(6)	7(2)	14(4)
3	3 8 13(1 2 3)	24(6)	8(2)	16(4)
4	4 9 14(4 5 6)	27(15)	9(5)	18(10)
...	...	...	...	...
8	3 8 13(7 8 9)	24(24)	8(8)	16(16)
...	...	...	...	...
15	5 10 15(13 14 15)	30(42)	10(12)	20(28)

From Fig. 2 and Table 1, it can be seen that the half bandwidth before optimization is 10, optimization The latter is only 3.

Example 2: Optimized sorting of spatial cable-stayed bridge nodes (Fig. 3).

A cable-stayed bridge model composed of space beam and cable elements, node compilation The number is according to the main beam - rigid arm 1 - rigid arm 2 -

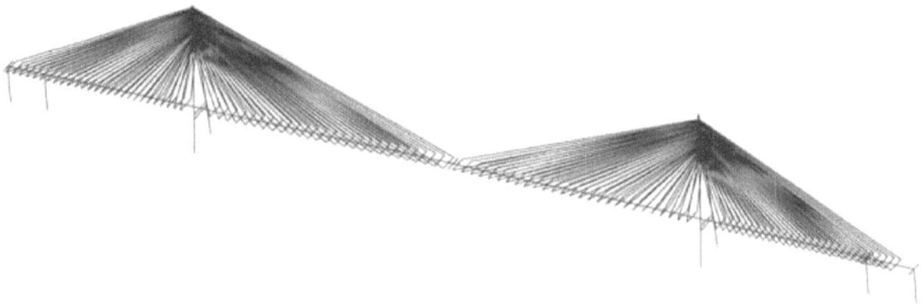

Fig. 3. Spatial cable-stayed bridge model

left bridge - right bridge tower - auxiliary Sequential modeling of auxiliary piers. In the past, the node optimization numbers were based on the rigid arm. - The law of Lasso-Lasso-stiff arm is completed manually, and the modeling process It is very complicated and error-prone.

The results optimized by the program are shown in Table 2.

Table 2. CPU Time Comparison Table (AMD3000 + 256M)

Structure	Freedom number	Half bandwidth	Static analysis time(s)
Original structure	3468	2208	5.953
After optimization	3468	164	0.328

As can be seen from Table 2, the optimization effect of the program on the total rigid bandwidth is very Significantly, and more importantly, the contour of the symmetric matrix is reduced, so that it can It greatly reduces memory usage and improves computing speed, especially for nonlinearity. And dynamic problems play a more significant role in solving them.

5 Conclusions

In finite element analysis, the global stiffness matrix is typically sparse and symmetric. By optimizing node numbering, the bandwidth of the stiffness matrix can be significantly reduced. This not only decreases the data storage requirements but also enhances computational speed. For this issue, this paper utilizes the AD algorithm and tailors it specifically to the characteristics of bridge structures. This optimization primarily relies on span criteria, as the geometric and load characteristics of bridge structures determine the layout of nodes and elements. Proper node numbering can effectively reduce matrix bandwidth, thus optimizing overall computational efficiency.

(1) Furthermore, this research also employs object-oriented programming techniques to develop a bandwidth optimization program. By using the STL library and integrating core object-oriented concepts such as inheritance, encapsulation, aggregation, and

polymorphism, the program effectively organizes and manages different classes using dynamic arrays and linked-list structures. This programming strategy not only enhances the flexibility and reusability of the program but also allows it to be used as an independent unit or to be encapsulated into a class, which can then be integrated into existing software frameworks.

(2) Extended to broader applications, this optimization method and program structure are not only applicable to bridge structures but can also be extended to other types of engineering structures, such as high-rise buildings or other large infrastructure projects. By reducing the consumption of computational resources, it is possible to accelerate the design process while ensuring accuracy. Additionally, the optimized node numbering strategy also enables more complex dynamic analyses and nonlinear calculations, bringing new tools and methodologies to the fields of engineering design and analysis.

Acknowledgments. Sichuan Provincial Natural Science Foundation, (No.:2023NSFC0389) Yunnan Provincial Science and Technology Plan Project, (No.:202305AF150138) Yunnan Prov. Sci. & Tech. Spec. Plan, (NO.: 202102AD080003) Yunnan Province Major Science and Technology Special Plan Project (202302AD080009).

China Railway Science and Technology Research and Development Project (KSWQ224013) Yunnan Province Innovation Guidance and Technology oriented Enterprise Cultivation Plan (No.:202204BQ040008).

References

1. Akhras, G., Dhatt, G.: An automatic node relabeling scheme for minimizing a matrix or network bandwidth. Int. J. Numeric. Methods Eng. **10**, 787–797 (1976)
2. Akhras, G.A.: finite element software for the analysis of two-dimensional continua. In: Second Sym. In App. of Solid Mechanics. McMaster Uni, vol. 9, pp. 4–9 (1974)
3. Grooms, H.R.: The Finite Element Method in Engineering Science. McGraw-Hill, London (1971)
4. Liu, W.X.: Solution of Large Sparse Linear Equations. National Defense Industry Press Edition Company, Beijing (1981)
5. Xing, Y.: Research on the optimal permutation for the nodes of finite element network. Chin. J. Comput. Mech. **16**(3), 366–369 (1999)
6. Wu, S.F.: Data structure optimized by grid coding. J. Changchun Univ. Sci. Technol. **22**(4), 44–47 (1999)
7. Yang, Z.J.: Fast solution algorithm for object oriented finite element method—part two: fast solution algorithm. J. Jilin Univ. **35**(2), 195–198 (2005)
8. Yu, M.H.: Finite Element Method and Object-Oriented Programming. Science Press, Beijing (2004)

Research on Intelligent Vibration Parameters of Assembled Lightweight T-Beam Members

Zuoqian Xu[1], Zhitian Wang[2(✉)], and Dading Sun[3]

[1] Anhui Jiaocheng Engineering Group Co, Hefei 230088, Anhui, China
[2] Anhui Province Transportation Planning and Design Institute Co, Ltd., Hefei 230088, Anhui, China
402072776@qq.com
[3] Anhui Jiaocheng Industrialized Construction Company Limited, Hefei 230088, Anhui, China

Abstract. With the evolution of urban development and the increasing demand for space utilization, the advantages of assembly construction are becoming more and more prominent, and gradually developing towards industrialization. But nowadays, prefabricated components are still uneven, loose and other quality defects, at the same time, the vibration step as an important part of the impact of the quality of prefabricated components, so control the effect of vibration, is conducive to the control of the entire prefabricated components of the final quality. In order to effectively control the precast T-beam concrete internal defects due to vibration as well as the apparent quality and other problems, the article takes the Anhui wisdom beam yard as an engineering example, and designs the vibration parameters when pouring light T-beam, and studies the influence of each vibration parameter on the quality of precast components. The parameters include the number of layers of materials in the vibration process, the concrete mix ratio of pouring, and the vibration frequency of attached vibrator. The results show that: the reduction of layering height, or more refined pouring, is conducive to the full exhaustion of each layer of concrete in the process of pouring and vibration, and the small bubbles can rise from the lower layer to the upper layer more smoothly and gradually in the process of vibration, thus reducing the phenomenon of small voids and holes; appropriately increasing the frequency of vibration is conducive to the reduction of the area of the large bubbles in the horseshoe portion of the lower part of the T-beam, and it also improves the strength of the whole.

Keywords: T-beam · vibration · fabrication · production quality · intelligences

1 Introduction

Precast concrete T-beam has the characteristics of large structural rigidity and convenient construction, which is commonly used in the construction of large highway bridges [1–4]. Since the T-beam is an anisotropic structure, and the sloping surface at the connection

The original version of the chapter has been revised. A correction to this chapter can be found at https://doi.org/10.1007/978-3-031-76102-7_51

B. Yuan et al. (Eds.): ICCASE 2024, SUCI, pp. 158–166, 2024.
https://doi.org/10.1007/978-3-031-76102-7_19

between the horseshoe and the web is not conducive to the discharging of air bubbles in the concrete during the vibration process, once the vibration process is unreasonable, it will lead to the occurrence of quality problems such as air bubbles and honeycomb in the whole precast component, which will affect the aesthetics and durability of the product [5, 6]. Therefore, precast concrete T-beams put forward high requirements for the control of concrete casting and vibration process.

At present, most of the precast concrete T-beam casting quality improvement aspects are focused on raw material design, template selection, maintenance process, etc. Qiu et al. [7] combined with the decomposition and arrangement characteristics of T-beam streaming production, researched the T-beam production process, special equipment for streaming production, selection of prefabrication yards, and difficulties in construction site arrangement, analyzed the energy consumption of the production process, and finally summarized the advantages of the T-beam industrialized production and Finally, it summarizes the advantages and efficiency of T-beam industrialized production and provides reference for judging the focus of the whole production process of T-beam. Mi [8] analyzed the appearance quality defects and appearance quality improvement of T-beam in bridge engineering, focusing on the impact of vibration on precast girder construction. Zhao [9] also analyzed the appearance quality control technology of precast concrete T-beam construction, and gave the suggestion that, in order to ensure that the appearance quality is up to the standard, the concrete mixing ratio should be optimized, and do a good job in formwork design and installation, strengthen vibration during the concrete pouring process, and take effective technical measures to maintain the concrete to avoid cracks. Strengthen the vibration in the process of pouring concrete, and take effective technical measures to maintain the concrete to avoid cracks. Liu et al. [10] compared and analyzed the advantages and disadvantages of each fabrication process in the process of concrete pouring, and sorted out the advantages of the whole layered fabrication in the vibration process. Li [11] analyzed the causes of air bubbles in the production process of T-beam, and effectively reduced and lowered the generation of air bubbles on the surface of concrete through the selection of raw materials, adjustment of admixtures, optimization of mixing ratios, and improvement of construction technology, etc. Liang [12] proposed measures to prevent air bubbles in the precasting of T-beam from the raw materials, the mixing time of concrete, and the changes in temperature respectively, and strictly controlled the vibration time to achieve the expected results. Vibrator vibration time and so on to achieve the expected results.

In summary, many scholars have analyzed the whole construction process of T-beam, and analyzed the factors such as bubble generation, surface unevenness and honeycomb of precast T-beam, focusing on the influence of concrete in the process of vibration, and clarified the influence of the mixing ratio and fabric on the final appearance of the precast beams, but there is a lack of research on the effect of vibration in the case of different parameters. Therefore, this paper takes a girder yard project example in Anhui Province as a basis, adopts the scaled-down modeling experimental method, and further analyzes the influence of the number of layers of materials in the vibration process and the different vibration frequency of the attached vibrator on the vibration effect through the optimized mixing ratio before.

2 Experimental Program

A full-size T-beam formwork was used for this test, with the beam length scaled down by 1:10 (2.5 m). The effect of reinforcement on the vibration process had to be taken into account throughout the test, and in order to better simulate the actual production conditions through the test, the scaled-down members were kept identical in cross-section to the original T-beam members. The cross-section is shown in Fig. 1.

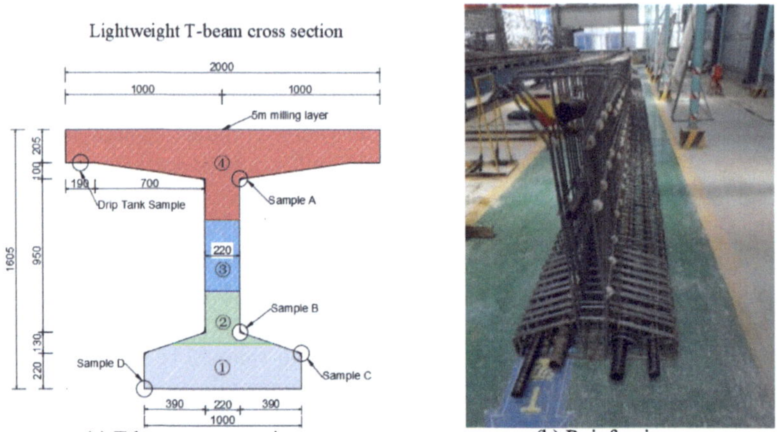

(a) T-beam cross section (b) Reinforcing cage

Fig. 1. Cross-section of the member

Table 1. Optimal mixing ratio(kg/m^3)

Clinker	Coal ash	Mineral powder	Sand	Rock (5-10mm)	Rock (10-20mm)	Water	Water reducer
346	44	53	691	322	854	168	5.1

The article through the previous test on the mix ratio optimization to get the optimal mix ratio as in Table 1, and as this test casting mix ratio, after the beam factory vibration process of the field investigation, combined with some existing research literature, proposed a few major factors that may affect the vibration effect: (1) the concrete pouring situation, (2) the height of the material or layer, (3) vibration frequency. The test protocols are summarized in Table 2, a total of three tests, organized in the article proposes five layers of fabric (20 cm + 25 cm + 40 cm + 45 cm + 30.5 cm) in the order shown in Fig. 1(a), the first and the second layers of fabric are only opened after the completion of the following row of vibrators, the third and the fourth layers of fabric are opened after the completion of the above row of vibrators, the vibrating time and frequency of vibrators are consistent with the original girder factory vibration process. The fifth layer of fabric is not considered in this study because it is mainly vibrated by manual insertion type vibrators.

Table 2. The test program and the specific process

Number	Vibration frequency	Vibrating time	Pouring and vibration process	Thickness of each layer of the section (cm)
1	130 Hz	50 s	After pouring the first layer, turn on the lower vibrator; after pouring the second layer, turn on the upper and lower vibrators; after pouring the third layer, turn on the upper vibrator only	First layer: 30; Second layer: 45; Third layer: 55
2	130 Hz		After pouring the first layer, turn on the lower vibrator; after pouring the second layer, turn on the lower vibrator; after pouring the third layer, turn on the upper vibrator only; after pouring the fourth layer, turn on the upper vibrator only	First layer: 20; Second layer: 25; Third layer: 40; Fourth layer: 45;
3	150 Hz		After pouring the first layer, turn on the lower vibrator; after pouring the second layer, turn on the upper and lower vibrators; after pouring the third layer, turn on the upper vibrator only	First layer: 30; Second layer: 45; Third layer: 55

The article focuses on evaluating the quality of prefabricated T-beams by measuring porosity and rebound strength. It was calculated by visually inspecting the size and number of bubbles and applying the following formula: Total porosity $= \frac{S_{\text{Large air bubble}} + S_{\text{Small air bubble}}}{S_{\text{lower horseshoe}}} \times 100\%$, where the area of large and small bubbles is estimated by taking the average radius of large bubbles (5 mm < r < 10 mm) and small bubbles (0 < r < 5 mm) to obtain the average area, The total area of the entire lower horseshoe is the product of the length of the T-beam member and the width of the horseshoe, The smaller the calculated value, the better. Finally, the strength of the T-beam after maintenance was measured by rebound meter. $S_{\text{lower horseshoe}} \approx 2500\times\ 400\ mm^3 = 1m^3$.

3 Analysis of Results

3.1 Appearance Quality Evaluation

The main purpose of this test is to calculate the number of large bubbles and small bubbles in the lower horseshoe part, so as to roughly estimate the area of all the bubbles, and compare it with the area of the whole horseshoe part to get the total void ratio of the horseshoe part. This index is used as the evaluation index of the appearance quality of T-beam members. The test data were recorded as shown in Table 3 below:

Table 3. Calculation results of bubble area

Beam No	Frequency (Hz)	Storey	Large air bubble (number)			Small air bubble number)			Total porosity (%)
			1	1'	Total	1	1'	Total	
1	130	3	185	125	310	450	580	1003	5.88
2	130	4	145	174	319	289	228	517	5.52
3	150	3	107	144	251	526	667	1193	5.11

a. Influence of the number of layers poured on the quality of appearance.

A comparison of the total porosity of the horseshoe under Test Beam 1 and Test Beam 2 reveals that the porosity of the bubbles decreased by 0.36% with the decrease in the height of each pouring layer, or the increase in the number of layers poured. The number of large bubbles did not decrease significantly, but mainly the number of small bubbles decreased significantly, and the number of small bubbles in test beam 2 decreased by about 48.4% compared to test beam 1. The reason may be that the reduction of layering height, or more refined pouring, is conducive to the full exhaustion of each layer of concrete in the process of pouring and vibration, and the small bubbles can rise more smoothly from the lower layer to the upper layer gradually in the process of vibration, which reduces the phenomenon of small voids and holes.

b. Effect of pouring frequency on appearance quality.

A comparison of the total porosity of the horseshoe under Test Beam 1 and Test Beam 2 reveals that the porosity of the bubbles decreased by 0.77% as the frequency of vibration increased. There was an 18.9% increase in the number of small bubbles but a significant decrease in the number of large bubbles by about 19% for test beam 2 compared to test beam 1. The reason for this phenomenon may be that the higher the vibration frequency, the large bubbles inside the concrete will re-split during the vibration process and form more tiny bubbles. It can be concluded that increasing the vibration frequency appropriately is beneficial to reduce the area of large bubbles in the lower horseshoe portion of the T-beam.

3.2 Evaluation of Rebound Strength

Concrete strength was measured by rebound meter with reference to Technical Specification for Testing Compressive Strength of Concrete by Rebound Method (JGJ/T 23-2011), and each concrete specimen was divided into upper and lower layers on average to examine the strength homogeneity of concrete in different vibration states.

a. The results of the rebound strength test of test beam 1 were recorded as shown in Table 4 below:

Table 4. Measurement results of rebound strength of test beam 1 (unit: MPa)

Measurement point	Survey area									
	1	2	3	4	5	1'	2'	3'	4'	5'
1	50	34	46	39	36	37	33	37	36	34
2	44	32	40	36	35	34	31	51	35	33
3	47	32	41	36	36	37	34	37	35	33
4	45	34	39	37	45	41	35	36	34	35
5	42	36	40	40	37	42	34	38	48	36
6	47	33	49	36	38	33	34	43	36	36
7	44	34	36	40	39	50	36	38	36	35
8	46	35	39	35	34	46	34	38	39	34
9	46	38	34	36	36	47	35	40	38	36
10	46	33	39	38	40	51	33	39	43	36
11	42	35	38	37	36	41	35	38	40	35
12	48	36	40	40	36	38	34	36	46	35
13	50	37	36	42	36	34	36	35	36	36
14	40	35	40	39	38	44	33	51	49	36
15	40	36	42	42	41	38	36	51	39	35
16	45	35	39	35	33	45	33	36	36	36
Average strength	45.1	34.7	39.9	38.0	37.3	41.1	34.1	40.3	39.1	35.1
Final Strength:(45.1 + 34.7 + 39.9 + 41.1 + 38.0 + 37.3 + 34.1 + 40.3 + 39.1 + 35.1)/10 = 38.5										
Standard deviation:$S=\sqrt{\frac{\sum f_{cu,i}^{2}-n(m_{f}^{c})^{2}}{n-1}}=3.39$										

b. The rebound strength test results of test beam 2 were recorded as shown in Table 5 below:

Table 5. Measured results of rebound strength of test beam 2 (in MPa)

Measurement point	Survey area									
	1	2	3	4	5	1'	2'	3'	4'	5'
1	41	36	35	40	44	37	35	38	47	40
2	49	36	46	43	43	36	35	43	46	38
3	42	35	36	45	38	43	34	50	43	42
4	53	34	36	44	36	48	36	42	42	37
5	45	35	36	40	35	39	34	37	36	36
6	51	36	35	40	36	46	35	44	42	41
7	41	36	37	40	40	40	35	48	36	37
8	46	36	36	50	52	45	35	38	42	44
9	42	36	35	40	47	42	34	44	51	37
10	46	38	36	44	38	46	33	36	36	37
11	55	37	38	44	45	45	33	37	42	38
12	46	34	36	41	47	42	36	37	38	37
13	50	35	41	58	39	41	35	37	47	38
14	39	38	38	39	38	38	34	36	52	41
15	42	36	38	42	42	39	34	36	38	38
16	44	37	37	51	36	36	35	40	38	50
Average strength	45.1	34.7	39.9	38.0	37.3	41.1	34.1	40.3	39.1	35.1
Final Strength:(45.6 + 35.9 + 37.3 + 43.8 + 41.0 + 41.4 + 34.6 + 40.2 + 42.3 + 39.4)/10 = 40.2										
Standard deviation: $S=\sqrt{\frac{\sum f_{cu,i}^2 - n(m_f^c)^2}{n-1}} = 3.47$										

c. The rebound strength test results of test beam 3 were recorded as shown in Table 6 below:

From the data in Tables 4, 5 and 6, it can be seen that except for the rebound strength of test area 2 and test area 2'(which are located at the horseshoe of the T-beam), which is only 35 MPa, the strength of the rest of the test areas is basically around 40 MPa. This also shows that the air bubbles at the horseshoe do affect the concrete strength in this area. The rebound strengths of test beam 2 and test beam 3 in both test area 2 and test area 2' are higher than that of test beam 1. This indicates that reducing the number of air

Table 6. Measurement results of rebound strength of test beam 3 (unit: MPa)

Measurement point	Survey area									
	1	2	3	4	5	1'	2'	3'	4'	5'
1	42	36	41	44	43	50	34	38	36	40
2	42	36	37	40	35	47	32	39	40	36
3	46	40	40	34	52	45	33	41	49	41
4	40	32	36	37	53	51	32	38	35	38
5	42	38	39	41	44	51	34	42	36	37
6	40	34	38	45	52	38	34	40	46	38
7	38	33	38	39	45	36	32	46	40	51
8	48	32	37	48	45	36	33	42	36	38
9	38	40	37	38	47	47	34	45	36	36
10	48	36	43	45	35	45	33	43	45	40
11	39	36	43	46	34	40	33	38	43	52
12	45	37	40	35	38	46	32	39	45	49
13	40	35	41	52	36	44	33	35	44	53
14	47	35	38	44	36	42	32	40	45	36
15	42	36	37	36	43	40	36	40	42	36
16	40	37	40	42	52	38	32	48	40	36
Average strength	42.3	35.8	39.1	41.6	43.1	43.5	33.1	40.9	41.1	41.1
Final Strength:(42.3 + 35.8 + 39.1 + 41.6 + 43.1 + 43.5 + 33.1 + 40.9 + 41.1 + 41.1)/10 = 40.2										
Standard deviation: $S=\sqrt{\frac{\sum f_{cu,i}^{2}-n(m_{f}^{c})^{2}}{n-1}}=3.32$										

bubbles at the horseshoe of the T-beam also improves the overall strength of the T-beam member.

4 Conclusions

(1) The smaller the layering height, or more refined pouring, the more conducive to the full exhaustion of each layer of concrete in the process of pouring and vibration, and the small bubbles can rise more smoothly and gradually from the lower layer to the upper layer in the process of vibration, which reduces the phenomenon of small voids and holes.

(2) The higher the vibration frequency, the large air bubbles inside the concrete will be re-split in the process of vibration, the formation of more tiny bubbles. Appropriately increasing the vibration frequency is conducive to reducing the horseshoe portion under the T-beam.

(3) Reducing the number of air bubbles at the horseshoe of the T-beam also improves the overall strength of the T-beam member.
(4) Intelligent optimization of fabric and vibration, strict control of parameters, and subsequent real-time monitoring and measurement through mixing and vibration will further improve the quality of prefabricated T-beam products and promote the development of prefabricated component intelligence.

References

1. Liu, H.Q.: Research on construction jointing program of large span single line railroad concrete continuous girder bridge. J. Railway Sci. Eng. **16**(03), 706–711 (2019). https://doi.org/10.19713/j.cnki.43-1423/u.2019.03.019. (In Chinese)
2. Abdulkadir, G., Metin, M.K.,Tolga, M.G.: Experimental investigation of flexural performance of T-section prefabricated cage reinforced beams with self-compacting concrete. Structures, **33**, 2190–2197 (2021). https://doi.org/10.1016/J.ISTRUC.2021.05.074
3. Sah, P.T., Lacey, W.A., Hao, H., et al.: Prefabricated concrete sandwich and other lightweight wall panels for sustainable building construction: State-of-the-art review. J. Building Eng. 89109391 (2024). https://doi.org/10.1016/j.jobe.2024.109391
4. Xu, T.L.: Key technology for construction of merging of prestressed concrete continuous girder bridge. Railway Constr. (2017). (In Chinese)
5. Bao, Y.J., Ding, M.B., Zhu, L.: Analysis of the influence of merging scheme on construction monitoring of multi-span continuous girder bridge. Railway Standard Des. (2016). https://doi.org/10.13238/j.issn.1004-2954.2016.03.018.(InChinese)
6. Yang, Y., Cao, X.L., Cui Jian, L., et al.: Influence of different merging methods on the bridge formation state of continuous girder bridges. Railway Constr. (2014). (In Chinese)
7. Qiu, S.L., Shen, L.C., Wang, Z.F., et al.: Research on prefabricated T-beam production technology based on ergonomics. Jiangsu Building Mater. (04), 10–12+15 (2023). (In Chinese)
8. Mi, Y.J.: Analysis of T-beam appearance quality defects and improvement focus in bridge engineering. Sichuan Building Mater. **48**(07), 137–138 (2022). (In Chinese)
9. Zhao, P.: Appearance quality control technology of precast concrete T-beam construction in bridge engineering. Transp. World **27**, 164–166 (2023). https://doi.org/10.16248/j.cnki.11-3723/u.2023.27.028. (In Chinese)
10. Liu, Z.Y., Zang, X.Q., Tang, Z.L., et al.: Key control measures of concrete casting and vibration quality of precast T-beam. Inner Mongolia Highway Transp. **04**, 5–8 (2022). https://doi.org/10.19332/j.cnki.1005-0574.2022.04.002. (In Chinese)
11. Li, Y., Hei, J.L., Wu, D.L., et al.: Research on the causes of air bubbles in T-beam concrete and preventive measures. Shanxi Constr. **47**(20), 131–133 (2021). (In Chinese)
12. Liang, Y.L.: A preliminary study on the causes of air bubbles and preventive measures in T-beam prefabrication. Sci. Technol. Innov. Appl. (20), 139–140 (2017). (In Chinese)

Wet Joint Tensile Test of Concrete Bridge Deck with Cross Arranged U-shaped Main Reinforcement

Gang Liu(✉), Congquan Wang, Songkai Jia, and Xuming Cui

CCCC No.3 Engineering Co., Ltd, Beijing 101102, People's Republic of China
15854436@qq.com

Abstract. In order to improve the construction progress and quality, precast concrete bridge panels are commonly used in composite beams, cable-stayed Bridges and arch Bridges, which are then transported to the site for installation and then joined with wet joints. To solve the complicated problem of wet joint of main reinforcement connection in bridge, a crosswise arranged U-shape main reinforcement is proposed in this work. The tensile experiment was carried out on the wet joint of crosswise arranged U-shape main reinforcement specimens with different overlapping lengths. And the results were compared with the wet joint with traditional main reinforcement lap weld. The final results show that the stress mode of the joint varies with the overlap length. With enough overlap length, the wet joint with crossing arranged U-shape main reinforcement can meet the stress requirements of practical projects.

Keywords: Engineering Structure · Wet Joint · U-Shape Main Reinforcement · Crosswise Arrangement · Tensile Experiment

1 Introduction

At present, the main connection forms of wet joint main bars include straight rib lap welding, sleeve connection [1–5], cross arrangement of raised steel bars at the end, U-shaped main rib lap welding, U-shaped main rib cross arrangement, etc. In order to ensure sufficient adhesion between the steel bar and the concrete, the cross-bar arrangement requires sufficient extension of the steel bar. The straight rib lap welding and U-shaped main rib lap welding have some disadvantages, such as difficulty in connecting two bars, heavy welding workload and difficult welding operation. The sleeve connection form requires high installation accuracy of steel bars, which is mainly used in the civil construction field. The U-shaped main bar cross arrangement form omits the on-site welding operation and speeds up the construction speed, which is very suitable for the on-site installation of prefabricated structures, and is the ideal connection form at present. In actual engineering, after the installation of prefabricated bridge deck panels is completed, the two U-shaped main bars cross each other and overlap within a certain range, making it most convenient to pour concrete. However, there is currently limited research on the tensile reliability of this structure, and the force mechanism and overlap length need to be further clarified.

B. Yuan et al. (Eds.): ICCASE 2024, SUCI, pp. 167–173, 2024.
https://doi.org/10.1007/978-3-031-76102-7_20

Therefore, this article takes the cross arranged U-shaped main reinforcement wet joint as the research object, and conducts tensile tests on specimens with three overlapping lengths of 350 mm, 450 mm, and 550 mm, as well as main reinforcement lap welding specimens. The displacement changes, steel strain changes, crack distribution and size of different specimens are compared and analyzed.

2 Experimental Plan

2.1 Test Piece Design

The test pieces are shown in Table 1 and Fig. 1, with a length of 2200 mm, a width of 1200 mm, and a thickness of 250 mm. The length of the middle tensile test section is 940 mm and the width is 400 mm. There are two types of main reinforcement connection forms for the middle tensile test section. The straight reinforcement lap length is 550 mm, and the U-shaped main reinforcement cross overlap length is 350 mm, 450 mm, and 550 mm. All steel bars are made of HRB400 with a diameter of 20 mm.

Table 1. Summary of specimen

Test piece name	Reinforcement connection method	Quantity	l1/mm	l2/mm
T1-350	Cross arrangement	2	350	925
T1-450		2	450	875
T1-550		2	550	825
T2-550	Lap welding	2	550	825

2.2 Material Performance Experiment

The compressive strength test of concrete adopts a standard 150 × 150 × 150 mm cube, and the strength is tested on 7 days, 28 days, and during the test period. The corresponding standard values are 27.4 MPa, 35.3 MPa, and 35.6 MPa. Six sets of steel bar tensile tests were conducted, with an average elongation of 23.8%, a yield point of 420.7 MPa, and a tensile strength of 544.7 MPa.

2.3 Loading Devices and Methods

The loading method is shown in Fig. 2. The specimen is placed horizontally, with two jacks arranged on both sides of the middle tensile test section, and the axis coincides with the center of the specimen's load-bearing wing. The loading classification is shown in Table 2. From the beginning of loading until the concrete cracks (theoretical cracking load 318.7 kN), until all main reinforcement reaches the tensile strength state (load 1026.6 kN), there are a total of 7 levels of load. Install a percentage displacement meter between the load-bearing wings to measure displacement, use a vernier caliper to measure crack width, and collect strain information using a uT7110Y static strain acquisition instrument.

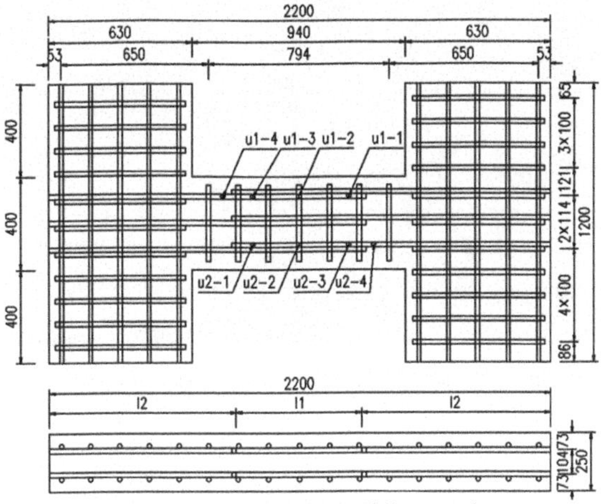

(1) Straight steel bar lap welding specimen

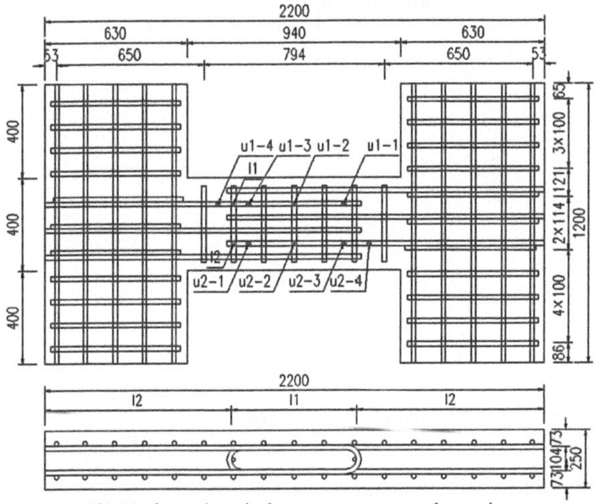

(2) U-shaped main bar cross-arranged specimen

Fig. 1. Demonstration of dimension and setup of reinforcement

Table 2. Load rate table

Loading classification	1	2	3	4	5	6	7
Load size /kN	100	200	300	400	600	800	1000

Fig. 2. Loading diagram of tensile experiment

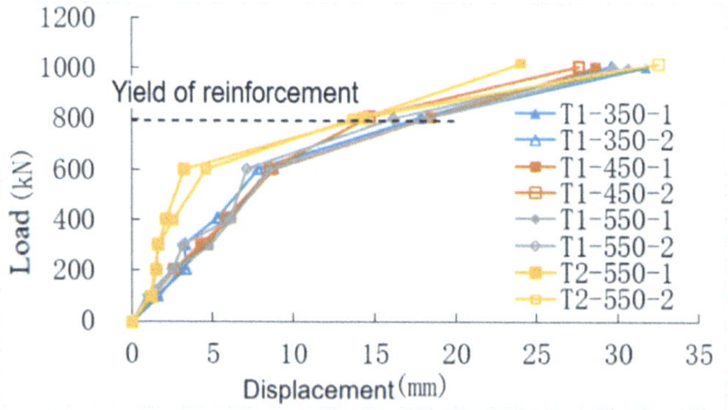

Fig. 3. Load-displacement curve

3 Experimental Results and Analysis

3.1 Load-Displacement Curve

The load displacement curve of the specimen is shown in Fig. 3. There is no obvious difference between the three U-shaped main bar specimens, the steel bar is basically linear before yielding, and the stiffness of steel bar drops sharply after yielding. The initial stiffness of the butt welded specimens with straight bars is greater than that of the U-shaped main bars, but the stiffness decreases obviously after the steel bars yield.

All specimens are loaded to a maximum of around 1000 kN, which is greater than the theoretical calculated diagonal compressive failure bearing capacity of concrete, indicating that the calculation method for bearing capacity under this working condition is smaller than the actual value.

3.2 Reinforcement Strain

The load strain curve of the main reinforcement is shown in Fig. 4. The strain changes at the ends of the overlapping area U1-1 and U2-1 of specimen T1-350 are the smallest, and the slope is basically linear. The strains in the overlapping zone intervals U1-2 and

U2-2 are basically the same as those in U1-3 and U2-3, and are always in an elastic state. The strain changes on the outer side of the overlapping area U1-4 and U2-4 are the largest, and they have already reached a plastic state during maximum loading.

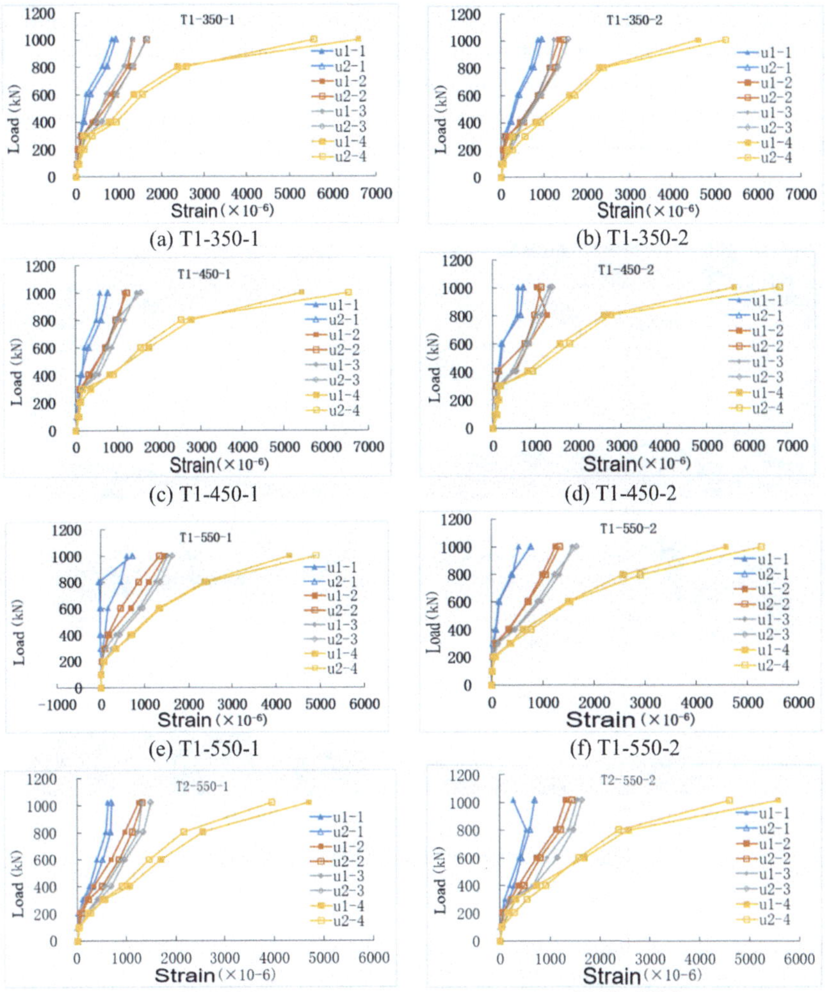

(a) T1-350-1 (b) T1-350-2

(c) T1-450-1 (d) T1-450-2

(e) T1-550-1 (f) T1-550-2

(g) T2-550-1 (h) T2-550-2

Fig. 4. Load-strain curve of main reinforcement

The strain of specimen T1-450 decreases sequentially from the outer side of the overlapping zone to the end. At maximum loading, only the outer side reaches a plastic state, while the rest are in an elastic state. The strain variation of specimen T1-550 is basically similar to that of specimen T1-450, with a more significant difference between the strains in the overlap zone U1-3, U2-3, and U1-2, U2-2. The strain variation pattern of specimen T2-550 is basically similar to that of specimen T1-550.

3.3 Distribution of Concrete Cracks

The transverse crack width at the end of the U-shaped main reinforcement on the outer side of specimen T1-350 is the largest, mainly because the outer U-shaped main reinforcement cannot reach equilibrium state, and the concrete at the end of the main reinforcement is directly under tension. Two longitudinal penetrating cracks appeared between the transverse cracks at the ends of the left and right U-shaped main bars. The main reason for this phenomenon is that after the transverse cracks in concrete exceed the outer main reinforcement, the tensile force of the main reinforcement produces an out of plane bending moment on the cracking surface, which affects the transverse tensile force of the concrete.

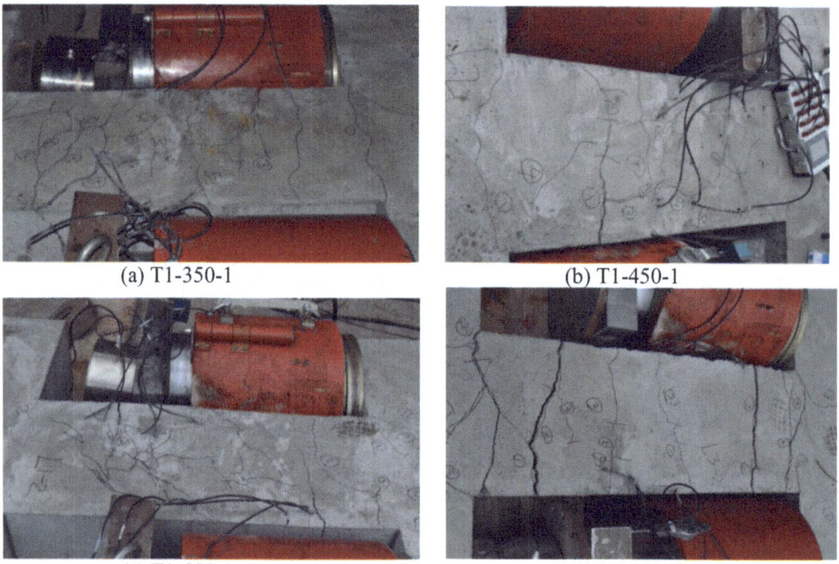

(a) T1-350-1 (b) T1-450-1 (c) T1-550-1 (d) T2-550-1

Fig. 5. Distribution of crack after experiment

The location and shape of cracks in specimens T1-450 and T1-550 are basically similar to those in specimens T1-350, but the transverse crack at the end of the U-shaped main reinforcement does not penetrate the entire cross-section. Test specimen T2-550 only has transverse cracks without longitudinal cracks, and the crack width at the overlapping end is the largest. The main reason for this phenomenon is that the main reinforcement of the specimen is a continuous construction of concrete that is only subjected to tension, and the overlapping area of the steel bars is larger than the outer side. The distribution of concrete cracks after loading is shown in Fig. 5.

4 Conclusion

- As the overlap length of the U-shaped main bar specimen becomes shorter, the longitudinal crack inclination becomes more obvious, indicating that the stress mode is closely related to the overlap length.
- The U-shaped main reinforcement specimens all showed transverse and longitudinal cracks, but the lap welded specimens only showed transverse cracks, indicating that the stress mechanisms of the two specimens are completely different.
- When the maximum load is applied to the U-shaped main bar specimen, the whole section of the main bar reaches the tensile strength state, which indicates that this structure fully meets the stress requirements and can be applied in practical engineering.

References

1. Alexander, A., Clifford, L., Bala, T.: Investigation of closure-strip details for connecting prefabricated deck systems. PCI J. **56**(3), 75–93 (2011)
2. Ling, N., Shuixing, Z., Fang, Y., et al.: Experimental study of shrinkage compensation concrete on wet joints of bridge deck. Concrete **11**, 39–41 (2009)
3. Lewis, S.: Experimental Investigation of Precast Bridge Deck Joints with U-Bar and Headed Bar Joint Details. University of Tennessee, Knoxville (2009)
4. Li-Lin, C., Ai-Qun, L., Hong-En, X., et al.: Experimental research on splicing joints of large plate assembly. J. Southeast Univ. (Natural Science Edition) **41**(06), 1253–1258 (2011)
5. Zhu, Y., Dehao, D.: Study on lap length of precast concrete beams with wet joints. Highway **60**(12), 89–93 (2015)

Analysis of Ultimate Strength of Steel Bridge Deck New Type with J-shape Stiffener

Congquan Wang(✉), Gang Liu, Songkai Jia, and Xuming Cui

CCCC No.3 Engineering Co., Ltd., Beijing 101102, People's Republic of China
wcq19831s@126.com

Abstract. To improve fatigue resistance of steel bridge deck, a new structure of J-shape stiffeners consisting of vertical line segment and bottom semicircle, realizing double-side welding at the connection of stiffeners and the roof, and the plate thickness and section height are the same as those of the U-shaped stiffeners. At the same time, the vertical the second moment of area of J-shape stiffeners is basically the same and shares the advantages of U-shaped stiffeners, which overcomes the defects prone to fatigue. In this paper, the bearing capacity, deformation and plastic behavior of J-shape stiffeners in a bridge deck has been studied by using ANSYS software. The results show that the area and moment of inertia of J-type stiffeners meet the requirements of existing steel structure codes, but the bearing capacity decreases slightly.

Keywords: Steel Bridge Deck · Fatigue Resistance · J-shaped Stiffeners · Ultimate Strength · Stability

1 Introduction

Steel bridge panel stiffeners can be divided into open stiffeners and closed stiffeners according to the section shape. The type of closed stiffener is U-shaped stiffener, which has high vertical stiffness and good stability, and is used in many long-span suspension Bridges, cable-stayed Bridges, continuous beams and other steel box girders. However, fatigue cracking of U-stiffened bridge panel is easy to occur due to welding residual stress, single-side welding, stress concentration and construction quality.

In order to solve the above problems, domestic and foreign experts have conducted a series of studies [1–5]. For example, by increasing the thickness of the plate, improving the welding process, and using bolt connections to improve the fatigue life of key parts, and adding ultra-high toughness concrete to enhance the stiffness of the bridge deck, but all are limited to U-shaped stiffeners. In this paper, a new type of structure is developed by changing the shape of stiffener structure to improve the fatigue performance. The stiffener adopts J-shape and is composed of vertical straight section and bottom semi-arc. The top plate and stiffener are welded on both sides. The plate thickness and height of J-type stiffener are the same as U-type stiffener, and the vertical moment of inertia is basically the same. J-type stiffener not only has the advantages of U-type stiffener, but also makes up the defects of fatigue easily.

B. Yuan et al. (Eds.): ICCASE 2024, SUCI, pp. 174–179, 2024.
https://doi.org/10.1007/978-3-031-76102-7_21

2 Structure of J-type Stiffened Steel Bridge Panel

The structure of the bridge panel is shown in Fig. 1 and Fig. 2. The minimum thickness of the stiffened plate is 14 mm, and the maximum thickness is 20 mm, which is increased according to a gradient of 2 mm. The distance between the stiffeners should be 300 mm, the height of the straight section should be 210 mm, and the radius of the circular section should be 40 mm. The minimum and maximum transverse spacer spacing are 1500 m and 3500 m, and are increased according to the gradient of 500 mm. In this paper, the flat steel box girder bridge panel is taken as the research object, the width is only considered 10 stiffeners with a width of 3000 mm, and the length direction is taken 6 spans considering the influence of end boundary conditions.

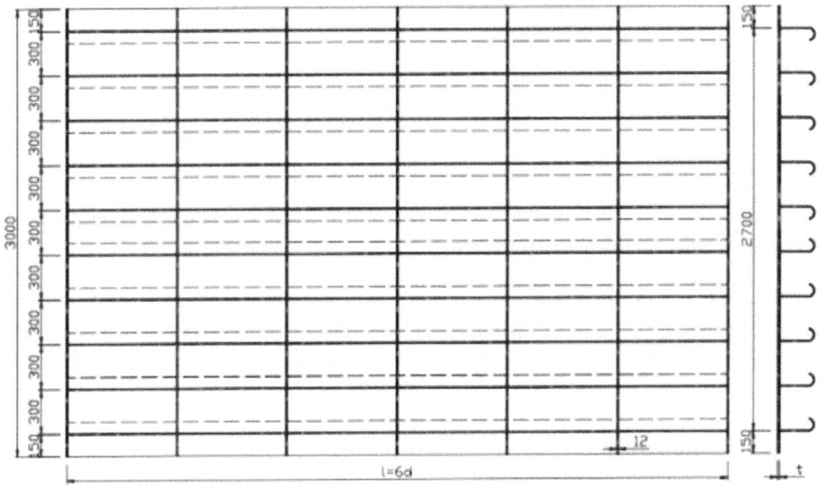

Fig. 1. Structure diagram of bridge panel

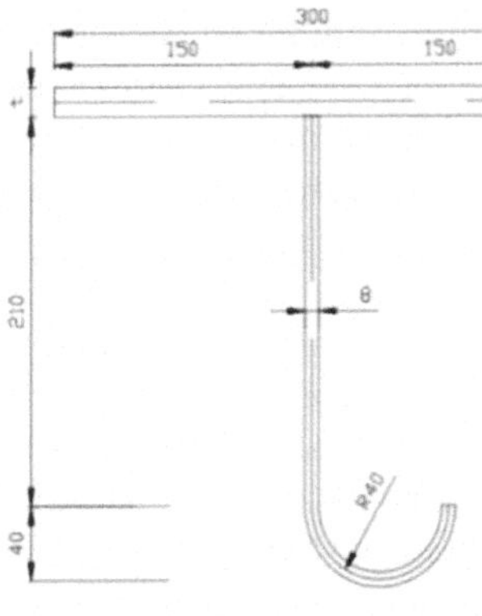

Fig. 2. Structure diagram of J-type stiffener

3 Finite Element Calculation

3.1 Finite Element Model

The calculation software adopts Ansys19.0, and all units adopt board unit shell181. The calculation model is shown in Fig. 3.

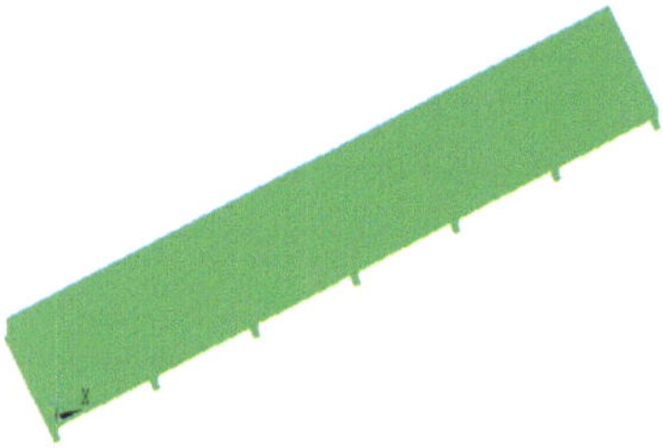

Fig. 3. Finite element calculation model

3.2 Stress-Strain Relationship of Steel

The steel adopts Q345, the elastic modulus of the material is 2.06 × 105 MP, Poisson's ratio is 0.3, the yield value is 345 MPa, and the hardening rate is 0.02 times of the elastic modulus. The material yield criterion is Von Mises isotropic reinforced yield criterion.

4 Finite Element Result Analysis

4.1 Load-Displacement Curve and Bearing Capacity

The load displacement curve of the main node is shown in Fig. 4.

In the elastic stage when the load is small, the slope of the curve is the same although the spacing of the transverse partition is different. When the residual stress reaches the yield point, the stiffness of the compression area begins to decrease, and the greater the distance between transverse partitions, the more obvious the stiffness decrease. The final ultimate bearing capacity continues to increase, and the greater the spacer spacing, the smaller the ultimate bearing capacity. The main reason for this phenomenon is that the slenderness ratio of the member increases with the increase of the transverse partition spacing, and the failure form changes from strength failure to instability failure. When the load is maximum, the displacement corresponding to the bridge panels of different widths is compared. The displacement of the bridge panels of d = 1500 mm is maximum, and the displacement of the bridge panels of d = 3000 mm is minimum. The displacement of the bridge panels of d = 2000 mm and d = 2500 mm is irregular.

The summary results of maximum load are shown in Table 1. The maximum load on a bridge deck with d = 1500 mm is 0.96–0.97, the maximum load on a bridge deck with d = 2000 mm is 0.87–0.89, the maximum load on a bridge deck with d = 2500

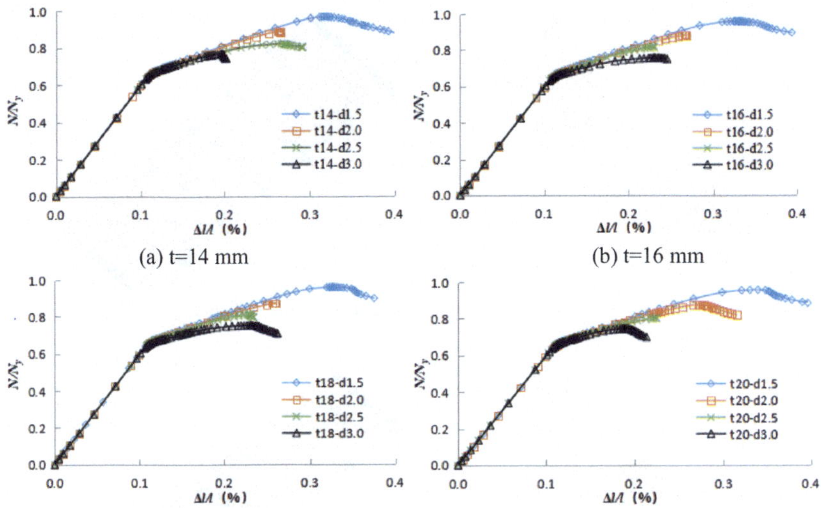

(a) t=14 mm (b) t=16 mm (c) t=18 mm (d) t=20mm

Fig. 4. Load-displacement curve

mm is 0.80– 0.82, and the maximum load on a bridge deck with d = 3000 mm is 0.74–0.76. It can be seen that the stiffening plate thickness changes have little influence on the maximum load when the transverse spacer spacing is the same, and the greater the transverse spacer spacing, the greater the maximum load reduction.

Table 1. Results of finite element analysis

Bridge panel	N/kN	N_y/kN	N/N_y	Bridge panel	N/kN	N_y/kN	N/N_y
t14-d1.5	235364	242643	0.97	t18-d1.5	273499	284895	0.96
t14-d2.0	215952		0.89	t18-d2.0	247859		0.87
t14-d2.5	198967		0.82	t18-d2.5	230765		0.81
t14-d3.0	177129		0.76	t18-d3.0	213671		0.75
t16-d1.5	253362	263919	0.96	t20-d1.5	293636	305871	0.96
t16-d2.0	234888		0.89	t20-d2.0	269166		0.88
t16-d25	216414		0.82	t20-d2.5	244697		0.80
t16-d3.0	201322		0.76	t20-d3.0	226344		0.74

4.2 Stress Distribution and Deformation

The stress distribution and deformation of the t = 14 mm bridge panel at the last loading step are shown in Fig. 5.

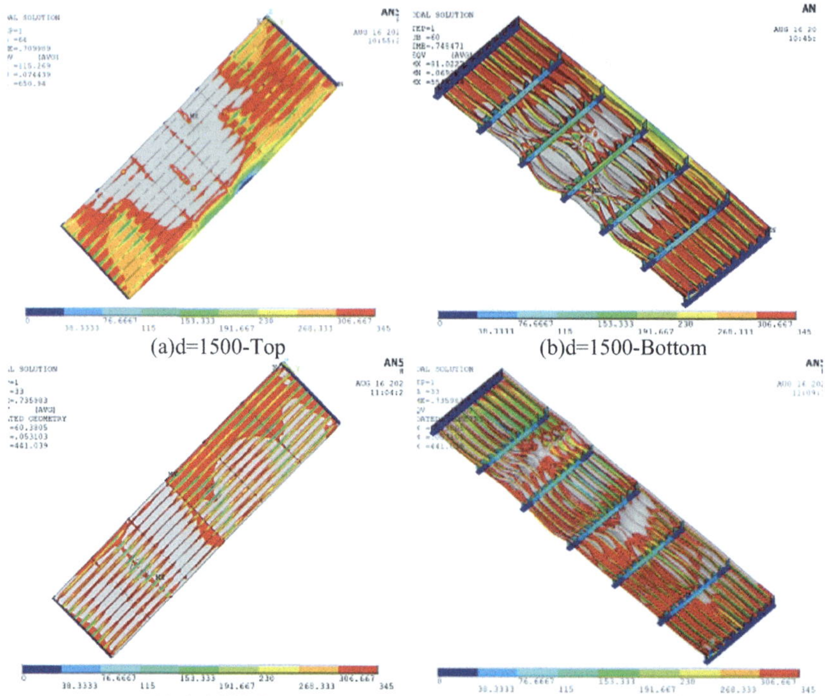

(a)d=1500-Top (b)d=1500-Bottom

(c) d=2000-Top (d) d=2000-Bottom

Fig. 5. Stress moire diagram (unit: MPa, t = 14 mm)

The d = 1500 mm bridge panel in the third and fourth span of the stiffener plate and stiffener most areas have reached the yield state, and the stiffener occurred transverse instability. The d = 2000 mm bridge panel third span stiffener and stiffener connecting small area did not reach the yield point, the rest of the area reached the yield point. The d = 2500 mm bridge panel is similar to the d = 2000 mm bridge panel, but the area of the stiffened plate does not reach the yield point is slightly increased. The area of d = 3000 mm bridge panel that has not reached the yield point continues to increase, and most of the second and fourth span stiffeners reach the yield state and lose stability to the inside. The stress distribution and deformation trend of t = 16 mm, t = 18 mm and t = 20 mm bridge panels are basically similar to those of t = 14 mm bridge panels.

5 Conclusion

Based on the finite element analysis and theoretical analysis, the following conclusions are drawn.

- When the distance between the transverse partitions is small, most of the stiffener plate and stiffener ribs can reach the yield state, but when the distance is larger, the local area of the connection between the stiffener plate and the stiffener is still in the elastic state, and the stiffener is both vertically and horizontally unstable.

- The support spacing of the J-shape bridge panel should not be greater than 1500 mm, less than 3000 mm of the conventional U-shape stiffener, and it is necessary to increase the support at the compression place.
- The bearing capacity of J-shape bridge panels still declines under the conditions of minimum stiffening area and stiffness of existing steel structure specifications. If the existing design method is unfavorable, it is necessary to consider the decline. The actual application of J-shape stiffeners in engineering is good.

References

1. Yunshi, Z., Faxiong, L., Feng, X., et al.: Analysis and control of fatigue crack of orthotropic steel bridge panel. Highw. Transp. Sci. Technol. **30**(8), 75–80 (2013)
2. Jia, L., Shang, X., Bu, X., et al.: Comparison and analysis of force character calculation and real bridge test of new STC steel bridge deck pavement system. China J. Highw. Transp. **27**(3), 39–44 (2014)
3. Liu, Y.: Mechanical Properties Analysis of New Steel box girder Orthotropic Bridge Panels. Nanjing: Southeast University (2015)
4. Lipeng, L., Liang, T.: Fine finite element analysis of hot spot stress of welded joint of steel bridge roof plate and U-stiffener. J. Highw. Transp. **35**(4), 58–66 (2018)
5. Liu, H., Luo, Y., Han, S., Zhang, Q.: Study on ultrasonic impact control method of residual stress field in double-sided welding of new U stiffener and roof. Steel Struct. **36**(10), 25–33 (2021)

Research on TBM Construction Technology of Small Clearance Crossing of Existing Track Lines at Soft and Hard Rock Interfaces

Shuxiang Zhang(✉), Wentao Liu, and Qingshi Li

No. Three Engineering Co., LTD., CCCC First Highway Engineering Group Co., Ltd., Beijing 100020, China
540718447@qq.com

Abstract. With the rapid development of urban rail transit, the municipal rail line network has shown a trend of complexity and densification, and the engineering construction conditions have become more and more complex, but due to the limitation of construction cost and geological topography, the design and planning spacing of some lines is extremely small, and the construction of the later tunnel may have a certain impact on the safe operation of the existing operation line. This paper takes the small clearance (structural clearance 3 m, excavation contour line and primary bolt clearance 15 cm) crossing the existing operation Line 9 at the interface of the upper soft and lower hard rock layer at the interface of the upper soft and lower hard rock layer in the middle of Chongqing Rail Transit Line 18 North Extension Project ~ end point, and theoretically analyzes the adverse effects of crossing construction under extreme working conditions. Through the development of targeted construction control measures, to ensure that the settlement and scope can be controlled within the effective range, and with the help of automatic monitoring means, to lay the foundation for smooth through the construction.

Keywords: small clear distance · large longitudinal slope · TBM · upper soft and lower hard · automatic monitoring

1 Introduction

Usually, when the distance between two tunnel lines is less than 1/2 of the tunnel diameter, it is called a small clearance tunnel, and during construction under this condition, the two tunnels will affect each other. Therefore, before the construction of the tunnel with small clear distance, the force and displacement of the tunnel structure should be checked and analyzed, and targeted control measures should be taken during construction to reduce the crossing risk as much as possible [1–5]. At the same time, the existing structure should be automatically monitored and targeted to grasp the deformation of the existing line at any time.

The section structure of Chongqing Rail Transit Line 18 project (TBM method construction) crosses the section structure of Line 9 at the soft and hard rock interface of the lower sandstone and upper sandy mudstone (mining method construction). The

B. Yuan et al. (Eds.): ICCASE 2024, SUCI, pp. 180–186, 2024.
https://doi.org/10.1007/978-3-031-76102-7_22

attitude control during TBM crossing process is difficult, and the construction process has high safety risks, which may affect the construction safety of Line 18 and the operation safety of Line 9.

2 Analysis of Crossing Impact

2.1 Project Overview

The TBM section of Chongqing Metro Line 18 from Daping West Station to the endpoint is 697 m long, with an outer diameter of 6.6 m and a width of 1.5 m for the lining ring. The minimum turning radius is 350 m and the maximum slope is 43 ‰. The tunnel is buried at a depth of 32 m to 95 m, and two single shield TBMs are used for tunnel construction. The geological condition of the crossing location is the interface between the lower sandstone and upper sandy mudstone soft and hard rock layers with a horizontal inclination of 30°.

The minimum clear distance between the tunnel body of Line 18 and the existing tunnel structure of Line 9 is only 3 m. Excavation construction may cause redistribution of surrounding rock stress, resulting in deformation of the surrounding rock, which may cause deformation or uplift of the tunnel structure of Line 9. When the TBM of Line 18 crosses Line 9, the outer edge of the cutterhead is only 15 cm away from the initial anchor rod of Line 9. In addition, the crossing position is located at the intersection of the upper soft and lower hard rock layers of the sandy mudstone in the lower sandstone, making it difficult to control the posture of the TBM. If the cutterhead touches the initial anchor rod, it will cause structural damage to the tunnel of Line 9.

2.2 Calculation and Analysis of Displacement Situation

The geological parameters of the crossing location are shown in the Table 1 below.

2.3 MIDAS-GTS NX Modeling Analysis

Three dimensional finite element analysis was conducted on the construction process of Line 18 tunnel using MIDAS-GTS NX universal finite element software, as shown in Fig. 1. The yield condition of rock materials adopts the Mohr Coulomb yield criterion.

According to the above Fig. 2, the maximum uplift value of the tunnel structure in the section of Line 9 is 1.25 mm, located at the arch position; The maximum horizontal displacement value is 0.22 mm, located at the arch waist position, and the displacement change does not exceed the specification requirements.

2.4 Structural Force Calculation

Considering the most unfavorable working conditions, all the loads on the section of the proposed Line 18 will be applied to the structure of the Line 9 section. At the same time, spring elements will be used to simulate the effect of surrounding rock around the tunnel to constrain the calculation of the stress on the Line 9 structure during the construction

Table 1. Rock and Soil Characteristics and Mechanical Parameters

Rock name	Sandstone		sandy mudstone	
	highly-weathered	Moderate weathering	highly-weathered	Moderate weathering
Weight (kN/m^3)	23*	24.9*	23.5*	25.9
Cohesion force (kPa)		1330*		500*
Internal friction angle (°)		40.6*		30.0*
Elastic modulus (MPa)		3020*		1100*
Deformation modulus (MPa)		2500*		780*
Poisson's ratio μ		0.30*		0.35*

Fig. 1. The Numerical Calculation Model

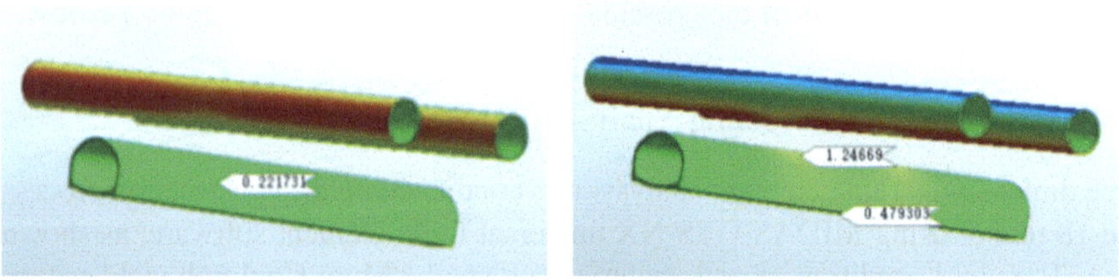

Fig. 2. Horizontal and vertical displacement

process. The calculation model is established as follows. The force cloud map of the section of Line 9 is calculated as Fig. 3.

According to the characteristics of the lining section and relevant specifications, when e_0/h_0 is less than 0.55, the crack width may not be checked. According to the simulation calculation results, the maximum bending moment during the crossing construction process occurs at the arch waist positions on both sides, with a size of 648.9 kN · m. The maximum axial force occurs at the arch foot position, with a size of 2672.1 kN, both of which are within the allowable bearing capacity of the lining section.

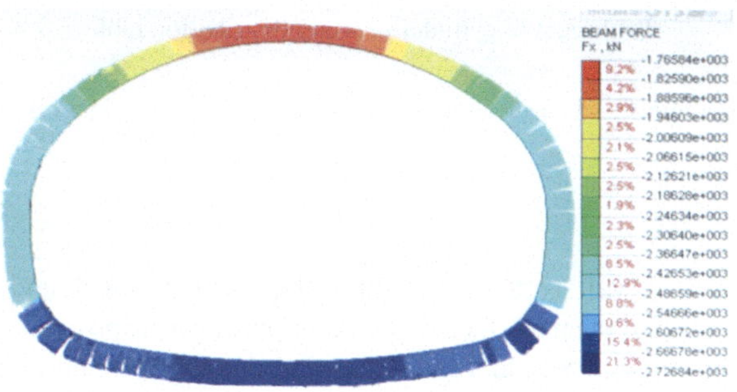

Fig. 3. Internal force diagram of lining structure cross-section

3 Construction Control Measures

According to theoretical calculation results, the impact of the crossing construction of Line 18 on the existing Line 9 is controllable, and its axial force, bending moment, and displacement values are all within the allowable range. However, considering the 15 cm limit distance between the TBM cutterhead and the initial anchor rod of Line 9 during tunnel construction, certain control measures need to be taken for the theoretical displacement and deformation of Line 9 during crossing construction, and automatic monitoring of the tunnel structure of Line 9 is required to further reduce crossing risks.

3.1 Excavation Posture Control

Due to the minimum distance of 15 cm from the outer edge of the TBM cutterhead to the initial anchor rod of Line 9, strict control of the TBM posture is required during construction. Considering various factors such as guidance measurement system errors and Line 9 construction errors, the TBM head posture needs to be pre lifted. During construction, the TBM posture should be strictly controlled within the range shown in the Table 2 below, while minimizing unnecessary correction.

Table 2. TBM attitude control table for crossing sections

Vertical deviation of shield head (mm)	Vertical deviation of shield tail (mm)
25–40	20–30

3.2 Excavation Parameter Control

(1) **Thrust control**

According to the geological conditions of the crossing section, the total thrust should be controlled between 10000 kN and 12000 kN. At the same time, considering the upper

Table 3. Oil cylinder pressure distribution table

Upper oil pressure (MPa)	Lower oil pressure (MPa)	Left and right oil pressure (MPa)
9–11	14–17	9–15

soft and lower hard, as well as the longitudinal slope of the tunnel, and in combination with the pressure ratio of the palm face oil cylinder partition proposed by He Xiangfan in the "Optimization Study of Shield Tunnel Crossing Upper Soft and Lower Hard Strata Palm Face Top Thrust", the pressure distribution during the crossing process should be as shown in the Table 3 below.

(2) **Speed control**

For the construction of upper soft and lower hard TBM, the formula for calculating the rotational speed of a single formation is no longer applicable and needs to be modified using the distribution coefficient of the formation. The specific data is shown in Table 4.

$$n_u = \sqrt{\frac{a * D}{1.2q_u^v}} \tag{1}$$

where n_u is the rotational speed of the cutterhead, q_u is the strength of the composite formation, β is the safety factor, α is the torque coefficient of the cutterhead, D is the diameter of the shield cutter head, v is the speed of shield tunneling.

The calculation of composite formation strength needs to be based on the formula:

$$q_u = Kq_{ui} + (1 - K)q_{uj} \tag{2}$$

where q_u is the speed of the cutterhead, K is the distribution coefficient of the strata, q_{ui} and q_{uj} are the strengths of soft and hard formations. The coefficient K depends on the ratio of the hard rock area to the soft rock area of the palm face, and is taken as 1 here.

Table 4. Calculation Parameter Table

β	α	D	q_{ui}	q_{uj}	v
0.9	15	6.88	12	30.1	2.1

After being substituted into the calculation, the rotational speed n_u = 2.3r/min was calculated. Due to the consideration of the excessive speed of the cutting edge, which caused sudden force changes and damage to the tool at the soft hard interface position, the rotational speed was selected between 1.8 r–2.0 r/min.

(3) **Propulsion speed**

Keep the propulsion speed constant and control it between 30 mm and 35 mm/min.

4 Analysis of Monitoring and Measurement Results

TBM completed the crossing construction directly above the 55–67 ring road from March 2, 2023 to March 4, 2023. During the crossing period, uninterrupted monitoring and measurement were carried out on the tunnel of Line 9. The monitoring data after the crossing was completed are shown in the Fig. 4 below.

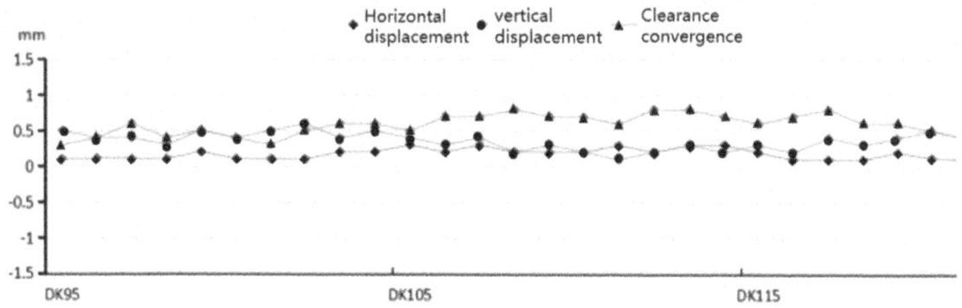

Fig. 4. Monitoring and Measurement Results of Line 9

5 Conclusion

To minimize construction risks, a detailed theoretical analysis was conducted before the construction of Chongqing Line 18 and Line 9, and multiple targeted measures were taken during the crossing process. And the following conclusions are drawn:

During the construction process, the maximum bending moment appears at the arch waist on both sides of the tunnel, and the maximum axial force appears at the arch foot, both of which are within the allowable bearing capacity of the lining section

In order to ensure the normal operation of Line 9 during the crossing period, the horizontal and vertical displacement, headroom convergence and other items of the tunnel structure of Line 9 were monitored, and automated monitoring and manual checking were adopted to meet the normal operation requirements.

After the completion of the crossing, the displacement and deformation of the tunnel structure of Line 9 were all less than the theoretical control values, providing some construction experience for similar working conditions.

During the construction process, always observe the excavation situation of the belt conveyor, and immediately report any non undisturbed soil on the belt conveyor. Report immediately when TBM excavation parameters fluctuate abnormally and torque and thrust suddenly increase. And timely synchronize grouting and secondary grouting to ensure that the gaps between the shield tail buildings are filled tightly.

References

1. Sun, H.: Research on the rock breaking mechanism of shield tunneling cutter at the interface of upper soft and lower hard strata. China University of Mining and Technology; China University of Mining and Technology (Jiangsu) (2019

2. Yihao, L., Xuchun, W., Minhe, L., et al.: Construction control technology for shield tunnels in upper soft and lower hard strata. Low Temp. Build. Technol. **46**(01), 114–118 (2024). https://doi.org/10.13905/j.cnki.dwjz.2024.1.026
3. He, X., Gao, F., Shen, X., et al.: Optimization of top thrust of shield tunnels passing through soft and hard strata. J. Underg. Space Eng. (2018)
4. Hongpeng, L., Tengteng, W., Linjie, Z., et al.: Research on construction control of shield tunnels undercrossing existing subway tunnels considering friction between shield shell and soil. Tunn. Constr. (Chinese and English) **41**(05), 729–739 (2021)
5. Qiu, G., Zhou, T., Wang, F.: Discussion on the construction of subway tunnel sections crossing existing line structures. Build. Technol. (009), 053 (2022)

Exploring Intelligent Construction Practices: A Case Study of Feilong Bridge

Kunqiu Zhang[1], Tuo Shi[1,2(✉)], Jiahai Chen[1], Zhigang Li[1], and Jiaqing Geng[1]

[1] Guangxi Road Construction Engineering Co., Ltd., Nanning, China
1179848174@qq.com
[2] Guangxi University, Nanning, China

Abstract. There are no web crack in the continuous rigid frame bridge with corrugated steel webs and the main beam deflection was small. Because of these advantages, this bridge type was widely used. Characterized by a continuous rigid frame bridge with corrugated steel web, Feilong Bridge's main span was 560 (100 + 180 + 180 + 100) m. The bridge featured double-legged thin-walled piers and single box composite box beams. Addressing challenges in main bean cantilever construction technology in precision and complexity, an intelligent construction method was proposed. Two core technologies were applied in this approach: an intelligent transportation system by the Beidou system and a lightweight intelligent bridge deck crane. This method had been demonstrated a remarkable construction speed of seven days per section for the main beam, with installation accuracy of the corrugated steel web cantilever reaching the millimeter level.

Keywords: Continuous rigid frame bridge · Waveform steel web plate · Intelligent construction

1 Introduction

The rapid expansion of bridge infrastructures in China had spurred the proliferation of rigid frame bridges, enhancing the bending resistance of the main beam and the crossing capacity of the bridges by connecting bridge span elements with piers and abutments [1]. Notably, rigid frame bridges spanning from 100 to 300 m have witnessed significant development in recent years in China. However, with increasing span lengths, there were some challenges such as the web plate cracking of the main beams and deflection issues in conventional concrete rigid frame bridges. In response, a solution had been proposed, which was to adopt corrugated steel web technology instead of using a concrete web, thereby forming a corrugated steel web rigid frame bridge [2]. This innovative bridge design fully capitalized on the material and structural advantages of steel-concrete composites, resulting in a more balanced and rational distribution of stress across the bridge structure [3]. Following the inception of the first corrugated steel web composite bridge in France, similar designs had flourished in Japan and China [4, 5].

B. Yuan et al. (Eds.): ICCASE 2024, SUCI, pp. 187–193, 2024.
https://doi.org/10.1007/978-3-031-76102-7_23

Cantilever construction technology was commonly employed for the main beams of corrugated steel web rigid frame bridges, with two primary construction methods, which refer to the whole box section method and the different part method, respectively [6]. Compared to the former, the latter leveraged the vertical bearing capacity of the installed corrugated steel web plates to increase the working face and facilitated open operation, thereby enhancing construction efficiency and accelerating construction speed [7, 8]. Also, traditional installation methods for corrugated steel web panels often involve cumbersome procedures, such as lifting the individual components to the design position through large arm tower cranes for connection and installation. Alternatively, components may be lifted to the top of the pier using tower cranes before transportation to the beam top for installation via a bridge deck crane [9–12]. However, there were some challenges in construction by the methods, such as poor construction accuracy, slow construction speeds, and weak intelligence [13–16].

Based on the different part method, the intelligent ship and intelligent crane were developed and manufactured in the Feilong bridge. These two kinds equipment were cooperated with each other to form a complete set of key construction technologies. They could realize the deep intelligence and automation of the main beam installation. The complete set of construction technology had been successfully applied in Feilong bridge. The construction speed of each section for the main beam was seven days, and the installation accuracy reached millimeter level, which could provide reference for similar bridge construction. Therefore, intelligent construction exploration and practice were great significance for the Feilong Bridge project.

2 Materials and Methods

The main spans of Feilong Bridge was $560(100 + 2 \times 185 + 100)$ m. It was constructed as a continuous rigid frame bridge with corrugated steel web. The main pier was constructed using C50 double limb thin-walled bridge piers, which supported by cap pile foundations. The type of main beam was a single box. The concrete grade of the top plates bottom plates and the inner lining web plate were C60, with the top plate ranging in thickness from 0.3 to 0.84 m and the bottom plate from 0.3 to 1.5 m. The web plate utilized 1800 type Q420qDNH weather-resistant steel waveform steel plate, with the thickness of the steel plate increasing gradually from the mid-span to the root, ranging from 14 to 28 mm.

A cantilever pouring construction scheme was adopted for the main beam construction of Feilong Bridge, whose box girders were divided into 19 sections. The each length of Sections 1 # to 13 # was 3.6 m, while each length of Sections 14 # to 19 # was 5.4 m. During the construction of each T-structure, the cantilever pouring was conducted symmetrically. To maintain structural integrity, the actual unbalanced deviation of the load on the cantilever at both ends should not exceed 1/4 of the weight of the beam segment. Strict control measures were implemented to ensure the T-structure remains in a balanced stress state. The construction was proceeded step by step from Section 1 # to Section 19 # until completion.

The bearing capacity of the corrugated steel web plate was fully optimized in the cantilever construction for the main beam of Feilong Bridge. The corrugated steel web plates were served as the load-bearing beam for hanging baskets during construction.

The different parts method was used in cantilever pouring, which allowed simultaneous construction of the top plate of section N-1, the bottom plate of section N, and the corrugated steel web plate of section N + 1. The schematic diagram of this construction using the different parts methods is depicted in Fig. 1.

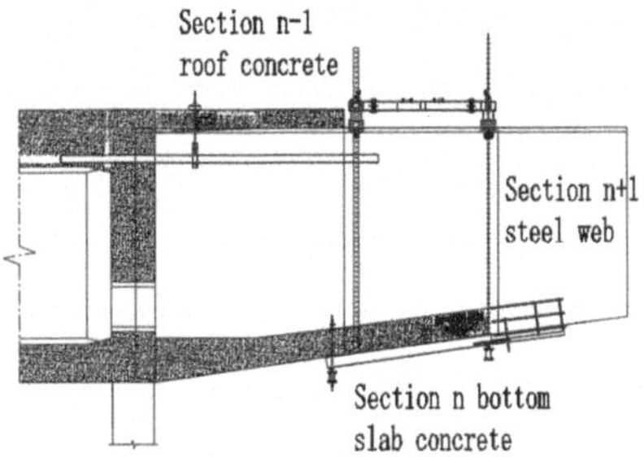

Fig. 1. Schematic diagram of construction using different parts methods

3 Results and Discussion

Considering the unique construction challenges posed by the location of Feilong Bridge above the Xijin Reservoir, an innovative waveform steel web intelligent transportation and hoisting system, based on Beidou technology, had been meticulously developed to streamline the construction of the main beam in the bridge. The schematic diagram of the intelligent construction is presented in Fig. 2, while the intelligent construction process diagram is depicted in Fig. 3.

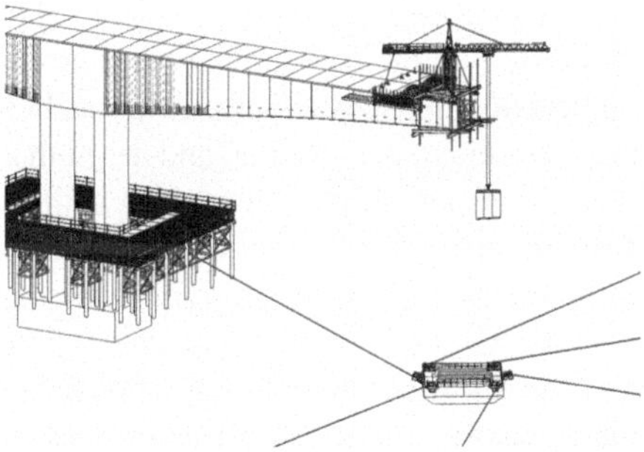

Fig. 2. Schematic diagram of intelligent construction in Feilong Bridge

3.1 Intelligent Rapid Transportation System for Corrugated Steel Belly Plates Based on Beidou Technologies

In response to on-site conditions, corrugated steel web units were transported by intelligent floating ship. This approach involved the installation of underwater wire rope

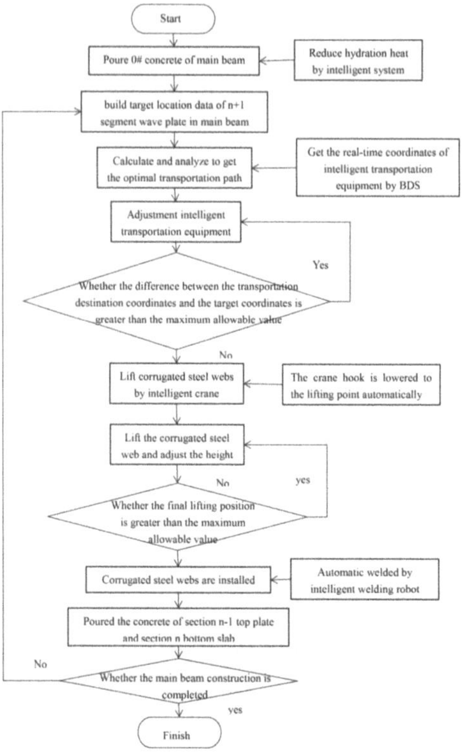

Fig. 3. Intelligent Construction Flow Chart of Feilong Bridge

anchors (two on each side) positioned strategically on both the upstream and downstream riverbeds, along with wire rope anchor points (in a total of six directions) on the steel platforms flanking both sides of the navigation hole. The coordination of six small winches were installed on the floating ship, connected to each anchor (point) via steel wire ropes enabled precise movement, positioning, and azimuth angle adjustment of the floating ship. This intelligent transport ship configuration is illustrated in Fig. 4.

The development of intelligent control systems and terminals, encompassing both software and hardware components, for floating ships in the pier and navigation sections, had facilitated intelligent control of collaborative movement and precise positioning of floating ships on the river surface. Furthermore, this system enabled the intelligent lifting of single components for waveform steel web plates on each side of the main beam section, initiating with a single-click click command.

3.2 Limit Location and Weight Intelligent Rotary Bridge Deck Crane

The crane and corrugated steel webs were designed to be integrated together, and a sophisticated limit location and weight intelligent rotary bridge deck crane was produced. Considering the bearing capacity of corrugated steel web plates, the crane was optimized by connecting base and steel web the plate to counter balance the crane's

Fig. 4. Intelligent transport ship

load. Additionally, the hoist was served as a counterweight for small tonnage lifting, eliminating the need for counterweight blocks. This design facilitated full rotation of the lifting arm during the lifting of smaller components, thereby enhancing the mechanization of the main box section construction. To ensure precise and controlled lifting operations, the lifting arm was set as the horizontal limit (fixed angle) while the lifting point served as the vertical limit (fixed force arm) during the lifting of corrugated steel web units. Anchoring the rear end of the lifting arm in the tight rear box section facilitated crane balance and significantly reduced crane weight. Further stability and lighter tower columns were achieved through the installation of a steel wire rope between the crane base and the top of the tower column, serving as a stabilizing link. The configuration of the intelligent crane is illustrated in Fig. 5.

Fig. 5. Intelligent crane

An intelligent control host terminal, accompanied by related systems, with both software and hardware components, had been developed for cranes, enabling intelligent control of dual limit and amplitude. This system featured automatic weight sensing capabilities, allowing for automatic full rotation lifting when lifting materials weighing less than two tons. In the event of lifting weights exceeding the rated lifting capacity (the maximum weight of the corrugated steel web unit), an alarm mechanism was triggered to halt lifting operations automatically. During the lifting of corrugated steel web units, precise control was maintained over the left and right swing of the lifting arm and the lifting point, ensuring alignment with the intended installation section's center plane. Additionally, the rear anchor was locked in advance to enhance operational stability.

Integrating the double limit and weight limit intelligent rotary bridge deck crane with the intelligent transportation system ensured seamless coordination in the construction process. Upon the successful transportation of the corrugated steel belly plated to the design position was transported by the transport ship, the crane automatically lowers the lifting hook to connect with the lifting point Subsequently, the crane lifted the corrugated steel belly plate unit to the designated height for installation. Figure 6 shows the construction site of Feilong Bridge.

Fig. 6. Construction Site of Feilong Bridge

4 Conclusion

Given the challenging construction environment, the construction of Feilong Bridge entailed a costly and complex process, involving tower transportation, beam top transportation, and the utilization of bridge deck cranes for the main beam construction. To enhance the precision of main beam construction and achieve a high level of automation, Feilong Bridge had pioneered the development of an intelligent transportation and hoisted system specifically tailored for corrugated steel web plates by Beidou technology. This system comprised two pivotal technologies: an intelligent rapid transportation system for corrugated steel belly plates based on Beidou technologies, and an intelligent rotating bridge deck crane featuring limit position weight constraints. Together, these technologies facilitated intelligent transportation and intelligent lifting operations for corrugated steel belly plates, thereby significantly improving construction efficiency and accuracy.

Notably, the construction speed of the main beam in Feilong Bridge had accelerated to seven days per section, while achieving an impressive installation accuracy of five millimeters for the corrugated steel web plate. These results were leading in similar projects. The intelligent construction methodology developed in this project could be served as a valuable reference for similar bridge construction projects.

Acknowledgments. This research was supported by the Key Technology Projects of the Chinese Transportation Industry (2022-ZD7-127) and the Chinese Postdoctoral Science Foundation (2023MD744182).

References

1. Jiang, R.J., Au, F.T.K., Xiao, Y.F.: Prestressed concrete girder bridges with corrugated steel webs: review. J. Struct. Eng. **141**(2), 04014108 (2015). https://doi.org/10.1061/(ASCE)ST.1943-541X.0001040
2. Li, H.J.: Review on special issues in prestressed concrete box girders with corrugated steel webs. J. Basic Sci. Eng. **26**(2), 440–454 (2018). in Chinese
3. Li, F.R., Yuan, B.: Design, research and application of prefabricated waveform steel web beam bridges. Bridge Constr. (002), 052 (2022). in Chinese
4. Chen, K.L., Liu, H.Y.: Key technologies for design and construction of waveform steel belled plate bridges in Japan. World Bridge **50**(4), 8–17 (2022). in Chinese
5. Nie, J.G., Zhu, Y.J., Tao, M.X., et al.: Optimized prestressed continuous composite girder bridges with corrugated steel webs. J. Bridg. Eng. **22**(2), 04016121 (2017). https://doi.org/10.1061/(asce)be.1943-5592.0000995
6. He, J., Li, X., Li, C., et al.: A novel asynchronous-pouring-construction technology for prestressed concrete box girder bridges with corrugated steel webs. Structures **27**, 1940–1950 (2020). https://doi.org/10.1016/j.istruc.2020.07.077
7. Xia, S.J., Niu, C.T., Zhang, L.F., et al.: Comparative analysis and research on asynchronous construction and traditional cantilever construction process of PC corrugated steel web composite box girder bridge. Ind. Archit. **53**, 371–374 (2023). in Chinese
8. Liu, F., Ai, J., Zhang, J.D., et al.: Application communication of multi working face construction Schemein box girder bridge with corrugated steel webs. Constr. Technol. **40**(23), 46–49 (2011). in Chinese
9. Liu, X.H., Wang, C.C., Zhang, W., et al.: Key technologies for the design of the main bridge of the Liangdu Xiangjiang bridge in Guizhou. Bridge Constr. **53**(S01), 9–15 (2023). in Chinese
10. Ren, H.C., Chen, L.P., Luo, Q.D., et al.: Analysis of cantilever construction of the main beam of Gexi bridge in Wentai expressway. Bridge Constr. (005), 051 (2021). in Chinese
11. Deng, W.Q., Liu, D., Feng, J., et al.: Experimental study on asynchronous construction segmental full scale model of box girders with corrugated steel webs. Bridge Constr. **49**(1), 53–58 (2019). in Chinese
12. Bao, Y.J., Yi, J.P, Xian, N.: Research on construction control of waveform steel web composite box beam bridges. Highway **68**(1), 7 (2023). in Chinese
13. Teng, B.Y., Wang, X.M., Lu, Z.Y., et al.: Intelligent construction control for the erection process of continuous rigid frame bridges with corrugated steel webs. J. Build. Sci. Eng. (004), 039 (2022). in Chinese
14. Wang, X., Miao, C., Wang, X.: Prediction analysis of deflection in the construction of composite box-girder bridge with corrugated steel webs based on MEC-BP neural networks. Structures **32**(3), 691–700 (2021). https://doi.org/10.1016/j.istruc.2021.03.011
15. Xu, Q., Wan, S.: Design and application of PC composite box girder bridges with corrugated steel webs. People's Communications Press, Beijing (2009). in Chinese
16. Zhou, X.H., Liu, J.P., Feng, L., et al.: Preliminary Study on Intelligent Building Technology and Its Application. China Architecture & Building Press, Beijing (2021). in Chinese

Bridge Pile Foundation and Ground Improvement

Experimental Study on Mechanical Properties of Segmental Precast Assembled Concrete T-section Beams

Jingchuan Xun[1], Jianqing Bu[2(✉)], Lingpei Meng[3], Zizhuo Zhang[3], and Zhibo Guo[2]

[1] China Construction Road & Bridge Group Corporation, Shijiazhuang 050001, Hebei Province, China

[2] School of Transportation, Shijiazhuang University of Railways, Shijiazhuang 050043, Hebei Province, China
bujq2004@163.com

[3] School of Civil Engineering, Shijiazhuang University of Railways, Shijiazhuang 050043, Hebei Province, China

Abstract. The division of sections makes multiple joints in the beam, and also prevents the longitudinal reinforcement from being continuously arranged, resulting in a discontinuous local mechanical behavior of the beam at the joint location. The number of key-joints at the joints and the position of key-joints are used as variable parameters for the study. Four test beams were designed, including one single, one double, and one combined key-joint T-section beam and one integral beam. The mechanical properties of each type of test beam were judged according to the load-deflection curves in the span of the static load test, the damage characteristics of the test beams, the law of concrete stress changes along the beam height, and the stress changes of prestressing strands during the test. The results show that increasing the number of key-joints (from single to double) can improve the load-carrying capacity of the test beam within a certain range. Under the same conditions, the deflection of the integral beam is the smallest, followed by double, combined, and single key-joint segmental T-section beam; the strain of concrete at both sides of the B-joint is the largest for single key-joint segmental T-section beam, followed by combined key-joint and double key-joint segmental T-section beam; and the strain at the span center position is the largest for integral T-section beam, followed by single key-joint, combined key-joint, and double key-joint segmental T-section beam.

Keywords: Highway bridges · key-joint · experimental studies · mechanical properties · epoxy-joints · segmental precast assembled concrete T-section beams

1 Introduction

Segmentation results in joints within the bridge span, and longitudinal reinforcement cannot be continuous, resulting in the discontinuous mechanical behavior of the beam. Therefore, it is necessary to study their mechanical properties. In 2005, J Turmo [1]

B. Yuan et al. (Eds.): ICCASE 2024, SUCI, pp. 197–211, 2024.
https://doi.org/10.1007/978-3-031-76102-7_24

et al. proposed a finite element analysis method (FEM) for segmental beams. In 2012, S Saibabu [2] found that the corners and deflections at individual joints were highly concentrated. Yuan Aimin [3, 4] found that the direct shear bearing capacity of shear keys is directly proportional to the area of the failure surface, and is independent of key-joint depth and key-joint pitch. Propose a formula for calculating the ultimate bearing capacity of segmental beams [6]. In 2016, Cao Zenghua [5] proposed a formula for calculating crack resistance. In 2018, Jiang Haibo [7] pointed out that as the shear span ratio increases, the shear bearing capacity of segmental beams decreases. In the same year, Lu Wenliang [8] discovered that an excessively thick adhesive layer would reduce the stiffness of the beam. In 2019, Yan Zeyu [9] found that the shear-bearing capacity of UHPC single bond epoxy-joints was much higher than the shear-bearing capacity of flat epoxy-joints of the same cross-sectional dimensions through tests; the positive stresses increased and the shear bearing capacity were all improved. The simulation data obtained by Yulin Zhang [10] is in good agreement with the experimental results.

At present, the research on the mechanical properties of prefabricated segmental beams mainly focuses on the direct shear performance of test block joints and the analysis of the mechanical properties of concrete beams under single load effects. The research on T-section beam is also a simple 3-segment beam, or even a 2-segment beam, with pre-stressed steel bars generally arranged externally. Based on this, the mechanical performance of segmental precast and assembled concrete T-section beam bridges is analyzed. We plan to use three 5-segment horseshoe-shaped T-section beams and one integral T-section beam for research and summarize the laws of deflection, strain, and ultimate bearing capacity at the selected position of the beam under four-point loading.

2 Experimental Protocol Design

The effects of design parameters such as the number (single key-joint, double key-joint) and position of key-joints on the mechanical properties of segmental precast assembled concrete T-section beams were investigated through static load tests on four test beams. The section was selected as a T-section beam with a horseshoe shape, with a length of 5 m, divided into 5 sections of 1 m each, a beam height of 0.555 m, a flange width of 0.5 m, and a beam rib width of 0.18 m. The concrete strength grade was C60, and the internal reinforcement was HRB400, Φs15.24 mm high-strength low-relaxation steel strand. Key-joints to be used single, double two, the angle of 45°. The detailed information is shown in Table 1.

Table 1. Test beam information

Serial number	Test beam number	Type of key-joint	Design Parameters
1	T1	/	/
2	T2	Single key-joint	Number of key-joints
3	T3	Double key-joints	Number of key-joints
4	T4	Combined key-joint	key-joint position

The specific section size and prestressing strand placement are shown in Fig. 1, 2, 3 and 4.

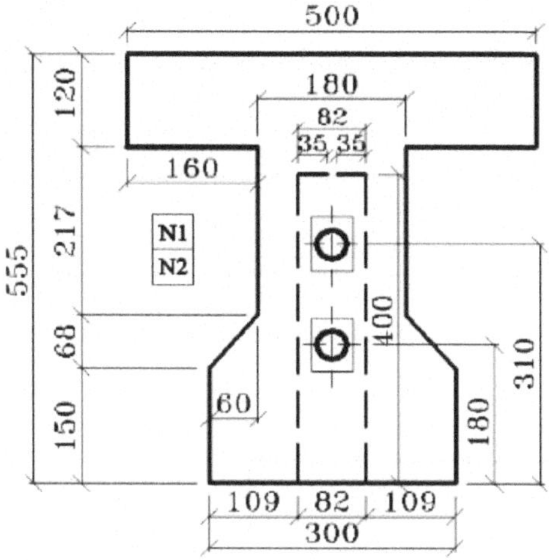

Fig. 1. End cross section (unit: mm)

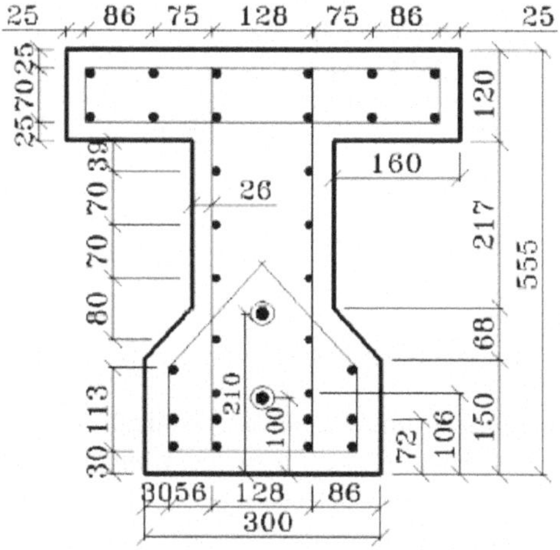

Fig. 2. Cross-section in the span (unit: mm)

The detailed dimensions of the key-joints of the segmental beam are shown in Fig. 5 and Fig. 6.

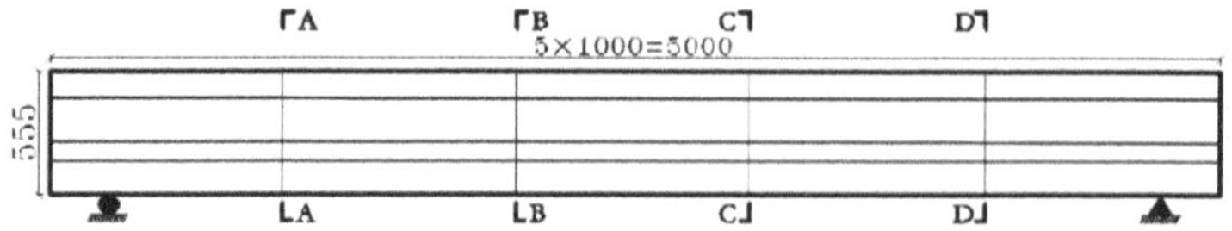

Fig. 3. Elevation of segmental precast assembled concrete T-section beam (Unit: mm)

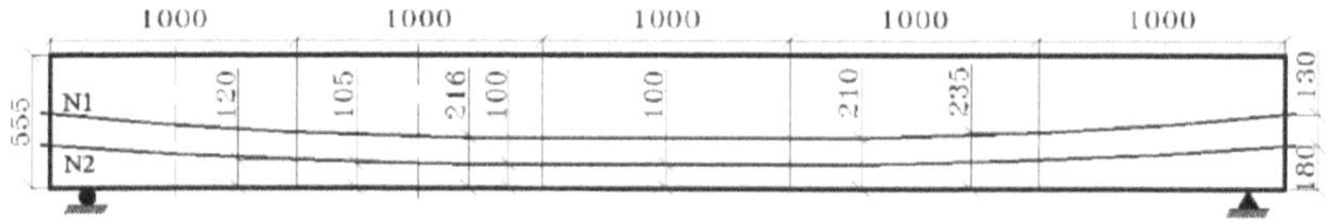

Fig. 4. Steel strand laying (unit: mm)

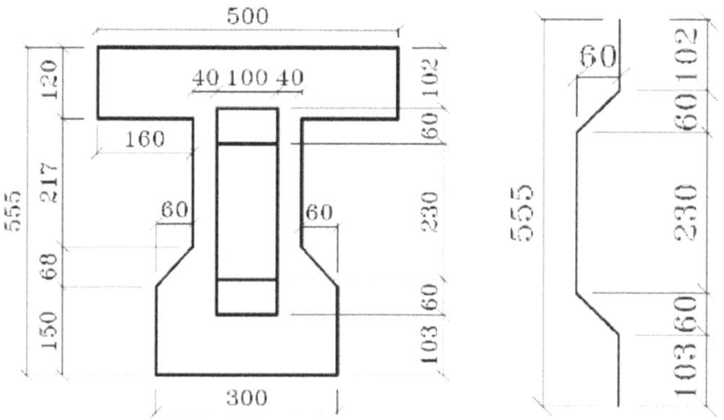

(a) Front (b) Side

Fig. 5. Single key-joint size (unit: mm)

3 Loading

Two-point loading was used to analyze the mechanical properties of the segmental precast assembled concrete T-section beams by recording and measuring the joint and mid-span deflection values, concrete strains on both sides of the joint and in the span, ultimate load capacity, and the location, development, and distribution of cracks under all levels of loading of the test beams. The loading cases are shown in Fig. 7 and Fig. 8.

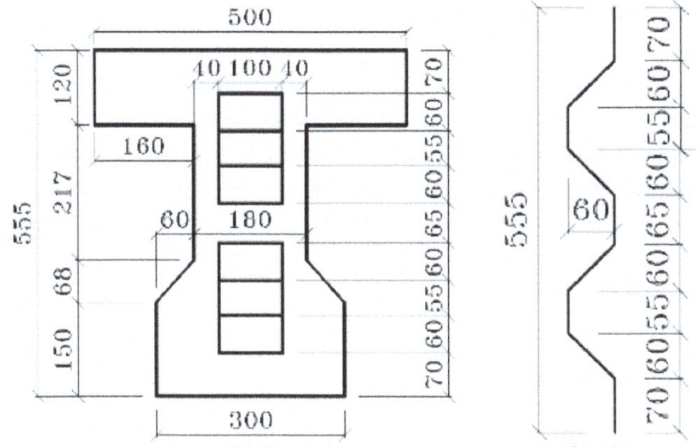

Fig. 6. Double key-joint size (unit: mm)

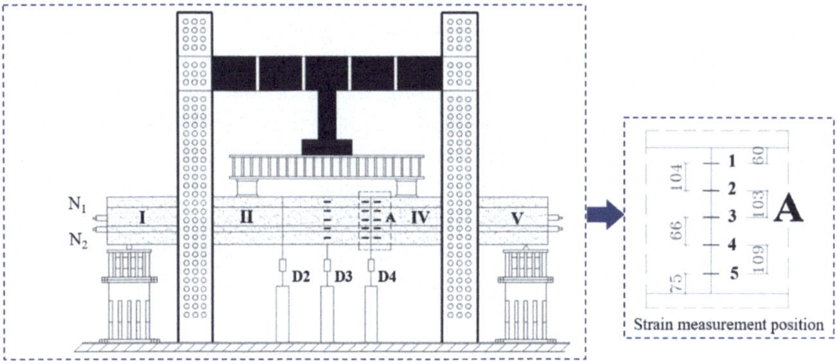

Fig. 7. Schematic diagram of test beam loading

Fig. 8. Actual test loading

4 Analysis of Test Results

To analyze the mechanical properties of segmental precast concrete T-section beams through static load test, the test phenomena should be observed and data recorded in time for the purpose of the study, including the ultimate load carrying capacity of the four test beams; damage pattern during the stressing process, i.e., location, characteristics and development of the first crack and other cracks; deflection at the mid-span position; strain variation law of concrete at the corresponding measurement points on both sides of the joints and in the mid-span, etc.

4.1 Damage Characteristics of the Test Beam

When TM1 was loaded, the first crack appeared when the external load increased to 27.4 t. The cracks gradually extended upward from small cracks; as the load increased, cracks appeared outside the two small distribution beams, and the distribution became denser and denser, and in general, it showed the distribution characteristics of "fine and dense". When the loading reached 72 t, TM1 reached yield and made a strange sound, which was very slight at first, and then the sound gradually increased. Considering the force process, it was determined that the internal longitudinal reinforcement of TM1 concrete lost its role and fractured. At the end of loading, the block concrete fell off and the load had reached 84 t. The stress concentration in the concrete of the small distribution beam near the side of the hinge support end was too large and small cracks appeared, and at this time the main crack in the span position developed to the wing edge position and continued to develop. When the strain at the edge of the concrete in the compressed zone reaches the ultimate compressive strain, a longitudinal horizontal crack appears in the compressed zone, and as the concrete in the compressed zone crushes, the displacement increases and the load no longer changes, declaring that the test beam TM1 has been damaged and the loading is completed. At this time, the anchor cable at the tension end of the prestressing reinforcement N2 position was subjected to 20.8 t of squeezing force, and the ultimate bearing capacity of the test beam was 87 t. The crack location, development condition, and damage form are shown in Fig. 9.

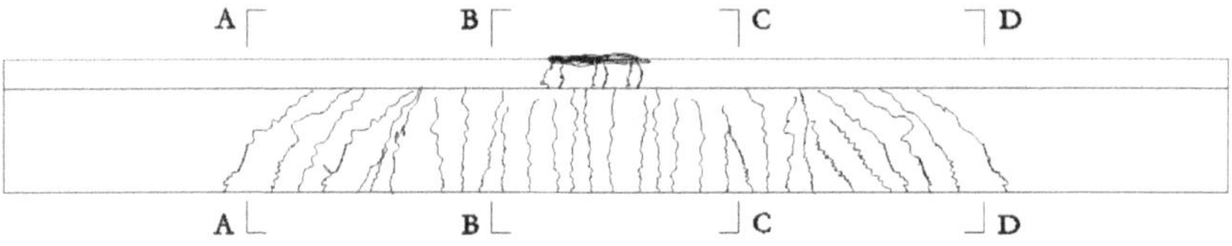

Fig. 9. Integral beam

TM2 When the joints are not cracked, the damage characteristics basically match with TM1, and the segmental precast assembled concrete T-section beams at this time can be treated as integral beams. When the load reaches 22.4 t, the first crack appears with the direction of meandering vertically upward. The load increases and other cracks appear in the beam, the starting point of the B and C joints is 2.2 cm from the B joint and

2.7 cm from the C joint, from the fine cracks gradually extend upward, at 30 t, the cracks around the B joint section extend to the joint B. At 72 t, the cracks get developed around the position of the flange of the C joint section, the glued structure of the C joint section at the flange is not seriously damaged, but the cracks began to develop around the concrete location of the C joint section, and at 73.9 t, the concrete in the compression zone at the flange location of the B joint section was crushed, announcing the destruction of the test beam TM2 and the end of loading. At this time, the prestressing steel N2 anchor cable was subjected to 23.8 t. The TM2 crack location, development condition, and damage characteristics are shown in Fig. 10.

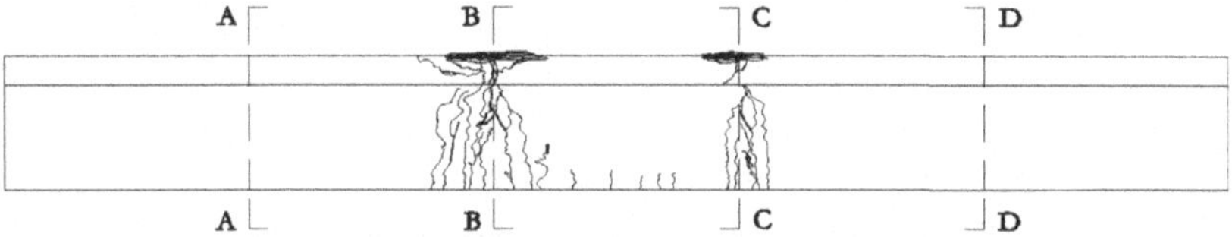

Fig. 10. Single-key-joint segmental beam

During the loading of TM3, the first crack appeared at 25.4 t, at 1.9 cm from the fixed support at the bottom span, and meandered vertically upward; at 34.2 t, the crack extended to joint C. However, with the increase of external load, the crack here extended directly through joint C, without directly damaging the adhesive structure at the joint, and cracked directly along the joint, as did the joint at B. When the external load reaches 74 t, the crack gets developed at the flange of the C joint section, and the rubber layer here starts to appear damaged when the external load reaches 76 t, the crack gets developed at the flange of the B joint section, and the rubber structure starts to appear damage, and the crack starts to develop at the B joint when the external load reaches 80.1 t, the rubber structure at the C joint is damaged, and the crushing of the concrete in the pressurized area announces the TM3 damage. At this time, the prestressing steel N2 anchor cable was subjected to 21.9 t. The location, development, and damage characteristics of the TM3 crack are shown in Fig. 11.

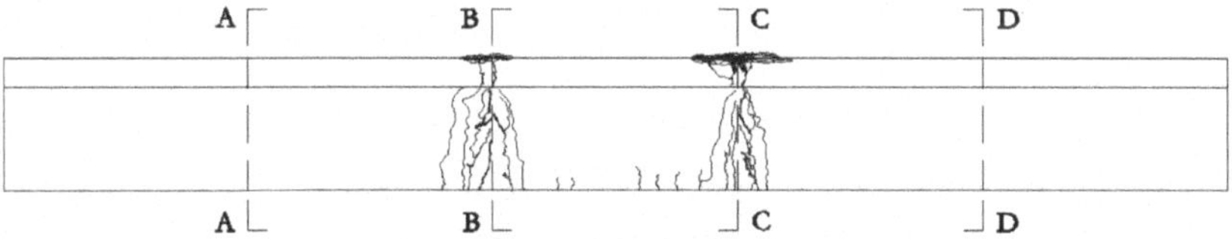

Fig. 11. Double-key-joint segmental beam

The first crack appeared when TM4 was loaded to 23.7 t. It was located at 1.1 cm from the bottom part of the beam in the span at the offset sliding support and meandered vertically upward; the starting point of joints B and C was at 2.1 cm from joint B and 3.2 cm from joint C. The crack gradually extended upward from the fine crack and reached 32.1 t when the crack extended to joint B. The load increased and the crack here extended directly through joint B extends. When the load reached 71 t, the crack was developed at joint B and the rubber structure began to be damaged, and when the load reached 72 t, the crack was developed at joint C and the rubber structure in the web part began to be damaged; when the external load reached 77.6 t, the rubber structure at joint B was damaged and the crushing of concrete in the pressurized area announced the destruction of TM4. At this time, the prestressing steel N2 anchor cable meter was subjected to 22.7 t pressure. The test beam TM4 crack location, development condition, and damage characteristics are shown in Fig. 12.

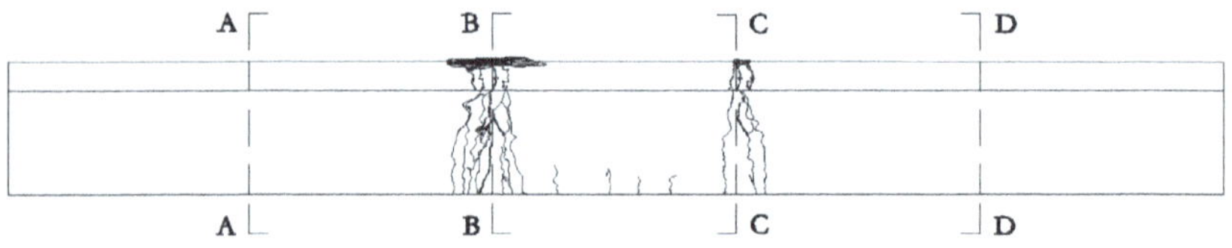

Fig. 12. Combination-key-joint segmental beam

4.2 Load-Deflection Curve

The load-deflection curves of the four T-section beams at the mid-span position are shown in Fig. 13.

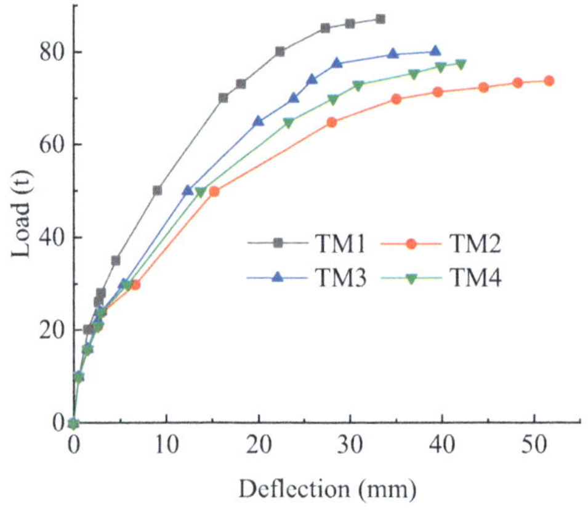

Fig. 13. Load-deflection curves at mid-span position

As can be seen from Fig. 13, the load-deflection curves of the four test beams at the mid-span position initially had similar slopes, and the stiffnesses differed after the concrete cracking, with TM1 being the largest, with a maximum deflection of 33.3 mm at the mid-span position; followed by the stiffness of TM3, with a maximum deflection of 39.3 mm; then TM4, with a maximum deflection of 42.1 mm; and finally TM2, with a maximum deflection of 51.7 mm.

4.3 Load - Prestressing Strand Tension

Because the experimental beam adopts upper and lower pipeline tensioning, compared to the upper side pipeline, the lower side pipeline tensioning range is larger and plays a greater role. Therefore, the lower side steel strand N2 is selected for analysis of the variation law of prestressed steel strands under external load. The detailed information is shown in Table 2.

Table 2. Information on the variation of N_2 values of steel strands with load

Serial number	Load/t	Prestressing steel strand N_2 tension value/t			
		TM1	TM2	TM3	TM4
1	0	8.0	8.1	8.0	8.0
2	10	8.5	8.8	8.7	8.7
3	20	9.5	9.8	9.7	9.6
4	26	11	11.8	11.3	11.5
5	28	11.3	12.7	11.9	12.1
6	35	12.7	14.1	13	13.8
7	50	15.2	18.2	16.3	17.1
8	70	17.2	20.8	18.8	20.3
9	73	17.4	23.6	19.2	21.7
10	80	18.9	–	21.9	–
11	85	20.1	–	–	–
12	86	20.5	–	–	–
13	87	20.8	–	–	–

From Table 2, it can be seen that under the same conditions, the tensile force of N2 has the smallest increase in the integral beam TM1, followed by the precast and assembled concrete T-section beam TM3, TM4, and TM2; At the initial stage of loading, the N2 value increases slowly, and as the load increases, the N2 tensile value of the lower prestressed pipeline steel strand increases at an accelerated rate; In the later stage of loading, the joint fails, and the bearing capacity is borne by the steel strand, quickly reaching the critical value. The faster the joint position fails, the faster the N2 tension value increases.

4.4 Load-Concrete Strain

Combined with the actual situation, with the development of cracks on the concrete surface during the loading process, some strain gauges were damaged. The data of these strain gauges were sorted out, so 10 t, 20 t, 26 t, 28 t, 35 t, 50 t, etc. were selected to carry out stress analysis along the beam height of the external surface of the concrete. The specific information is shown in Fig. 14.

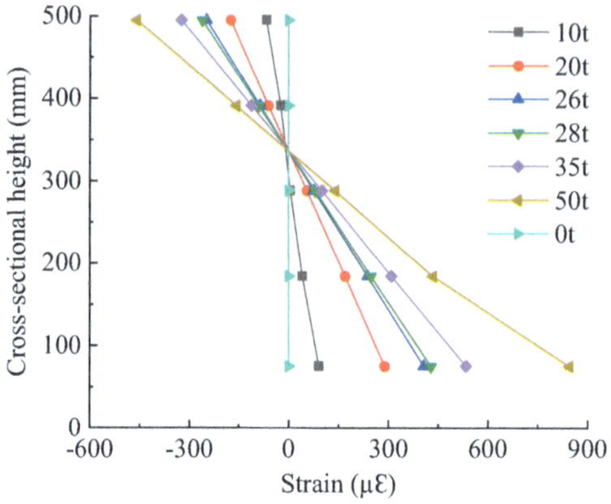

Fig. 14. TM1 span mid-section along the beam height strain

From Fig. 14, it is known that the strain range is increasing as the load increases, and the strain pattern of the concrete outside TM1 is consistent with the assumption of a flat section. The strain variation pattern of TM1 at 2 m is shown in Fig. 15.

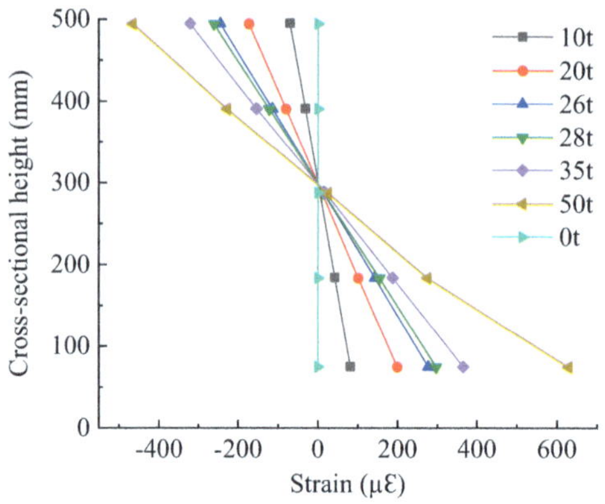

Fig. 15. High strain along the beam at TM1 2m

From Fig. 15, it can be seen that the strain variation pattern of TM1 at 2 m is roughly the same as that at the mid-span position, but the strain variation range at 2 m is relatively smaller compared to the mid-span position. This is because, in the integral T-section beam, the mid-span position has the largest deflection change and strain change throughout the entire loading process. The strain of TM2 concrete at the B joint is shown in Fig. 16.

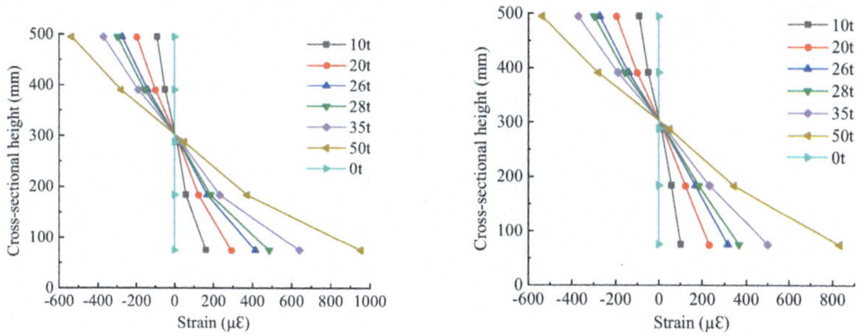

(a) B joint near the side of the span

(b) B joint near the side of the end

Fig. 16. TM2 B joint

From Fig. 16, it can be seen that the strain of the concrete on both sides of the B joint section in TM2 is greater on the side near the mid-span compared to the side near the end, resulting in an increase in load and a smaller rate of increase on the side near the end compared to the side near the mid-span. The concrete strain of TM2 at the mid-span section is shown in Fig. 17.

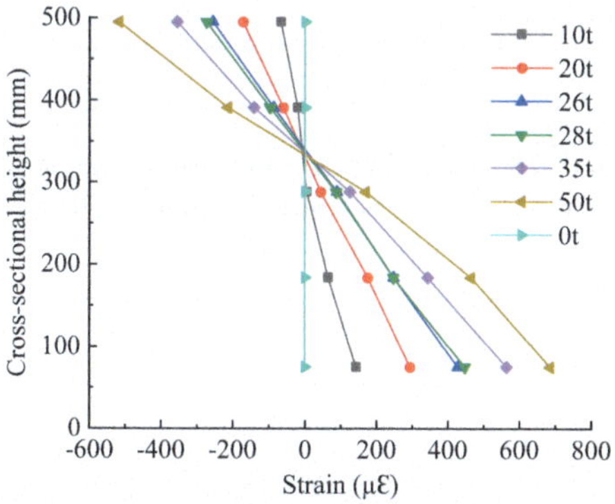

Fig. 17. Spanning cross-section

The strain variation pattern of TM3 at the B joint and mid-span section is shown in Fig. 18.

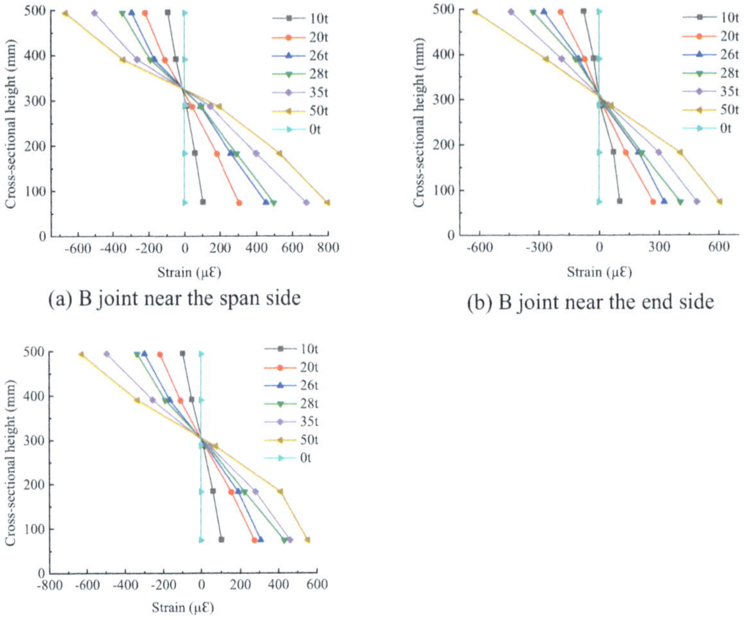

(a) B joint near the span side

(b) B joint near the end side

(c) Span section

Fig. 18. TM3 concrete strain

From Figs. 18 (a) and (b), it can be seen that the concrete strain of TM3 on both sides of the B joint and at the mid-span is similar to that of TM2, but the value is smaller. In the later stage of loading, the neutral axis shifts relatively upwards, and the slope between strain and section height increases. The concrete strain of TM4 at the B and C joint sections is shown in Fig. 19.

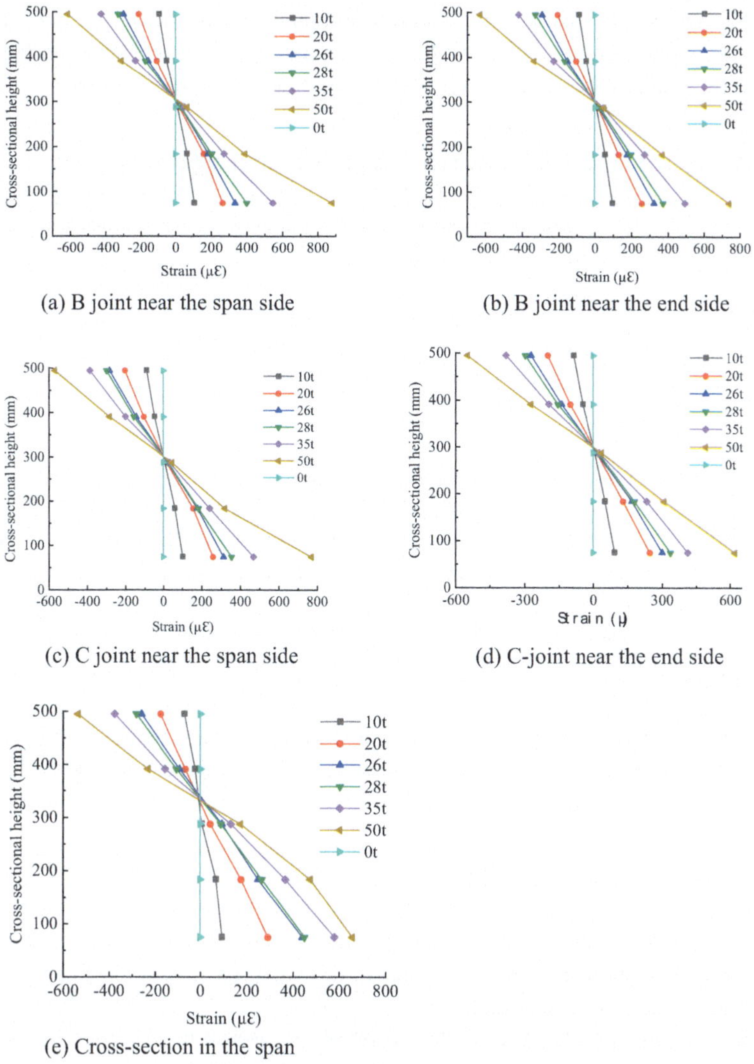

Fig. 19. Test beam TM4 concrete strain

From Fig. 19, it can be seen that during the initial stage of TM4 loading, the strain in concrete at each joint and mid-span section of TM4 is similar to that of TM1 and TM2. As the load increases, cracks appear, and the absolute value of strain on the side near the end of joints B and C is relatively smaller than that on the side near the mid-span of joints B and C.

4.5 Ultimate Load Carrying Capacity

When the static load test is loaded, when there are cracks running through the upper and lower parts of the test beam, resulting in the failure of the concrete in the compression area or the inability to increase the load during displacement loading, it indicates that the test beam has been damaged and has no loading significance. Based on the above situation, the ultimate bearing capacity of the test beam is shown in Table 3.

Table 3. The ultimate load capacity of test beam in the static load test

Serial number	Test beam number	Ultimate bearing capacity/t
1	TM1	87
2	TM2	73.9
3	TM3	80.1
4	TM4	77.6

The maximum ultimate bearing capacity of TM1 is 87 tons, followed by TM3, 80.1 tons, TM4, 77.6 tons, and finally TM2, 73.9 tons. Compared with TM1, TM2 to TM4 have a smaller ultimate bearing capacity, and at the same time, the stress on the double-keyed shear key is more uniform and reasonable than that on the single-keyed shear key.

5 Conclusions

The influence of key-joint mesh, key-joint position, and beam manufacturing process parameters on the mechanical properties of concrete T-section beams is studied through static load tests. The main conclusions obtained are:

(1) The precast and assembled concrete T-shaped beam with double-keyed key-joint segments has a more reasonable stress than the single key-joint beam, and the final failure occurs at the joint position. The cracks around the single key-joint appeared earlier than the double key-joint, and the overall beam ultimately failed at the mid-span position, indicating that the joint is the weak part of the segmental precast assembled concrete T-section beam.
(2) Under the same conditions, the deflection of the overall test beam is the smallest, followed by a double, combined, single key-joint segmental T-section beam. In the T-shaped beam with combined key-joint, the deflection of the single key-joint is greater compared to the double key-joint at the symmetrical position of the span. Increasing the number of key-joint (from single to double) increases the bearing capacity within a certain range.
(3) Under the same conditions, the strain on both sides of the B joint on the concrete surface of the beam body is the highest in the single key-joint segment type T-section beam, followed by the combination key-joint and double key-joint segment type T-section beam. In the segment-type T-section beam, the strain on the side near the mid-span of the B joint is greater than the strain on the side near the end; At the mid-span position, the strain of the integral T-section beam is the highest, followed by single key-joint, combined key-joint, and double key-joint segmented T-section beam.

Acknowledgments. This research was supported by the National Key R&D Program of China (Project No. 2021YFB2600605, 2021YFB2600600), and the Key R&D Program of Hebei Province (Project No. 19275405D).

References

1. Turmo, J., Ramos, G., Aparicio, A.C.: FEM study on the structural behavior of segmental concrete bridges with unbonded prestressing and dry joints: Simply supported bridges. Eng. Struct. **27**(11), 1652–1661 (2005)
2. Saibabu, S., Srinivas, V., Sasmal, S., Lakshmanan, N., Iyer, N.R.: Performance evaluation of dry and epoxy jointed segmental prestressed box girders under monotonic and cyclic loading. Constr. Build. Mater. **38**(C), 931–940 (2013)
3. Yuan, A., He, Y., Cheng, L., Qian, S.: Experimental study on the shear performance of segmental prefabricated bridge adhesive joints. J. Chongqing Jiaotong Univ. (Nat. Sci. Ed.) **33**(06), 22–26+33 (2014)
4. Yuan, A., Wu, W., Sun, D., Dai, H.: Bending performance test of precast box girders with mixed in-body and out-body beam sections. J. Chang'an Univ. (Nat. Sci. Ed.) **35**(05), 73–81+103 (2015).https://doi.org/10.19721/j.cnki.1671-8879.2015.05.011
5. Zeng-Hua, C.: Calculation of crack resistance of prefabricated assembled bridges with glued joint sections. Railway Archit. **12**, 21–23 (2016)
6. Aimin, Y., He, Y., Hang, D., Dasong, S.: Experiments on the force performance of segmental precast box girders under different loading methods and beam distribution ratios. J. Chang'an Univ. (Nat. Sci. Ed.) **36**(01), 58–68 (2016). https://doi.org/10.19721/j.cnki.1671-8879.2016.01.009
7. Jiang, H., Li, Y., Xiao, J., Chen, Y., Fang, Z.: Shear performance test of precast segmental dry jointed extracorporeal prestressed concrete simply supported beams. Chin. J. Highw. **31**(12), 188–195 (2018).https://doi.org/10.19721/j.cnki.1001-7372.2018.12.018
8. Lu, W.-L., Ma, X.-L., Zheng, Q.-Q.: Study on the thickness of epoxy-joints of segmental prefabricated glue joint members. Railway Archit. **59**(08), 39–41 (2019)
9. Zeyu, Y.: Section prefabricated assembled UHPC adhesive joints shear performance test and finite element analysis. Highw. Eng. **44**(06), 228–233 (2019). https://doi.org/10.19782/j.cnki.1674-0610.2019.06.043
10. Zhan, Y.,et al.: Experimental and numerical investigations on shear performance of key tooth joints of the precast concrete segmental bridge under repeated loading. Constr. Build. Mater. **351**, 128794 (2022)

Analysis of the Influence of Supporting Structure Stiffness on Deformation of Deep Foundation Pit

Dan Miao(✉) and Chushao He

Guangzhou Guangjian Construction Engineering Testing Center Co., Ltd., GuangDong Engineering Technology Research Cencer of Building Health Monitoring and Security Early Warning, Guangzhou, China
279766509@qq.com

Abstract. Explore the influence of support structure stiffness on foundation pit deformation using the inner support with an underground diaphragm wall configuration. Finite element analysis (FEM) assesses deformation patterns due to varying support stiffness. Results reveal significant deformation variations within the support structure and its effect on surrounding strata due to differing stiffness. Top displacement inadequately represents actual stress and deformation state. Prioritize deep horizontal displacement monitoring with surface horizontal displacement control. Support stiffness changes primarily impact deformation magnitude within the support structure, without affecting deformation patterns. It provides insights into support structure stiffness's role in governing foundation pit deformation, with implications for project design and monitoring, enhancing our understanding of support systems' interactions with the geological environment.

Keywords: foundation pit support · supporting stiffness · foundation pit monitoring · displacement deformation · FEM

1 Introduction

In the context of escalating urban land space constraints and the rapid expansion of super high-rise buildings and underground rail transit, the proliferation of deep foundation pits has become increasingly prevalent. As the excavation depth of these foundation pits deepens, so does the associated construction risk. In urban cores, the foundation pit's vicinity is marked by complex environmental conditions, and excessive deformation of the pit can jeopardize the proper functioning of surrounding pipelines, roads, and other structures [1]. Effectively enhancing the control of deep foundation pit deformation has consequently emerged as a pressing and imperative challenge to address.

The factors that cause the deformation of the foundation pit mainly include support form, geological conditions, foundation pit dewatering, excavation depth, and excavation method [2–4]. Xu et al. [5] based on Clough system stiffness, proposed a mobilized strength design synthetic system stiffness to explore the relationship between the displacements of deep foundation pit and the stiffness. In engineering practice, Terzaghi

B. Yuan et al. (Eds.): ICCASE 2024, SUCI, pp. 212–226, 2024.
https://doi.org/10.1007/978-3-031-76102-7_25

and Peck et al. [6] proposed the total stress method, which adopts the theory of the overall stability of the soil mass excavated by the magnitude of the support load and the actual excavation. E. Pujades et al. [7] reviews the effects of groundwater during excavation and points out that the prediction of surface subsidence caused by groundwater changes and the problems of consideration of non-Darcy flows in the design of dewatering systems need to be improved. Liu et al. [8] proposed an improved MSD method to modify the deformation mechanism of the foundation pit system. Zhang et al. [9] pointed out that deformation is the main result for the load of support and should be paid more attention to in the tunnel engineering design. Wei et al. [10] proposed a shield segment hydraulic loading system to explore the internal force and the law of the excavation deformation. Ma et al. [11] utilized finite element method to establish a three-dimensional numerical model to simulate the excavation process. Parameter sensitivity analysis was used to study three stiffness parameters. Wang et al. [12] indicated that different stiffness would lead to different deformation modes of foundation pits. Some existing analytical method were used to simulate pit deformation and the horizontal displacement curves of a neighboring single pile are obtained. Cui et al. [13] simulated four different excavation schemes and found that increasing the overall stiffness of the main structure in excavation area can reduce the additional deformation. In light of these developments, the finite element method is employed to analyze and study the patterns of foundation pit deformation induced by support structure stiffness. The research outcomes hold significant implications for foundation pit design, monitoring, and early warning systems.

2 Methodology

In practical engineering, it is impossible to compare and analyze the deformation of the foundation pit under various influencing factors, and it is also impossible to obtain relevant measured comparative data. Parameter sensitivity analysis is an effective method for structure simulation and comparative analysis. To study the deformation of deep foundation pit under different supporting structure stiffnesses, the following analysis method is adopted:

- FEM is used to analyze the deformation of the foundation pit and the accuracy and reliability of the parameters used in the numerical analysis are determined by comparing with the measured results.
- The relevant influencing factors are adjusted respectively to simulate the deformation of the foundation pit, based on the finite element model. Under different conditions, the influence of each factor on the deformation of the foundation pit is compared and analyzed.

3 Practical Engineering Project

3.1 Introduction of Project

The foundation pit area of this project is about 18000 m^2, and the excavation depth is about 17.9 m. The foundation pit adopts the support form of an underground diaphragm wall with three reinforced concrete supports in the pit. The underground diaphragm wall on the side of the main building is 1000 mm, and the effective length is about 37.8 m. The section sizes of the three internal supports are 1000 mm × 700 mm, 1300 mm × 800 mm, and 1200 mm × 800 mm, respectively. The vertical pile adopts a 4L 160 × 16 steel lattice column, and the section is 460 mm × 460 mm.

3.2 Finite Element Model

The Mohr-Coulomb model is used for the constitutive relationship of the soil element, and the linear elastic model is used for the supporting structure. The specific indicators are shown in Table 1 and Table 2. To consider the nonlinear relationship between the relative displacement and stress on the pile-soil contact surface, the contact surface element is used in the elastic-plastic finite element analysis to simulate the friction characteristics between the structure and the soil.

According to the actual size of the foundation pit, the finite element analysis model is established. The plane triangular 15-node element is used to divide the grid. The horizontal constraints are applied on both sides of the boundary, and the fixed constraints are applied on the bottom boundary. Considering the symmetry of the model, only the left half of the graph is drawn, as shown in Fig. 1.

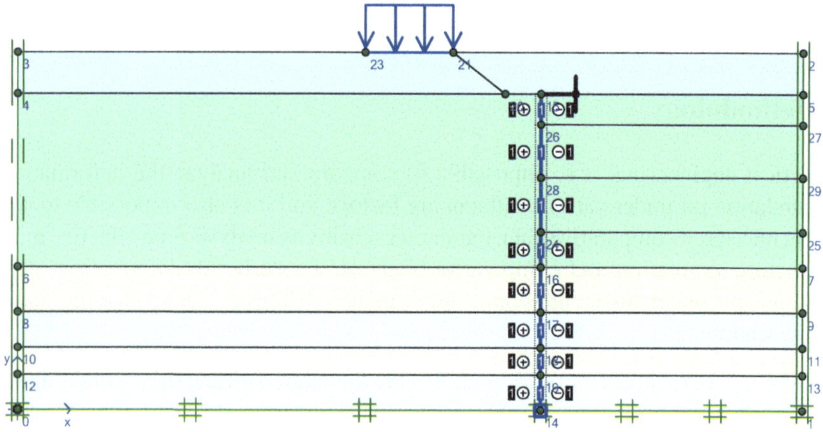

Fig. 1. The diagram of the foundation pit FEM model.

Table 1. Physical and Mechanical Properties of Soil

Soil Layer	Natural Unit Weight	Elastic Modulus E	Poisson Ratio V	Cohesion C	Internal Friction Angle φ	Permeability Coefficient K_x	Contact surface Coefficient R_{inter}
	$kN \cdot m^{-3}$	kPa	–	kPa	°	$m \cdot d^{-1}$	
Mixed backfill soil	18.0	12000	0.21	1.0	22.0	0.100	0.65
Clayey silt	18.3	7500	0.24	4.0	22.0	0.001	0.62
Silty clay	16.7	5000	0.24	13.0	11.0	0.001	0.63
Silty clay-1	18.0	8000	0.27	13.0	15.5	0.001	0.58
Silty clay-2	18.2	8000	0.28	14.0	21.5	0.001	0.64
Silty clay-3	19.8	8000	0.32	42.0	16.0	0.001	0.62
Reinforced soil	20.4	50000	0.18	150	25.0	0.000	0.65

Table 2. Properties of Supporting Structure

Supporting Structure	Axial Stiffness EA	Bending Stiffness EI	Depth D	Unit Weight W	Poisson Ration V	Support Spacing L_S
	$kN \cdot m^{-1}$	$kN \cdot m^2 \cdot m^{-1}$	m	$kN \cdot m^{-1} \cdot m^{-1}$	–	m
Underground diaphragm wall	3.00 × 107	2.5 × 106	1.0	10	0	/
Support-1	2.10 × 107	/	/	/	/	10
Support-2	3.12 × 107	/	/	/	/	10
Support-3	2.88 × 107	/	/	/	/	10

3.3 Model Validation

To verify the model, the calculated value of the deep horizontal displacement at the supporting pile is compared with the measured value, as shown in Fig. 2. It can be seen from Fig. 2 that the measured value is in good agreement with the calculated value. The measured value tends to decrease at the top of the pile, but the calculated value does not decrease significantly. This is because the numerical calculation cannot accurately reflect the installation time of the support, which leads to the late support effect of the support. In the range below 2.5 m, the maximum error between the measured value and the calculated value is about 5.58 mm, and the error rate is about 16%, indicating that the calculation model adopted in this paper can reflect the deformation characteristics of the actual foundation pit excavation.

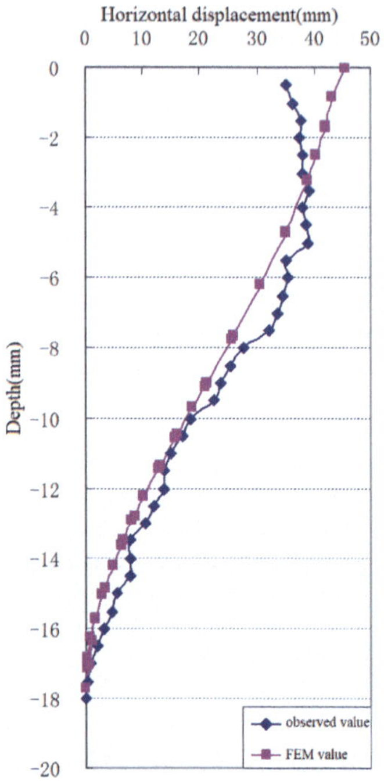

Fig. 2. The variation curve of inclinometer data with depth.

3.4 Parametric Analysis

After the finite element model validation, the stiffness of diaphragm wall and the stiffness of inner support are adjusted respectively to simulate the deformation of the deep foundation pit, and the influence of each factor on the deformation of the foundation pit is compared and analyzed. Under the premise of keeping D of the supporting structure unchanged, the axial stiffness and bending stiffness of the underground diaphragm wall are adjusted accordingly. The detail of the adjustment is shown in Table 3 and Table 4 for the stiffness of diaphragm wall and stiffness of inner support respectively.

Table 3. Stiffness Calculation Comparation of Underground Diaphragm Wall

No.	Axial Stiffness EA	Bending Stiffness EI	Depth D	Unit Weight W	Poisson Ration V
	kN·m^{-1}	kN·m^{2}·m^{-1}	m	kN·m^{-1}·m^{-1}	–
1	3.0 × 107	2.5 × 106	1.000	10	0
2	6.0 × 106	5.0 × 105	1.000	10	0
3	3.0 × 106	2.5 × 105	1.000	10	0
4	1.5 × 108	1.25 × 107	1.000	10	0
5	3.0 × 108	2.5 × 107	1.000	10	0

Table 4. Stiffness Calculation Comparison of Supporting Structure

No.	Axial Stiffness EA			Support Spacing L_S
	kN			m
	Support-1	Support-2	Support-3	–
1	2.10 × 107	3.12 × 107	2.88 × 107	10.0
2	4.20 × 106	6.24 × 106	5.76 × 106	10.0
3	2.10 × 106	3.12 × 106	2.88 × 106	10.0
4	1.05 × 108	1.56 × 108	1.44 × 108	10.0
5	2.10 × 108	3.12 × 108	2.88 × 108	10.0

4 Analysis and Results

4.1 Stiffness of Underground Diaphragm Wall

The calculation results of the underground diaphragm wall under different stiffness conditions are shown in Table 5. Figure 3 shows the relationship between the bending stiffness and the maximum horizontal displacement of the diaphragm wall.

It could be seen that with the increase of the stiffness of the underground diaphragm wall, the horizontal displacement of the underground diaphragm wall decreases continuously, and the settlement of the soil at the edge of the foundation pit and the uplift of the bottom of the foundation pit also decrease obviously. The increase in the stiffness of the underground diaphragm wall plays a significant role in reducing the deformation of the foundation pit. In this sample case, when the bending stiffness is less than 2.5 × 106 kN·m^{2}·m^{-1}, the horizontal displacement of the underground diaphragm wall develops rapidly with the decrease of the stiffness of the underground continuous wall. When the bending stiffness is greater than 2.5 × 106 kN·m^{2}·m^{-1}, the horizontal displacement of the underground diaphragm wall develops slowly with the decrease of the stiffness of the underground diaphragm wall. When the bending stiffness of the underground diaphragm wall is reduced from 2.5×107 kN·m^2·m^{-1} to 2.5×105 kN·m^2·m^{-1}, the horizontal displacement of the underground diaphragm wall increases by 1.2 times. Therefore, the corresponding horizontal displacement control standards should be formulated for different supporting structure stiffness.

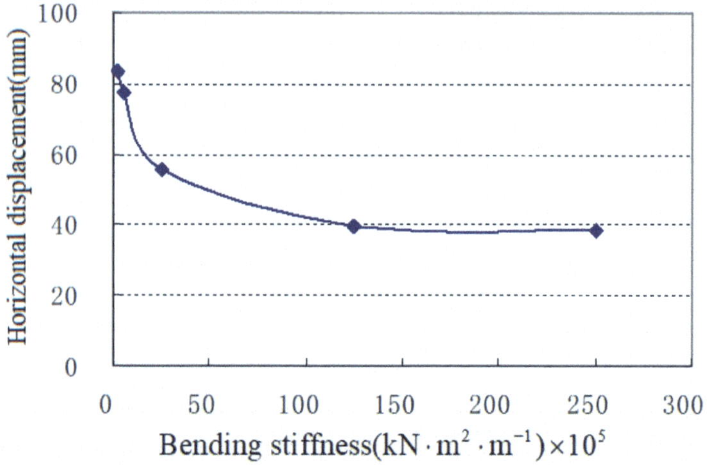

Fig. 3. The relationship curve between bending stiffness and maximum horizontal displacement of underground diaphragm wall.

Table 5. Stiffness calculation comparison of underground diaphragm wall

No	Horizontal Displacement	Maximum Horizontal Displacement	Heave of Pit Bottom	Settlement of Pit Edge	Axial Force of Support kN·m^{-1}		
	mm	mm	mm	mm	Support-1	Support-2	Support-3
1	55.6	69.1	418.2	20.9	155.0	429.7	243.6
2	77.3	89.7	421.6	22.8	117.8	395.8	348.8
3	83.5	94.4	422.8	22.5	115.8	374.5	388.0
4	39.6	49.8	414.6	20.1	278.5	347.8	153.6
5	38.3	45.9	413.5	19.3	315.5	307.6	123.6

Figure 4 shows the changes in the bending stiffness and the axial force of the support of the underground diaphragm wall. It is shown that with the increase of the stiffness of the underground diaphragm wall, the axial force of the support also changes obviously. The axial force of support-1 increases with the increase of the bending stiffness. The change of the axial force of support-2 is relatively small, and it also decreases with the increase of the bending stiffness. The axial force of support-3 decreases with the increase of the bending stiffness, which indicates that with the increase of the stiffness of the underground diaphragm wall, the center of gravity of the support force distribution tends to move up gradually, and the stress characteristics of the three supports are significantly different.

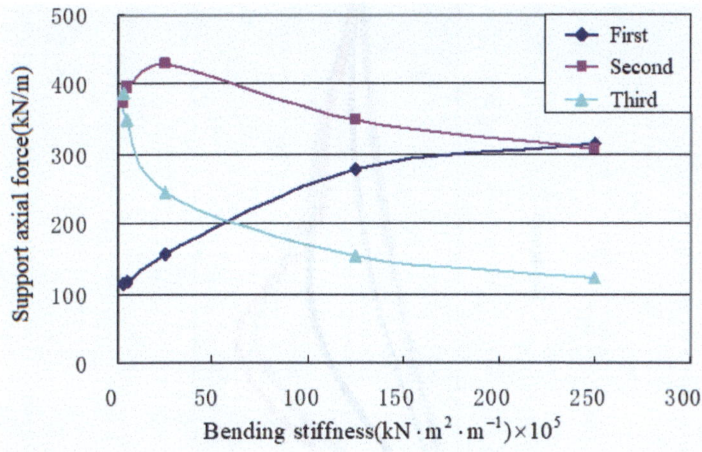

Fig. 4. The relationship curve between bending stiffness and axial force of support.

The curve of horizontal displacement of the underground diaphragm wall with depth under different stiffness conditions is shown in Fig. 5. With the decrease of the stiffness of the diaphragm wall, the change of the surface displacement is not obvious, and the change of the horizontal displacement is only 7.9 mm. However, below the depth of 2.6 m (support-1), the deformation under different stiffness conditions is different. The stiffness of No.4 and No.5 is larger, and the change of the diaphragm wall shows the characteristics of rigid change. The change along the depth is approximately linear, while No.1, No.2, and No.3 show flexible changes. In the range of depth of 2.6 m (support-1) and 18.0 m (the bottom of the foundation pit), the diaphragm wall under the three stiffness conditions has obvious extrusion deformation characteristics. Below the bottom of the foundation pit, under the condition of No.2 and No.3 stiffness, the underground diaphragm wall is further squeezed out, so the corresponding horizontal displacement control standards should be formulated for different stiffness. In addition, when the stiffness of the diaphragm wall is small, the displacement at the top of the supporting structure cannot truly reflect the stress and deformation characteristics of the support. The deep horizontal displacement control should be the main control, supplemented by the surface horizontal displacement control.

The curves of soil settlement of the foundation pit side wall with distance under different stiffness conditions are shown in Fig. 6. When the stiffness of the diaphragm wall is small, the soil settlement increases first and then decreases with the increase of the distance from the pit edge. When the stiffness of the diaphragm wall is large, the soil settlement decreases with the increase of the distance from the edge of the pit. Due to the use of three internal supports in the project, the stress of the three internal supports has also changed significantly under the condition that the stiffness of the ground wall

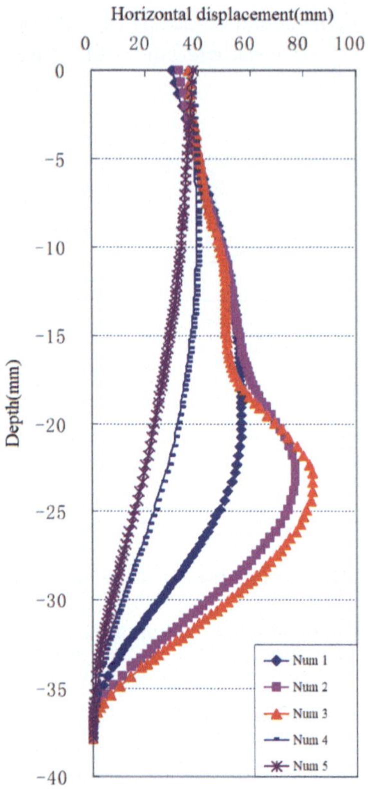

Fig. 5. The variation curve of inclinometer data with depth.

is decreasing, as shown in Fig. 4. Therefore, the distribution of surface settlement is different. For the case of large stiffness (No.4 and No.5), the surface settlement decreases with the increase of the distance from the edge of the foundation pit, while the stiffness is small (No.1, No.2, and No.3), and the maximum surface settlement occurs 15 m away from the foundation pit.

4.2 Stiffness of Inner Support

Figure 7 is the relationship curve between the axial stiffness and the maximum horizontal displacement of the diaphragm wall. The calculation results of the support under different stiffness conditions are listed in Table 6 and the displacement shadow diagram under the minimum axial stiffness condition is shown in Fig. 8.

It is said that the stiffness of support also plays a very important role in the deformation of the foundation pit. When the axial stiffness of the three internal supports changes in the same proportion, the horizontal displacement of the underground diaphragm wall decreases with the increase of the axial stiffness of support. When the axial stiffness of the first support is greater than 2.1×107 kN·m^{-1}, the maximum horizontal displacement of the underground diaphragm wall develops slowly. When the axial stiffness of

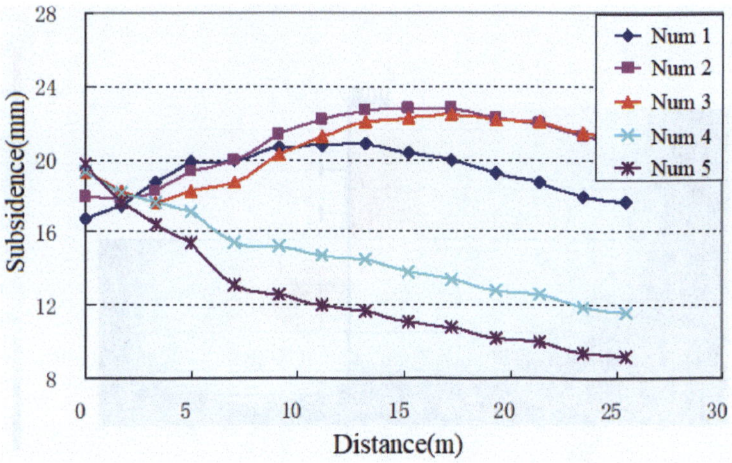

Fig. 6. The variation curve of subsidence with distance from the edge of the foundation pit.

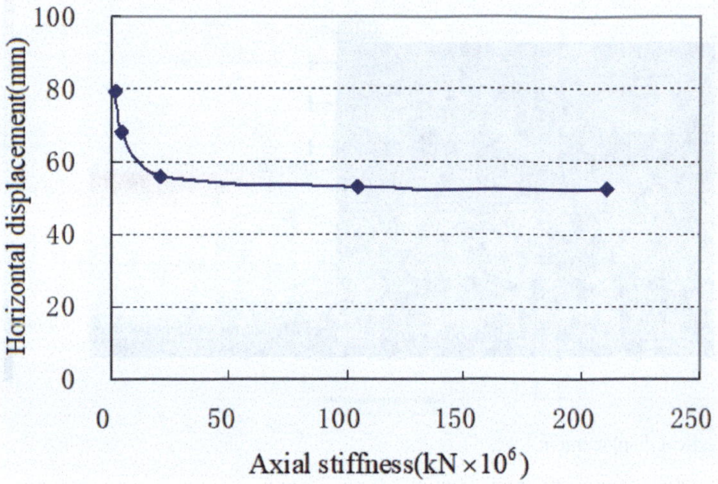

Fig. 7. The relationship curve between axial stiffness and maximum horizontal displacement of the underground diaphragm wall.

underground diaphragm wall is reduced from 3.0 × 108 kN·m^{-1} to 3.0 × 106 kN·m^{-1}, the horizontal displacement of the underground diaphragm wall increases by 0.5 times. The change in support stiffness also has a similar effect on the settlement of the foundation pit. In general, the stiffness of the underground diaphragm wall has a more obvious effect on controlling horizontal displacement.

Figure 9 shows the change in the axial stiffness and axial force of the support. It could be seen that the axial force of the first support decreases with the increase of the axial stiffness, and the axial force of support-2 and support-3 increases with the increase

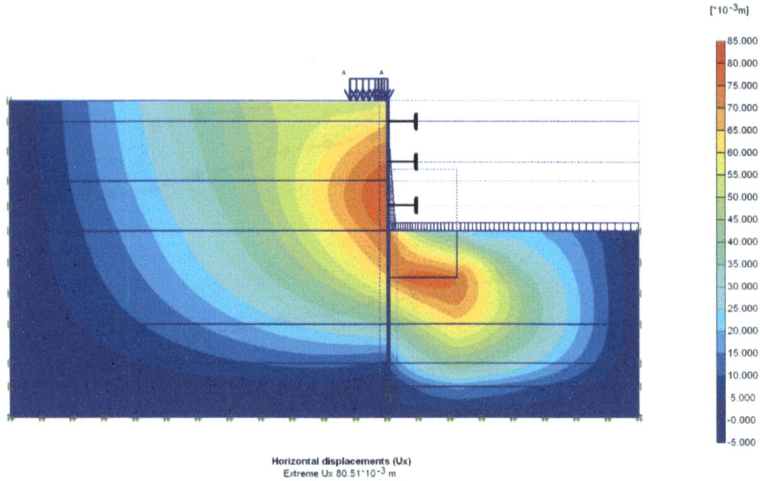

(a) Horizontal displacement

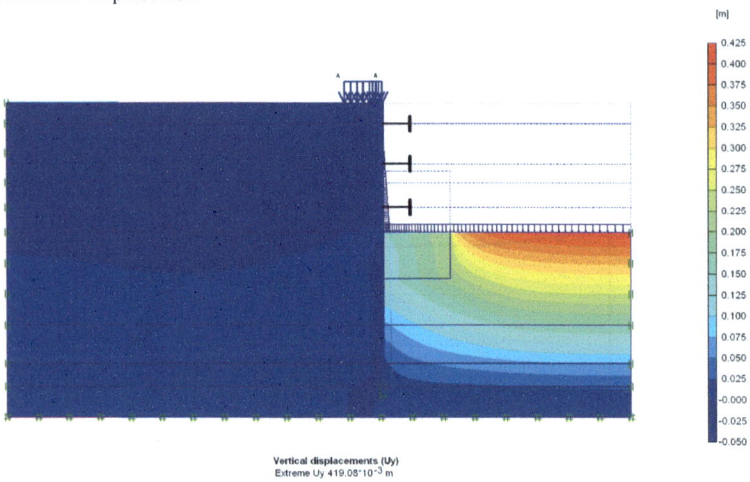

(b) Vertical displacement

Fig. 8. The diagram of Num 3 displacement.

of the axial stiffness. It shows that with the increase of the axial stiffness of the support, the center of gravity of the support force distribution would have a tendency to gradually move down, which is different from the law shown by the change of the stiffness of the underground diaphragm wall.

The curve of horizontal displacement of the underground diaphragm wall with depth under different support stiffness conditions is shown in Fig. 10. The change of support stiffness can only cause the change of horizontal displacement of the underground diaphragm wall. The maximum change occurs at the top of the pile. The displacement of the pile top under the condition of minimum stiffness (No.3) is about 2 times that under

Table 6. Stiffness calculation comparation of support

No	Horizontal Displacement	Maximum Horizontal Displacement	Heave of Pit Bottom	Settlement of Pit Edge	Axial Force of Support kN·m^{-1}		
	mm	mm	mm	mm	Support-1	Support-2	Support-3
1	55.6	69.1	418.2	20.9	155.0	429.7	243.6
2	68.5	75.9	418.8	25.5	204.9	325.1	177.2
3	79.6	80.5	419.1	29.5	203.6	275.5	140.6
4	53.1	66.1	417.8	19.2	109.6	492.8	267.0
5	52.6	66.2	417.7	18.9	100.7	505.1	269.7

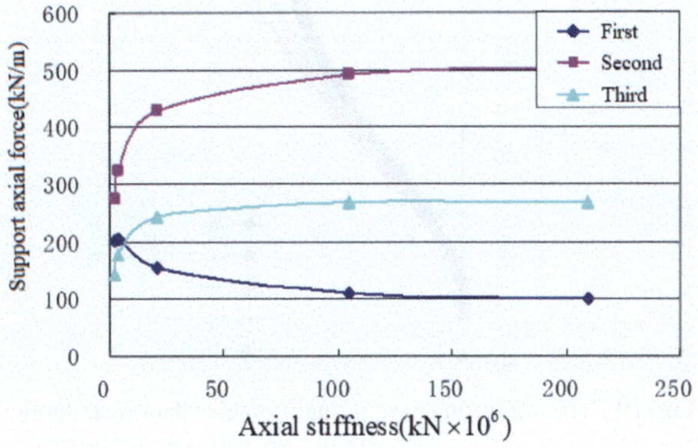

Fig. 9. The relationship curve between axial stiffness and axial force of support.

the condition of maximum stiffness (No.5), but the change of support stiffness could not cause the change of deformation characteristics of the underground diaphragm wall.

The curves of soil settlement of the foundation pit side wall with distance under different support stiffness conditions are shown in Fig. 11. The settlement characteristics under various conditions are the same. The maximum settlement occurs between 10 ~ 15 m from the edge of the foundation pit. The surface settlement under the condition of minimum stiffness (No.3) is 1.56 times that under the condition of maximum stiffness (No.5).

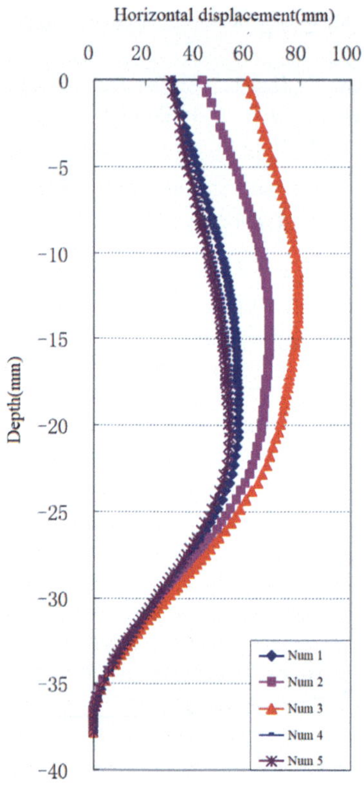

Fig. 10. The variation curve of inclinometer data with depth.

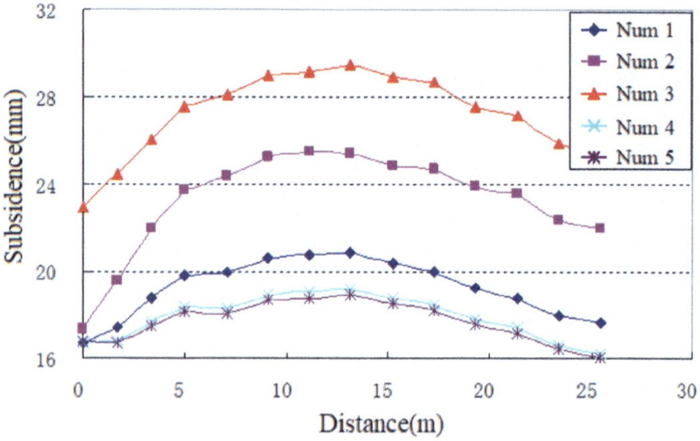

Fig. 11. The variation curve of subsidence with distance from the edge of the foundation pit.

5 Conclusion

In this paper, the deformation of foundation pit is studied around the supporting structure stiffness and the main conclusions are as follows:

- Deformation characteristics of the supporting structure itself and its influence on foundation pit deformation vary significantly under different support structure stiffness conditions. Greater stiffness results in rigid deformation characteristics in the supporting pile, whereas lower stiffness tends to induce flexible deformation traits. Notably, the influence of the supporting structure on the surrounding soil also varies with differing stiffness levels. To simulate foundation pit deformation, relevant influencing factors are adjusted individually within a finite element model, enabling comparison and analysis of their respective impacts.
- Monitoring the horizontal displacement of the pile top is insufficient for accurately capturing the deformation characteristics and trends of the supporting structure. Instead, emphasis should be placed on deep horizontal displacement monitoring, with supplementary control measures for surface horizontal displacement.
- Changes in support stiffness solely impact the deformation magnitude of the supporting structure and do not alter the deformation patterns or laws governing the supporting structure.
- In foundation pit monitoring, it is advisable to establish distinct monitoring and control standards tailored to different support stiffness conditions.

Acknowledgments. Research on the safety state of bearing capacity of foundation pit retaining structure based on monitoring deformation data (NO: 2023KJ43).

References

1. Lei, G., Gong, X.N.: Analysis of lateral displacement law of deep foundation pit support in soft soil based on improved MSD method. Adv. Civil Eng. **s**, 5550214 (2021)
2. Zhang, Y.G., Wang, Y.H., Zhao, Y.Y.: A case study on the deformation of metro foundation pit in silt stratum in North China. Shock Vibr. **2021**, 7454596 (2021)
3. Zhang, X.H., Wei, G., Lin, X.B., Xia, C., Wei, X.J.: Transverse force analysis of adjacent shield tunnel caused by foundation pit excavation considering deformation of retaining structures. Symmetry-Basel **13**, 1478 (2021)
4. Zhou, Y., Wang, K.: Application of improved WOA-exponential power product model in foundation pit deformation prediction. J. Water Resour. Water Eng. **31**, 233–239 (2020)
5. Xu, H., Sun, P., Cui, W., Sun, Y.: MSD synthetic system stiffness and deformation prediction of support structure for deep excavation in Nanjing subway. J. China Univ. Min. Technol. **47**, 907–912 (2018)
6. Terzaghi, K., Peck, R.B.: Soil Mechanics in Engineering Practice. Wiley, New York (1948)
7. Pujades, E., Jurado, A.: Groundwater-related aspects during the development of deep excavations below the water table: a short review. Undergr. Space **6**, 35–45 (2019). https://doi.org/10.1016/j.undsp.2019.10.002
8. Liu, M., Fang, Q., Zhang, D., Hou, Y.: Prediction of transient deformation due to excavation based on improved MSD method. Chin. J. Rock Mech. Eng. **37**, 1700–1707 (2018)

9. Zhang, D., Fang, H., Chen, L., Sun, Z.: Stiffness design theory for tunnel-support system. Chin. J. Rock Mech. Eng. **40**, 649–662 (2021)
10. Wei, G., Feng, F.F., Cui, Y.L., Wang, X.Q., Diao, H.G., Wu, Y.: Research on the influence of foundation pit excavation on the lateral force and deformation of side shield tunnels based on full-scale experiments. Tunn. Undergr. Space Technol. **140**, 105236 (2023)
11. Ma, H.F., Hao, X., Yang, Y.P., Li, H., Yi, L.X.: Study on parameters of modified Mohr-Coulomb model for water-rich soft soil in Tianjin, China. Sn Appl. Sci. **5**, 187 (2023)
12. Wang, Z., Guo, B.L., Wei, G., Hu, C.B., Zhang, L.S.: Simplified deformation analyses of a single pile subjected to different deformation modes of the pit enclosure structure. Int. J. Geomech. **23**, 04023014 (2023)
13. Cui, X.Y., Li, Z.P., Guo, F., Han, X.G., Song, J.P.: Displacement analyses of main structure of parallel pit excavation and analysis of countermeasures. Appl. Sci.-Basel. **13**, 3706 (2023)

Seismic Response Analysis of Bridge Pile Foundations in Saline Frozen Soil Sites

Xiaodong Lai(✉)

Lanzhou Jiaotong University, Lanzhou 730030, Gansu, China
1310049032@qq.com

Abstract. This article studies the seismic simulation analysis of bridge pile foundations in saline frozen soil sites, focusing on the acceleration time history and peak acceleration amplification coefficient of the foundation soil and bridge pile foundations. From the perspective of dynamic analysis, the seismic response of the site and bridge structure is studied, and the following conclusions are drawn: the seismic response of the foundation soil and bridge pile foundations is different, and the seismic response of the bridge pile foundation is relatively obvious, The analysis of the peak acceleration amplification coefficient also confirms this rule. In subsequent seismic research, the focus should be on seismic analysis of bridge structures.

Keywords: Bridge pile foundation · Seismic performance · Quasi static test · Finite element analysis · Permafrost regions

1 Introduction

The impact of permafrost on bridge structure damage has been confirmed by actual earthquakes [1, 2]. Earthquakes are frequent in western China and are also the main area of permafrost distribution [3–5]. In recent years, major earthquakes that have occurred in the Qinghai Tibet region include the 7.1 magnitude earthquake in Yushu, Qinghai in 2010 [6], the 7.4 magnitude earthquake in Maduo, Qinghai in 2021 [7], and the 6.2 magnitude earthquake in Jishishan, Gansu in 2023 [8], all causing extensive damage to building structures and surface damage. In permafrost areas with frequent seismic activity, the likelihood of pile foundation bridges being damaged by earthquakes is very high. Due to the difference between frozen soil and conventional thawed soil, bridge pile foundations are affected by complex external environmental factors, especially the influence of external temperature changes [9]. After an earthquake occurs in permafrost areas, bridge pile foundations may enter a nonlinear state under strong seismic action [10], making the dynamic response of bridge structures particularly complex [11]. Therefore, the study of seismic resistance of bridge pile foundations in permafrost regions is a key and complex issue in the field of bridge seismic resistance.

Since the 2001 Kunlun Mountain earthquake with a magnitude of 8.1 and the 2002 Alaska earthquake with a magnitude of 7.9 in the United States, seismic resistance in permafrost regions has gradually come into people's attention, realizing the importance

B. Yuan et al. (Eds.): ICCASE 2024, SUCI, pp. 227–232, 2024.
https://doi.org/10.1007/978-3-031-76102-7_26

of the presence of permafrost on the seismic response of bridges [12]. With the deepening development of transportation engineering in cold regions of China, more and more pile foundation bridge structures have emerged in permafrost regions. As an important component of lifeline engineering such as railways and highways, bridge structures need to be given sufficient attention to their seismic design theory and related research.

2 Establishment of Finite Element Calculation Model

This article uses the finite element calculation software ABAQUS for simulation and analysis. The dynamic calculation model of soil characterizes the basic stress-strain relationship of soil under dynamic loads. How to reasonably describe the deformation and stress of soil under earthquake action is the core problem of studying soil dynamics. The deformation of soil under dynamic loads usually includes two parts: elastic deformation and plastic deformation, and soil is usually in the stage of elastic-plastic deformation. Under dynamic loads, soil not only has the characteristics of elastic-plastic behavior, but also has the characteristics of viscosity. Soil can be regarded as a viscoelastic plastic body with elasticity, plasticity, and viscosity. Under cyclic dynamic loads, the stress-strain relationship of soil exhibits three characteristics: nonlinearity, hysteresis, and cumulative deformation. Especially under strong seismic conditions, the soil exhibits strong nonlinearity and plastic deformation, which are significantly different from static loads.

2.1 Model Parameter Settings

In order to reduce the influence of boundary effects during the modeling process, the soil width is taken as 3 m, the height is 2 m, the upper 40 cm is the seasonal frozen soil layer, and the lower 110 cm is the thawed soil layer; The diameter of the pile foundation is taken as 15 cm, the pile length is 150 cm, and the pile foundation is buried at a depth of 140 cm. The counter clockwise numbering of the upper right pile is 1 #, 2 #, 3 #, and 4 #, respectively; In terms of model unit division, both concrete and soil use three-dimensional solid elements (C3D8R), while steel bars use truss elements (T3D2). The contact between the side and bottom of the pile and the soil is achieved through surface to surface contact, which means a hard contact is used in the normal direction and a "penalty" contact is used in the tangential direction. The "penalty" contact is similar to setting a friction coefficient. In addition, the contact between the steel reinforcement and the pile foundation is embedded. In terms of boundary conditions, the bottom of the soil restricts its translational and rotational displacement in three directions, while the side of the soil restricts displacement parallel to the coordinate axis. In terms of constitutive model, the concrete constitutive model adopts the concrete damage plasticity model, the steel reinforcement constitutive model adopts the bilinear constitutive model, and the soil constitutive model adopts the Mohr Coulomb model. The ABAQUS finite element calculation model and stress cloud diagram are shown in Fig. 1. The mechanical parameters of the simulated soil are obtained through static triaxial tests of saline frozen soil.

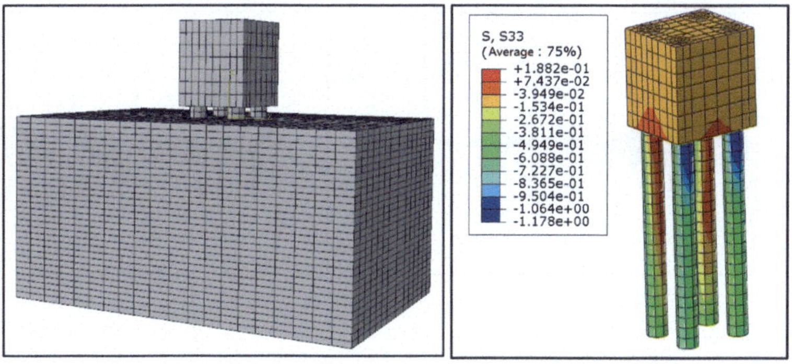

Fig. 1. Schematic diagram of calculation model and stress cloud map of pile foundation

To investigate the influence of soil mechanical properties in saline frozen soil areas on the seismic performance of bridge pile foundations, based on a validated finite element model, the influence mechanism of changes in soil mechanical properties after temperature, initial moisture content, and compaction degree of frozen saline soil on the seismic performance of bridge pile foundations was studied. The mechanical parameters of frozen soil were obtained through static triaxial tests of saline frozen soil, with a dry density of 1.75 g·cm^{-3} and a moisture content of 14%, as shown in Table 1.

Table 1. Mechanical parameters of soils with different variables

Test conditions	Temperature/°C	Confining pressure/kPa	Elastic modulus/kPa	Cohesio-n/kPa	Internal friction angle/(°)
1	2	50		120.62	19.56
2	2	250			
3	2	500	36419.57		
4	−5	50		507.39	24.52
5	−5	250			
6	−5	500	392814.92		

2.2 Seismic Wave Selection and Input

This article selects the seismic wave as the El Centro wave and adjusts it to the corresponding seismic peak acceleration of 0.3 g, as shown in Fig. 2. To improve simulation efficiency, the selected seismic wave is truncated for the first 15 s and input along the bridge direction.

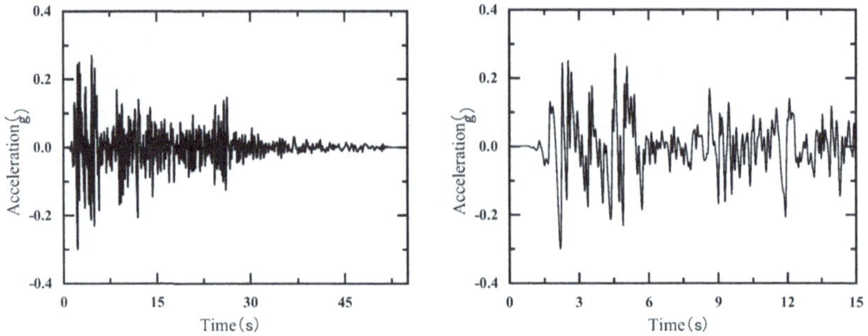

Fig. 2. Schematic diagram of El Centro time history curve

3 Seismic Response Analysis of Pile Foundation

3.1 Acceleration Time History

The calculated time history curves at the bottom and top positions of the pile when the input peak PGA is 0.3 g are shown in Fig. 3. According to the stress cloud map of the model, it can be seen that the foundation of the bridge group piles is subjected to symmetrical forces, so the focus is on analysing 1 # and 3 #; Numerical calculation can reflect the dynamic response law of the model system and reflect the main characteristics of the waveform curve. Under the condition of PGA = 0.3, the dynamic response values and time history waveforms of the calculated results for 1 # and 3 # are basically consistent. As the seismic wave propagates from the bottom to the top, the dynamic response values of the calculated waveform gradually increase, indicating that the energy dissipation capacity of the pile foundation at different positions of the calculation model is equivalent and good under high-intensity earthquake conditions.

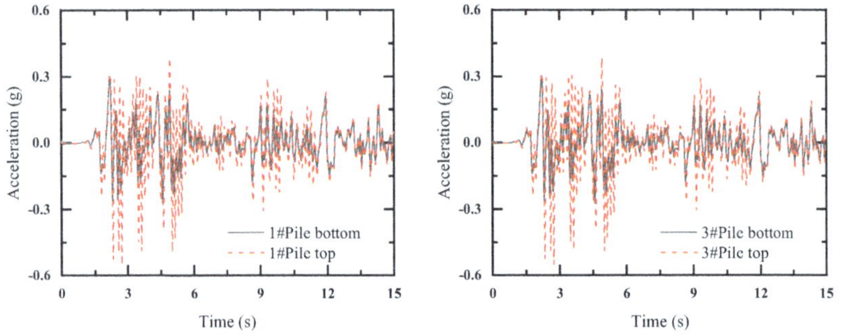

Fig. 3. Acceleration time history of pile foundation

3.2 Acceleration Peak Amplification Factor

Table 2 compares the variation curves of the acceleration peak amplification coefficients of the foundation soil, pile body, and upper structure with height (depth) when the calculation model PGA is 0.3, with a ground height of 0cm, negative for the soil downwards, and positive for the soil upwards.

As the lower depth decreases, the acceleration response is continuously amplified, and the amplification factor for the calculation of the foundation soil and pile body on the surface is about 1.35. Due to the presence of frozen soil layers, it can be observed that the amplification factor amplification has decreased; The acceleration force amplification trend of the pier cap is more obvious, and the acceleration amplification coefficient of the pier cap structure is above 1.7. Due to the lack of constraints at the free end, the amplification coefficient increases significantly. Among them, the amplification coefficients of 1 # and 3 # have the same trend of change. It is found that in the group pile foundation, the seismic response of the pile foundation at different positions is more consistent, and the force is balanced. The specific changes are shown in Fig. 4.

Table 2. The amplification coefficient of each position of the model

Depth/(cm)	70	40	10	0	−10	−30	−60	−90	−120	−150
Foundation soil	–	–	–	1.357	1.342	1.321	1.243	1.115	0.998	1
1#Pile	1.716	1.631	1.552	1.427	1.421	1.367	1.295	1.172	0.982	–
3#Pile	1.694	1.622	1.537	1.476	1.433	1.386	1.284	1.167	0.978	–

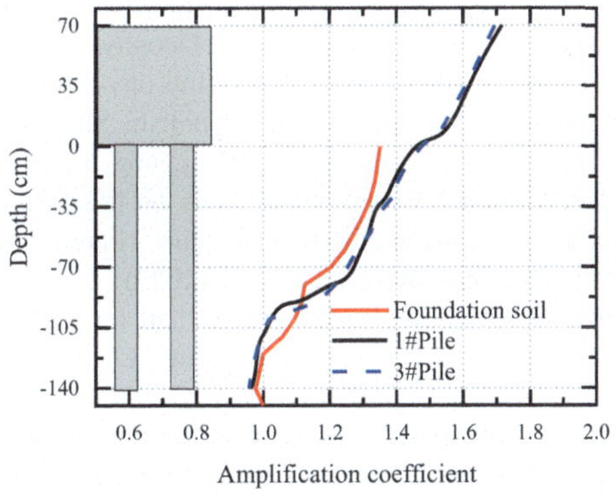

Fig. 4. Peak amplification factor of foundation soil and pile foundation

4 Conclusion

The article analysed the seismic response of bridge pile foundations in frozen saline soil sites and draws the following conclusions: 1. The seismic response of frozen saline soil sites is different from that of ordinary sites. Due to the presence of frozen soil layers, the acceleration response is weakened, and this effect is more significant on the foundation soil than the acceleration response of bridge structures; 2. By comparing and analysing the peak acceleration amplification coefficients of the foundation soil and pile foundation, it was found that the peak acceleration amplification coefficient of the foundation soil is relatively small compared to the peak amplification coefficient of the structure, with a maximum amplification coefficient of around 1.35. On the other hand, the peak acceleration amplification coefficient at the free end of the pile foundation increases significantly, due to the lack of constraints from the foundation soil at the free end, resulting in a more pronounced seismic response.

References

1. Wang, W., Zhang, X., Chen, X., et al.: Study on dynamic interaction between bridge pile and soil with permafrost effect: status and review. J. Glaciol. Geocryol. **42**(04), 1213–1219 (2020)
2. Chen, X., Gao, F., Wu, S.: Effect of frozen soil layer on seismic response of bridges. J. Eng. Mech. (03), 120–125 (2007)
3. Ran, Y., Li, X., Cheng, G., et al.: Distribution of permafrost in China: an overview of existing permafrost maps. J. Permafrost Periglac. Process. **23**(4), 322–333 (2012). https://doi.org/10.1002/ppp.1756
4. Deng, Q., Cheng, S., Ma, J., et al.: Seismic activities and earthquake potential in the Tibetan Plateau. J. Chin. J. Geophys. **57**(07), 2025–2042 (2014)
5. Liu, S., Zhang, J.: Review on physic-mechanical properties of warm frozen soil. J. Glaciol. Geocryol. **34**(01), 120–129 (2012)
6. Lin, A., Rao, G., Jia, D., et al.: Co-seismic strike-slip surface rupture and displacement produced by the 2010 Mw 6.9 Yushu earthquake, China, and implications for Tibetan tectonics J. Geodyn. **52**(3–4), 249–259 (2011). https://doi.org/10.1016/j.jog.2011.01.001
7. Lan, Z., Qi, L., He, T., et al.: An analysis of the co-seismic deformations and fault slip models of the 2021 Qinghai MaMW7.5 earthquake. Chin. J. Geophys. **52**(3–4), 249–259 (2022)
8. Zhuang, W., Cui, D., Hao, M., et al.: Geodetic constraints on contemporary three-dimensional crustal deformation characteristics in the Laji Shan-Jishi Shan tectonic belt. J. Geodesy Geodyn. **14**, 589–596 (2023). https://doi.org/10.1016/j.geog.2023.03.006
9. Zhu, S., Zhao, H., Wei, Y., et al.: Experimental study on triaxial mechanical properties of high−temperature frozen loess under different moisture content and confining pressure in Yili, Xinjiang. J. Northwest. Geol. **56**(05), 140–150 (2023)
10. Zhang, X., Yu, S., Wang, W., et al.: Seismic failure mechanism and fragility of pile foundation with elevated cap of railway bridge in permafrost region. China Civil Eng. J. **55**(07), 77–89 (2022)
11. Shelman, A., Tantalla, J., Sritharan, S., et al.: Characterization of seasonally frozen soils for seismic design of foundations. J. Geotech. Geoenviron. Eng. **140**(7), 04014031 (2014). https://doi.org/10.1061/(ASCE)GT.1943-5606.0001065
12. Wotherspoon, L.M., Sritharan, S., Pender, M.J., et al.: Investigation on the impact of seasonally frozen soil on seismic response of bridge columns. J. Bridge Eng. **15**(5), 473–481 (2010). https://doi.org/10.1061/(ASCE)BE.1943-5592.0000075

Analysis of the Load-Bearing Characteristics of Bridge Piles and Monopiles by Improved Saline Soil Under Freeze-Thaw Cycle

Mingzhao Cui(✉)

Lanzhou Jiaotong University, Lanzhou 730070, Gansu, China
1845526506@qq.com

Abstract. In order to explore the effect of saline soil improvement on the monopiles of pile foundation, the soil shear strength parameters under different conditions were obtained through static triaxial test. The ABAQUS software was used to understand the influencment of soil salt deterioration and the change of soil shear strength parameters before and after the improvement on the bearing capacity, axial force, side friction resistance and settlement of pile foundation, and raw the following conclusions: When the freeze-thaw cycle grows, the deterioration of pile foundation is intensified, and the side friction of pile gradually decreases and becomes stable after 10 freeze-thaw cycles. After adding the modifier, both fly ash modified soil and cement modified soil can effectively inhibit the amount of subsidence of the bridge monopiles, and the improvement effect is 14.5% and 40.6%, respectively. It provides reference for evaluating the bearing capacity of pile foundation of improved soil on salt-frozen environment.

Keywords: Numerical analysis · Lift saline soils · Pile-frozen soil interaction · Sedimentation

1 Introduction

Saline soil [1] is widely distributed in China, including in cold, less water, more salt areas, the eastern coastal areas and the Northeast Plain. The western China cold and arid areas belong to the seasonal frozen soil area, which is characterized by severe winter, hot summer and obvious temperature difference. The variation of seasonal temperature difference will cause the soil to freeze and thaw repeatedly, resulting in salt crystallization and delixiation in the salty soil, and eventually Physical, chemical, biological reactions of the saline soil will change significantly. Bridge pile foundation serving in complex geological environment is vulnerable to saline soil erosion and deterioration of freeze-thaw cycle. Resulting in reduced bearing capacity, shortened service life and other phenomena. So far, many scholars have used field tests, laboratory tests and numerical simulations to study the load bearing characteristics of pile foundation under freeze-thaw cycles in saline soil areas. Cheng Zhuo [2] determined the optimal content of fly ash through freeze-thaw cycle tests. The freeze-thaw cycle is an important factor leading to

B. Yuan et al. (Eds.): ICCASE 2024, SUCI, pp. 233–240, 2024.
https://doi.org/10.1007/978-3-031-76102-7_27

the change of physical properties of saline soil. Li Ying Tao [3] established finite element analysis by considering pile-soil interaction. It provides reference value for pile foundation design. Wang Fuchuncai [4] used Abaqus to explore the reaction characteristics of saline soil corrosion to piles in a bridge On this basis, Utkin [5] pointed out that the warming climate and the increase of active layer thickness will further weaken the load-bearing performance of concrete piles Jia Yanmei et al. [6] create a finite virtual model of contact by using ansys finite element simulation software and found that the cohesive force of frozen soil is the main source of the bearing capacity of pile foundation. Luan Hong [7] considered the relationship between pile side freezing strength, initial water content, soil particle composition and temperature at a certain temperature, and proposed the transfer mechanism of pile-soil interaction. Based on the pile-soil interaction model, Ponomarev [8] deduced the prediction model of the bearing capacity of frozen soil pile foundation, atmospheric temperature and seasonal thawing layer thickness. Crowther [9] conducted a numerical simulation of the relationship between a single pile under lateral load and frozen soil, and compared the measured values with the calculated values Churkin [10] used Geostudio software to conduct numerical simulation research on building deformation caused by foundation frost heave in cold areas.

Much of the current research is on the performance of bridge monopiles in permafrost environments, and few studies on the damage of bridge substructure under the coupling effect of freeze-thaw cycle deterioration and saline soil. With global warming, permafrost degradation speeds up, and salt-freeze coupling deterioration has an increasing impact on the bearing capacity of pile foundation. By adding a reasonable proportion of amendments, the durability and strength of saline soil in the freeze-thaw cycle can be greatly improved. It is of great significance to study the variation of bearing capacity of improved saline soil pile foundation under different conditions by using the parameters of soil measurement in the chamber, and it also provides reference direction for the reinforcement of bridge pile foundation in the future.

2 The Establishment of Finite Element Calculation Model

2.1 Virtual Models Arrangement

Monopile foundation of a railway bridge in the frozen soil area is taken as the prototype, the indoor model test is single pile foundation, and the scale ratio of 1:10 pile top is 10 cm higher than the soil surface, and the indoor model test is a single pile foundation, with a scale ratio of 1:10, the pile top is 10 cm higher than the soil surface, Fig. 1 shows the size and environment of the monopile and rebar, the model construction method is bored pile, and the test conditions are shown in Fig. 1. In the test loading and unloading system, slow load maintenance method and step by step loading method are adopted, and the holding time is increased step by step with 8 kN to 120 min. Data are collected once at 5 min, 10 min, 15 min, 30 min, 45 min, 60 min, 90 min and 120 min respectively. When the termination loading condition determined by the specification is reached, the loading will stop and the unloading will start. The unloading phase adopts step-by-step unloading.

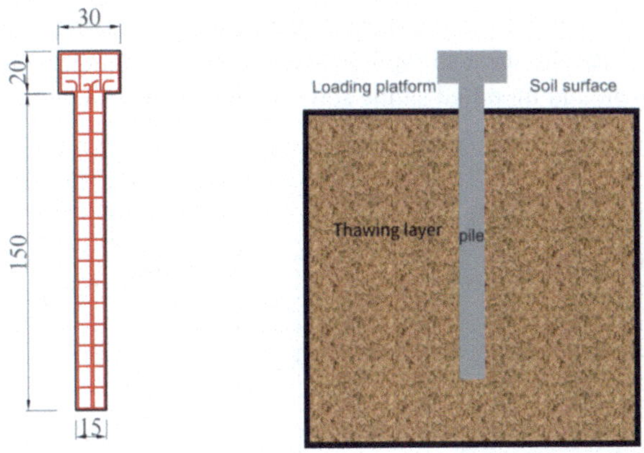

Fig. 1. Model reinforcement drawing and model layout drawing. (unit: cm).

2.2 Calculation Assumption

1. All materials are homogeneous elastic, continuous, and isotropic.
2. Without considering the effect of cement hydration heat on the soil around the pile, the geometric size of the soil is expanded to reduce the boundary effect.
3. C30 concrete CDP model was adopted for pile material properties, molar Coulomb plastic model was adopted for soil, and three-fold line model was adopted for rebar.
4. The pile-soil contact surface adopts "penalty" constraint, the normal direction is "hard" contact, and the interaction between reinforcement and concrete adopts embedment.

2.3 Establishment of Geometric Model

As shown in the left figure of Fig. 2, ABAQUS was used to establish a connection model with a scale of 1:1 with the model test and the pile, reinforcement and cushion cap In the radial model of soil mass, 20 times the diameter and length of the bottom of single pile and the calculated thickness of soil layer were taken as 2 times the buried depth of foundation to avoid boundary effect. CDP damage model of C30 concrete was adopted for pile foundation, and the density was 2400 kg/m^3. The soil with an elastic modulus of 29.7 GPa was created using the Mohr-Coulomb model, and the three-fold model is used for rebar. All constraints are applied to the bottom of the soil model, and the displacement parallel to the coordinate axis is constrained on the side of the soil. The interaction between the monopile and the saline soil is achieved by defining the surface-to-surface contact. The pile body is set as the main plane, the soil body as the slave plane, the Coulomb friction model is adopted in the tangential direction, and the hard contact is adopted in the normal direction. A coupling point is set at the top of the cap to apply vertical load, and the settlement curve of pile top load can be obtained by output coupling point reaction. In terms of load step setting, in-situ stress balance is calculated first, and then the single pile foundation is activated to simulate the actual construction situation of the project. As shown in Fig. 2 right, the accuracy of the model was verified by the finite element numerical analysis which was consistent with the field test data.

According to the laboratory tests, the shear strength indexes of saline soil under different freeze-thaw cycles and the shear strength indexes and soil parameters of modified saline soil with 8% cement and modified saline soil with 8% fly ash were obtained, as shown in Table 1.

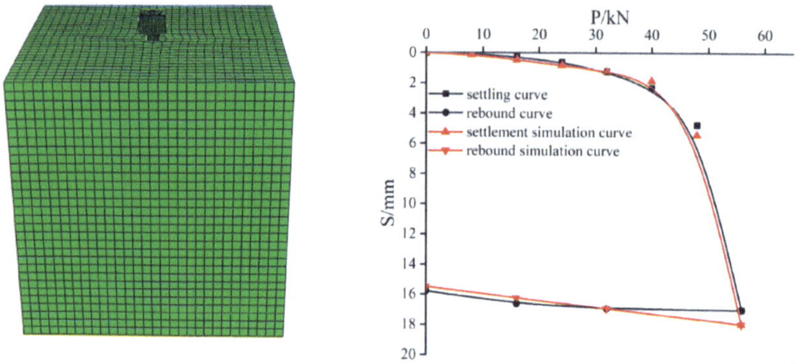

Fig. 2. The establishment and self-proof of finite element model

Table 1. Soil mechanics parameters of saline soil.

Soil type	Cohesion/kPa	Internal friction angle/(°)	Elasticity modulus/MPa	Poisson's ratio	Number of freeze-thaw cycles
Unimproved saline soils	41.23	27.15	1.8	0.44	0
Unimproved saline soils	39.26	27.95	1.8	0.44	1
Unimproved saline soils	38.56	28.16	1.8	0.44	6
Unimproved saline soils	38.12	28.62	1.8	0.44	10
Unimproved saline soils	38.02	28.65	1.8	0.44	15
Cement improved soil	53.62	29.52	1.83	0.46	0
Fly ash improved soil	52.15	28.68	1.82	0.45	0

3 Simulation Results and Analysis

3.1 Axial Force and Stress Distribution of Pile Body

The axial stress extracted from the center of the pile is drawn along the depth under different graded loads in Fig. 3 left. The axial stress of a single pile along the buried depth of the pile foundation decreases with the buried depth of the pile foundation, and a focusing phenomenon occurs at the pile bottom. Due to the lateral damping effect of the pile and soil contact embedding, the downward transfer process of the graded load applied to the top of a monopile is gradually cancelled, resulting in a smaller load on the pile bottom. The stress distribution of pile foundation under graded load 56 kN is shown in Fig. 3 right.

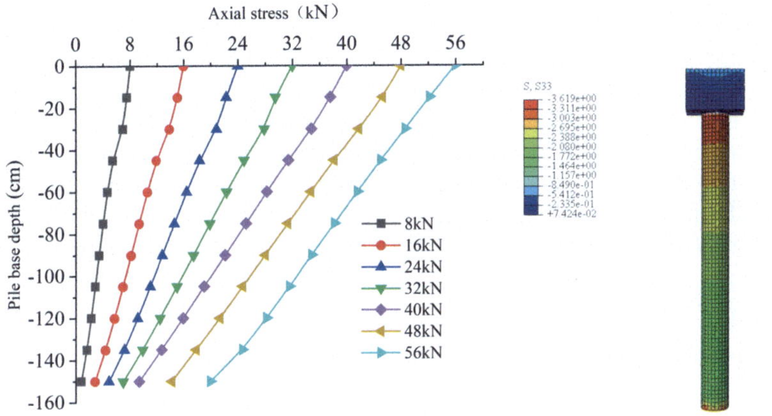

Fig. 3. Axial force of pile foundation and pile stress nephogram diagram

3.2 Pile Side Friction Resistance

The friction generated by the contact between the monopile and the saline soil is shown in the Fig. 4 left, and the calculated lateral friction resistance of pile foundation under different loading grades is shown in the figure. From the perspective of the whole loading process, Frictional resistance in the direction of burial depth gradually increases with the add of pile top load, and the lateral friction resistance increases significantly in the early loading stage, and Lateral damping curve of a bridge monopile are far apart under different loading grades.

When the load reaches 56 kN, the lateral friction resistance of bridge monopile increases slowly, and the lateral friction curve is relatively dense, because the resistance of bridge monopile is the bonding force between loading of bridge monopiles and saline soil at the initial stage. With the increase of load, the relative sliding occurs between pile and soil, and the lateral friction resistance gradually transitions from the bonding force to the friction force between pile and soil. At this time, the lateral friction resistance

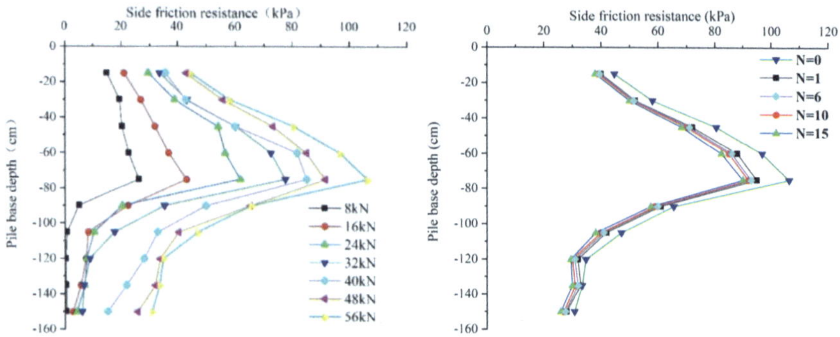

Fig. 4. Comparison of pile side friction resistance before and after freeze-thaw

gradually becomes stable and the load gradually transfers to the pile bottom. It can be seen from Fig. 4 right that based on 56 kN graded load, under different freeze-thaw cycles, the friction between the pile and the saline soil of the pile changes with the vertical direction of the bridge monopiles, that is, with the add of the buried depth of the pile foundation, the lateral friction resistance gradually increases, reaches the maximum at the buried depth of 70 cm, and then gradually decreases. As the number of freeze-thaw cycles increases, soil will deteriorate, and the friction between pile and soil will gradually decrease (N represents the number of times the saline soil has been frozen and thawed).

3.3 Improved Settlement Curve of Single Pile in Saline Soil

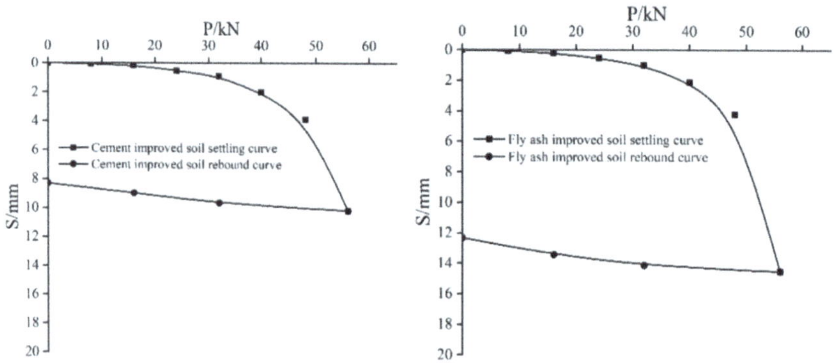

Fig. 5. P-S curves of different improved saline soils

The P-S curve reflects the law between the bridge monopile and the settlement of the foundation when it is subjected to the upper load. The load on pile foundation under non-frozen soil and permafrost conditions and the corresponding settlement curve are shown in Fig. 5. On the whole, the P-S curve of pile foundation shows a slow decline in both frozen saline and non-frozen saline soils. Under the condition of non-frozen soil, when the load applied on the bearing table is less than 32 kN, the slope of the P-S curve is at a constant datum, and the relationship between load and pile foundation settlement is approximately linear. It can be seen from the Fig. 5 that under the unimproved saline soil, the maximum settlement of a single pile is 17.02 mm after various graded loads, and the settlement of a single pile changes after adding modifier. The maximum settlement of single pile of fly ash modified saline soil is 14.52 mm, which is increased by 14.5%. The maximum displacement of single piles of bridges is 10.02 mm, which increases 40.6%. Therefore, the addition of amendment has obvious inhibition effect on pile foundation settlement. Figure 6 are stress cloud maps of cement modified saline soil and fly ash modified saline soil, respectively.

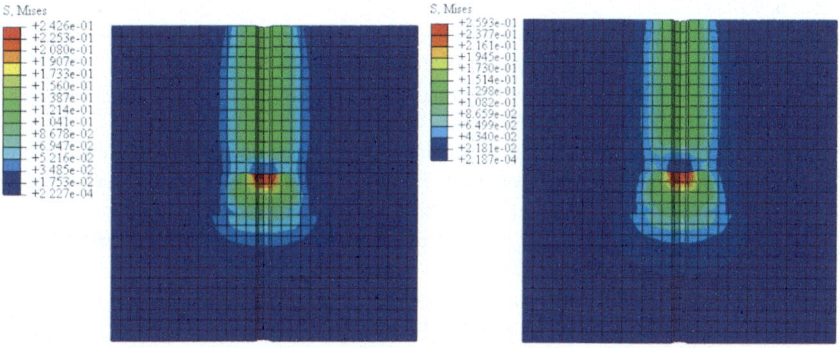

Fig. 6. Stress cloud map of cement and fly ash improving saline soil

4 Conclusion

In this paper, the difference of load bearing performance of pile foundation before and after freeze-thaw cycle and improvement is compared by numerical simulation, and the influence of soil parameters on load bearing performance of pile foundation (including load settlement curve, lateral friction resistance curve and axial force curve) is analyzed. The following five conclusions were reached:

(1) Axial stress of the pile foundation decreases with the buried depth of the pile foundation, and the focusing phenomenon occurs at the pile bottom. The downward transfer process of superstructure loads is gradually cancelled due to the exerting of pile side friction resistance, resulting in the partial load of the bridge pile bottom becoming smaller.
(2) Freeze-thaw cycle deterioration can weaken load-bearing characteristics of the bridge monopile in the vertical direction, resulting in increased settlement, and reach a stable state when the freeze-thaw cycle is 10 times.
(3) The improvement of saline soil can significantly increase the load-bearing characteristics of piles and improved soil foundations and significantly inhibit the settlement of pile foundation. The improvement effect of cement on saline soil reached 40.6%

(4) The deterioration of Freeze-thaw cycle can affect the decrease of bearing capacity of saline soil pile foundation pile foundation. The method of appropriately increasing pile length or pile diameter can make up for the problem of red bearing capacity of pile foundation caused by freeze-thaw cycle deterioration
(5) The bearing effect of the pile on the superstructure can be greatly improved after salt improvement, which provides ideas for the strengthening direction of bridge pile foundation

References

1. Wang, L., Gan, H., Yu, F.L.: Salted soil and its development in Northwest China. J. Hydraul. Eng. **6**, 90–94 (2018)
2. Cheng, Z., Hou, Z., Gang, H.: Study on effect of freeze-thaw cycle on shear resistance of fly ash modified saline soil. China Foreign Highw. **43**(03), 205–211 (2023)
3. Li, Y.T., Yuan, L.: Study on strength formation mechanism of cement improved saline soil. Highway 164–168 (2013)
4. Wang, F.C., Yao, X.H.: Numerical simulation of Effect of salt marsh corrosion on vertical ultimate bearing capacity of highway bridge pile foundation. Highway **62**(01), 60–66 (2017)
5. Utkin, V.S, Sushev, L.A.: The reliability analysis of existing reinforced concrete piles in permafrost regions. Int. J. Comput. Civil Struct. Eng. **13**(2), 64–72 (2017). https://doi.org/10.22337/2587-9618-2017-13-2-64-72
6. Jia, Y.M., Guo, H.Y.: Finite element analysis of pile-frozen soil interaction in permafrost region. Chin. J. Rock Mech. Eng. 3134–3140 (2018)
7. Luan, H.: Permafrost region of bridge pile foundation freeze strength test study. J. Urban Road Bridge Flood Control 114–117 (2011). 16799/.2011.03.042
8. Ponomarev, V.D., Fedoseev, Y.G., et al.: Field investigations of negative friction forces in thawing soils. Soil Mech. Found. Eng. **15**, 241–245 (1978). https://doi.org/10.1007/bf02111768
9. Crowther, G.S.: Lateral pile analysis frozen soil strength criteria. J. Cold Reg. Eng. **29**(2), 04014011 (2015). https://doi.org/10.1061/(ASCE)CR.1943-5495.0000078
10. Churkin, S.V., Nikitin, A.V., et al.: Deformation of building on pile foundation due to frost heave. Jpn. Geotech. Soc. Spec. Publ. **2**(39), 1420–1423 (2016). https://doi.org/10.3208/jgssp.KAZ-13

Influence of Expansive Agent on the Mechanical Properties of Ultra-High-Strength Concrete Filled Steel Tube

Yangyueye Cao and Dianyi Song(✉)

Undergraduate School, National University of Defense Technology, Changsha 410072, Hunan, China
cy3yeah@163.com

Abstract. Ultra-high-strength concrete filled steel tube (UHSCFST) specimen is a composite formed by filling steel tube with ultra-high-strength concrete (UHSC) and it has a great load-bearing capacity. However, the shrinkage of UHSC is relatively high compared to normal concrete and hence UHSC tends to debond from the tube in UHSCFST. UHSC and the steel tube are only "mechanically" combined together, making it difficult to realize their joint work and hindering the high performance of UHSCFST. In this study, 5%, 10% and 15% UEA expansive agents are added to UHSC to reduce the shrinkage. The influences of UEA dosage on the compressive strengths of UHSC and UHSCFST, as well as the bond strength between the tube and the concrete are analysed. The experimental results show that adding 10% UEA leads to only 3.2% increase of the UHSC compressive strength but 17.3% increase of the UHSCFST compressive strength thanks to the combined effects of the expensive agent and the tube confinement. Moreover, 15% UEA results in the highest pushout force, which is 29.1% higher compared to that without UEA. The study confirms that adding expansive agent contributes to improving the co-working properties of the steel tube and UHSC in UHSCFST.

Keywords: Ultra-high-strength concrete · Concrete filled steel tube · Expansive concrete · Mechanical properties

1 Introduction

Concrete is one of the most commonly used building materials in the world and ultra-high-strength-concrete (UHSC) is a type of concrete with a compressive strength over 100 MPa. With the increase of the strength, the brittleness of UHSC also increases. Filling steel tube with UHSC, i.e., creating ultra-high-strength concrete filled steel tube (UHSCFST) specimen, helps to overcome this problem and can change the damage mode from brittle to ductile destruction [1, 2]. Moreover, the confinement of the steel tube makes the core UHSC under a three-dimensional stress state, which contributes to the higher strength of the specimen. These advantages of UHSCFST makes it a promising material in construction field. However, the shrinkage of UHSC is also relatively high compared to normal concrete. Consequently, UHSC tends to debond from the tube in UHSCFST [3]. UHSC and steel tube are only "mechanically" combined together, making it difficult to realize their joint work and hindering the high performance of UHSCFST.

B. Yuan et al. (Eds.): ICCASE 2024, SUCI, pp. 241–248, 2024.
https://doi.org/10.1007/978-3-031-76102-7_28

Expansive agent can reduce the shrinkage of UHSC and improves the synergistic performance of steel tube and UHSC. For example, Li et al. [3] analysed the effects of CaO and MgO based expansive agents on the performance of concrete filled steel tubes and confirmed that expansive agents restrained the separation between the core concrete and steel tube. Chen et al. [4] investigated the mechanical performance of expansive concrete filled steel tube columns under compression and proposed two calculation methods of the ultimate bearing capacity. Tam et al. [5] studied the performance of recycled aggregate concrete filled steel tubular stub columns with expansive agent and found that the columns with expansive agent have higher strength, ultimate strain, and ductility than the reference columns.

It is noteworthy that the concrete studied by the above research are under 70 MPa and there are very few studies focus on UHSCFST filled with concrete with strengths greater than 100 MPa. On the other hand, UHSC with a relatively high strength usually has the characteristics of low water/cement ratio, rapid development of early strength, and larger shrinkage. It would be expected that the compensating shrinkage effect of expansion agents on UHSC is different from that on low-grade concrete [6] and the co-working properties of the tube and UHSC in UHSCFST still needs to be further investigated.

In this study, UEA expansive agent is added to UHSC of C100 to reduce its shrinkage and thus improve the synergistic performance of steel tube and UHSC in UHSCFST specimens. The influences of the expansive agent dosage, namely 5%, 10% and 15%, on the compressive strengths of the UHSC material and the UHSCFST specimen are investigated and discussed. Furthermore, in order to evaluate the bond strength between the steel tube and the UHSC, pushout experiments of UHSCFST specimens are conducted. This study has important scientific significance and engineering applications.

2 Materials and Experimental Methods

2.1 Materials and Mix Design

Recipes of the UHSC are presented in Table 1. The raw materials include Portland Cement CEM I 52.5 R, standard sand, coarse aggregates with diameters of 5–10 mm, tap water, superplasticizer and UEA expansive agent. The properties of the steel tube used in the study are given in Table 2.

Table 1. Recipes of the UHSC.

No.	Cement	Sand	Coarse aggregate	Water	Superplasticizer	Expansive agent
1	100	100	80	19.0	0.4	0
2	100	100	80	19.5	0.4	5
3	100	100	80	20.0	0.4	10
4	100	100	80	20.4	0.4	15

Table 2. Properties of the steel tube.

Q235	Tensile strength	Yield strength	Elastic modulus	Elongation
	416 MPa	247 MPa	198 GPa	25%

2.2 Mixing and Casting Procedure

UHSC cubes and UHSCFST specimens are produced in the study. The cubes are used for experimenting the compressive strength of the UHSC material with different amounts of expansive agent, and the UHSCFST specimens are used for the compressive experiments of the UHSCFST and the pushout experiments.

The mixing procedure is given below: dry mix of cement and sand for 1 min; add UEA expansive agent with 75% water and mix for about 2 min; add the remaining water with the superplasticizer incorporated and mix for another 2 min; subsequently add the coarse aggregates and mix for around 5 min until they are well distributed in the mixture.

Steel cubic moulds of 100 × 100 × 100 mm^3 and steel tube of 70 × 70 × 250 mm^3 are used to cast the UHSC cubes and the UHSCFST specimens, respectively. The thickness of the steel tube is 2.5 mm. The bottom of the tube is carefully wrapped with plastic film to avoid UHSC leakage from the bottom during the casting procedure. There are two types of UHSCFST specimens in the study: one for the compressive experiment, in which the steel tube is completely filled with UHSC; and the other one for the pushout experiment, in which the UHSC is filled to a height of 200 mm, i.e., a space of 50 mm is left, so that the concrete can slide inside the tube during the pushout experiment. The UHSC cubes are cured in the steel moulds and the UHSCFST specimens do not require demoulding. They are covered with plastic films on top and cured in the air at room temperature of about 20 °C and 65% humidity for 27 days.

2.3 Experimenting Methods

The compression and pushout experiments of the designed specimens are conducted with universal testing machines. The experiment setups are given in Figs. 1 and 2. In the pushout experiments, the UHSCFST specimen is put in the middle of the machine, with the 50 mm high empty space located at the bottom of the tube. A rectangular steel block is placed on the top of the UHSCFST specimen to push the core UHSC to slip inside the tube. The length of the steel block is slightly smaller than the inner cross-section length of the steel tube. For the compression experiment of UHSC, displacement control under a rate of 0.2 mm/min is adopted. In the case of UHSCFST compressive and pushout experiments, load control is adopted and the rate is 2 kN/s. The load stops when the displacement of the loading head reaches 40 mm.

Fig. 1. UHSC under compression.

a) b)

Fig. 2. UHSCFST under a) compression and b) pushout.

3 Experimental Results and Discussion

3.1 Compressive Strength of UHSC

The compressive strengths of the designed UHSC cubes at the age of 28 days are given in Table 3. It can be obtained from the table that adding 10% and 15% UEA leads to the improvement of the strength. More specifically, the compressive strengths of UHSC-10 and UHSC-15 are about 3.2% and 9.2% higher than that of group UHSC-0, respectively. The increase in compressive strength with the addition of UEA is also observed in references [7, 8], which analysed the properties of high-strength-concrete (HSC) with various UEA dosages. It is found that UEA can react with cement and generate C-S-H and calcium alumina, filling the pores in concrete and reducing the porosity [9]. The denser material structure is conducive to improve the concrete strength. Furthermore, the expansion also compensates for the shrinkage stress of the UHSC, thereby increasing its strength [8]. However, the specific trend of strength enhancement in this study differs from that in reference [7], where adding 10% UEA increases the compressive strength while the addition of 14% UEA reduces the strength. The reason may be the different curing conditions. The UHSC cubes in this study are cured in steel moulds, whereas the HSC cubes in reference [7] are demoulded and then cured for 28 days. It is obvious that the steel mould provides a circumferential confinement on UHSC, so that it is under a three-dimensional stress state, thus increasing its compressive strength. Thanks to this confinement effect, the large amount of UEA in group UHSC-15 results in a higher circumferential stress on the cube and therefore its compressive strength does not decrease in this study. Furthermore, Table 3 also shows that the compressive strength of group UHSC-5 is smaller compared to that of the reference group without UEA. This may be due to the slightly higher water/cement ratio of UHSC-5 and the small dosage of UEA in this group, which makes its positive effect less pronounced.

Table 3. Compressive strengths of the UHSC.

No.	Identification	Expansive agent dosage (%)	Compressive strength (MPa)
1	UHSC-0	0	110.0
2	UHSC-5	5	103.0
3	UHSC-10	10	113.5
4	UHSC-15	15	120.1

3.2 Compressive Strength of UHSCFST

Damage modes of the UHSCFST specimens with and without UEA are shown in Fig. 3. It seems that adding UEA has no obvious influence on the damage mode. Table 4 gives the compressive strengths of the UHSCFST specimens. Compare to the effects on increasing the compressive strength of UHSC (within 10%), the effects of UEA on the strength of UHSCFST is much more prominent. To specify, in comparison to UHSCFST-C0, adding 10% and 15% UEA results in 17.3% and 19.5% higher compressive strengths of the UHSCFST, respectively. This more significant effect of the expansive agent on the compressive strength of UHSCFST can be attributed to the stronger confinement provided by the steel tube. However, with regard to UHSCFST-C5, its compressive strength is slightly lower than that of the UHSCFST-C0. This is mainly due to the lower strength of the core UHSC with 5% UEA. Moreover, this small amount of UEA produces an insignificant expansion that is not sufficient to offset the shrinkage of the core UHSC, which prevents the full realization of the confinement effect of the steel tube. Nevertheless, it is still noteworthy that the difference in strength between specimens with 0% and 5% UEA is smaller in UHSCFST than in UHSC, which further confirms the effect of tube confinement on increasing the compressive strength.

3.3 Pushout Force of UHSCFST

The pushout process of the UHSCFST can be mainly described below. (1) During the initial phase of the pushout experiment, there is almost no sliding between the core UHSC and the steel tube. Chemical bonding force between the UHSC and the tube is the most important factor in this phase. (2) As the load increases, the UHSC-tube interface starts to failure begins. The UHSC starts to slide inside the tube, during which

a) b)

Fig. 3. Damage modes of UHSCFST with a) 0% and b) 10% under compression.

Table 4. Compressive strengths of the UHSCFST.

No.	Identification	Expansive agent dosage (%)	Compressive strength (MPa)
1	UHSCFST-C0	0	138.3
2	UHSCFST-C5	5	137.0
3	UHSCFST-C10	10	162.3
4	UHSCFST-C15	15	165.3

the mechanical occlusion force plays an important role [10]. (3) After the load reaches its maximum value, the UHSC slips significantly. The effects of chemical bonding force and mechanical occlusion force decrease in this phase, while the friction between the UHSC and the inner wall of the steel tube contributes most to the bond strength.

Figure 4 illustrates the pushout force versus UHSC slip obtained from the UHSCFST pushout experiments. For the UHSCFST specimens containing 0% or 5% UEA, the curves drop sharply after the pushout forces reach the peak, indicating a weak bond between the core concrete and the steel tube. This also confirms that in UHSCFST with 5% UEA, the confinement effect of the steel tube is not fully utilized, and it is an important reason for its lower compressive strength (see Table 4). On the other hand, the addition of 10% and 15% UEA effectively improves the ductility of the UHSCFST specimens. Especially in the case of UHSCFST with 15% expansive agent, a hardening trend of the curve can be observed in the figure, which indicates a remarkably enhanced bond strength between UHSC and the steel tube.

The maximum pushout forces of UHSCFST specimens with various UEA dosages are listed in Table 5. The influence of the expansive agent on the maximum pushout force is similar to its effects on the compressive strengths described above. Compared to the reference group without UEA, the addition of 5% UEA has a slight negative effect on the pushout force. This is because the expansion of the core UHSC in UHSCFST-P5

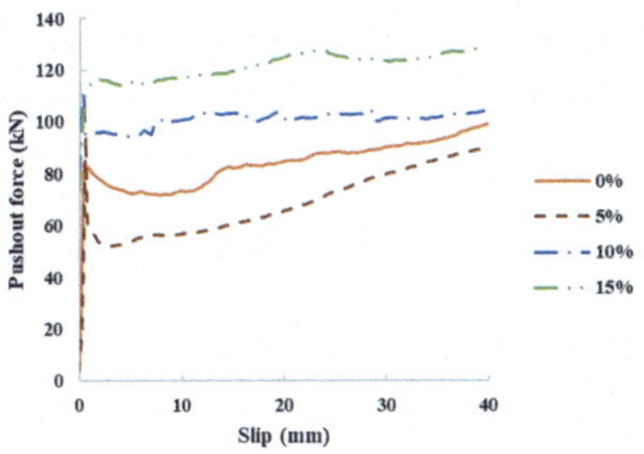

Fig. 4. Pushout force- slip relation of UHSCFST.

is insufficient to compensate its shrinkage, which weakens the bond between the UHSC and the tube. In contrast, higher amounts of UEA, namely 10% and 15%, increase the maximum pushout force by 15.8% and 29.1%, respectively. This is mainly associated with the fact that the larger the amount of UEA, the greater the expansion of the core UHSC, resulting in a more pronounced confinement on the UHSC by the steel tube.

Table 5. Pushout forces of the UHSCFST.

No.	Identification	Expansive agent dosage (%)	Maximum pushout forces (kN)
1	UHSCFST-P0	0	99.6
2	UHSCFST-P5	5	97.7
3	UHSCFST-P10	10	115.3
4	UHSCFST-P15	15	128.6

4 Conclusions

This study investigates the influences of UEA dosage (5%, 10% and 15%) on the mechanical properties of UHSC and UHSCFST. The compressive strengths and bond strength between the steel tube and the core UHSC are analysed. The following conclusions can be drawn from the study:

(1) The effect of UEA on the compressive strength of UHSCFST is more significant than that on UHSC. Adding 10% UEA leads to 3.2% increase of the UHSC strength but 17.3% increase of the UHSCFST strength thanks to the combined effects of the UEA and the tube confinement.
(2) The addition of 15% UEA results in the highest pushout force, which is 29.1% higher compared to that without UEA. A hardening trend of the pushout force-slip curve is observed, indicating the remarkably improved co-working performance of the steel tube and the UHSC.

The UHSCFST proposed in this study helps to fully utilize the mechanical properties of both the steel tube and the UHSC. Thereafter, the thickness of the steel tube can be reduced at the same level of confinement, which in turn saves material. This study can provide a reference for the engineering design and boarder application of UHSCFST in the construction field.

Acknowledgments. Authors wishing to acknowledge the financial support from the National Natural Science Foundation of China (No. 52208297), the Scientific Research Project of National University of Defense Technology (No. ZK21-50) and the Natural Science Foundation of Hunan Province, China (No. 2021JJ30776).

References

1. Liew, J., Xiong, M., Xiong, D.: Design of concrete filled tubular beam-columns with high strength steel and concrete. Structures **8**, 213–226 (2016)
2. Mirmomeni, M., Heidarpour, A., Zhao, X., Al-Mahaidi, R., Packer, J.: Size-dependency of concrete-filled steel tubes subject to impact loading. Int. J. Impact Eng **100**, 90–101 (2017)
3. Li, H., Wang, Y., Wang, Y., Liu, J., Tian, Q.: Effect of CaO and MgO based expansive agent on deformation and mechanical properties of concrete-filled steel tubes. Constr. Build. Mater. **250**, 118723 (2020)
4. Chen, Y., Cao, G., Chen, Z., Hu, J.: Research on mechanical performance and bearing capacity calculation method of expansive concrete filled steel tube columns under axial compression. Build. Struct. **44**, 36–40 (2014). (In Chinese)
5. Tam, V., Tao, Z., Evangelista, A.: Performance of recycled aggregate concrete filled steel tubular (RACFST) stub columns with expansive agent. Constr. Build. Mater. **272**, 121627 (2021)
6. Zhou, X., Guo, K., Li, X., Chen, J., Ji, J.: Application research of expansion agent in high stone powder content of manufactured sand C80 concrete filled steel tube. World Build. Mater. **38**, 63–72 (2017). (In Chinese)
7. Li, Y., Zhou, S., Yang, M.: Influence of UEA on the properties of cement mortar and high performance concrete. Concrete **5**, 12–15 (2001). (In Chinese)
8. Song, C., Ma, S., Shi, H.: The low temperature super early high strength concrete prepared by UEA expensive agent instead of early strength agent. New Build. Mater. **6**, 12–14 (2015). (In Chinese)
9. You, B., Chen, F., Han, L., Yan, H.: Investigation on the long-term properties of UEA cement mortar and concrete. J. Chin. Ceram. Soc. **28**, 314–318 (2000). (In Chinese)
10. Sun, B., Zhang, J., Chen, W., Zeng, S., Zhang, Z.: Influencing factors on bond-slip performance of recycled aggregate concrete filled steel tubes under repeated push-out. J. Build. Mater. **26**, 970–979 (2023). (In Chinese)

Research on the Influence of Concrete Shrinkage on the Stress Behavior of Bridge Piers

Shenglong Yan(✉), Jian Wang, Xiaoliang Zhai, and Yuncheng Feng

CCCC First Highway Consultants Co., Ltd., Xian 710065, Shaanxi, China
303634536@qq.com

Abstract. Concrete shrinkage is the key factor affecting the mechanical properties of continuous rigid bridge, especially on the bridge piers. Taking the Tianning Gully Bridge as the engineering background, the influence of concrete shrinkage on the stress behaviour of bridge piers was studied in the present work. Abutment height and cross-section size, which were considerd as the most significant impact factors on the abutment stiffness, were selected as control variables and the variable parameters of the whole bridge were analyzed. The results showed that the concrete shrinkage effect was more sensitive to the stiffness of the abutment, which increased with the increase of abutment stiffness, showing a parabolic trend. Concrete shrinkage on the double-limbed pier stress was slightly larger than the single-limbed pier, when the pier height was less than 60 m, this trend was more obvious. When the pier height was greater than 120 m, the level of influence on the two tends to be the same.

Keywords: concrete shrinkage · rigid structures · stiffness · bridge piers · influence law

1 Introduction

Continuous rigid bridge is a multiple super-static structure. Dahui Gong et al. found that the shrinkage effect of concrete often leads to structural elongation and shrinkage, which generates secondary internal forces on the bridge superstructure and substructure [1–4], and then affects the overall internal force distribution of the whole structure, especially the force on the main pier.

Research by Qiang Yu and others has shown that the shrinkage effect of concrete can cause uniform expansion and contraction deformation of the structure [5, 6]. For the continuous rigid system, the size of the structural secondary internal force caused by the concrete shrinkage, and the constraint stiffness of the abutment is closely related. Generally speaking, the smaller the stiffness of the bridge pier, the less obvious the concrete shrinkage effect, and the smaller the secondary internal force generated by it in the structure [7, 8]. However, by the abutment bearing capacity and the maximum double cantilever construction control, the abutment size can not be reduced indefinitely. Therefore, a reasonable determination of the size of the side abutment can not only effectively reduce the impact of concrete shrinkage secondary internal forces on the structure, but also take into account the needs of the structural capacity at the same time.

B. Yuan et al. (Eds.): ICCASE 2024, SUCI, pp. 249–257, 2024.
https://doi.org/10.1007/978-3-031-76102-7_29

2 Rigid Bridge Pier Force Characteristics

2.1 Types of Rigid Bridge Piers

For the cross-section type of rigid bridge, the completed bridge is mainly composed of two major types of single-limb piers and double-limb piers as shown in Fig. 1, and individual piers will also be used in the combination of piers [9].

Single-limbed thin-walled pier means that the entire pier consists of a cross-section, which can be hollow or solid. For high pier span continuous rigid structure, single-limb box section has a strong bending and torsion stiffness, making it be a more widely used pier cross-section form. But the box section has a large longitudinal thrust stiffness, it has poor adaptability to the longitudinal deformation of the structural system, Double-limb thin-walled piers due to its relatively small stiffness in the direction of the bridge, can be well adapted to the longitudinal deformation of the bridge, is currently more often used in the form of a bridge pier.

2.2 Thrust Stiffness of Rigid Bridge Piers

Because of its two end boundaries are consolidated, the internal force distribution of rigid bridge pier is most significantly affected by its thrust resistance stiffness under the action of concrete shrinkage [10, 11]. Since the stiffness of the main girder is large relative to the pier, the lower end of the pier of a rigid bridge can be regarded as cemented at both ends, the upper end is cemented with the main girder and the lower end is cemented with the foundation. From the theory of structural mechanics, it can be seen that under the external force at the top of the pier, the thrust stiffness K_{tl} of the single-limb pier and K_{t2} of the two-limb pier can be calculated by formula (1) and (2):

$$K_{t1} = \frac{12EI}{h^3} \tag{1}$$

$$K_{t2} = \frac{12EI}{h^3(1+2\rho_0)} \tag{2}$$

where E is the modulus of elasticity of concrete, I is the flexural moment of inertia of the section, h is the height of the pier, ρ_O is the overall stiffness coefficient of the double-limbed pier.

As seen from the above equation, for the concrete structure, its modulus of elasticity E is certain, therefore, the main parameters affecting the thrust stiffness of the bridge pier are the height of the pier h and the bending moment of inertia of the pier's own cross-section I. Therefore, the influence of these two variables on the mechanical properties of piers is the main research object of this paper.

3 Finite Element Model

3.1 Project Overview

Tianning Gully Bridge is located in Qingyang City, Gansu Province, with an average annual relative humidity of 55% in the project area. The main bridge adopts (95 + 3 × 180 + 95) m variable section prestressed concrete continuous rigid bridge, the

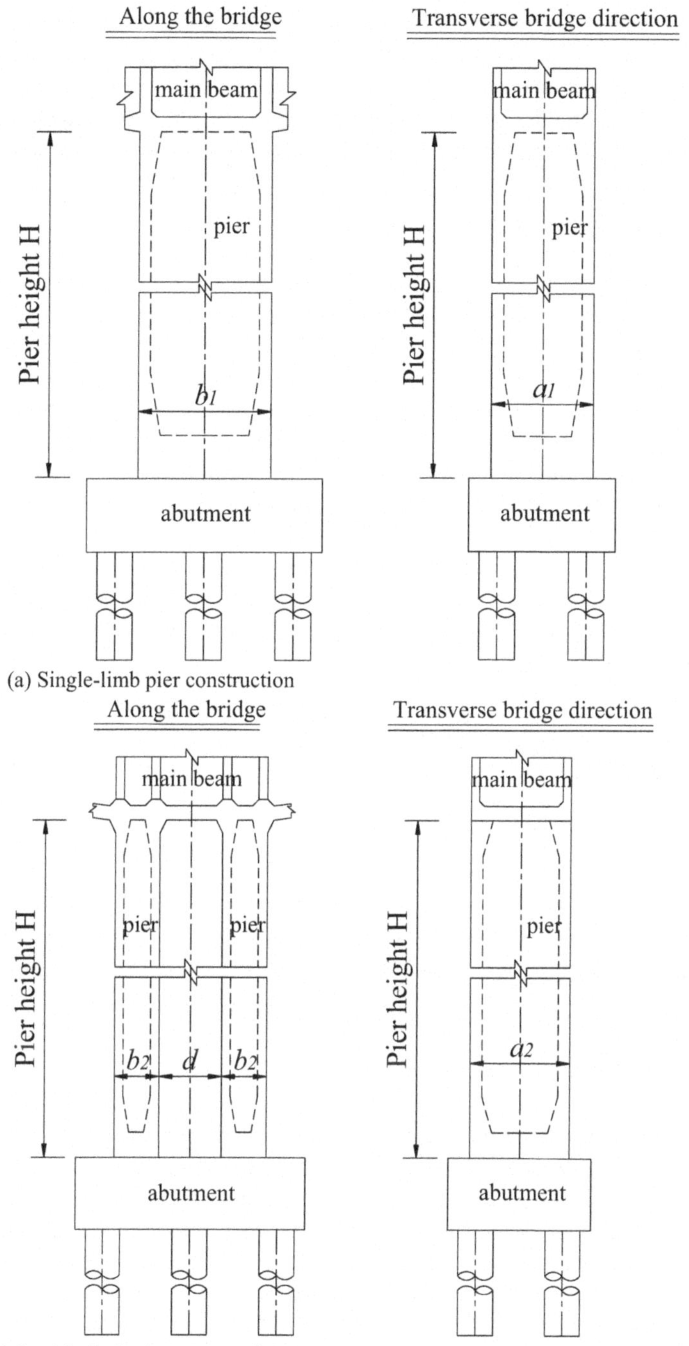

(b) Double-limb pier construction

Fig. 1. Schematic illustration of bridge pier structure.

main girder adopts single box and single chamber section, configured with three-way prestressing system. The root girder height is 11.0 m, the span girder height is 4.0 m, the piers adopts single-limbed hollow and thin-walled piers and the group pile foundation with diameter of 2.5 m, the maximal pier height is 163 m, and the maximal pile length is 96 m, Tianning Gully Bridge is the first high pier in China's loess area when completed.

3.2 Finite Element Model

The space finite element software was used to establish a three-dimensional model of the whole bridge, the main girder and piers were simulated using beam units, according to the cantilever pouring method to consider the impact of the construction stage and the age of the concrete. Main pier bottom boundary according to the "m" method was considered as the role of foundation pile soil, according to the node spring imposition. The top of the pier and the main beams were connected by the cementation and the transition pier according to the actual change of bearing stiffness for simulation. The finite element model of the whole bridge was shown in Fig. 2.

(a) full Bridge model (b) bridge pier cross-section model

Fig. 2. Finite element model.

4 Variable Parameter Analysis

In order to study the variation law of stress conditions on bridge piers under concrete shrinkage effect, a variable parameter analysis was conducted based on a real bridge with a span of 3 × 180 m. From the previous section, it can be seen that the size of the rigid structure pier thrust stiffness affects the size of the secondary internal force generated by the concrete shrinkage effect directly, so the pier thrust stiffness of the most significant impact of the pier height h and the section of the bending moment of inertia I were taking as variables in the variable parameter analysis. The cross-section of the moment of inertia I was directly controlled by the geometric dimensions.

For single-limb piers, the cross-section size was $a_1 \times b_1$, for double-limb piers, it's the cross-section size was $a_2 \times b_2$. a_1, a_2 were the size of the abutment along the bridge to the size of b_1, b_2. It can be seen from the analysis that the constraint effect of side pier was the strongest for rigid structural systems. Therefore, the height of side pier was taken as a variable in the variable parameter analysis, and h = 40 m, 60 m, 80 m, 100 m, 120 m and 140 m were taken respectively. In order to make the analysis results more reliable, the pier height should be increased according to the structural requirements, and the pier size should be changed accordingly according to the principle of equivalent abutment stiffness. The transverse dimensions of the pier were the same as the bottom width of the main beam, which were 6.6 m. The longitudinal widths of the single pier were a1 = 4.0 m, 6.0 m, 8.0 m and 10.0 m, and the longitudinal widths of the double pier were a2 = 2.5 m, 3.0 m, 3.5 m and 4.0 m. The effect of concrete shrinkage on the stress of the bridge abutment.

The development of concrete shrinkage effect is a complex process with a long time period, and the shrinkage strain of concrete is generally calculated according to the following formula:

$$\varepsilon_{cs}(t, t_s) = \varepsilon_{cs0} \cdot \beta_s(t - t_s) \tag{3}$$

where t is the age of concrete at the moment of calculation consideration (d), t_s is the age of concrete at the beginning of shrinkage (d), $\varepsilon_{cs}(t, t_s)$ is the total strain of shrinkage at the calculated age of t, ε_{cs0} is the nominal shrinkage factor, and β_s is the coefficient of the development of shrinkage with time. According to Eq. (3), the development curve of shrinkage strain with time for different grades of concrete from C20 to C80 can be obtained, as shown in Fig. 3.

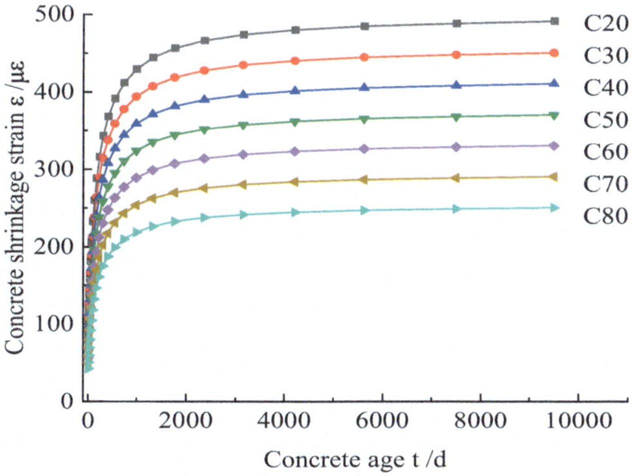

Fig. 3. Development curve of concrete shrinkage strain with time.

As it was shown in Fig. 3, the variation trend of shrinkage strain of different grades of concrete was basically the same. Both developed rapidly in the early stage, stabilized in the later stage, and reached the end of the basic development of shrinkage strain in about 10 years (3650 days). At the same time, it can be seen that the lower the concrete grade was, the greater the total shrinkage strain was. Therefore, in the study of the influence of concrete shrinkage on the force of bridge piers, the total shrinkage effect of concrete in 10 years (3650 days) was taken as the control index, and the variable parameters of different pier heights and different pier sizes were analyzed. Under the effect of concrete shrinkage for ten years, by changing the type and height of the main pier of Tianning Gully bridge, the stresses at the top and bottom of the pier under corresponding working conditions were calculated as shown in Table 1. From the table, it can be seen that the stress of the pier decreased with the increase of the pier height under the action of the 10-year concrete shrinkage effect.

Table 1. Comparison of stresses under ten-year shrinkage of concrete for different pier heights (MPa).

pier height H/m	Single-limb pier stress (a1 = 6.6 m, b1 = 8.0 m)		Double limb pier stress (a1 = 6.6 m, b2 = 3.5 m)	
	top of pier	the bottom of a pier	top of pier	the bottom of a pier
40	7.27	19.42	17.6	15.12
60	6.61	12.37	9.15	8.24
80	5.15	8.23	5.4	4.97
100	3.96	5.79	3.51	3.26
120	3.07	4.26	2.44	2.28
140	2.42	3.25	1.78	1.66

Under the action of ten-year shrinkage-creep effect of concrete, the variable parameter analysis was carried out with the pier height H and the downstream bridge dimensions $b1$ and $b2$ as variables, and the corresponding results were shown in Figs. 4 and 5.

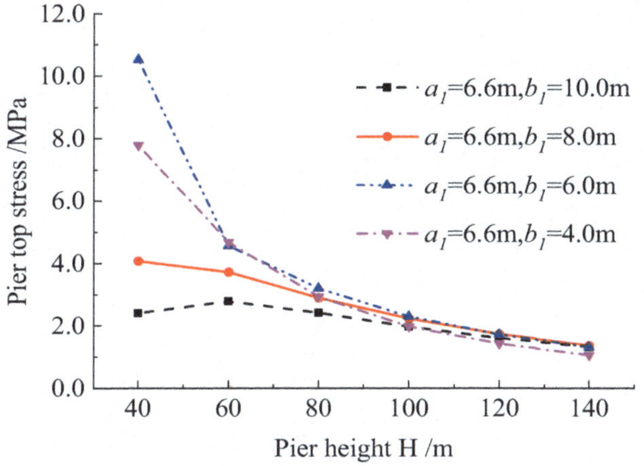

(a) Stress changes at the top of single-limb piers

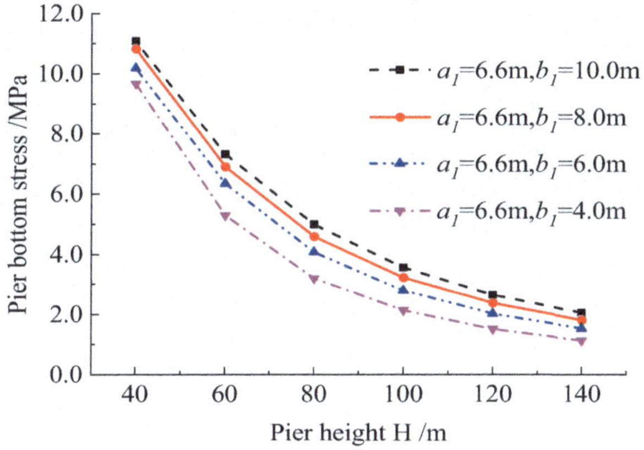

(b) Stress changes at the base of single-limb piers

Fig. 4. Shrinkage stresses in concrete of single-limb piers versus pier height and abutment size.

As shown in Fig. 4, the shrinkage stress of the concrete at the top and bottom of the double-limb pier is inversely proportional to the change of the pier height. The shrinkage stress of concrete at the top and bottom of single-limb pier varies with the change of pier height. The stress at the bottom of pier decreases sharply with the change of pier height. The stress at the top of pier decreases with the change of pier height, although it decreases, but the rate of decrease is slow. Especially when the pier height is less than 60 m, the stress at the bottom of pier changes more discretely.

With the increase of pier size, the overall trend of concrete shrinkage stress was the same, and both of them changed in direct proportion. When the pier height was less than 60 m, the pier size was more sensitive to the shrinkage stress of concrete. When the pier height was greater than 120 m, the influence rate of pier size on the shrinkage stress of concrete abutment decreases and tends to be stable.

At the same time, the effect of concrete shrinkage and creep on the stress level at the top and bottom of the pier was basically the same. Under the influence of linear stiffness,

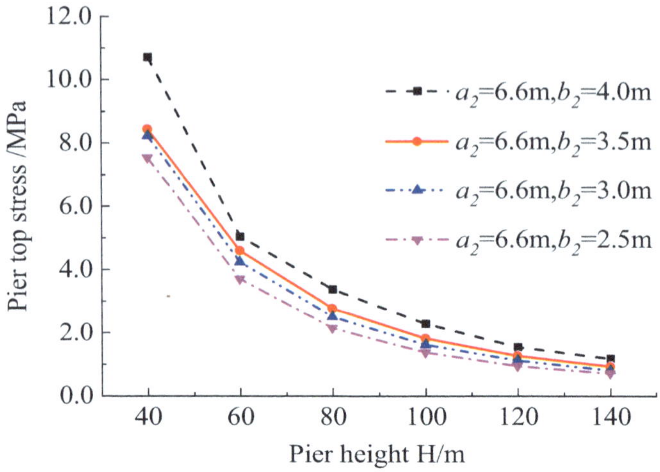

(a) Stress changes at the top of a double-limbed pier

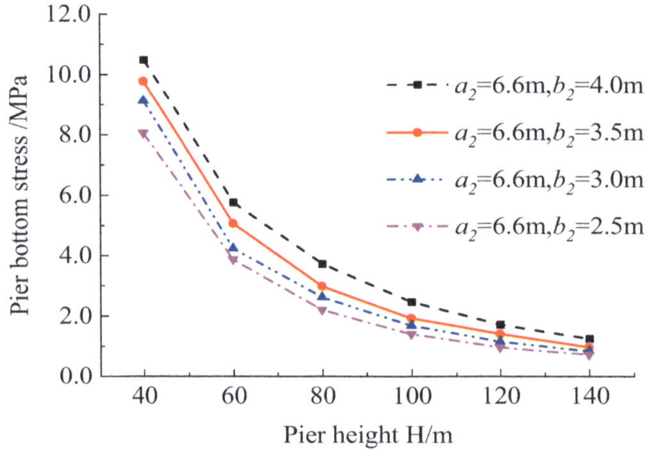

(b) Stress changes at the base of a double-limbed pier

Fig. 5. Shrinkage stresses in concrete of double-limb pier versus pier height and abutment size.

the influence of pier height change on the bottom of the pier was greater than that on the top of the pier, and the change of pier size was no longer obvious when the pier height was greater than 120 m.

5 Conclusions

The effect of concrete shrinkage on the stress of main pier of continuous rigid frame bridge often controls the design of the structure. This paper took Tianning Gully bridge as the background, through the numerical analysis of a large number of variable parameters, the law of the effect of concrete shrinkage on rigid pier was obtained, it can provide a certain reference for the design of bridges.

(1) The calculation results showed that the concrete shrinkage effect on the stress of double-limbed pier was slightly larger than the single-limbed pier.
(2) The effect of concrete shrinkage on the stress at the top and bottom of double-limb pier was the same, both of which increased with the increase of pier stiffness and showed a parabolic trend.
(3) When the pier height was less than 60 m, the concrete shrinkage had a great influence on the force of rigid pier. When the pier height was greater than 120 m, the thrust stiffness of the pier became smaller, the influence of concrete shrinkage on the side force of the pier also became smaller, and the stress level of the pier tended to be basically stable. At this time, the concrete shrinkage load was no longer the controlling factor of the pier force.

References

1. Pan, H., Chu, W., Dai, G.: Analysis of temperature effect of reinforced concrete support under compressive bending load. Build. Struct. **43**(18), 70–73 (2013)
2. Ariyawardena, N., Ghali, A., Elbadry, M.: Experimental study on thermal cracking in reinforced concrete members. Struct. J. ACI **94**(4), 60–67 (1997)
3. Gong, D.: Study on the effect of abutment height difference on the force and deformation of three-span continuous rigid bridge. Constr. Technol. **51**(12), 50–54 (2022)
4. Foust, N.R.: Statistical Models of the Lambert Road Bridge: Changes in Natural Frequencies Due to Temperature. Utah State University (2014)
5. Li, H., Zhan, J.: Temperature stress analysis of main bridge pier columns of Huaihe River Bridge. World Bridge **02**, 45–48 (2010)
6. Li, M.: Research on design parameters of continuous rigid double thin-walled short pier. China Water Transp. **21**(06), 127–128 (2021)
7. Chore, H.S., Shelke, N.L.: Prediction of compressive strength of concrete using multiple regression model. Struct. Eng. Mech. **45**(6), 837–851 (2013)
8. Ruoyu, S.: Research on the arrangement of hollow thin-walled pier for continuous rigid bridge. Value Eng. **42**(22), 72–74 (2023)
9. Chen, Z.: Study on longitudinal stiffness of rectangular hollow-double thin-walled combined piers for rigid bridges. J. Huazhong Univ. Sci. Technol. (Nat. Sci. Ed.) **49**(01), 116–121 (2021)
10. Liu, X.: Research on High Pier Stiffness Matching Problem of High Pier Multi-span Continuous Rigid Bridge. Chang'an University (2015)
11. Shan, C.: Analysis of the influence of pier size and foundation stiffness on the force and self-vibration characteristics of high-pier multi-span continuous rigid bridge. Railw. Constr. **04**, 13–16 (2016)

Study on Fracture Characteristics of Nanomaterials in Concrete

Fulong Li[1], Xianpeng Wang[2](✉), Ya Liu[1], and Jianjun Zheng[1]

[1] School of Civil Engineering, Hainan Vocational and Technical College, Hainan, Haikou, China
[2] Belarusian National Technical University, Civil Engineering, Minsk, Belarus
wxp-phd@bntu.by

Abstract. The wide range of applications of nanomaterials in the field of science and technology has drawn attention to the potential impact of nanotechnology in the field of concrete. Studying the fracture behavior of concrete under load is essential to ensure the reliability of the structure under extreme conditions.In this paper, the three-point bending test of nano-silica concrete is carried out, and the fracture toughness and fracture energy are calculated based on the "Hydraulic Concrete Fracture Test Regulations" and ASTM specifications, and the effects of nano-silica content on the fracture properties of concrete are obtained, respectively. The results show that the addition of silica can significantly increase the fracture energy of nano-silica concrete, and with the increase of nano-silica, the load of nano-material concrete increases first and then decreases, and the fracture toughness increases first and then decreases. With the increase of nano-silica, the average effective crack length decreases first and then increases. In addition, the influence mechanism of nano-silica content on fracture characteristics was explored, and the optimal content of nano-silica content was obtained. This paper can provide a certain reference for construction engineering.

Keywords: Silica · Fracture energy · Fracture toughness · Effective crack length

1 Introduction

The wide range of applications of nanomaterials in the field of science and technology has drawn attention to the potential impact of nanotechnology in the field of concrete. Concrete is one of the most commonly used materials in construction engineering, especially in bridges, buildings and basic structures, making the improvement of its performance an important research direction in the field of civil engineering. However, traditional concrete has certain limitations in terms of impact resistance and fracture toughness, so researchers are always looking for new ways to improve the performance of concrete. The introduction of nanomaterials provides new possibilities for improving the performance of concrete. Nanomaterials, such as nanoparticles, carbon nanotubes, oxide nanoparticles, etc., are introduced into concrete matrices with their special physical and chemical properties. This introduction aims to improve the overall performance of concrete by changing its microstructure and properties inside. For example, the addition of nanomaterials can enhance the compressive strength, improve durability, and improve electrical

B. Yuan et al. (Eds.): ICCASE 2024, SUCI, pp. 258–266, 2024.
https://doi.org/10.1007/978-3-031-76102-7_30

conductivity of concrete. In civil engineering, impact fracture characteristics are a key performance indicator. When a structure is subjected to impact load, its impact resistance and fracture toughness are directly related to the stability and safety of the structure. By enhancing the performance of concrete, nano-concrete is expected to improve the resistance of building structures to natural disasters such as earthquakes and storms. The improved performance of nano-concrete can improve the efficiency of construction process, reduce maintenance costs, and extend the service life of the structure. The study of nano-concrete offers the possibility of new applications, such as in electromagnetic shielding, self-healing concrete, and smart materials. Overall, nanoconcrete research has important implications for advancing building materials science, improving existing infrastructure, and addressing sustainability challenges. Therefore, studying the fracture behavior of concrete under load is essential to ensure the reliability of the structure under extreme conditions.

Wang Jie et al. studied the effect of nanomaterials on the mechanical properties of concrete by adding silica nanoparticles to concrete. They found that as the silica nanoparticle content increased, the compressive strength and fracture toughness of the concrete increased. This indicates that the introduction of nanoparticles can effectively improve the impact fracture characteristics of concrete [1]. Zhu Yunsheng et al. studied the effect of nanomaterials on the conductivity and impact fracture properties of concrete by adding carbon nanotubes to concrete. They observed that the addition of carbon nanotubes significantly improved the conductivity of the concrete and enhanced its impact toughness to some extent [2]. Gao Yaqing et al. used alumina nanoparticles to modify concrete, and studied the effects of different alumina nanoparticle contents on the impact fracture characteristics of concrete. The results show that an appropriate amount of alumina nanoparticles can significantly improve the impact resistance of concrete while maintaining high compressive strength [3]. Huang Kun et al. studied the impact properties of reinforced concrete using nano-titanium dioxide particles. They found that the introduction of nano-titanium dioxide significantly improved the impact resistance of concrete and increased its fracture toughness [4]. Xu Zhidong et al. studied the influence of nano-iron oxide particles on the fracture characteristics of concrete by introducing them. They found that an appropriate amount of nano-iron oxide particles could significantly improve the impact toughness of concrete and improve the compressive properties of concrete to a certain extent [5]. Wang Lingling et al. added diatomaceous earth nanoparticles to concrete to study their effects on the mechanical properties and impact toughness of concrete. The results show that the introduction of diatomaceous earth nanoparticles improves the impact toughness of concrete and increases its compressive strength [6]. Cui Jia et al. studied the effect of nanomaterials on the microstructure and impact properties of concrete by adding nano-calcium silicate. It has been found that an appropriate amount of calcium nanosilicate can significantly improve the fracture toughness of concrete and improve its overall impact resistance [7]. Li Xiaoying et al. studied the effect of nano-zinc oxide particles on the conductivity and impact toughness of concrete. The results show that the addition of nano-zinc oxide particles not only improves the conductivity of concrete, but also significantly enhances its impact toughness [8, 9]. Yu Haoxuan et al. elaborated on the characteristics of nanofibers, and analyzed the influence and mechanism of non-metallic nanofibers on the compressive strength, tensile strength, flexural strength and durability of concrete [10].

The purpose of this study is to systematically study the fracture characteristics of concrete under load by introducing nanomaterials. By gaining an in-depth understanding of the mechanism of interaction between nanomaterials and concrete, we hope to provide a scientific basis for the application of nanoconcrete in practical engineering. The significance of this research is to expand the way of concrete performance improvement and promote the research and application of nanotechnology in the field of civil engineering.

2 Materials and Experimental Methods

In this paper, one of the reasons for choosing three-point bending for fracture testing is that this test method can assess the bending properties and toughness of the material. In a three-point bending test, the sample is placed between two support points and then a force is loaded in the middle, simulating the behavior of the material while bending. This method is commonly used to assess the fracture strength and fracture toughness of materials. The test process involves applying a load under standard conditions and measuring the deformation and fracture behavior of the material. The process of three-point bending concrete fracture test is as follows: preparing concrete specimen; Before the test, the specimen was soaked and cured for a certain time to ensure that the concrete reached the required strength and toughness; Install the support and loading devices of the testing machine to ensure that the spacing and loading points are set according to the standard specifications; Place the specimen on the support point and load the intermediate point. Gradually apply the load to simulate the stress of concrete during bending; During the loading process, the load and deformation of the specimen are recorded. The data will be used for subsequent analysis; Continue loading until the specimen breaks. Observing fracture patterns is crucial to understanding the toughness and bending properties of concrete. With the recorded data, the toughness, bending strength and other performance indexes of concrete are calculated. These indicators are very important for engineering design and quality control. The three-point bending test can provide the performance information of concrete in practical use, and has guiding significance for structural design and construction engineering. A three-point flexure standard sample was prepared by adding different contents of nano-silica (1%, 3%, 5%, 7%, and 9%), and then subjected to a three-point flexure test [11, 12]. After the experiment, the effects of different contents of nano-silica on the fracture energy GF and fracture toughness of concrete were studied.

3 Analysis of Results

3.1 Fracture Toughness

As shown in Fig. 1, With the increase of silica content of nanomaterials, the fracture toughness increases first and then decreases within a certain range. For example, when the nano-silica content increases from 1% to 9%, the roughness increases from 1.82 to 2.12

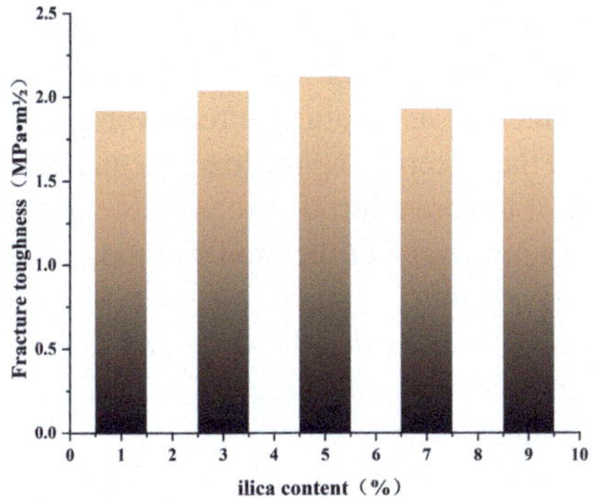

Fig. 1. Variation of fracture toughness

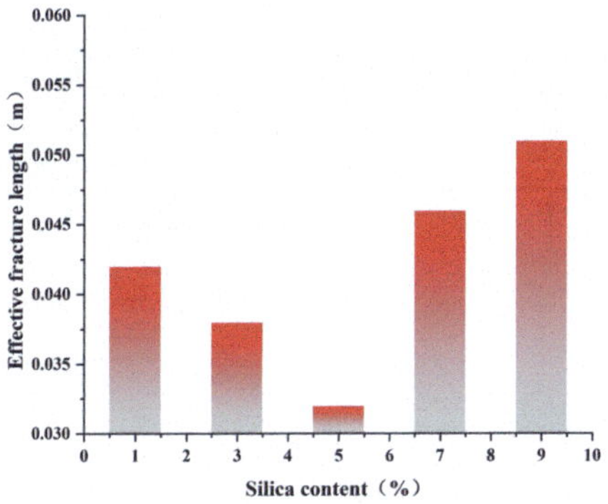

Fig. 2. Variation of effective crack length

and then decreases to 1.76. This phenomenon may be affected by many factors, and needs to be analyzed from the perspectives of material microstructure, fracture development, and engineering applications. As shown in Fig. 3, As the silica content increases, the microstructure of nanomaterial concrete may undergo complex changes. In the initial stage, an appropriate amount of nano-silica particles fills the pores of the concrete, which improves the compactness of the concrete, thereby increasing the compressive strength of the material. However, when there are too many nanoparticles, it may lead to aggregation between the particles, forming locally concentrated intensity regions, which in turn promote the formation of fissures. This may be one of the reasons why the load increases first and then decreases. During the loading process, the development of cracks inside the concrete has a significant impact on the overall performance. In the initial stage, an appropriate amount of nano-silica particles may inhibit the propagation of cracks to a certain extent and improve the toughness of concrete. As shown in Fig. 2, However, when there are too many nanoparticles, it may lead to concentrated propagation of cracks, which makes the fracture toughness of concrete decrease. Therefore, the development of fractures is one of the important factors affecting the change of fracture toughness. From an engineering point of view: Combined with the load-displacement curve and the fracture toughness trend, the optimal ratio can be found by optimizing the nano-silica content to improve the overall performance of the concrete. This requires detailed material design and experimental verification in real-world engineering. In view of the trend of increasing first and then decreasing the load, it can be considered to improve the compressive performance of concrete through reasonable nanomaterial allocation, and at the same time adopt other material modification methods, such as fiber reinforcement, to enhance the durability of concrete. Considering the cost and production difficulty of nanomaterials, it is necessary to consider the economy and feasibility in engineering practice. There may be an economic advantage of achieving the best performance in terms of nanomaterial content.

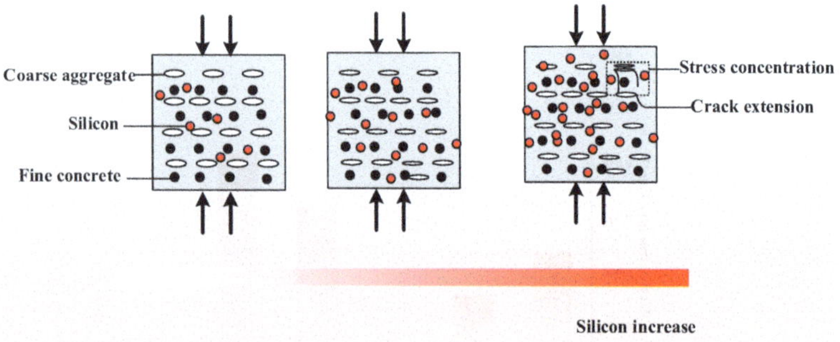

Fig. 3. Mechanism of silica effect on fracture toughness

With the increase of nanomaterials, the average effective crack length decreases first and then increases. For example, when the content of nano silica increases from 1% to 9%, the effective crack length decreases from 0.042 to 0.031 and then increases to 0.0526. This change can be influenced by a number of factors. The right amount of nano-silica particles may fill the pores of the concrete, improving the compactness of the concrete, thereby inhibiting the formation and propagation of cracks. This can lead to a decrease in the effective crack length during the initial loading phase. However, when there are too many nanoparticles, it may lead to the concentration of particles in a certain area, forming a region of high local intensity. This may trigger the concentrated propagation of the fracture to a certain extent, resulting in a gradual increase in the effective fracture length. In addition, a series of chemical reactions may occur between the nano-silica particles and the hydration products in the cement matrix. These reactions may have altered the internal structure of the concrete to a certain extent, affecting the formation and propagation of cracks. The different phases of the chemical reaction can lead to changes in the effective crack length. In view of the trend of effective crack length, it is necessary to comprehensively consider the mechanical properties of concrete, including compressive strength and toughness. In practice, a balance needs to be found so that the concrete improves crack development without sacrificing other important properties. By carefully designing the ratio of nano-material concrete, the structure of concrete can be optimized and its durability can be improved. The control of effective crack length helps to slow down the rate of deterioration of concrete structures during use.

3.2 Fracture Energy

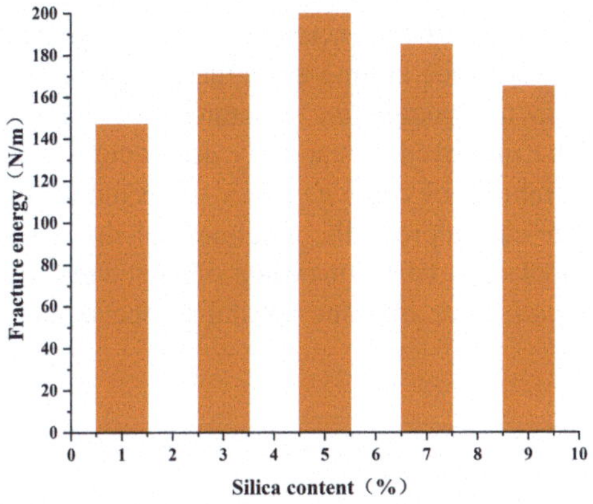

Fig. 4. Fracture energy variation law

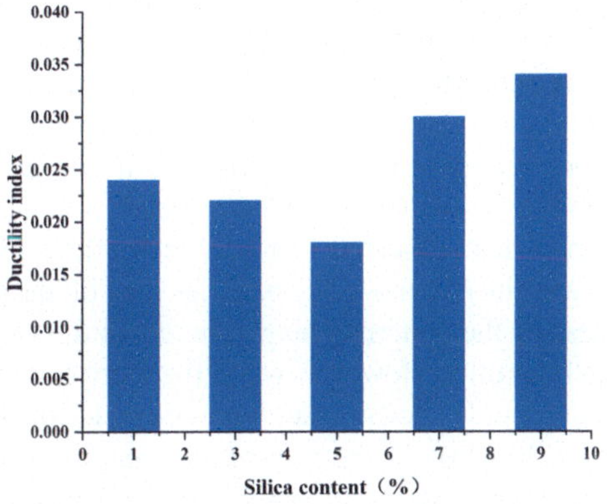

Fig. 5. Ductility index variation law

As shown in Fig. 4, With the increase of nano-silica materials, the fracture energy first increases and then decreases. For example, when the nano-silica content increases from 1% to 9%, the fracture energy increases from 148 to 196 and then decreases to 165. This change can be influenced by a variety of factors. As the silica particles increase, a chemical reaction with the cement hydration product may occur to form a substance with higher strength. This may lead to an increase in the initial tensile properties of the nanomaterial concrete, which in turn increases the fracture energy. However, when the reaction gradually becomes saturated, the excess particles may cause local aggregation to a certain extent, forming a fracture-sensitive area, resulting in a decrease in fracture energy. Initially, the right amount of silica particles may fill the pores in the concrete, improve the compactness of the concrete, help slow the crack propagation, and thus increase the fracture energy. However, when there are too many particles, it can lead to the aggregation of particles, forming a source of cracks, increasing the sensitivity of the concrete and reducing the fracture energy. In the project, the content of nanoparticles can be reasonably adjusted to find an optimal ratio. This requires a significant performance improvement at the beginning of the increase in particle content, and the ability to control the decline in performance in time when there is an excess of particles. In view of the change trend of fracture energy, the structure of concrete can be optimized and its fracture energy can be improved by carefully designing the ratio of concrete. This helps to improve the durability of concrete when subjected to force The optimization of nanomaterial concrete can improve the toughness of the structure and increase its resistance to external shocks and vibrations when subjected to force. This has important practical application significance in some special projects, such as seismic regional architectural design.

3.3 Ductility Index

As shown in Fig. 5, With the increase of nano-silica, the ductility index first decreases and then increases. For example, when the nano-silica content increases from 1% to 9%, the ductility index decreases from 0.024 to 0.0175 and then increases to 0.034. With the increase of nano-silica particles, it may cause a chemical reaction with cement hydration products to form compounds with higher strength. This can improve the initial tensile properties of the concrete to some extent, but too many particles may lead to the formation of cracks, which can reduce the overall ductility. As the reaction progresses, a certain chemical equilibrium may be reached, and the distribution of excessive particles in the concrete may tend to be uneven, which may lead to non-uniform deformation of the concrete structure when subjected to force, thus affecting the ductility index. Initially, an appropriate amount of nano-silica particles may fill the pores in the concrete, improve the compactness of the concrete, help slow down the crack propagation, and thus improve the overall ductility. However, when there are too many particles, it can lead to the aggregation of particles, forming a source of cracks, increasing the sensitivity of concrete and reducing ductility. In the project, the content of nanoparticles can be adjusted to find an optimal ratio. An initial increase in content can improve some of the properties of concrete, while a moderate decrease in content can prevent the accumulation of excessive particles and improve the overall ductility. In engineering, the ductility of the overall structure can be improved by optimizing the structural design of concrete and considering the content and distribution of nanomaterials. This is especially important for structures that need to resist shock and vibration, such as bridges and buildings in seismic areas. Considering the mechanical properties of nanomaterial concrete, it can be used in the seismic design to improve the ductility of the structure in the event of an earthquake and reduce the structural damage. The rational use of nanomaterials can improve the durability of concrete, reduce the incidence of cracks and damage, and thus increase the service life of the overall structure. By comprehensively considering the properties of nano-material concrete, it can be better applied to engineering practice, give full play to its excellent mechanical properties, and improve the safety and durability of concrete structures.

4 Conclusion

Nano-silica concrete is a new kind of building material, which has many excellent properties, such as high strength, durability and crack resistance. Understanding the fracture behavior of nano-silica concrete can guide studying the fracture characteristics is helpful to design more reasonable structures, reduce the formation and expansion of cracks, and improve the safety and stability of buildings. The formulation and preparation process of the material can be optimized to improve its cracking resistance and prolong its service life. The detailed conclusions are as follows:

(1) With the increase of nano-silica, the fracture toughness of nano-material concrete first increases and then decreases. With the increase of nano-silica, the average effective crack length decreases first and then increases. The results provide a new idea for the application of nanomaterials in concrete.
(2) With the increase of nano-silica, the fracture energy first increases and then decreases. This is due to the increase in silica particles, which may undergo a chemical reaction with cement hydration products to form a stronger substance. This may lead to an increase in the initial tensile properties of the nanomaterial concrete, which in turn increases the fracture energy. However, when the reaction gradually becomes saturated, the excess particles may cause local aggregation to a certain extent, forming a fracture-sensitive area, resulting in a decrease in fracture energy.
(3) With the increase of nano-silica, the ductility index decreases first and then increases. This is due to the fact that as the number of nano-silica particles increases, it may cause a chemical reaction with cement hydration products to form compounds with higher strength. This can improve the initial tensile properties of the concrete to some extent, but too many particles may lead to the formation of cracks, which can reduce the overall ductilit.
(4) The future research direction is to deeply understand the fracture mechanism of nano-silica concrete, including the formation and propagation mechanism of cracks, as well as the fracture toughness of the material, etc. The effects of different shapes, sizes and contents of nano-silica fillers on the fracture performance of concrete can also be explored to optimize the fracture characteristics and strength.

Acknowledgments. This article is one of the phased results of the school-level project of Hainan Vocational and Technical College "Research on Impact Fracture Characteristics of Nanoconcrete Materials" (HZY (QN) 2024–4).

References

1. Wang, J., Fu, R., Dong, H.: Carbon nanofibers and PVA fiber hybrid concrete: abrasion and impact resistance. J. Build. Eng. **80**, 107894 (2023)

2. Yunsheng, Z., Yuzhen, X., Xigang, D., et al.: Effect of nano-SiO_2-CSH composite regulator on the properties of cement concrete. J. Chongqing Jiaotong Univ. Nat. Sci. Ed. **42**(11), 57–62 (2023)
3. Effect of nano SiO_2 and Al_2O_3 on the properties of fly ash base polymers. Transportation World, 2023, (32):11–13. https://doi.org/10.16248/j.cnki.11-3723/u.2023.32.016
4. Huang, K., Xie, J., Feng, Y., et al.: Axial impact behaviors of UHPC: the roles of nanomaterials and steel fibres. Constr. Build. Mater. **384**, 131396 (2023)
5. Zhidong, X., Junpeng, M., Zhixin, W., et al.: Effect of nano-silica on physical and mechanical properties and hydration microstructure of mineral powder-cement system. Concrete **04**, 69–77 (2023)
6. Lingling, W., Baiyan, C., Zhiqiang, W., et al.: Study on physical and mechanical properties of photocatalytic concrete. Concrete **04**, 29–35 (2020)
7. Jia, C., Chenxin, N., Xusheng, Z., et al.: Effect of nano-hydrated calcium silicate on concrete performance of C60 shield segments. New Build. Mater. **50**(10), 22–48 (2023)
8. Xiaoying, L.I., Jun, L.I., Zhongyuan, L.U., et al.: Preparation and performance of antibacterial active powder concrete. J. Wuhan Univ. Technol. **42**(02), 14–18 (2020)
9. Dinesh, A., Parthiban, V., Parvathi, N. S., et al.: 2023 Crack-bridging and strengthening prospects of nanofibers in the cement composite–a review. Mater. Today: Proceedings
10. Yu, H., Wang, X., Nikolaevich, K. S., et al.: 2023 Mechanical properties and environmental protection analysis of non-metallic fibers//2023 9th International Conference on Architectural, Civil and Hydraulic Engineering (ICACHE 2023). Atlantis Press, 830–842
11. Dong, H., Wang, J.: Carbon nanofibers and polyvinyl-alcohol fiber hybrid-reinforced high-performance concrete: mechanical property, chloride penetration resistance, and material characterization. Constr. Build. Mater. **399**, 131891 (2023)
12. Yang, Y., Xing, Y., Fang, K., et al.: Influence of dynamic load and water on energy accumulation and dissipation in sandstone. Sci. Rep. **13**(1), 22010 (2023)

Study on the Influence of Mineral Evolution on Structural Strength of Loess During Collapsibility

Zhechao Zhu(✉), Deren Liu, Yafang Guo, and Yan Fei

Lanzhou Jiaotong University, Lanzhou 730070, Gansu, China
1227116730@qq.com

Abstract. With the implementation of the western development strategy, the continuous promotion of the transportation strategy, the construction and operation of a series of high-speed railroad projects, such as the Yinxi High-speed Railway and the Yinlan Railway, higher requirements have been put forward for the engineering and construction technology in loess areas. To study the variation rule of the mineral composition evolution on the structural strength influence situation with time in the process of loess wet subsidence. The test results show that the structural strength of loess will be obviously reduced after being exposed to water, which can be divided into three stages of steep decline, slow decline and slow rise in different time. The change of mineral composition of loess from the beginning of water immersion to 24 h is mainly the dissolution of easily soluble salts, and the structural strength decreases rapidly; from 24 h to 336 h, the dissolved encapsulated colloidal particles slip and fill in between the skeleton particles, and the structural strength decreases slowly. From 336 h to 672 h within the influence of feldspar hydration, the structural strength rises slowly and the microstructure is gradually stabilized. The structural strength was power function correlated with the change of calcite content from the beginning to 168 h. The microstructural observation shows that the pore structure shows a change process from destabilization to stabilization. It can be seen that the evolution of loess mineral composition and loess structural strength show consistency, and the microstructure shows salt dissolution and regular changes in particle size and pore space, and mineral hydration and dissolution is an important reason for the change of loess structural strength.

Keywords: mineral evolution · Structural strength · Microscopic characteristics · Collapsible loess

1 Introduction

Under the arid and semi-arid environmental conditions in Northwest China, Northwest China loess shows some unique mineral composition and structural characteristics, which are characterized by the development of vertical joints, obvious pore structure, weak cementation and strong wet subsidence [1, 2]. Wet subsidence is a unique engineering geological property of loess, which is a serious threat to the construction process and

B. Yuan et al. (Eds.): ICCASE 2024, SUCI, pp. 267–275, 2024.
https://doi.org/10.1007/978-3-031-76102-7_31

the safety of buildings due to the wet deformation of loess in contact with water, which has caused many serious engineering problems [3, 4]. With the continuous promotion of China's western development strategy, the development and construction of transportation infrastructure construction, introduction of new energy sources, urban renovation and other projects in Northwest China have put forward more stringent requirements for the engineering construction in loess areas [5].

Gao Guorui [6] proposed that particle morphology, connection form and arrangement are the three major factors affecting the wet subsidence of loess. Based on the study of loess microstructure evolution, Zheng Wanpeng [7] et al. found that the content of medium and small pores was the key structural factor determining the collapsibility of loess. Li [8] observed and quantitatively analyzed the microstructures of loess samples after various mechanical tests (consolidation-wet subsidence test, triaxial test) and found that under the action of force, the microstructural evolution is related to the level of stress and the path of stress.The study of wet subsidence in terms of the structural properties of in situ unsaturated loess and its mechanical properties showed that there are large differences in the structural properties of soils from one place to another, which affect their mechanical properties differently [9], and the wet subsidence situation varies. However, the existing structural theory [10] explains the mechanism of wet subsidence of loess from the structural features such as the morphology, contact relationship, arrangement of skeleton particles in the soil, the type of cement and the endowment state, etc., but it fails to illustrate the wet subsidence situation of loess well from the effect of the evolution of loess mineral compositions on the microstructures, especially for the feldspars in loess, the solubilization of carbonate minerals in contact with water, and the change of hydration reaction over time on the microstructure has not been studied much. Therefore, this paper investigates the evolution of minerals in the process of loess wet subsidence and tests the microstructural changes of loess through indoor experiments, which will help to understand the wet subsidence and deformation of loess, to predict and assess the risk of wet subsidence before engineering construction, and to provide guidance on prevention and control techniques.

2 Overview of the Experiment

2.1 Selection of Soil Samples

Test selected Yongdeng County G312 line Fu Bitter Road near the soil sampling site wet subsidence loess, the use of mechanical excavation and artificial soil sampling combined with sampling, sampling depth of about 10m, to obtain the original soil samples transported back to the laboratory into the test. The test can be known as sampling loess basic physical properties indicators shown in Table 1.

2.2 Test Methods

2.2.1 Straight Shear Test

Loess specimen preparation was carried out in accordance with GB /T50123–2019 Standard for Geotechnical Test Methods [11]. For the same soil samples prepared as-is and

Table 1. Indicators of basic physical properties of loess

Natural moisture content/%	Liquid limit/%	Plastic limit/%	Plasticity index/%	Optimum moisture content/%	Maximum dry density/(g/cm^3)
4.69	26.4	14.4	12	13.32	1.76

remodeled ring knife samples, spray humidification method was used to humidify to the optimal moisture content of the soil samples and sealed and stored. After reaching the set time, the tests were carried out using ZJ-type strain-controlled straight shear apparatus, totaling eight parallel groups of tests with set time points of 0 h, 1/30 h, 1 h, 12 h, 24 h, 168 h, 336 h, and 672 h, respectively.

2.2.2 X-Ray Diffraction Test

The test was carried out by Bruker D8 ADVANCE X diffractometer, and the mineral samples were selected to be dissolved at different times, sieved through a 200-mesh sieve, and the samples were filled into the sample carrier by the positive-pressure method, compacted and flattened, and then gently swept away by a soft brush to remove the surrounding excess samples, and then placed the sample carrier into the diffractometer to be analyzed, so that the X-diffraction spectra of the samples were obtained at different times, and the mineral composition and content of the samples were obtained by Jade6.0 software analysis and processing. The X diffraction patterns of different samples at different times were obtained, and the mineral composition and content of the samples were analyzed and processed by Jade 6.0 software.

2.2.3 Scanning Electron Microscopy Tests

The test was carried out using a ZSISS GeminiSEM 500 scanning electron microscope. During the drying process of soil samples, moisture transportation and the increase of capillary tension will cause the "wet inside and dry outside" state of soil samples, which will lead to shrinkage and deformation. Freeze-drying method can greatly reduce the contraction and deformation of the sample when drying, the tissue structure can be better preserved, and spend a short time. In order to better obtain the soil structure in the natural state, the freeze-drying method was chosen to dry the test soil samples in this study. The dried soil samples were cut into cubes with flat surfaces, sprayed with gold and then subjected to scanning electron microscopy, to obtain the magnified images of primary loess at different times of water immersion, 500 times, 1000 times, and 2000 times, and the image of 500 times was chosen to analyze the microstructural characteristics of primary loess for comparative observation.

3 Analysis of Test Results

3.1 Analysis of the Evolution Process of Mineral Composition in Contact with Water

The changes of mineral composition content at different times of water immersion are shown in Fig. 1, which shows that the mineral composition content of in situ loess at different times of water immersion is significantly different, with the highest quartz content (42.6–44.8%), sodium feldspar (16.7–3.2%), ilmenite (10.5–15.0%), calcite (6–8.8%), montmorillonite (2.9–7.6%), plagioclase (5.2–5.9%) followed by dolomite (2.1–.1%), kaolinite (0.3– 3.7%), and hematite (0–2.1%) in the least amount.

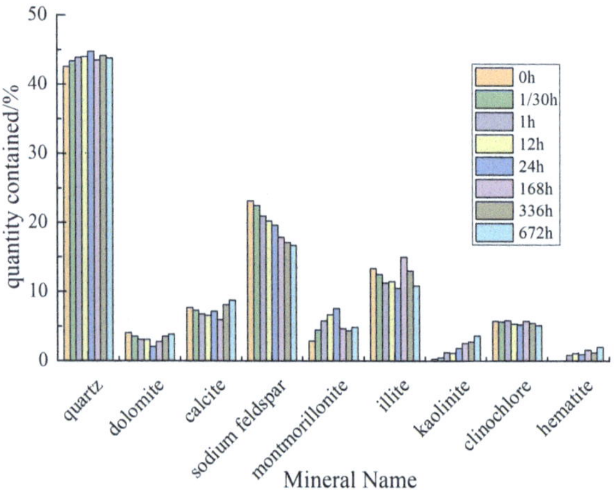

Fig. 1. Variation curve of mineral content with water immersion time

Hydrolysis of sodium feldspar in loess occurs during the process of water immersion, resulting in a decrease in content with time. When feldspar particles are hydrolyzed, they absorb water to remove surface cations and combine with hydrogen ions in water to produce kaolinite ($Al_2Si_2O_5(OH)_4$) and sodium montmorillonite ($NaAl_2[(Si,Al)_4O_{10}][OH]_{2n}H_2O$) to be deposited in the pore space of the colluvium, resulting in loess The overall trend of particle size after dissolution by water immersion becomes smaller, and makes montmorillonite and kaolinite content increase. So that the feldspar content decreases its hydrolysis reaction equation as shown in Eqs. (1) and (2):

$$Na_2[AlSi_3O_8] + 4H_2O \rightarrow Al_2Si_2O_5(OH)_4 + 2Na^+ + 4OH^- \quad (1)$$

$$Na_2[AlSi_3O_8] + H_2O \rightarrow Na^+ + NaAl_2\left[(Si, Al)_4O_{10}\right][OH]_2nH_2O \quad (2)$$

Dissolution of dolomite and calcite in loess occurs under submerged water conditions. When the water molecules come in contact with the dolomite in the loess, the water molecules exchange ions (e.g. calcium ions Ca^{2+}) with ions on the surface of the dolomite. This leads to the partial dissolution of carbonate, which reduces its content, and after the reaction reaches 168 h, the ion concentration in the loess reaches a dynamic equilibrium inhibiting the continued hydrolysis of carbonate, and the content of carbonate rises slowly.

3.2 Analysis of the Process of Structural Strength Change in Loess

The strength of unsaturated loess mainly consists of two parts: friction strength and cohesion strength [9], in this test, the structural strength is defined as the linkage strength produced by the cementing material at the point between the coarser grains and the magnitude of the strength lost in the destruction of the natural structure of loess, the magnitude of the structural strength of loess can be expressed by the difference in the shear stresses, qs, of the in situ loess and the corresponding (same gravity, same moisture content) remolded loess at the time of destruction of its natural structure.

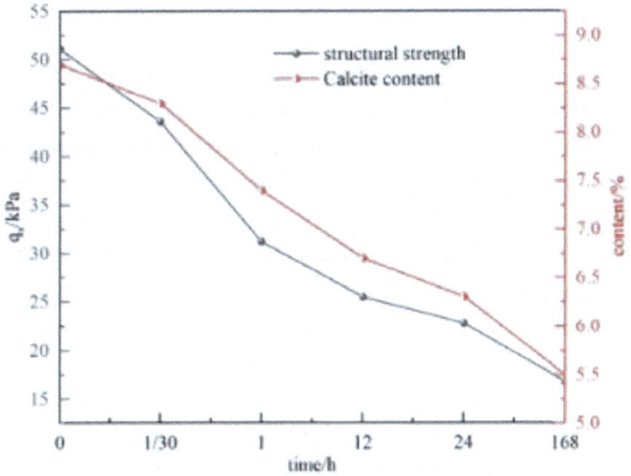

Fig. 2. Curve of structural strength and calcite content with time

The measured structural strength and the corresponding time points are plotted in the rectangular coordinate system, as shown in Fig. 2, it can be seen that the structural strength of loess varies with time for the trend of decreasing and then increasing, and the decrease in a short period of time is much larger than that in a long period of time. For example, in the humidification only 24 h loess, the structural strength decreased by 30.34kPa, the decrease reached 59.4%, while in the humidification 336 h after loess, the structural strength of loess compared with 24 h, decreased by 4.33kPa, the decrease was only 20.9%. At 672 h, the structural strength of loess increased compared with that at 336 h. The reason is: loess particles are distributed with a lot of calcite and dolomite as the representative of carbonate and other soluble salts, or in the form of flocculent wrapped around part of the small soil particles attached to the surface of large particles; or in the form of large particles with a mosaic distribution between the mineral particles, the formation of a cemented structure to enhance the bonding strength of the particles, the infiltration of water makes the soil particles on the surface of the mineral components,

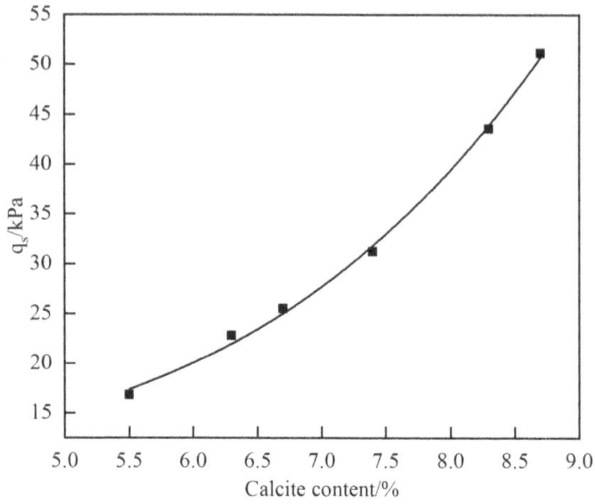

Fig. 3. Curve of structural strength of loess versus calcite content

dissolved or slipped to the soil in the large and medium pore space, and then the soil structure strength is increased. The water infiltration makes the mineral components on the surface of soil particles, dissolve or slip into the large and medium pores in the soil, so that the soil structure collapse, and these salt minerals hydrate quickly, in a short time can make the structural strength drop dramatically, and then due to the reduction of pore space, the mosaic structure increases, the impact of loess particles between the particles-particle contact area and friction between the particles, which slows down the decline of the structural strength or even make the structural strength rise slowly.

And from Fig. 2 can be seen that the structural strength of loess and calcite content in 168 h change quite, the change of structural strength is mainly to the dissolution of salt minerals hydration slip is mainly, the longer the experience time by feldspars and other minerals, the greater the influence of the relationship between the two is smaller, so it can be done in the qs-calcite content in the 168 h relationship curve, as shown in Fig. 3.

It is known that there is a power function relationship between the structural strength of unsaturated loess the change of calcite content within 168 h of humidification, which can be described as Eq. (3):

$$q_s = An^{\alpha} + C \tag{3}$$

where qs is the structural strength of unsaturated loess (kPa); n is the calcite content (%), and A, α and C are parameters determined by the nature of the soil. Regression analysis yielded the parameters A = 0.00709, α = 3.98985, and C = 11.01363 for the loess used in the test, with a correlation coefficient R = 0.9979.

3.3 Characterization of the Microstructure of the Loess Leaching Process

Figure 4 shows the scanning electron microscope images of loess at different times of water immersion. From the beginning of water immersion to 24 h, along with the

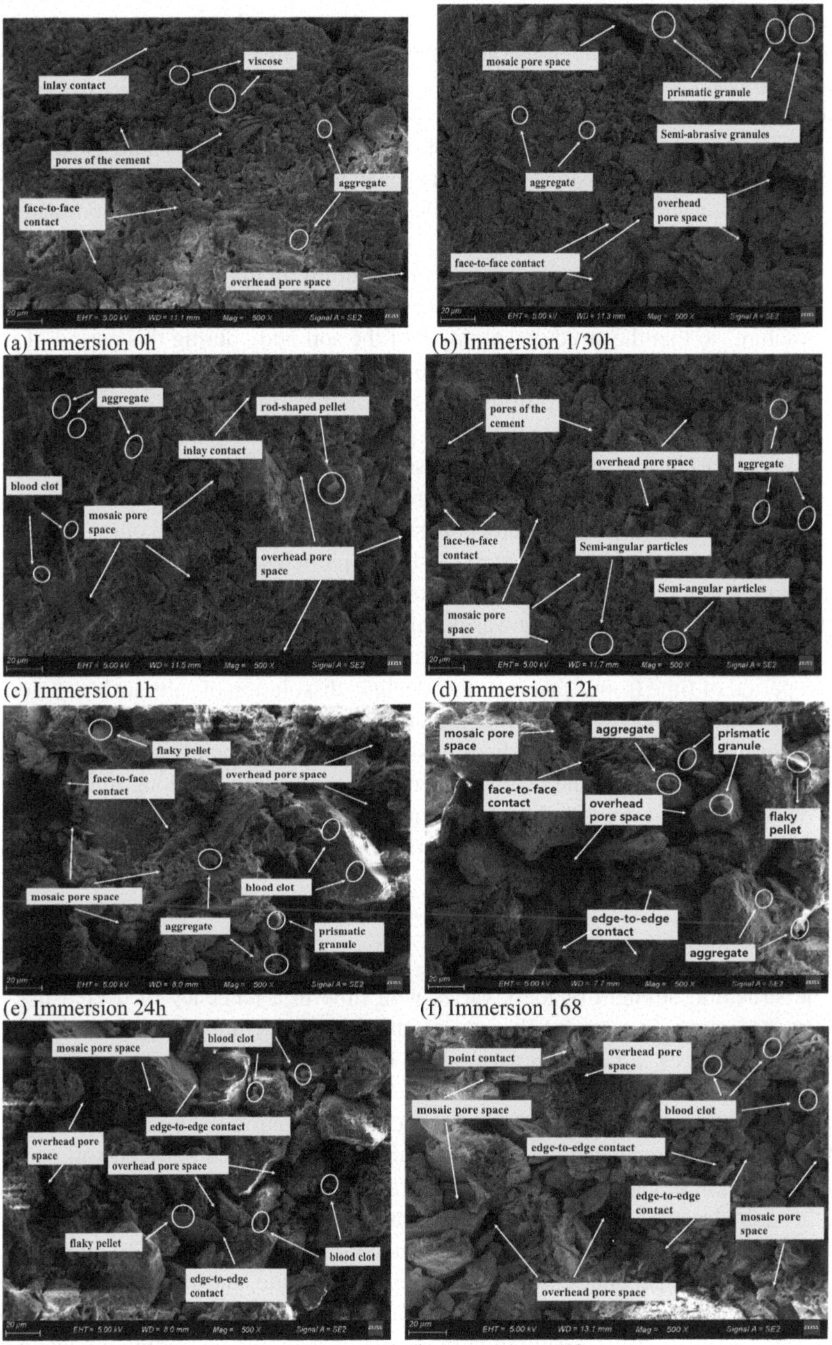

(a) Immersion 0h
(b) Immersion 1/30h
(c) Immersion 1h
(d) Immersion 12h
(e) Immersion 24h
(f) Immersion 168
(g) Immersion 336h
(h) Immersion 672h

Fig. 4. SEM images of loess microstructure under different submergence duration dissolution of carbonates and soluble salts, the agglomerates begin to decrease, and due to the dissolution of these agglomerates, fine powder grains composed of tiny minerals and detrital particle groups such as quartz or feldspar begin to appear. The pores of the cement begin to decrease, the mosaic pores begin to increase, and the structural strength is rapidly becoming smaller. The outline of the loess skeleton begins to be clear. Within 24 to 336 h of water immersion, small clots began to increase, the angles became clear, the form of connection between the particles began to mosaic contact is dominated by the shelf pores and mosaic pores coexist, and the pore space of the cement became less. Dissolution of agglomerates intensified, resulting in clay minerals, finer grained particles set, interspersed between the skeleton particles, to enhance the stability of the soil structure, so that the structural strength of the soil body during this period of time to decline slowly. In the submerged 336 to 672 h, the scaffold pore space due to the hydrolysis of clay minerals slip filling and become less, the number of pores reduced, back to mosaic pore space is dominated, a large number of hydrolyzed clay minerals and colloidal particles to fill the pore space so that the strength of the soil body structure during this period of time increased slightly.

The reason for the above phenomenon is: calcite and dolomite and other minerals in loess are attached to the surface of quartz and other particles in the form of agglomerates, dissolution of agglomerates in the process of water immersion and dissipation, and the position between the particles that was originally a water-soluble agglomerates become pores, and the form of contact becomes unstable, resulting in the structural strength of a short period of time from a fast to a slow decline; dissolution of salts such as dolomite and other minerals, feldspar and other minerals wrapped up in the carbonate dissolved in it. After the contact with water, hydrolysis occurs, clay minerals and other colloidal particles filled with slip to the overhead pore space, resulting in a long time under the slow rise in structural strength. It can be seen that the effect of water immersion time on the microstructure of primary loess is very obvious.

4 Conclusion

1) The structural strength of loess varies with time in a tendency of decreasing and then increasing. In terms of the rate of change of structural strength, it is divided into three stages: a rapid decrease in 24 h, a slow decrease in 24 h-336 h and a slow increase in 336 h-672 h; meanwhile, it can be seen that the structural strength of loess and the content of calcite show a power function relationship in 168 h. In loess, dolomite and calcite as carbonate are mostly attached between soil particles in the form of agglomerates and agglomerates, and the dissolution of carbonate and other soluble salts at the beginning of leaching destroys the agglomerate structure of loess, leading to a rapid decrease in the structural strength of loess. The hydrolysis of sodium feldspar generates montmorillonite and kaolinite, and after the destruction of the agglomerate, the wrapped feldspar and clay minerals are hydrolyzed and slipped, and the pores are filled, resulting in the effect of rounding and filling up, leading to a slow decline in the structural strength of the loess to a slow increase in the trend of corresponding to the phenomenon of wet subsidence of the loess from the rapid wet subsidence to the stable state.

2) microstructural analysis showed that the process of water immersion in the loess in the agglomerate dissolution dissipation phenomenon, between the particles was originally a water-soluble agglomerate position into a pore, dolomite and other salt minerals dissolved, inside the wrapped feldspar and other minerals in the dissolution of carbonate water contact occurs hydrolysis, clay minerals filled with slippage to the shelving pore space, resulting in a long period of time under the structure of the intensity of the structure will go through three phases.

References

1. Nima, V., Alireza, T.: Laboratory Investigation of Plant Root Reinforcement on the Mechanical Behaviour and Collapse Potential of Loess Soil. Eur. (2022). https://doi.org/10.1080/19648189.2020.1715848
2. Zhang, Y., Liu, D.R., Wang, X.: Study on surface deformation regularity of deep loess site of China-Lanzhou Railway. J. Railway Sci. Eng. **19**(09), 2461–2469 (2022)
3. Xu, S.H., Liu, D.R., Wang, X.: Experimental study on macro and micro parameters of large thickness collapsible loess in Lanzhou new district. J. Railway Sci. Eng. **19**(07), 1918–1926 (2022)
4. Giomi, I., Francisca, F.M.: Numerical modeling of the oedometrical behavior of collapsible loess. Geotech. Geol. Engi. **40**(5), 1–12 (2022)
5. Cheng, F.Q.: Research on microstructure of in-situ loess based on field immersion test. Railway Constr. Technol. **01**, 20–23 (2023)
6. Gao, G.: Formation and development of the structure of collapsing loess in China. Eng. Geology (2018). https://doi.org/10.1016/0013-7952(88)90029-4.
7. Zheng, W.P., Dou, H., Liang, W.: Collapsibility mechanism based on microstructure evolution of loess. Highway **68**(08), 322–325 (2023)
8. Li, P., Xie, W.L., Park, R.Y.S.: Microstructural evolution of loess soils from the Loess Plateau of China. Catena (2019). https://doi.org/10.1016/j.catena.2018.10.006.
9. Liu, H.S., Ni, W.K., Yan, B.: Study on the relationship between structural strength and collapsibility of loess. Rock Soil Mech. **03**, 722–726 (2008)
10. Wei, Y.N., (2019) Evolution of three-dimensional microstructure and collapsibility mechanism of loess under water action [D]. Chang 'an University.
11. GB/T 50123–2019, Standard for geotechnical test methods [S].

Study on Post-Construction Settlement of Soft Foundation Treatment Considering Secondary Consolidation

Dan Miao[1,2](✉) and Chushao He[1,2]

[1] Guangzhou Guangjian Construction Engineering Testing Center Co., Ltd., Guangzhou 510080, China
1285842180@qq.com

[2] GuangDong Engineering Technology Research Center of Building Health Monitoring and Security Early Warning, Guangzhou 510080, China

Abstract. The calculation of the settlement after the soft foundation treatment is directly related to the unloading time of the soft foundation treatment. In order to study the post-construction settlement characteristics of soft foundation treatment under secondary consolidation, the numerical models of soft foundation without consideration of secondary consolidation and under secondary consolidation were established by MC (Mohr-Coulomb) constitutive and SSC(soft soil creep) constitutive construction, respectively, and the influencing factors of secondary consolidation were proposed γ and the existing settlement calculation formula was modified, and the results showed that: (1) The secondary consolidation has a significant effect on the sedimentation rate, and the sub-consolidation is only about 0.5 times of the non-secondary consolidation sedimentation rate, and it is recommended that the sub-consolidation impact factor should be 0.5–0.8. (2) The sedimentation rate calculated by the method proposed in this paper is similar to the measured data and has good applicability.

Keywords: soft base treatment · secondary consolidation · numerical simulation · sedimentation rate · Impact factor

1 Introduction

In recent years, the post-treatment settlement of soft foundation has become a concern. Due to inadequate treatment of soft foundation, early unloading leads to excessive settlement in the later stage of foundation, resulting in frequent accidents of cracking of superstructure. Researchers at home and abroad have carried out extensive and in-depth research on the post-treatment settlement of soft foundation.

In terms of unloading standard of soft foundation treatment, sedimentation rate method is the most used unloading standard at present, which is simple to measure and has strong operability. Liu Jifu et al. [1] deduced the relationship between post-construction settlement and settlement rate, and the formula parameters were simple and practical. In terms of post-construction settlement prediction of soft foundation, Shi

B. Yuan et al. (Eds.): ICCASE 2024, SUCI, pp. 276–289, 2024.
https://doi.org/10.1007/978-3-031-76102-7_32

Boyi et al. [2] conducted a high pressure consolidation test for yellow soil settlement, and established a settlement prediction model based on the test results. Zhou Fei [3] used the combination of drainage consolidation and lightweight embankment to treat the normal consolidated soil into super-consolidated soil, which greatly reduced the creep coefficient of the soil and effectively reduced the secondary consolidation settlement. Lin Chen [4] established a calculation method for the settlement of the main consolidation considering the β variation of consolidation parameters. Chen Haiying et al. [5] combined Duncan-Zhang and generalized Huke law to calculate soil lateral deformation, carried out high-pressure consolidation creep test, and substituted the relevant parameters into the component model for nonlinear improvement. The results showed that the final settlement of soil calculated by the improved model was highly consistent with the field monitoring data. Xie Jiehui et al. [6] proposed a variable weight combined prediction model based on the sum of squares of errors and reciprocal absolute value of errors, which is more robust than a single prediction model in predicting the actual soft soil settlement. Wang Bolin et al. [7] optimized the combination of Logistic and Gompertz prediction models and found that their prediction accuracy was higher than that of a single model. In recent years, the secondary consolidation has attracted more and more attention. The secondary consolidation settlement is mainly soil creep. How to determine the start time of secondary consolidation and the amount of secondary consolidation settlement is the key issue. Hong Yi et al. [8] applied the compression characteristics and secondary consolidation characteristics of deep-sea diatomite, and found that the compression index of deep-sea diatomite was about 0.6–1.2 larger than that of soft soil in shallow sea on land, and diatomite still had long-term volume deformation in the creep stage, and did not show the trend of convergence and stabilization. Chen et al. [9] proposed a calculation method of soil settlement and consolidation considering short-term and long-term creep, and conducted finite element analysis with viscoelastic constitutive considering creep limit. Finally, combined with practical projects, the accuracy and efficiency of the simplified method was verified. Abidhan et al. [10] proposed an improved Harris hawks optimization (IHHO) algorithm, which is efficient and practical in predicting soil settlement. Li Guoliang et al. [11] improved the empirical formula for the calculation of secondary consolidation settlement proposed by Buisman.

In order to study the settlement law of soft foundation under the action of secondary consolidation, this paper uses numerical simulation to simulate the settlement deformation of soft foundation after treatment under the treatment of secondary consolidation and without secondary consolidation by using two different constitutive models, and combines engineering examples to carry out research, and on the basis of the existing relevant settlement calculation formula, the influencing factor of secondary consolidation is proposed, and the calculated results are in good agreement with the measured data.

2 Settlement Calculation Theory

According to the one-dimensional consolidation theory of Terzaghi [12], the average degree of consolidation of foundation soil can be expressed as:

$$\overline{U_t} = 1 - \alpha e^{-\beta t} \tag{1}$$

In the formula, the drainage consolidation coefficient is taken according to Table 1 in various cases.

Table 1. Values of consolidation coefficients for drainage [13]

Drainage	α	B
Radial drainage	1	$8C_h/F(n)d_e^2$
Vertical drainage	8/π2	$\pi C_v/4H^2$
Radial and vertical	$8/\pi^2$	$8C_h/F(n)d_e^2 + \pi C_v/4H^2$

If n loads have been applied to the homogeneous foundation at time t, the settlement at time t should be calculated as follows [14]:

$$S_t = \sum_{i=1}^{n} (S_{\infty i} - S_{Di})(1 - \alpha e^{-\beta t_i}) \tag{2}$$

where $S_{\infty i}$, S_{Di} are the final settlement and the instantaneous settlement under a certain stage of loading.

Then the sedimentation rate at time t is:

$$\dot{S} = \frac{dS_t}{d_t} = \alpha\beta \sum_{i=1}^{n} (S_{\infty i} - S_{Di}) e^{-\beta t_i} \tag{3}$$

And the remaining principal consolidation settlement at time t is:

$$S = (S_{\infty i} - S_{Di})\left(1 - \overline{U_t}\right) = \alpha \sum_{i=1}^{n} (S_{\infty i} - S_{Di}) e^{-\beta t_i} \tag{4}$$

The relationship between the sedimentation rateat time t and the remaining principal consolidation settlement under multi-level loads can be obtained by Eq. (3) and Eq. (4).

$$\dot{S} = \beta S \tag{5}$$

β can be back calculated according to the following formula according to the settlement data:

$$\beta = \frac{1}{\Delta t} \ln \frac{S_2 - S_1}{S_3 - S_2} \tag{6}$$

where $\Delta t = t_3 - t_2 = t_2 - t_1$.

According to the definition of secondary consolidation, when the ultra-static pore water pressure in the soil is close to zero, when the ultra-static pore water pressure in the soil is dissipated to zero, the main consolidation is completed, and the subsequent settlement is caused by the secondary consolidation, so that the secondary consolidation settlement can be removed from the remaining settlement, and the remaining is the remaining main consolidation settlement, which can be applied to Eq. (5), that is, a more reasonable sedimentation rate unloading control standard can be formulated. Specifically, it can be expressed as follows:

$$\lambda^* = \frac{C_c}{2.3(1 + e_{init})} \tag{7}$$

$$\lambda^* = \frac{C_c}{2.3(1 + e_{init})} \tag{8}$$

Among them, Sa is the amount of secondary consolidation settlement; The consolidation duration ranges from t_1 to $\lambda^* = \frac{C_c}{2.3(1+e_{init})}$; Δu is the ultra-static pore water pressure.

3 Case Studies

3.1 Project Overview and Modeling

This project is a seawall project, the width of the bottom of the proposed seawall is 50.0 m, the two sides of the embankment are sloped according to the slope of 1:2, the lower part of the embankment is in-situ silt, the soil quality is poor, the thickness of the silt layer is about 25.0 m, the lower part of the silt layer is a medium dense sand layer, and the groundwater level is located 2.0 m below the original mud surface. The design height of the embankment is 5.0 m, considering that the lower silt layer will produce a large consolidation deformation, the design stacking height is 7.2 m, the embankment body is used as the loading material, the lower soft soil layer is reinforced, the silt layer is equipped with a plastic drainage board, the plastic drainage board adopts SPB-B type, and the square arrangement is carried out according to the spacing of 1.0 m, and the depth of laying is 20.0 m.

In order to fully consider the secondary consolidation characteristics of soft soil, the SSC model is used for the constitutive relationship of the silt soil layer in the finite element calculation, and the MC model is still used for the soil filling, and the calculation parameters are shown in Table 1. Here, point A (located on the original mud surface, 1.5 m away from the center of the roadbed) is taken as the research object, and the specific location is shown in Fig. 1, focusing on the analysis of the settlement deformation of the point.

The Mohr-Coulomb model is used for the constitutive relationship of the soil element, the plastic drainage board of the vertical drainage channel is simulated by the drain unit in the program, and a layer of drain unit is also set between the sand cushion and the original silt layer to simulate the horizontal drainage channel. The physical and mechanical properties of the soil used in the calculation are shown in the Table 2.

Considering the symmetry of the converted plane strain problem, only the right half of the problem was analyzed, the first layer of silt was taken as the analysis object, the depth range was taken as 25.0 m, and the horizontal influence range was taken as the range of 25.0 m outside the slope angle, and the finite element calculation model as shown in Fig. 2 was established, and the plane triangle 15-node element was used to divide the grid. The left and right vertical boundaries have only vertical displacement,

Table 2. The physical and mechanical properties of the earth were calculated

Soil	Natural severe/γ	Elastic modulus/E	Poisson's ratio/v	Cohesion Force/c	Nemo Rub the corners/φ	Permeability coefficient	
						K_x	K_y
	$kN \cdot m^{-3}$	kPa	-	kPa	°	$m \cdot d^{-1}$	$m \cdot d^{-1}$
Fill the soil	17.5	2500	0.3	5.0	24.0	1×10^{-3}	1×10^{-3}
silt	14.0	/	/	6.0	15.0	9.6×10^{-6}	5.2×10^{-5}

the right boundary is impervious boundary, the left boundary is permeable boundary within the depth range of the drainage plate, the rest is impervious boundary, and the bottom boundary has neither horizontal displacement nor vertical displacement, which is impervious boundary.

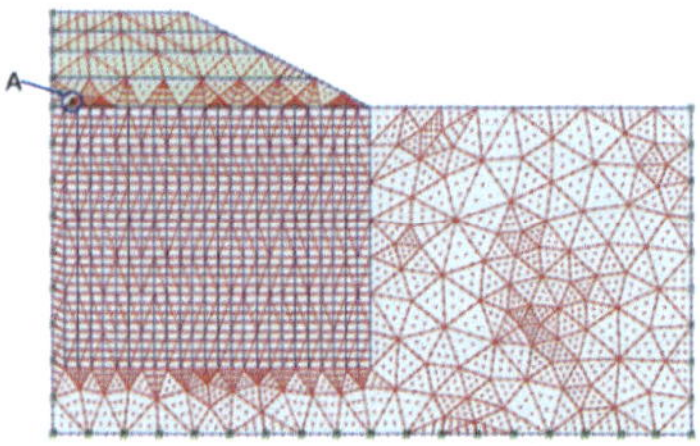

Fig. 1. Schematic diagram of the location of the analysis point

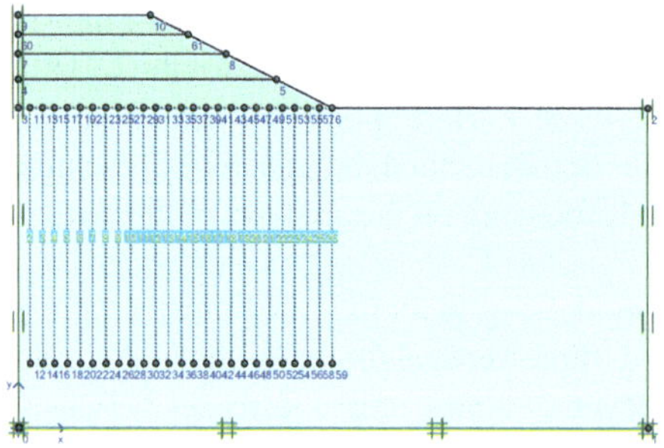

Fig. 2. Schematic diagram of the calculation model

3.2 Degree of Consolidation

There are various methods to define the consolidation time in PLAXIS consolidation calculation, such as the total multiplier, the incremental multiplier, and the minimum pore pressure, where the minimum pore pressure method refers to the soil consolidation process until the superhydrostatic pressure value on all nodes is less than the preset P-STOP value. The final settlement was calculated by the minimum pore pressure method and the total multiplier method, where the P-STOP was defined as 10.0 kPa and the total multiplier consolidation time was defined as 3650 days.

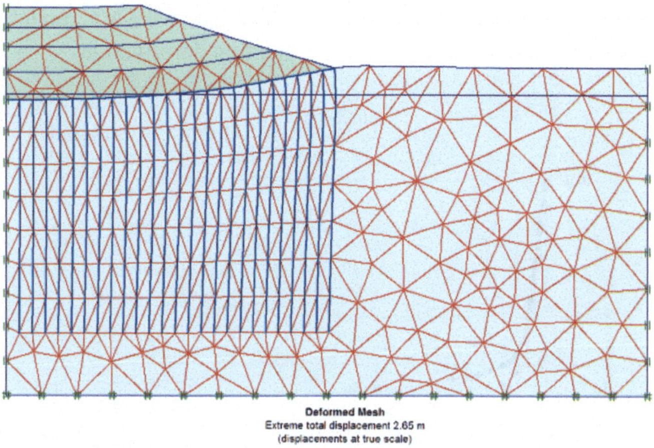

a. Total multiplier method

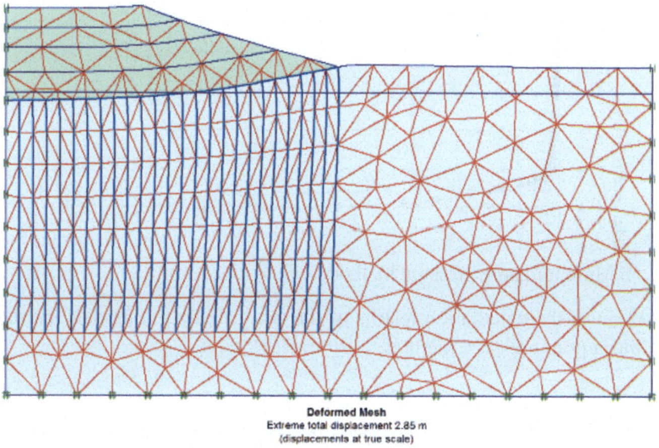

b Minimum pore pressure method

Fig. 3. Schematic diagram of settlement deformation (total multiplier method)

The final settlement deformation of the two methods is shown in Fig. 3, In the two cases, the final settlement of point A is 2.45 m and 2.60 m respectively, and the consolidation time of the minimum pore pressure method reaches 11940 days, indicating that after 120 days of dead load, a certain amount of settlement will be generated, and this settlement will occur slowly over time, the longer the time, the higher the proportion of secondary consolidation in the settlement, even after 3650 days of drainage consolidation, it still produces a settlement of 0.15 m, which shows that it is unreasonable to ignore the influence of secondary consolidation in the post-construction settlement.

As shown in Fig. 4, there is no significant difference between the two types in the early stage of construction, the difference occurs in the late loading stage, the later settlement of the SCC model is significantly greater than that of the MC model, and the settlement of 0.70 m is still generated after 120 days of dead load, which includes the remaining main consolidation settlement and also includes the secondary consolidation settlement, and the consolidation time is longer than that of the MC model, which is in line with the characteristics of secondary consolidation deformation.

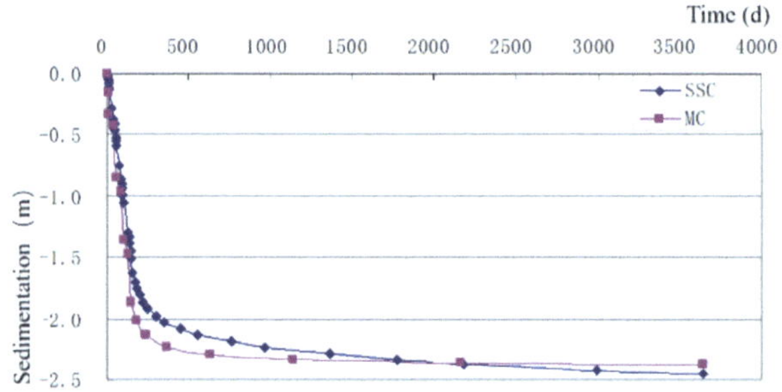

a. Total multiplier method

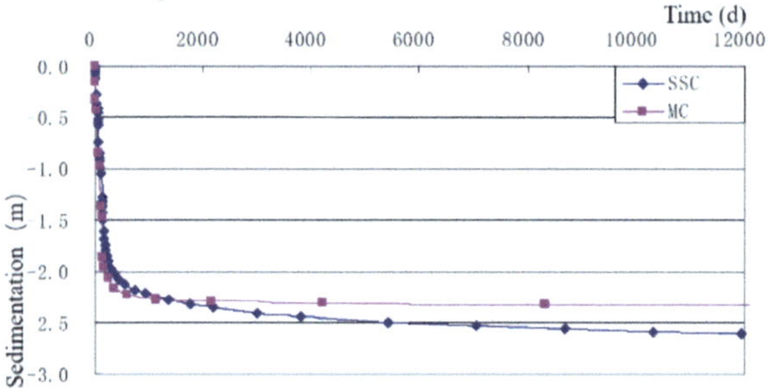

b Minimum pore pressure method

Fig. 4. Comparison of the calculation results of the two calculation models. a. Total multiplier method. b Minimum pore pressure method.

Under the premise of considering the secondary consolidation deformation, the various levels are loaded with sedimentation. The calculated value is compared with the final settlement (taken as 2.60 m) to obtain the consolidation table shown in Table 3.

Because the stacked material itself is the embankment body in this project, there is no problem of unloading, but for most of the soft foundation problems treated by the preloading method, the consolidation degree and post-construction settlement are often used as the requirements for designing unloading, and the consolidation degree and post-construction settlement of the soil are the best criteria for judging the unloading time, so the consolidation degree of the soil should be fully considered when formulating the unloading control standard. It can be seen from Table 3: when the fourth level of loading is completed, the degree of consolidation of soil has reached 73.9%, when the dead load

Table 3. Loading consolidation at all levels (considering secondary consolidation)

Level 1		Level 2		Level 3		Level 4	
sedimentation	Degree of consolidation	Sedimentation	Degree of consolidation	sedimentation	Degree of consolidation	sedimentation	Degree of consolidation
0.318	12.2%	0.852	32.8%	1.386	53.3%	1.922	73.9%
First-stage intermittent		Second-stage intermittent		Level 3 intermittent		Fourth-level intermittent	
sedimentation	Degree of consolidation	sedimentation	Degree of consolidation	sedimentation	Degree of consolidation	sedimentation	Degree of consolidation
0.428	16.5%	0.98	37.7%	1.5	57.7%	2.153	82.8%

is 120 days, the degree of consolidation of the soil is 82.8%, basically reaching or close to the standard of 80% or 85% of the consolidation degree set by the project, and the post-construction settlement of 120 days of dead load is about 0.45 m, which is obviously larger, if the post-construction settlement is considered 0.3 m, the degree of consolidation needs to reach 88.5%, and the consolidation time needs to reach 630 days, which will greatly extend the number of days of construction, Increase the construction cost of the project. When the dual control standards of consolidation and post-construction settlement are adopted, a reasonable control value should be formulated in combination with the actual situation of the project.

3.3 Sedimentation Rate

Equation (5) is the theoretical basis for formulating a reasonable standard for sedimentation rate. Post-construction settlement not only includes the main consolidation settlement, but also includes the secondary consolidation settlement, if only the remaining main consolidation settlement is used as the designed post-construction settlement to substitute Eq. (5) will obtain a large settlement rate, in other words, that is, the project has not yet reached the design requirements, the operation of unloading may be carried out in advance, thereby affecting the quality of engineering construction.

In the engineering construction monitoring, the measured settlement can not be divided into the main consolidation settlement and the secondary consolidation settlement, which brings great difficulty to accurately determine the settlement rate, Eq. (8) from the definition of secondary consolidation, the initial conditions for judging the occurrence time of secondary consolidation t1 are given, based on the above analysis, we take the settlement after the ultra-static pore water pressure at point A dissipates to 0.5 kPa as the secondary consolidation part, and the secondary consolidation settlement in this project is 0.28 m, and the secondary consolidation is added to the SSC model analysis. It can be obtained as shown in Table 4.

Explanation: 1. When not considering the impact of β, take $\beta = 0.01$; 2. γ is defined as the secondary consolidation influencing factor.

It can be seen from Table 4 that the secondary consolidation has a significant impact on the analysis of the settlement rate, the sedimentation rate of the secondary consolidation is only about 0.5 times that of the non-secondary consolidation, and the sedimentation rate of the secondary consolidation is closer to the measured results at 270 days, so the influence of the secondary consolidation should be fully considered when estimating

Table 4. Comparative analysis table

Condition	Time to consolidation (d)									
	270		300		350		400		500	
	Post-work settlement	Sedimentation rate	Post-work settlement	Sedimentation rate	Post-work settlement	Sedimentation rate	Post-work settlement	Sedimentation rate	Post-work settlement	Sedimentation rate
	m	mm/d	m	mm/d	m	mm/d	m	mm/d	m	mm/d
Not considered Secondary consolidationM_1	0.67	6.7	0.63	6.3	0.59	5.9	0.55	5.5	0.50	5.0
consider Secondary consolidationM_2	0.39	3.9	0.35	3.5	0.31	3.1	0.27	2.7	0.22	2.2
$\gamma = M_2/M_1$	0.58		0.56		0.53		0.49		0.44	

the control value of the unloading sedimentation rate by using the post-construction settlement.

Based on the above comparative analysis results, in order to facilitate the application in the project, it is recommended to adopt the following calculation formula in the project:

$$\dot{S} = \gamma \beta S_0 \quad (9)$$

where γ is the influencing factor of secondary consolidation, which is used to reflect the influence of secondary consolidation on the settlement rate, and it is recommended to take the value of 0.5–0.8; S_0 is the remaining main consolidation settlement, and the design value of post-construction settlement can be directly used in the project.

In order to better analyze the influence of the influencing factor of secondary consolidation on the sedimentation rate control standard, the same method was used to study point B (located 8.0 m below the original mud surface and 7.5 m away from the center of the subgrade) and point C (located 13 m below the original mud surface and 12.5 m away from the center of the subgrade), and the following conclusions can be obtained.

Figure 5 shows that the influence factor of secondary consolidation is related to the consolidation time, it can be seen that the secondary consolidation influencing factor is gradually decreasing with the increase of consolidation duration, in other words, the influence of secondary consolidation on the sedimentation rate unloading control standard is more and more significant with the increase of consolidation duration, which is also in line with the theoretical analysis results, with the increase of consolidation duration, the remaining main consolidation settlement is getting smaller and smaller, while the secondary consolidation settlement remains unchanged, and its proportion in the post-construction settlement is increasing, so its influence is more and more obvious, in addition, The influencing factors of secondary consolidation and the consolidation time basically showed a linear change trend.

Figure 6 shows the relationship between the influencing factor of secondary consolidation and depth, it can be seen that the influencing factor of secondary consolidation gradually decreases with the increase of depth, that is, the influence of secondary consolidation on the deep soil is more obvious, and the change trend is more significant within the depth of 8.0 m, and the change tends to slow down below the depth of 8.0 m.

To sum up, the following value method can be used for the secondary consolidation influencing factor γ, the large value is taken for the short consolidation time, the

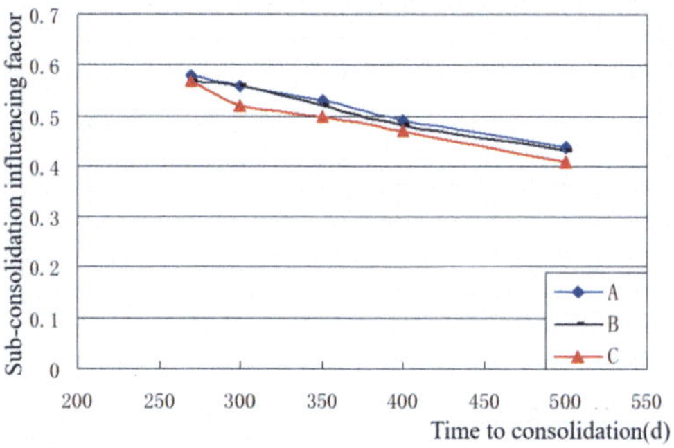

Fig. 5. The relationship between the influencing factor of secondary consolidation and the consolidation time

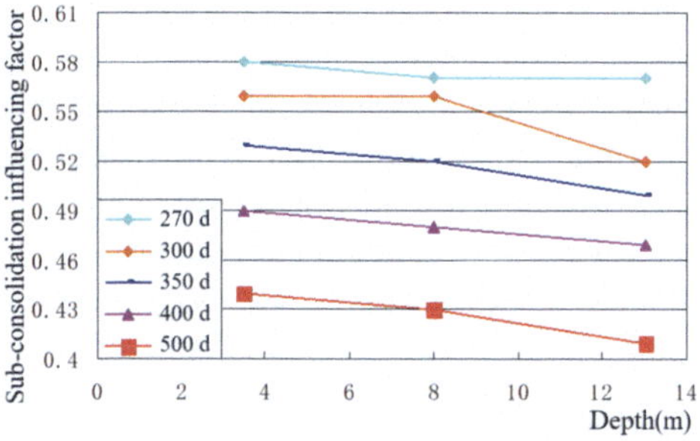

Fig. 6. Relationship between the influencing factor and depth of secondary consolidation

small value is taken for the long consolidation time, the small value is taken for the reinforcement depth, and the large value is taken at the surface.

Equation (9) can also be expressed as follows:

$$\dot{S} = \gamma\beta S_0 = \gamma\beta(1 - \overline{U}_t)S_\infty \tag{10}$$

It shows that the sedimentation rate of unloading is determined by the remaining main consolidation settlement and the average degree of consolidation without considering the influence of γ and β, and a large number of engineering practices show that the indicators of post-construction settlement and consolidation proposed by the design data are not consistent, that is, it may appear that the consolidation degree requirements may be reached, but the post-construction settlement does not meet the requirements, or the post-construction settlement meets the requirements, but the consolidation degree does not meet the requirements, which will cause the difficulty of formulating the settlement rate, and it is recommended to draw the sedimentation rate–post-construction settlement–consolidation degree relationship curve, and combined with the actual situation of the project for comprehensive consideration. Taking this project as an example, the relationship curve of sedimentation rate–post-construction settlement–consolidation degree as shown in Fig. 7 is obtained.

Figure 7 gives the sedimentation rate corresponding to the settlement of 0.3 m and the degree of consolidation of 85%, it can be seen that the results obtained by the two are not

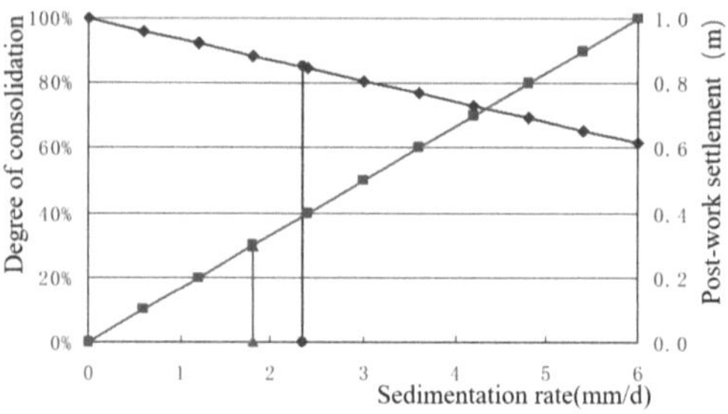

Fig. 7. Relationship curve of sedimentation rate–post-construction settlement–degree of consolidation

consistent, the corresponding settlement rates are 1.8 mm/d and 2.34 mm/d respectively, there is a certain difference, if the standard of the sedimentation rateshould be selected according to the consolidation time, engineering characteristics, etc. Figure 7 is obtained without considering γ and β, in fact, in the process of drainage consolidation, the above two parameters are constantly changing, and the relationship curve of sedimentation rate–post-construction settlement–consolidation degree cannot be straight, so there are still many uncertainties in the sedimentation rateat the time of unloading according to theoretical analysis.

3.4 Discussion of β Values

B is related to a variety of factors such as reinforcement scheme parameters, geological conditions, etc., and is accompanied by the continuous change of drainage consolidation process, and it is impossible to accurately obtain the changes of relevant parameters in the construction monitoring process, so the β obtained according to the indoor testC_h, $C_v$is a fixed value, and it is unreasonable, and this method is not recommended in construction monitoring. However, the β value is also insufficient to estimate the value according to the measured settlement amount and pore pressure value, and the β value has a great relationship with the time interval Δt taken, and the β value obtained by different time interval Δt is also different, which may lead to different monitoring personnel, using the same method, but obtaining different settlement and unloading standards, theoretically, the larger the time interval Δt, the higher the accuracy of the β value [11].

Taking the dead load period (150–270 days) as the analysis period, the SSC model was used to obtain the sedimentation curve as shown in Fig. 8, and the same time interval Δt was set for the rear analysis, and the calculated curve was fitted. According to the fitting curve, the time interval t is 6 days and 8 days, respectively, and the method of backcalculation according to the sedimentation rate and post-construction settlement amount, the β value variation curve as shown in Fig. 9 is obtained.

As can be seen from Fig. 9, the variation law of the calculation results of the three methods is basically the same, the curves of the time interval of 6 days and the time interval of 8 days basically coincide, and the value of the time interval of 8 days is slightly greater than the value of the time interval of 6 days. The result of the back-calculation is significantly smaller than the estimated value, and the difference between

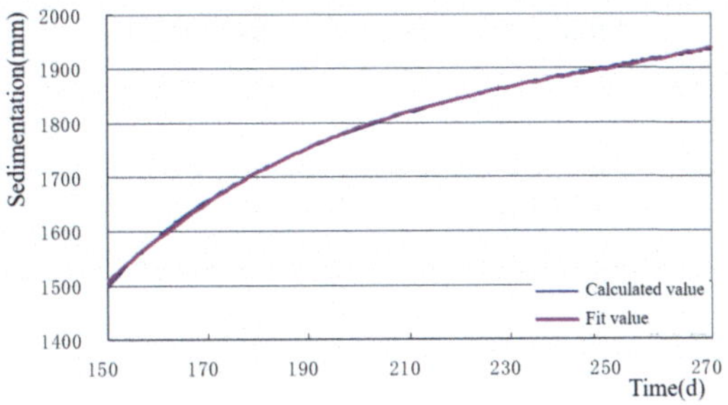

Fig. 8. Sedimentation curve with time

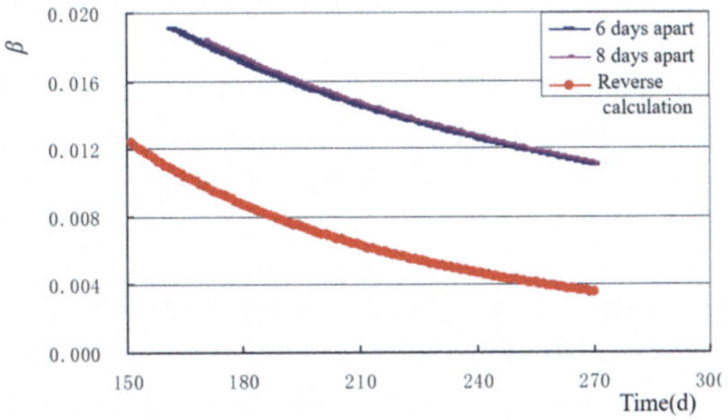

Fig. 9. Variation curve of dead load period value with time

the two cases is basically between 0.0075–0.0085, that is, the difference is smaller with the extension of the consolidation time. Therefore, it is recommended that the β value be taken as follows:

$$\beta = \beta_1 - k \tag{11}$$

Among them, β_1 is taken according to Eq. (6), k is the adjustment coefficient, the value is between 0.0045–0.0085, the consolidation time is long, the small value is taken, and the consolidation time is short, the large value is taken.

Substituting Eq. (11) into Eq. (9), the relationship between the sedimentation rate of the secondary consolidation settlement and the sedimentation rate of the β value and the post-construction settlement can be obtained as follows:

$$\dot{S} = \gamma\beta S_0 = \gamma(\beta_1 - k)S_0 \tag{12}$$

According to Eq. (12), the relationship between the sedimentation rate of this project and the post-construction settlement can be obtained, as shown in Table 5.

It can be seen from table 5 that the sedimentation rate at 270 days of drainage consolidation is about 2.0 mm/d, which is similar to the measured value of 1.7 mm/d, indicating that the correction method adopted in this paper can better reflect the actual situation of the project. What is proposed in this paper is not a specific value, but a method that has better versatility and certain generalization significance.

Table 5. Calculated sedimentation rate

Time to consolidation (d)										
condition	270		300		350		400		500	
	Post-work settlement	Sedimentation rate	Post-work settlement	Sedimentation rate	Post-work settlement	Sedimentation rate	Post-work settlement	Sedimentation rate	Post-work settlement	Sedimentation rate
Consider secondary consolidation	m	mm/d	m	mm/d	m	mm/d	m	mm/d	m	mm/d
	0.67	2.0	0.63	1.1	0.59	0.9	0.55	0.7	0.50	0.6

4 Conclusion

The later settlement of the SCC model considering the secondary consolidation is significantly greater than that of the MC model, and the consolidation time is longer than that of the MC model, which is in line with the consolidation characteristics of soft soil. When the dual control standards of consolidation and post-construction settlement are adopted, a reasonable control value should be formulated in combination with the actual situation of the project. The secondary consolidation has a significant effect on the settlement rate, and the sedimentation rate of the secondary consolidation is only about 0.5 times that of the non-secondary consolidation settlement rate, and it is recommended that the value of the secondary consolidation impact factor is 0.5–0.8, with a small value in the early stage of consolidation and a large value in the later stage of consolidation. The sedimentation rate obtained by the calculation method modified by the secondary consolidation impact factor tends to be consistent with the field measured values, and has good applicability.

Acknowledgments. Research on the safety state of bearing capacity of foundation pit retaining structure based on monitoring deformation data (NO:2023KJ43).

References

1. Jifu, L., Xinhua, C.: Calculation of residual settlement of soft soil embankment using sedimentation rate method. Chin. J. Geotech. Eng. **25**(2), 233–235 (2003). (in Chinese)
2. Boyi, S., Wankui, N., Yanhui, W.: et al. Rock Soil Mech. 37(7), 1963–1968. (2016) (in Chinese)
3. Fei, Z.: Research on one-dimensional creep characteristics of soft clay and method of reducing post-construction settlement of highway. Zhejiang University, 2022. (in Chinese)
4. Chen, L.: Research on soft soil subgrade settlement prediction and preloading design method based on measured data. Changsha University of Science and Technology, 2022. (in Chinese)
5. Haiying, C., Yongfen, R., Haitao, D., et al.: Calculation of rheological nonlinear settlement of peat soil foundation. Chin. J. Earthquake Eng. **45**(06), 1324–1332 (2023). (in Chinese)
6. Jiehui, X.: Settlement deformation mechanism and disease prevention and control countermeasures of Binhai soft foundation expressway. South China University of Technology, 2022. (in Chinese)
7. Bolin, W., Wenjie, M., Wang, X., et al.: Research on application of optimal combination prediction model in high fill settlement. Chin. Civil Eng. J. **52**(S1), 36–43 (2019). (in Chinese)
8. Yi, H., Bowen, Z., Menghao, Y., et al.: Rock Soil Mech. 44(S1)m 268–276 (2023) (in Chinese)

9. Chen, Z.J., Feng, W.Q., Yin, J.H., et al.: (2021) A new simplified method for calculating short-term and long-term consolidation settlements of multi-layered soils considering creep limit. Comp. Geotech. https://doi.org/10.1016/j.compgeo.2021.104324.
10. Bardhan, A., Kardani, N., Alzo'ubi, A.K., et al.: 2022 J. Rock Mech. Geotech. Eng. 14 5 1588 1608.
11. Guoliang, L., Chengbin, Y., Chenglong, B. et al.: Research on the improvement of the calculation method of secondary consolidation settlement. In: Proceedings of the 2022 Industrial Construction Academic Exchange Conference of MCC Construction Research Institute Co., Ltd. (in Chinese)
12. K. Terzaghi (translated by Xu Zhiying). Theoretical Soil Mechanics. Geological Publishing House, Beijing (1960) (in Chinese)
13. Yan, L.: Vacuum Drainage Preloading Method to Strengthen Soft Soil. China Communications Press, Beijing (2002). (in Chinese)
14. Changsheng, Z., Boyou, Z., Guonan, L., et al.: Estimation of remaining settlement and unloading time by sedimentation rate. J. Earth Sci. Environ. **27**(4), 28–32 (2005). (in Chinese)

Effect of Lateral Load on Vertical Bearing Capacity of Helical Piles—Part II. Theoretical Study

Bing Feng[1], Zhihao Sun[2,3], Qing LÜ[2,3(✉)], Ma Zhu[1], Jianjun LÜ[1], Jian Huang[4], and Weiming Ma[2]

[1] Shaoxing Daming Electric Power Design Institute Co. Ltd., Shaoxing 312000, Zhejiang, China
[2] College of Civil Engineering and Architecture, Zhejiang University, Hangzhou 310058, Zhejiang, China
lvqing@zju.edu.cn
[3] Center for Balance Architecture, Zhejiang University, Hangzhou 310058, Zhejiang, China
[4] State Grid Shaoxing Power Supply Company, Shaoxing 312000, Zhejiang, China

Abstract. The utilization of helical pile foundations in transmission towers is increasingly prevalent. This foundation must endure both vertical and lateral forces imposed by the transmission towers. However, there is insufficient theoretical research and calculation methodology to assess the impact of lateral loads on the vertical bearing capacity of helical piles. This paper explores the influence of lateral load on the vertical bearing capacity of helical piles through theoretical analysis. A practical calculation approach for the vertical load-bearing capacity of helical piles under combined loading conditions is proposed. This method considers the impact of lateral loads on both the pile shaft and pile helix vertical bearing capacity, using distinct discount factors denoted as k_1 and k_2, respectively. When a small lateral load is applied, it affects the bearing capacity of the shaft initially. The discount factor k_1 is used to adjust the shaft bearing capacity. When a substantial lateral force is exerted, resulting in the rotation point of the helical pile descend, it impacts the load-bearing capacity of the helix. The discount factor k_2 is utilized to modify the load-bearing capacity of the helix. The values for k_1 and k_2 are derived from numerical model tests reported in a companion paper. The effectiveness of the proposed computational approach is validated by comparing it to experimental outcomes documented in the existing literature.

Keywords: helical pile · combined loads · bearing capacity calculation · theoretical study · discount factor

1 Introduction

The author's companion paper [1] detailed numerical model tests on helical piles subjected to varied uplift-lateral load ratios, aiming to offer insights into the impact of lateral forces on the vertical load-bearing capacity of helical piles. The findings suggest a decrease in the vertical load-bearing capacity of helical piles under concurrent

B. Yuan et al. (Eds.): ICCASE 2024, SUCI, pp. 290–299, 2024.
https://doi.org/10.1007/978-3-031-76102-7_33

lateral forces. The effect of lateral loading on the compressive capacity of the helical piles surpasses its influence on the uplift capacity. Upon reaching 80% of the vertical load, the vertical load-bearing capacity experiences an approximately 50% reduction. Nevertheless, the current specifications [2–4] lack a calculation method and relevant provisions for vertical load-carrying capacity of helical piles subjected to combined loads. There is limited research on the on the load-bearing performance of helical piles under combined loading [5–8], and no scholars have proposed a method to calculate the vertical load-carrying capacity of helical piles subjected to combined loads. This paper investigates the influence of lateral loading on the vertical load-bearing mechanism of helical piles through theoretical research, providing a practical approach to determine the vertical load-carrying capacity of helical piles under combined loading. The method utilizes numerical model test results from the companion paper [1].

2 Theoretical Analysis of the Effect of Lateral Load on Vertical Bearing Capacity of Helical Pile

According to the classification criteria of flexible piles, semi-rigid piles, and rigid piles in the Chinese code [9], the pile shaft in the model experiments can be regarded as a rigid pile. Under the lateral load, the helical pile was inclined, as shown in Fig. 1. At this time, the soil on the right side of the upper part of the rotation point of the helical pile was pressurized, and the soil on the left side was loose from the shaft. The left-side soil would be gradually converted from earth pressure at rest to active earth pressure. When an uplift or compression load was applied to the pile, the friction force provided to the shaft by the left-side soil would be reduced because the active earth pressure is much smaller than the earth pressure at rest. Similarly, if the left-side soil at the lower part of the rotating point was pressurized and the right-side soil was loosened with the shaft, then the friction force provided to the shaft by the right-side soil would be reduced. According to the experimental data in the literature [10], when the pile top displacement exceeds 1.5‰ pile length, the soil on the side of the pile reaches the ultimate active earth pressure. The limit passive earth pressure is reached when the pile top displacement exceeds 15% of the pile length. The displacement requirement for the soil to reach the limit passive earth pressure is much larger than the active earth pressure. With the compressive deformation of the soil, the change from earth pressure at rest to limit passive earth pressure can be considered a linear [11]. Therefore, the value of passive earth pressure was basically unchanged, and the lateral friction resistance provided by the zone of passive earth pressure was basically unchanged.

As the lateral load increases, the soil on the bearing side of the pile shaft transitions from earth pressure at rest to the ultimate passive earth pressure state, while the soil on the non-bearing side transitions from earth pressure at rest to the ultimate active earth pressure state. Achieving the ultimate passive earth pressure state requires significantly greater pile top displacement compared to achieving the ultimate active earth pressure. Consequently, as the lateral load increases, the soil on the non-load-bearing side reaches the ultimate active state before the soil on the load-bearing side reaches the ultimate passive state. Due to the lower active earth pressure compared to the earth pressure at rest, the axial side resistance of the pile on the non-bearing side decreases. Despite the increase in shaft resistance on the bearing side, it is substantially smaller than the decrease on the non-bearing side, resulting in an overall reduction in the pile shaft resistance. The process can be reflected by multiplying the frictional resistance of the shaft by a discount factor of k_1.

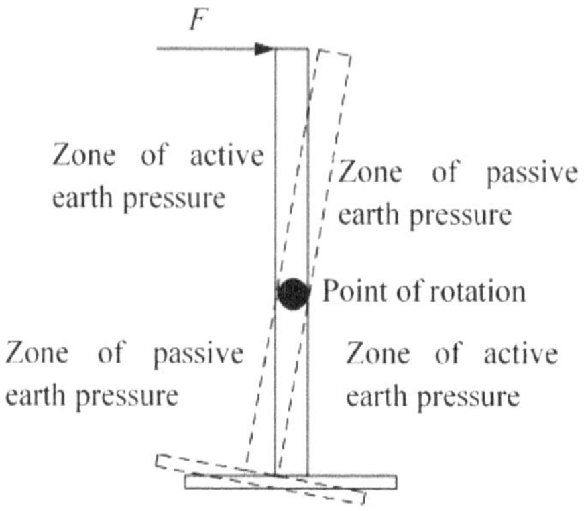

Fig. 1. Schematic diagram of helical pile inclination

For the pile helix, the failure modes are shown in figs. 2 and 3. Zone I is the elastic zone, and zones II and III are the plastic equilibrium zones. When the lateral load is relatively small ($R_h \leq 0.5T_t$ or $R_h \leq 0.5T_c$), the slip surface of the soil has not developed to the position of the pile helix. Therefore, it is assumed that there is no effect of the lateral load on the failure mode of soil in the helix position. That is, the contribution of the helix to the vertical bearing capacity is unchanged when the lateral load is small. In addition, some inclination of the helix under lateral loading occurs, but this inclination is calculated to be extremely small (about 0.23° for $R_h = 0.5T_t$) and negligible.

3 Calculation of Uplift Capacity of Helical Piles Under the Influence of Lateral Loading

The theoretical analysis presented relies on two assumptions: (1) the pile is rigid, and (2) the soil is sandy. In practical applications, some helical piles with relatively low slenderness ratios may be considered rigid [12], while certain transmission lines are situated in sandy soil layers [13]. The calculation methods derived below are based on theoretical analysis results. These methods are particularly effective in engineering designs that align with the assumptions.

According to the Chinese code for the design of helical pile foundations for overhead transmission line [2], the formula for calculating the uplift capacity of a single helical pile without considering the lateral load is as follows.

$$\begin{aligned} Q_{tu} = Q_{sh}^{t} + Q_{t} = \pi b(L - L_0)q_s^t + \frac{\pi B^2 H \cos\varphi}{4}\gamma' \\ + \frac{\pi B\psi\left(H^2 - H_t^2\right)}{2}\gamma + \pi B\left(H - H_t^2\right)c \end{aligned} \tag{1}$$

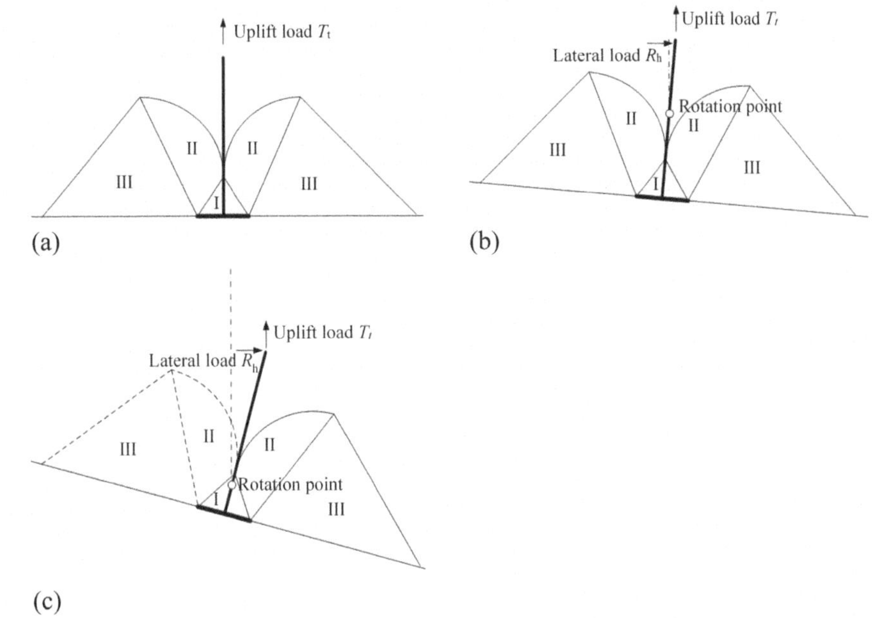

Fig. 2. Failure mode of helix under combined uplift and lateral loads: (a) single uplift load, (b) combined uplift and lateral loads (for $R_h \leq 0.5T_t$), (c) combined uplift and lateral loads (for $R_h \geq 0.5T_c$).

where Q_{tu} is the uplift capacity of the helical pile, Q_t is the uplift capacity of the helix, Q_{sht} is the uplift capacity of the pile shaft, B is the diameter of the helix, H is the helical depth, φ is the inclination angle of the helical pile, and ψ is the coefficient of the influence of lateral pressure, H_t is the depth of the top surface of the upper part of the helix, when $H > \lambda B$, it is taken as $H\text{-}\lambda B$, λ is the influence coefficient of the height of the shear soil column, c is the cohesion of the soil within the range of λB in the upper part of the helix; b is the outer diameter of the pile shaft, L is the length of buried in the soil, L_0 is the sum of L_1 and L_2, L_1 is the drastic disturbances zone depth of the upper part of the pile shaft, L_2 is the soil thickness that moves with the helix in the vicinity of the helix, L_2 can be taken as $0.5\lambda B$ and q_{st} is the lateral friction resistance of pile shaft.

The equation contains two parts: the first part is the bearing capacity provided by the shaft, and the second part is the bearing capacity provided by the helix. From the above analysis, the lateral load will cause the pile shaft lateral friction resistance to decrease. Therefore, the first part of the equation is multiplied by the discount factor k_1, and the revised formula for the uplift capacity of the shaft is shown below.

$$Q_{sh}^{t} = k_1 \pi b (L - L_0) q_s^t \tag{2}$$

where k_1 is the pile shaft bearing capacity reduction factor.

As the lateral load increases, the point of rotation continues to move downward until it reaches the helix position. At this point, the pile shaft and the soil on the side where lateral loads are exerted on can be considered to be basically completely disengaged.

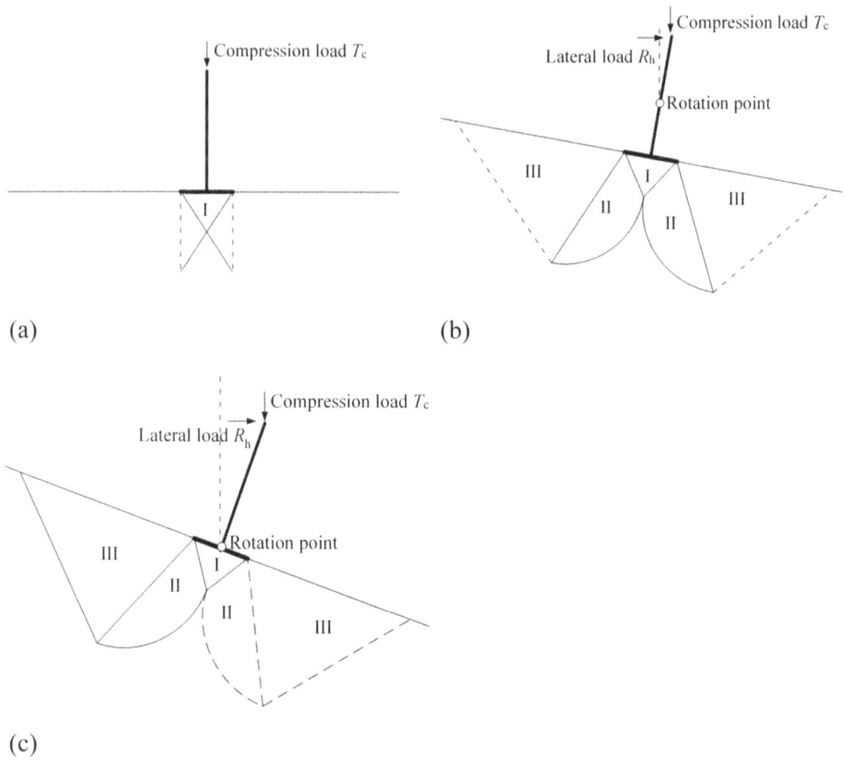

Fig. 3. Failure mode of helix under combined compression and lateral loads: (a) single compression load, (b)combined compression loads and lateral loads (for $R_h < 0.5T_c$), (c)combined compression loads and lateral loads (for $R_h < 0.5T_c$).

The bearing capacity of the shaft is reduced to half of its original capacity and does not continue to be reduced, after which the helix bearing capacity begins to gradually decrease. The second part of the equation is corrected by multiplying it by a reduction factor k_2, and the corrected formula for the helix uplift capacity is shown below.

$$Q_t = k_2 \pi B \left[\frac{BH \cos\varphi}{4} \gamma' + \frac{\psi\left(H^2 - H_t^2\right)}{2} \gamma + \left(H - H_t^2\right) c \right] \tag{3}$$

For the single uplift loading condition, the parameter values of the theoretical calculation are shown in Table 1. The values of the uplift capacity of the pile obtained by the theoretical calculation and numerical simulation (reported in the companion paper [1]) are shown in Table 2. When no lateral load is applied, the theoretical values are very close to the numerical simulation results, indicating that the theoretical calculation parameters are reasonable. After the lateral load is applied, the uplift capacity of the helix obtained from numerical simulation is used to back-calculate the values of k_1 and k_2, and the results are shown in Table 2. When $R_h = 0.5\,T_t$, the shaft bearing capacity has not been reduced to 50%. The lateral loads have not yet affected the helix bearing capacity, which coincides with the analysis of the helix slip surface in the companion paper (the slip surface analysis is in Sect. 3.2.1.).

Table 1. Theoretical calculation parameter values of uplift capacity of helical pile under single uplift loading.

B (m)	λ	φ (°)	ψ (°)	c (kPa)	Γ (kN·m^{-3})	L (m)	L_1 (m)	L_2 (m)	b (m)	q_s^{t} (kPa)
0.09	4	0	0.65	0	18	0.6	0.12	0.18	0.048	37

Table 2. Values of uplift capacity and discount factor k for helical piles under different load combinations.

	Theoretical calculation (N)	Numerical simulation (N)	Reduction factor k_1	Reduction factor k_2
$R_h = 0$ N	1 575	1 564	1	1
$R_h = 0.2T_t$	1 546	1 546	0.97	1
$R_h = 0.5T_t$	1 133	1 136	0.56	1
$R_h = 0.8T_t$	820	821	0.5	0.56
$R_h = 1.0T_t$	680	686	0.5	0.32

4 Calculation of Compression Bearing Capacity of Helical Piles Under the Influence of Lateral Loading

The failure modes of helical piles under combined lateral and compression loads are similar to those under combined lateral and uplift loads. When the lateral load is small ($R_h \leq 0.5T_c$), the failure mode of the helix is unchanged, and the bearing capacity of the shaft decreases with the increase of the lateral load, which is corrected using the discount factor k_1. When the lateral load is large ($R_h \geq 0.5T_c$), the bearing capacity of the helix is corrected using the discount factor k_2 and making $k_1 = 0.5$.

According to the Chinese code [2], the formula for calculating the compression bearing capacity of a single-helical pile without considering the effect of lateral load is as follows.

$$Q_{cu} = Q_c + Q_{sh}^c = \frac{\pi B^2 q_p}{4} + \pi b(L - L_0)q_s^c \tag{4}$$

where Q_{cu} is the compression bearing capacity of the helical pile, Q_c is the compression bearing capacity of the helix, $Q_{sh}{}^{c}$ is the compression bearing capacity of the shaft, q_p

is the ultimate end resistance of the soil in the lower part of the helix, and the meanings of the remaining parameters are the same as those in Eq. (1).

When the lateral load is considered, similar to the case for the combination of uplift and lateral loads, the calculation formula is described below.

$$Q_{sh}^{c} = k_1 \pi b(L - L_0) q_s^c \tag{5}$$

$$Q_c = k_2 \frac{\pi B^2 q_p}{4} \tag{6}$$

The values of the theoretical calculation parameters are shown in Table 3. The values of the compression bearing capacity of the pile obtained by theoretical and numerical simulation are shown in Table 4. For a single compression load, the theoretical calculation results are very close to the numerical simulation results, indicating that the theoretical calculation parameters are reasonable. After applying lateral load, the compression bearing capacity of the helical pile obtained by numerical simulation is used to back-calculate k_1 and k_2 values, and the results are shown in Table 4. The pile shaft bearing capacity has been reduced to 50% at $R_h = 0.5T_c$, i.e., the lateral load has started to have an effect on the helix bearing capacity, which coincides with the analysis of the helix slip surface in the companion paper (the slip surface analysis is in Sect. 3.2.2.).

Table 3. Theoretical calculation parameter values of compression bearing capacity of helical pile under single compression loading.

B (*m*)	q_p (*kPa*)	b (*m*)	L (*m*)	L_1 (*m*)	L_2 (*m*)	$q_s{}^c$ (*kPa*)
0.09	135	0.048	0.6	0.12	0.18	37

Table 4. Values of uplift capacity and discount factor k for helical piles under different load combinations.

	Theoretical calculation (*N*)	Numerical simulation (*N*)	Reduction factor k_1	Reduction factor k_2
$R_h = 0\ N$	1 814	1 871	1	1
$R_h = 0.2T_c$	1 678	1 682	0.81	1
$R_h = 0.5T_c$	1 103	1 106	0.5	0.69
$R_h = 0.8T_c$	774	773	0.5	0.31
$R_h = 1.0T_c$	623	621	0.5	0.13

5 The Applicability and Limitation Analysis of the Calculation Method

Through the above analysis, it can be found that the reduction coefficient k can reflect the change in soil sliding surface under combined loads. The discount factor k obtained by numerical simulation and theoretical analysis in this paper can be used to calculate the vertical bearing capacity of helical piles under different lateral-vertical load ratios. The values of the discount factor can be referred to in Tables 2 and 4.

The comparison between measured and calculated results is presented in Table 5, which includes data obtained from both the theoretical formulations proposed in this paper and those in Zhao's study [14]. Zhao's method incorporates the disparity in earth pressure on both sides of the pile under combined loading conditions, leading to an accurate prediction of the pile's uplift capacity. However, this method overlooks the influence of lateral loads on the helix for piles equipped with helical configurations. As a consequence, the estimated bearing capacity tends to surpass the measured value, especially under significant lateral loads. Notably, when employing the calculation method introduced in this paper, the ratio of theoretical to measured bearing capacity typically ranges from 0.94 to 1.10. The calculated bearing capacity is in great agreement with the measured bearing capacity. This indicates that the computational method in this paper can be used to predict the vertical bearing capacity of helical piles under combined loads.

Differences in the simulation method of combined loading led to the observed gap between the calculated and actual results. Prior studies (references [15] and [16]) typically utilize oblique loading to simulate both uplift and lateral loading simultaneously. In contrast, the approach in this paper involves the separate application of uplift and lateral loads within the numerical model. Despite the equal magnitude of the combined force applied to the pile, a minor discrepancy emerges as the lateral displacement of the pile increases.

Foundation designers in practical engineering applications can determine the bearing capacity of helical piles under combined loads using Eqs. 1–6 provided in this paper. The values of k_1 and k_2 can be selected from tables 2 and 4, respectively. For clay foundations, it is imperative to adjust the discount factors k_1 and k_2, with k_1 necessitating a proportionate increase. Similarly, slight adjustments are needed for k_1 and k_2 in loose sand conditions.

Although the proposed calculation method can offer a simple and effective means to determine the vertical load-bearing capacity of helical piles subjected to combined lateral and vertical loads, there are some limitations. This study is based on rigid helical piles, which may be flexible when the slenderness ratio of the pile shaft is too large in actual projects. The pile shaft is basically not affected below the rotation point, so the vertical bearing capacity of the shaft will be less affected by lateral loading. The calculation results obtained by using the method in this paper will be conservative. In addition, the study in this paper is based on sandy soils. In cohesive foundation soils, the pile displacement (s_a) when the soil on the non-bearing side reaches the ultimate active earth pressure is significantly greater than in sandy soil foundations, with the former being approximately 10 times greater than the latter. Additionally, the reduction in frictional resistance on the non-bearing side of the pile is typically less significant.

Table 5. Comparison of test bearing capacities with calculated bearing capacities from Eqs. 1–6.

Reference	L (m)	B (m)	φ (°)	γ $(kN{\cdot}m^{-3})$	R_h/T_t	$Q_C{}^a$ (N)	$Q_C{}^b$ (N)	$Q_A{}^c$ (N)	$Q_C{}^a/Q_A{}^c$	$Q_C{}^b/Q_A{}^c$
Lin et al. [15]	0.4	0.2	33.1	17	0.41	473	508	434	1.09	1.17
Lin et al. [15]	0.8	0.2	33.1	17	0.41	1033	1127	931	1.10	1.21
Hu et al. [16]	0.6	0.3	39	17	1	1387	2418	1477	0.94	1.63
Hu et al. [16]	0.9	0.3	39	17	1	2490	4684	2586	0.96	1.81
Hu et al. [16]	0.9	0.3	39	17	0.41	4256	4684	4197	1.01	1.11

[a] Computed capacity using the method in this paper

[b] Computed capacity using Zhao's method

[c] Actual capacity

Therefore, the discount factor k_1 proposed in this paper needs to be proportionately increased.

Future investigation is warranted to explore various aspects, such as assessing the method's applicability to different soil types, investigating the mechanical properties of longer piles or those with different helix configurations under combined loads, and integrating the method into a more comprehensive analytical model.

6 Conclusions

Based on the results of numerical simulation studies regarding helical piles under combined loads (as detailed in the companion paper), the mechanical properties of helical piles under combined loads were further explored through theoretical analysis. This investigation offers an efficient methodology for calculating the vertical bearing capacity of helical piles subjected to combined loads. The primary conclusions are as follows.

(1) Under simultaneous lateral and vertical loading, when the lateral force is small, the reduction in the vertical bearing capacity of the helical pile is primarily attributed to a decrease in the bearing capacity of the pile shaft. That can be considered by multiplying the bearing capacity of the shaft by the discount factor k_1.
(2) Under simultaneous lateral and vertical loading, the rotational point of the helical pile drops to the vicinity of the helix when the lateral load is great, which in turn causes a change in the helix failure mode as well as a decrease in the bearing capacity. This process can be considered by multiplying the helix bearing capacity by a discount factor of k_2.

(3) The results obtained in this paper offer insights into the failure mode and calculation of vertical bearing capacity of helical piles under simultaneous vertical and lateral loads. However, the failure modes and bearing capacity calculation methods of different types of helical piles in different soil layers still need further study.

Acknowledgments. This work received financial support from the State Grid Zhejiang Electric Power Co., Ltd. Collective Enterprise Technology Project (CG091201-DMSJY) the State Grid Zhejiang Electric Power Co., Ltd. Infrastructure Construction New Technology Research Project Based on Engineering (SGTYHT/23-JS-003).

References

1. Feng, B., Sun, Z. H., Lü, Q., et al.: Effect of lateral load on vertical bearing capacity of helical piles—Part I. Numerical simulation. In: 8th International Conference on Civil Architecture and Structural Engineering. pp. 26–47. Guangzhou (2024)
2. State Grid Corporation of China: Code for Design of Helical Anchor Foundation of Overhead Transmission Line: Q/GDW 10584–2022. China Electric Power Press, Beijing (2022)
3. International Code Council: International Building Code: 2021. ICC Publications, Lllinois (2021)
4. The Institute of Electrical and Electronics Engineers: IEEE Guide for Transmission Structure Foundation Design and Testing: IEEE Std 691. The Institute of Electrical and Electronics Engineers Inc, New York (2001)
5. Al Baghdadi, T.A., Brown, M.J., Knappett, J.A., et al.: Effects of vertical loading on lateral screw pile performance. Proc. Inst. Civ. Eng. Geotech. Eng. **170**(3), 259–272 (2017)
6. Cheng, P., Guo, J., Yao, K., et al.: Numerical investigation on pullout capacity of helical piles under combined loading in spatially random clay. Mar. Georesour. Geotechnol. **41**(10), 1118–1131 (2023)
7. Venkatesan, V., Mayakrishnan, M.: Behavior of mono helical pile foundation in clays under combined uplift and lateral loading conditions. Appl. Sci. **12**(14), 6827 (2022)
8. Venkatesan, V., Mayakrishnan, M., Shukla, S.K.: Effect of pile spacing in helical pile groups in soft clays under combined loading. Mar. Georesour. Geotechnol. **42**(4), 1–21 (2023)
9. Ministry of Communications of the People's Republic of China: Code for Pile Foundations of Harbour Engineering: JTJ 254–98. China Communications Press, Beijing (1999)
10. Fang, Y.S., Ho, Y.C., Chen, T.J.: Passive earth pressure with critical state concept. J. Geotech. Geoenviron. Eng. **128**(8), 651–659 (2002)
11. Fan, X.Z., Xu, C.J., Liang, L.J., et al.: Analytical solution for displacement-dependent passive earth pressure on rigid walls with various wall movements in cohesionless soil. Comput. Geotech. (2021). https://doi.org/10.1016/j.compgeo.2021.104470
12. Salem, T.N., Elkhawas, N.M., Elnady, A.M.: Lateral response enhancement of a conventional offshore monopile using skirts of helical piles and pipe piles with stiffeners. Innov. Infrastruct. Solut. **6**(3), 16 (2021)
13. Bittar, E., Lehane, B.M., Blake, A., et al. (CPT-based design method for helical piles in sand. Can. Geotech. J. 16 2023)
14. Zhao, C., Liu, F., Qiu, Z., et al.: Study on bearing behavior of a single pile under combined vertical and lateral loads in sand. Chin. J. Geotech Eng. **37**(1), 183–190 (2015)
15. Lin, Z., Hu, W., Zhao, P., et al.: Model test study on inclined pull-out bearing characteristics of flat circular anchor in sand. Rock Soil Mech. **42**(11), 3059–3068 (2021)
16. Wei, H., Yao, Y., Kai, L.S., et al.: Experiment study on inclined pullout bearing characteristic of screw anchor pile of photovoltaic bracket. Acta Energ. Sol. Sin. **43**(12), 50–61 (2022)

Effect of Lateral Load on Vertical Bearing Capacity of Helical Piles—Part I. Numerical Simulation

Bing Feng[1], Zhihao Sun[2,3], Qing LÜ[2,3,3](✉), Ma Zhu[1], Jianjun LÜ[1], Jian Huang[4], and Weiming Ma[2]

[1] Shaoxing Daming Electric Power Design Institute Co. Ltd., Shaoxing 312000, Zhejiang, China
[2] College of Civil Engineering and Architecture, Zhejiang University, Hangzhou 310058, Zhejiang, China
lvqing@zju.edu.cn
[3] Center for Balance Architecture, Zhejiang University, Hangzhou 310058, Zhejiang, China
[4] State Grid Shaoxing Power Supply Company, Shaoxing 312000, Zhejiang, China

Abstract. Helical pile foundations are increasingly utilized in transmission line projects. The unique nature of transmission line loads necessitates that helical pile can provide adequate vertical bearing capacity under combined lateral and vertical loads. To investigate the influence of lateral load on the vertical bearing capacity of helical piles and establish a calculation methodology for their vertical bearing capacity, numerical simulations were conducted using the finite element software PLAXIS 3D. These simulations were based on previously documented indoor experiments found in the literature. After the numerical model was validated, numerical static load experiments were performed on helical piles under varying lateral-vertical load ratios. The load-vertical displacement curves and soil slip surfaces were obtained (as reported in this paper). Additionally, a theoretical study on the calculation method for the vertical bearing capacity of helical piles under combined loads was undertaken and is reported in a companion paper. The numerical simulation study yielded the following findings: Under combined loads, the vertical bearing capacity of helical piles decreases, with a more pronounced reduction in the compressive bearing capacity. Moreover, with an increase in lateral load, the main soil slip surface is shifted from the position of the pile helix to the upper part of the pile shaft. Furthermore, the soil slip zone at the position of the shaft increases in both size and depth.

Keywords: Helical pile · Combined loads · Failure mode · Load-bearing mechanism · Numerical simulation

1 Introduction

Lately, the demand for electricity has surged significantly in China. The construction of transmission lines, encompassing various voltage levels, including high voltage and extra high voltage, is presently underway. Helical piles, as an economic and eco-friendly

B. Yuan et al. (Eds.): ICCASE 2024, SUCI, pp. 300–312, 2024.
https://doi.org/10.1007/978-3-031-76102-7_34

form of foundation, are increasingly utilized for transmission tower foundations. They offer advantages such as ease of construction, reusability, recyclability, and the ability to promptly generate bearing capacity [1]. Power transmission towers endure wind load and conductor load, resulting in significant vertical and lateral loads on the tower foundation. Additionally, the dead weight of the transmission tower generates compression loads. Hence, helical piles must accommodate both vertical and lateral loads when utilized as foundations for transmission towers [2, 3].

Trofimenkov et al. [4], Mitsch et al. [5], and Mooney et al. [6] each proposed distinct methods—the independent plate bearing method and cylindrical shear method—to calculate the vertical bearing capacity of helical piles. Mitsch et al. [5] and Ghaly et al. [7] proposed methods for calculating the bearing capacity of shallow and medium-depth helixes, respectively, based on the burial depth of the helix of the helical pile. These methods, with amendments, continue to be incorporated into national codes [8–10] and are extensively employed in calculating the vertical bearing capacity of helical piles. Furthermore, in recent years, researchers have utilized in-situ experiments, numerical simulations, and indoor experiments to investigate the effects of helix diameter [11], spacing [11, 12], and load sharing ratio between the pile helix and shaft [13] on the vertical bearing capacity of helical piles, building upon the aforementioned studies. These studies further contribute to the accuracy of the vertical bearing capacity calculation of helical piles.

Studies have demonstrated the influence of lateral loading on the vertical bearing capacity of helical piles [14]. However, there is a scarcity of research addressing the calculation method for the vertical bearing capacity of helical piles under combined lateral and vertical loads. During this study, finite element analysis software PLAXIS 3D was initially employed to simulate indoor vertical static load experiments of helical piles outlined in the literature. Following the validation of the model, numerical static load experiments for helical piles were conducted under various lateral-vertical load ratios. Ultimately, the calculation method for the vertical bearing capacity of helical piles under combined loads was derived through theoretical analysis. This paper concentrates on presenting the results of numerical static load experiments, aiming to explore the load-bearing mechanism and failure mode of helical piles under combined loads. The companion paper [15] delves into the impact of lateral loads on the vertical bearing capacity of helical piles using theoretical analysis, with the primary goal of providing a practical method for calculating the vertical bearing capacity under combined loads.

2 Establishment and Verification of Finite Element Calculation Model

To examine the impact of lateral force on the vertical bearing capacity and failure mode of helical piles, numerical simulation was employed as the research method in this study. Numerical simulation was conducted using PLAXIS 3D, a robust geotechnical finite element analysis software developed by PLAXIS B.V., Netherlands. PLAXIS is widely utilized for studying the force and mechanical characteristics of various pile foundations [16–18]. The finite element numerical simulation method can accommodate various complex geometries (e.g., inclined anchor plates and tapered anchor heads).

By discretizing the continuum into small cells, it accurately simulates the interaction between the pile foundation and surrounding soil, including nonlinear and elastic-plastic behaviours. The finite element method offers a comprehensive suite of post-processing tools for visualizing the deformation of the pile foundation and the displacement field of the soil, facilitating a thorough understanding and analysis of the problem under investigation. Due to their easier simulation compared to in-situ experiments, this study is based on indoor experiments of helical piles with a single plate conducted by George et al. [19]. The static load experiments on helical piles are simulated using PLAXIS 3D finite element software.

2.1 Finite Element Model Establishment

The soil layer parameters and helical pile mechanical parameters are the same as those in the literature of George et al. [19]. The Mohr-coulomb 10-node 4-plane body unit is used to simulate the soil. Plate unit and linear elastic isotropic material are used to simulate the pile helix and pile shaft, and the interface unit is used to simulate the pile-soil interaction. Detailed parameters are presented in tables 1 and 2. Installation effects are simulated by applying volume strain to the soil body at the pile shaft location to simulate cavity expansion. The value of volume strain is set to 25%, according to the literature [19]. Loading is carried out by applying a face load at the upper part of the pile shaft. The specific simulation steps are shown in table 3. After the model is established, meshing is carried out, and 23,383 cells and 36,369 nodes are formed. The model meshing is shown in Fig. 1.

Table 1. Changes in soil characteristics according to relative density.

Soil name	Relative density	Unit weight γ (kN·m^{-3})	Elastic modulus E (kN·m^{-2})	Poisson's ratio v	Internal friction angle φ (°)	Interface reduction coefficient R_{int}
Medium dense sand	50%	18	6000	0.35	33	0.4
Loose sand	30%	15	4000	0.30	30	0.2

Table 2. The geometric and mechanical parameters of piles.

Shaft diameter D_s (mm)	Helix diameter D_h (mm)	Installation depth H (m)	Pitch t (mm)	Elastic modulus E (kN·m^{-2})	Poisson's ratio v	Unit weight γ (kN·m^{-3})
48	90	0.6(0.8)	40	200e6	0.15	75.5

Table 3. Steps of numerical simulation.

Step 1	Step 2	Step 3	Step 4
Generate initial stress field	Cavity expansion	Deactivate the soil in the cavity, activate the plate element and interface element	Activate elements and interface element

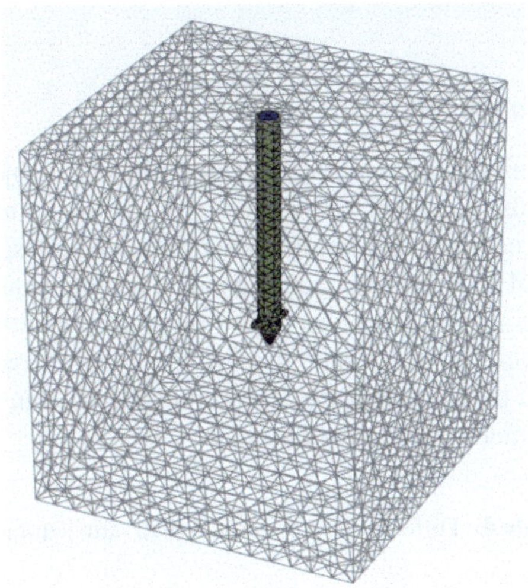

Fig. 1. Mesh discretization

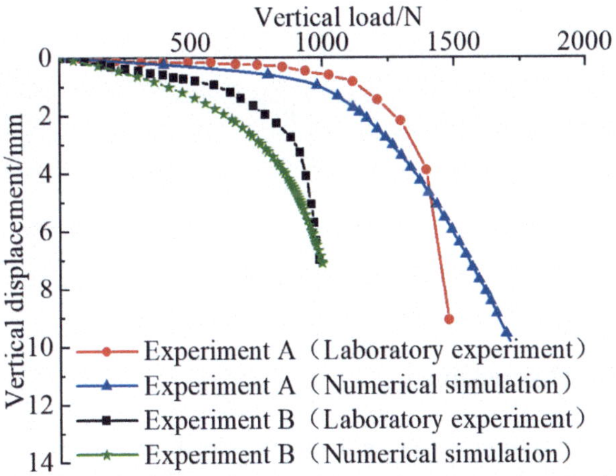

Fig. 2. Comparison of simulated and measured load-displacement curves.

2.2 Finite Element Model Validation

In order to verify the validity of the model, the model calculation results were compared with the load- displacement curves in the literature [19]. The comparison results are shown in Fig. 2. The experiment A represents the compression experiment of a helical pile in medium-dense sand with a 0.8 m burial depth. The experiment B represents the uplift experiment of a helical pile in loose sand with a 0.8 m burial depth. The change rule of the load-displacement curves obtained by numerical simulation in this paper is consistent with the measured curves, which verifies the validity of the current numerical modeling.

2.3 Loading Condition Settings

In real transmission line foundation projects, helical piles are subjected to simultaneous lateral and vertical loads in specific proportions. To assess the impact of lateral force on the vertical bearing capacity and failure mode of helical piles, numerical static load tests with various lateral and vertical load ratios must be conducted. On the basis of the numerical model of the above compression load experiments with the pile helix buried at a depth of 0.6 m in medium-dense sand, five different lateral-vertical load ratios were set up for the combined load experiments. The setups, detailed in table 4, encompassed a total of ten sets of numerical static load tests.

Table 4. Different lateral-vertical load ratio settings.

Load condition	R_h[a]/T_t[b]	Load condition	R_h/T_c[c]
1	0	6	0
2	0.2	7	0.2
3	0.5	8	0.5
4	0.8	9	0.8
5	1.0	10	1.0

[a] Lateral load
[b] Uplift load
[c] Compressive load

3 Results and Discussion

To establish the calculation method for the vertical load capacity of helical piles under combined loading, it is essential to analyze the impact of lateral load on both the vertical ultimate load capacity and failure mode of helical piles. Subsequently, numerical calculation results were extracted to acquire vertical load-displacement curves for helical piles subjected to various lateral-vertical load ratios. These curves facilitated the analysis of how much lateral load influences the vertical load capacity of helical piles. Additionally, numerical calculation results were utilized to generate Incremental displacement maps depicting soil behavior surrounding the helical pile. These maps were then employed to analyze the vertical failure mode of the helical pile induced by lateral and vertical combined loads.

3.1 Influence of Lateral Loading on Vertical Ultimate Bearing Capacity

Figures 3 and 4 show the uplift and compression load-displacement curves of the pile under combined loads. Under uplift and lateral combined loads, the load-displacement curves of the piles always show a sharply down curve type, while under compression-lateral combined loads, the load-displacement curves of the piles show adjustment curve type at the beginning. However, with the increase in lateral load, the load displacement curve gradually changed from adjustment curve type to sharply down curve type. This indicates that the lateral load changes the damage mode of the helical pile under compression load, from punching shear failure to local shear failure to general shear failure.

For sharply down load-displacement curves, the load value corresponding to the onset point of significant steepening is taken as the vertical ultimate bearing capacity in accordance with a Chinese code [20]. For the adjustment load-displacement curves, the load value at which the displacement reaches 0.15 times the diameter of the pile shaft is taken as the ultimate bearing capacity according to the recommendation of Khazaei et al. [21]. According to Fig. 3, the vertical ultimate bearing capacity of the helical pile under different lateral loads was obtained, as shown in table 5. The uplift and compression bearing capacity of the helical pile decrease when the lateral loads are applied at the same time. When the lateral load R_{h} is equal to 0.2, 0.5, 0.8, and 1.0 times the vertical load, the uplift bearing capacity decreases by 1.8%, 28%, 48%, and 57%, respectively, and the compression bearing capacity decreases by 10.1%, 40.8%, 59%, and 67%, respectively. The vertical bearing capacity of the helical pile decreases gradually with the increase in lateral load. When the lateral load is equal to 0.8 times the vertical load, the vertical bearing capacity decreases by about half. The effect of lateral load on compression bearing capacity is more significant, which is mainly due to P-Δ effect.

3.2 Influence of Lateral Loading on Vertical Failure Mode of Helical Piles

3.2.1 Influence of Lateral Loading on Uplift Failure Mode

The soil incremental displacement maps of the final calculation step in the numerical model are extracted, as shown in Fig. 4. The incremental displacement of the soil in the final calculation step can reflect the failure mechanism of soil slip under the ultimate load. Under the condition of a single uplift load, the slip surface of the soil mainly occurs at the location of the pile helix, and the shape of the slip surface is semicircular arc. When the lateral load was applied to 0.2 times the uplift load, the slip surface began to appear at the upper part of the pile shaft, but the main slip surface still occurred at the pile helix position. Continuing to increase the lateral load, when the lateral load reaches 0.5 times

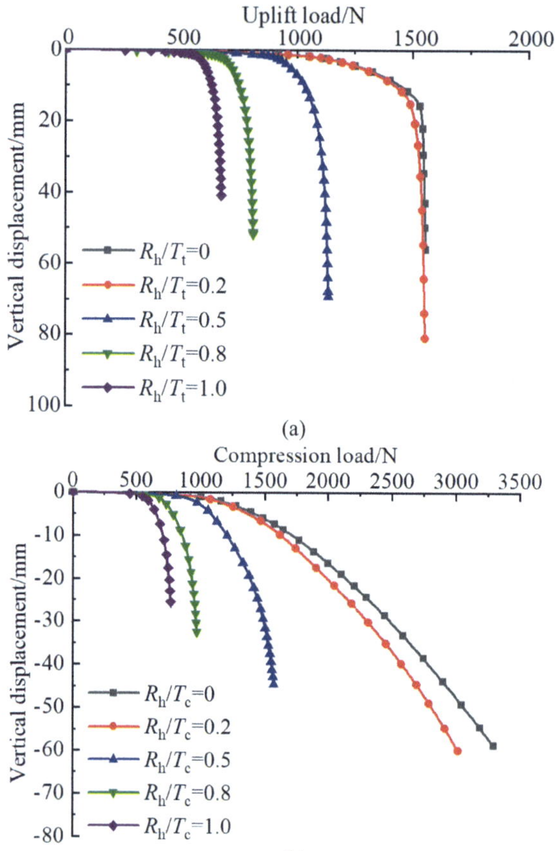

Fig. 3. (a) Combined loads-uplift displacement curves. (b) Combined load-compression displacement curves.

Table 5. Vertical ultimate bearing capacity under different lateral loads.

	Uplift bearing capacity (N)	Compression bearing capacity (N)
$R_h = 0N$	1 575	1 871
$R_h = 0.2T$	1 546	1 682
$R_h = 0.5T$	1 133	1 106
$R_h = 0.8T$	820	773
$R_h = 1.0T$	680	621

the uplift load value, a large soil slip failure has occurred at the upper part of the shaft. The incremental value is larger than that of the lower pile helix location, indicating that the main slip surface begins to shift to the upper part of the shaft. The slip surface still exists in the lower part of the pile helix, and the lower slip surface is distributed along the pile helix without being missed. However, with the further increase in lateral load value, the soil slip surface near the pile shaft becomes larger and deeper, and the slip surface at the pile helix appears to be missing on the upper right side.

Under left lateral load, soil displacement on the left side of the pile shaft significantly exceeds that on the right side, with the left slip surface being a primary contributor to pile failure. According to the experimental data in the literature [22], the pile top displacement needed for sand soil to reach the limit active state is significantly smaller than for the limit passive state, at 1.5‰ L (at 1.2 mm in this study). The displacement required for soil to reach the limit passive state is 15% L (at 120 mm in this study). Thus, the soil slip surface on the left side of the shaft is essentially penetrated when $R_h = 0.5T_t$. The soil has essentially reached its limit active state, resulting in a substantial reduction in pile side friction. Furthermore, the slip surface at the helix position gradually diminishes, signifying a reduction in the soil's resistance range and a consequent decrease in the bearing capacity of the helix. Moreover, the bearing capacity of the helix decreases further when $R_h = 0.8T_t$ and$1.0T_t$. The uplift bearing mechanism variation of helical piles under different lateral and uplift load ratios will be further analyzed in the companion paper. Quantitative calculations of vertical bearing capacity will utilize the numerical data from this paper.

3.2.2 Influence of Lateral Loading on Compression Failure Mode

The soil incremental displacement map in the *final calculation* step of the numerical model under the combined lateral and compression load is extracted and shown in Fig. 5. Under the single compression load, the soil slip surface occurs at the pile helix location, and the shape of the slip surface is also semicircular arc. However, different from the single uplift load, the slip surface mainly occurs in the lower part of the pile helix. When the lateral load reaches 0.2 times the value of the compression load, the slip surface starts to appear at the upper part of the pile shaft, which is different from the case of uplift loading. The slip surface at the upper part of the shaft has been the main slip surface at this time. This indicates that a small lateral load has an effect on the failure mode of the helical pile under combined lateral and compression load. This is mainly due to the fact that under the combined lateral compression loads, the downward pressure produces additional moments that cause the pile to rotate more. As the lateral load continues to increase, the slip surface area at the pile shaft location increases and becomes deeper. When $R_h = 0.5T_c$, the slip surface at the pile shaft has extended to the pile helix position, and the slip surface on the right side of the lower side of the pile helix starts to be missing.

Loading to the maximum load at $R_h = 0.2T_c$ reveals an evident slip surface on the left side of the pile, indicating that the soil has essentially reached the limit active state. Consequently, the soil on the left side of the pile exhibits a significant decrease in lateral frictional resistance. Upon loading to the maximum load at $R_h = 0.5T_c$, the slip surface of the soil near the pile helix location diminishes considerably. Additionally, the compression resistance offered by the soil at the pile helix experiences a notable decrease. Furthermore, the compression resistance provided by the soil at the pile helix diminishes further when loading occurs at $R_h = 0.8T_c$ and $1.0T_c$. The variation in the compression bearing mechanism of helical piles under different lateral and compression

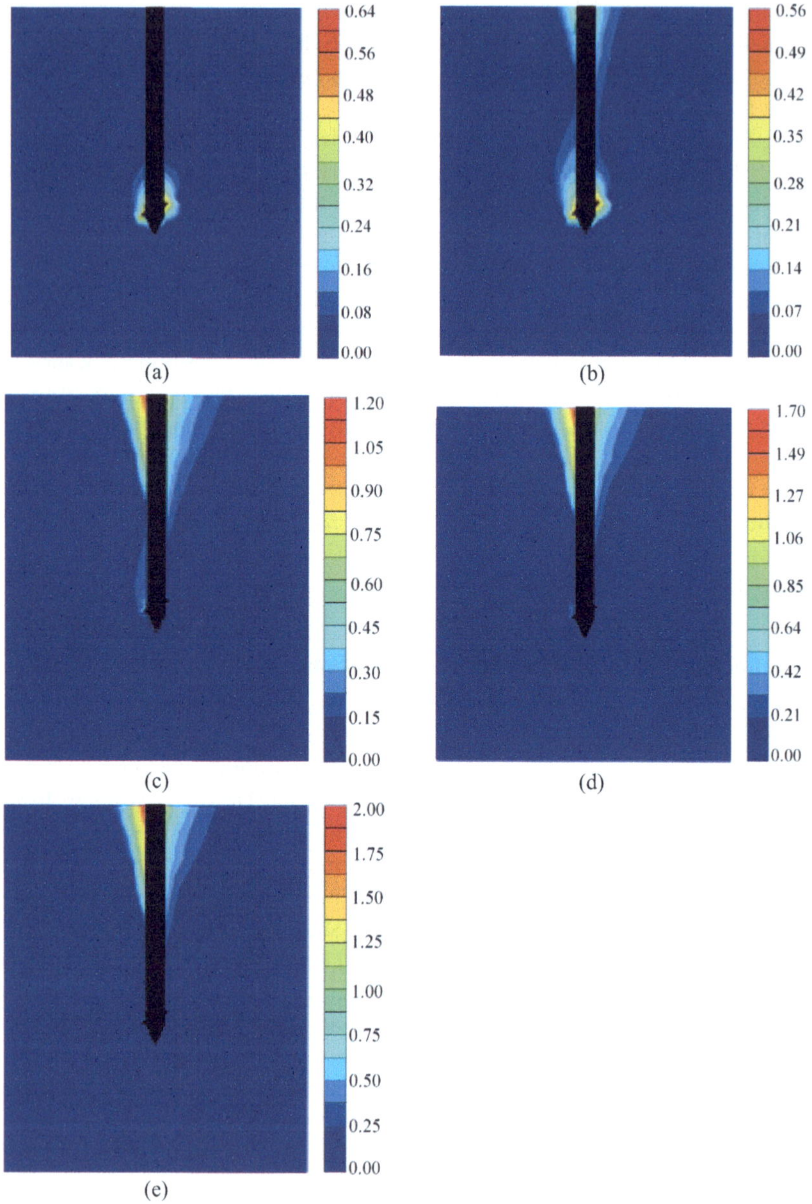

Fig. 4. Soil incremental displacements maps under lateral and uplift combined loads (unit: mm): (a) $R_h = 0$N, (b) $R_h = 0.2T_t$, (c) $R_h = 0.5T_t$, (d) $R_h = 0.8T_t$, (e) $R_h = 1.0T_t$.

loading ratios will be further analyzed in the companion paper. Quantitative calculations of the compression bearing capacity will be based on the numerical data presented in this paper.

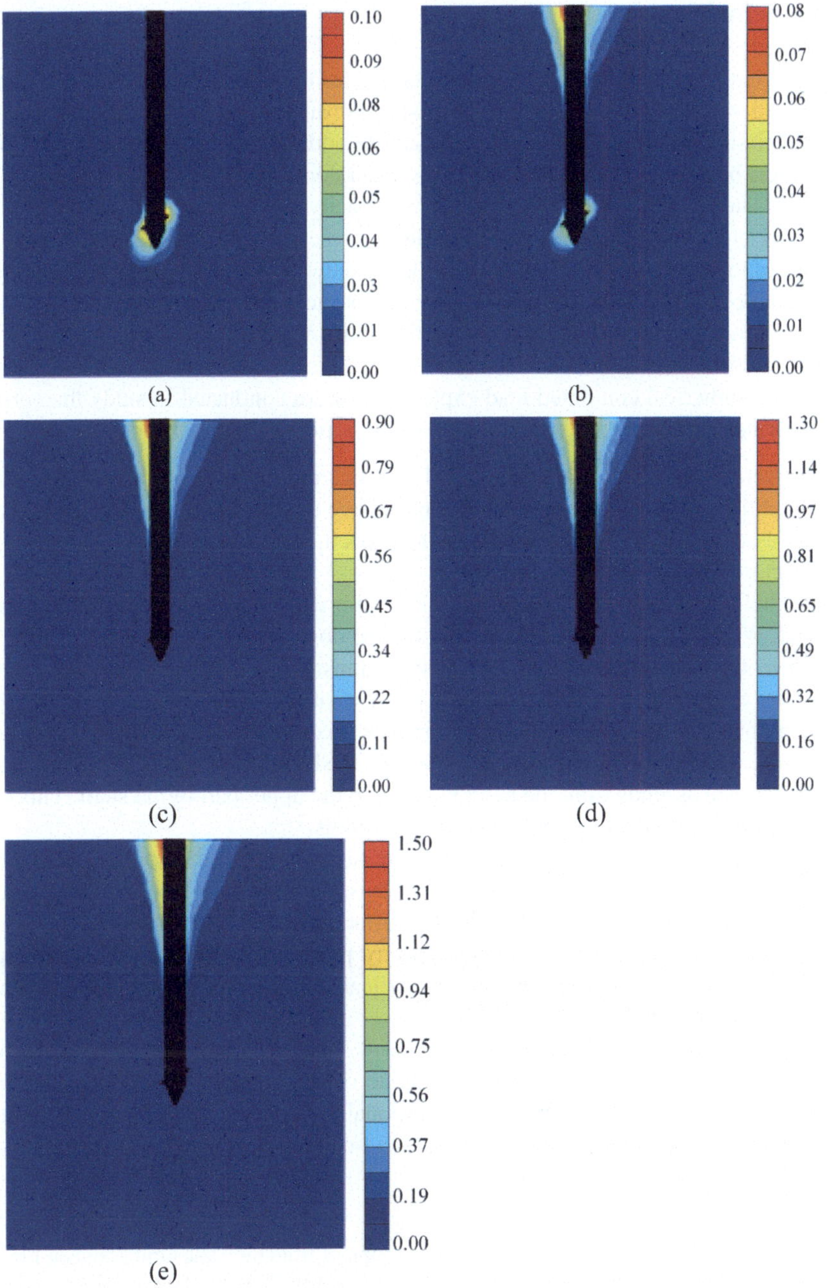

Fig. 5. Soil incremental displacements maps under lateral and compression combined loads (unit: mm): (a) $R_h = 0N$(b), $R_h = 0.2T_t$, (c) $R_h = 0.5T_t$, (d) $R_h = 0.8T_t$, (e) $R_h = 1.0T_t$.

4 Conclusions

The nature of transmission line loading necessitates that helical pile endure combined lateral and vertical loads. With the increasing prevalence of helical pile foundations in transmission lines, there is an urgent need to address the challenge of calculating the vertical bearing capacity of helical piles under combined loads. To comprehend the impact of lateral force on the vertical bearing capacity and failure mode of helical piles, this paper employs numerical simulation as the research methodology. Based on the findings of this paper, the calculation method for the vertical bearing capacity of helical piles under combined load is derived through theoretical investigation in the companion paper. In this paper, numerical analyses software PLAXIS 3D was used to emulate the indoor experiments of helical piles in the existing literature. Following model validation, ten sets of numerical combined load experiments were conducted to study the vertical load-bearing mechanism and failure mode of helical piles. The conclusion are as follows:

(1) The vertical bearing capacity of helical piles decreases after a lateral load is applied. The influence of lateral load on the compressive bearing capacity of the helical pile surpasses its impact on the uplift bearing capacity, primarily attributable to the P-Δ effect.
(2) The helical pile presents a punching shear failure mode under a single compressive load. When subjected to combined lateral and downward pressure loads, the failure mode of the helical pile undergoes a change, transitioning into a general shear failure.
(3) Under a singular vertical load, the slip surface causing the failure of the helical pile forms at the helix position. Upon the application of lateral load, the slip surface also emerges at the upper part of the pile shaft. As the lateral load increases, the principal slip surface ascends from the helix position to the upper part of the shaft. This shift occurs earlier under compression load conditions.
(4) As the lateral and vertical load ratios increase, the soil on the non-bearing side of the pile shaft tends toward the ultimate active state as the pile approaches failure. Concurrently, the area of the soil slip surface around the pile helix diminishes. Consequently, the vertical bearing capacity provided by the pile shaft and helix decreases with increasing lateral and vertical load ratios compared to the single vertical load condition.
(5) Future research should delve deeper into investigating the variations in the vertical load-carrying mechanism and failure mode of helical piles exposed to combined loads across diverse soil conditions. Additionally, exploring the impact of the relative stiffness of helical piles on their vertical load-carrying mechanism and failure mode under combined loads is warranted.

Acknowledgments. This work received financial support from the State Grid Zhejiang Electric Power Co., Ltd. Collective Enterprise Technology Project (CG091201-DMSJY) the State Grid Zhejiang Electric Power Co., Ltd. Infrastructure Construction New Technology Research Project Based on Engineering (SGTYHT/23-JS-003).

References

1. Feng, S.J., et al.: Field tests of micro screw anchor piles under different loading conditions at three soil sites. Bull. Eng. Geol. Env. **80**(1), 127–144 (2021)
2. Venkatesan, V., Mayakrishnan, M.: Behavior of mono helical pile foundation in clays under combined uplift and lateral loading conditions. Appl. Sci. Basel **12**(14), 1–23 (2022)
3. Kurian, N.P., Shah, S.J.: Studies on the behaviour of screw piles by the finite element method. Can. Geotech. J. **46**(6), 627–638 (2009)
4. Trofimenkov, J. G., Mariupolskii, L. G.: (1965) Screw piles used for mast and tower foundations. In: Soil Mech & Fdn Eng Conf Proc. Toronto. pp. 328–332
5. Mitsch, M. P., Clemence S P (1985) Uplift capacity of helix anchors in sand. In: Uplift Behavior of Anchor Foundations in Soil. New York. pp. 26–47
6. Mooney, J. S., Adamczak, S., Clemence, S. P.: (1985) Uplift capacity of helical anchors in clay and silt. In: Uplift Behavior of Anchor Foundations in Soil. New York. pp. 48–72
7. Ghaly, A., Hanna, A., Hanna, M.: (1991) Uplift behavior of screw anchors in sand. I: dry sand. J. Geotech. Eng. 117(5): 773–793
8. The Institute of Electrical and Electronics Engineers: IEEE Guide for Transmission Structure Foundation Design and Testing: IEEE Std 691. The Institute of Electrical and Electronics Engineers Inc, New York (2001)
9. International Code Council: International Building Code: 2021. ICC Publications, Lllinois (2021)
10. State Grid Corporation of China: Code for Design of Helical Anchor Foundation of Overhead Transmission Line: Q/GDW 10584–2022. China Electric Power Press, Beijing (2022)
11. Chen, B., Zhu, Z.Q., Feng, B., Li, P.F., Ding, S.J.: In-situ test research on uplift bearing capacity of screw anchor in soft ground of mire area. Ind. Constr. **50**(05), 71–74 (2020)
12. Qu, S.Z., Guo, Y.H., Wang, Y., Zhang, B., Zhang, J.M., Sun, Q.: In-situ test and numerical simulation of the uplift mechanical performance of screw anchor with large anchor plate in silty clay. Chin. J. Rock Mech. Eng. **39**(S2), 3655–3668 (2020)
13. Shi, D.D., Yu, K., Mao, Y.Y., Yuan, Y., Hao, D., Hu, W.: Experimental study on the uplift behavior and soil deformation characteristics of the double-blade screw anchor in loose sand. Rock Soil Mech. **43**(11), 3059–3072 (2022)
14. Cheng, P., Guo, J., Yao, K., Chen, X.J.: Numerical investigation on pullout capacity of helical piles under combined loading in spatially random clay. Mar. Georesour. Geotechnol. **41**(10), 1118–1131 (2022)
15. Feng B, Sun Z H., Lü Q., Huang J J.: (2024) Effect of lateral load on vertical bearing capacity of helical piles—Part II. Theoretical study. In: 8th International Conference on Civil Architecture and Structural Engineering. Guangzhou
16. Deendayal, R., Muthukkumaran, K., Sitharam, T.G.: Analysis of laterally loaded group of piles located on sloping ground. Int. J. Geotech. Eng. **14**(5), 580–588 (2020)
17. Jamil, I., et al.: Analysis and design of piled raft foundation taking into account interaction factors. Adv. Civ. Eng. **2022**, 1334136 (2022)
18. Amornfa, K., Quang, H., Tuan, T.V.V.: Effect of groundwater level change on piled raft foundation in Ho Chi Minh City, Viet Nam using 3D-FEM. Geomech. Eng. **32**(4), 387–396 (2023)
19. George, B.E., Banerjee, S., Gandhi, S.R.: Helical piles installed in cohesionless soil by displacement method. Int. J. Geomech. **19**(7), 04019074 (2019)
20. Ministry of Housing and Urban-Rural Development of the People's Republic of China: Technical Code for Testing of Building Foundation Piles: JGJ 106–2003. China Building Industry Press, Beijing (2014)

21. Khazaei, J., Eslami, A.: Post-grouted helical piles behavior via physical modeling by FCV. Mar. Georesour. Geotechnol. **35**(4), 528–537 (2016)
22. Fang, Y.S., Ho, Y.C., Chen, T.J.: Passive earth pressure with critical state concept. J. Geotech. Geoenviron. Eng. **128**(8), 651–659 (2002)

Detection Method for Surface Corrosion Level of Profiled Steel Plates Based on the Sliding Window Method

Peng Lu[1,2], Qian Li[3], Jian Wang[4(✉)], and Tiansong Zhao[4]

[1] Central Research Institute of Building and Construction Co., Ltd., Beijing 100088, China
[2] Inspection and Certification Co., Ltd., MCC Group, Beijing 100088, China
[3] CABR Technology Co., Ltd., Beijing 100088, China
[4] Tianjin Chengjian University, Tianjin 300384, China
avonlea@163.com

Abstract. Profiled steel sheet is a commonly used material in enclosure structures, and the service environments for these structures are often challenging. Regular inspections for the corrosion level of profiled steel sheets are necessary to ensure the safety of the structure. Currently, common detection methods mainly rely on visual inspections. A sliding window method for detecting the corrosion level of profiled steel sheets is proposed in this study, with the aim of saving time and manpower and promoting the intelligence of structural inspections. The detection method involves segmenting the images using a sliding window. The trained network is then utilized for corrosion level detection, achieving an accuracy of over 80%. Intelligent visual detection of the corrosion level of profiled steel sheets has been achieved in this study, providing an intelligent approach for assessing the corrosion degree of profiled steel sheets.

Keywords: Profiled Steel Sheet · Surface Corrosion · Damage Detection · Sliding Window Method

1 Introduction

Profiled steel sheet is a common type of metal sheet, often used as walls or roofs in protective structures. Due to its light weight, high strength, and ease of construction, it is widely used in industrial plants, public buildings, and other construction structures [1, 2]. Profiled steel sheets are generally used in environments with strong corrosion, such as industrial atmospheres and marine atmospheres. Corrosion damage is a common type of damage to profiled steel materials. Corrosion damage represents an early stage of deterioration, and failure to address rust in a timely manner can lead to structural failure, engineering accidents, and economic losses [3, 4]. Therefore, conducting regular inspections for corrosion damage in structures is crucial [5, 6]. Traditional inspections are typically visual inspections conducted by trained professionals. However, the inspection process is limited by the skill level of the personnel, and it comes with certain risks. The results and efficiency of the inspection may not be well guaranteed [7].

B. Yuan et al. (Eds.): ICCASE 2024, SUCI, pp. 313–318, 2024.
https://doi.org/10.1007/978-3-031-76102-7_35

In order to better address the limitations of traditional detection methods and achieve intelligent corrosion level detection for profiled steel sheets, a sliding window method for detecting the corrosion level of profiled steel sheets is proposed in this paper. The sliding window method combines image processing and machine learning technology to achieve precise identification and quantitative analysis of corrosion on the surface of profiled steel plates. By employing sliding window method, the method can effectively capture corrosion features in different regions and utilize deep learning algorithms to automatically assess the degree of corrosion. This approach not only enhances the accuracy and reliability of detection but also significantly improves detection efficiency, bringing about a significant technological breakthrough in the field of corrosion detection for profiled steel plates. In Sect. 2, the development of a corrosion level detection method based on the sliding window approach is detailed. In Sect. 3, the accuracy of the detection method is verified, and the impact of sliding window size on the detection results is discussed.

2 Method

To address the need for intelligent corrosion level detection in existing profiled steel structures, establishing a corrosion level detection network specifically designed for profiled steel plates is crucial. Images of profiled steel plates with varying degrees of corrosion are obtained through neutral salt spray accelerated corrosion tests. Based on the different surface corrosion morphologies, the corrosion levels are categorized into four classes: no corrosion, slight corrosion, moderate corrosion, and severe corrosion. These images are compiled to create a dataset for deep learning. The Swin-Transformer network architecture is chosen as the backbone feature network for corrosion level recognition. The network structure is illustrated in Fig. 1. After training, the network can accurately identify the corrosion levels of profiled steel plates.

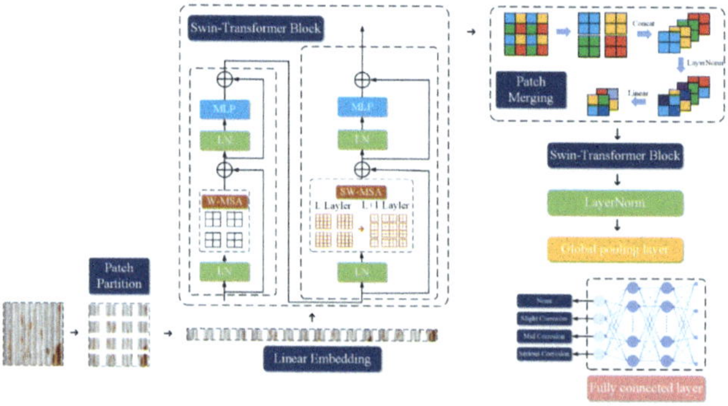

Fig. 1. Swin-Transformer network structure diagram

The sliding window method is a commonly used detection technique in computer vision. The process involves moving a window across the original image for segmentation. Before the entire original image is scanned, the window moves with a fixed size of

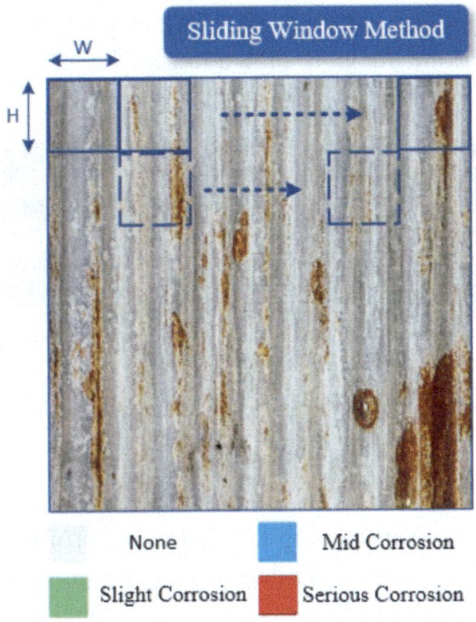

Fig. 2. Workflow diagram of the sliding window method, where H and W represent the height and width of the sliding window

pixels both horizontally and vertically, thus analyzing the image. At each position, the image information within the window is extracted, and then further corrosion level detection is conducted based on this extracted image information. By continuously moving the window over the image and repeating this process, the entire image can be comprehensively analyzed. The sliding window method is chosen as the detection tool for corrosion level in this paper, the segmented images are used as input, and a pre-trained corrosion level detection network is employed for identification. The recognition results are added to the original image in the form of transparent blocks of different colors, where no rust is represented as the original image; slight corrosion is represented in green, mid corrosion in blue, and serious corrosion in red. The detection process of the sliding window method is illustrated in Fig. 2. Finally, the resulting images are reassembled to complete the corrosion level detection of the profiled steel plate.

3 Results

The selected images for the validation of the detection method are taken from the 7th day of the experiment. The apparent condition of the pressed steel plates is mainly mid corrosion, with a small amount of slight corrosion and serious corrosion. There are a total of 10 images.

Selecting a subset of images for validation of the detection method. The apparent condition of the pressed steel plates is mainly mid corrosion, with a small amount of slight corrosion and serious corrosion. There are a total of 10 images. The images have a resolution of 6000 × 4000. To explore the impact of choosing different sliding window sizes on the detection results, three window sizes are selected: 150 × 100, 300 × 200, and

600 × 400. The number of cut images is 100, 400, and 1600, respectively. An example of the detection result for one image is shown in Fig. 3.

(A)150×100 (B)300×200 (C)600×400

Fig. 3. (A) – (C) Detection results for three sliding window sizes (A)150 × 100 (B)300 × 200 (C)600 × 400

As shown in Fig. 3, when inspecting the surface of profiled steel plates, the detection network can accurately assess the degree of corrosion, ensuring the accuracy of the results while overcoming the limitations of traditional inspection methods based on the skill level of inspectors. This detection method described in this paper enables rapid and precise identification and quantitative evaluation of the corrosion status on the surface of profiled steel plates, thereby assisting engineers and technicians in promptly implementing necessary maintenance and repair measures to safeguard the normal operation of protective structures. However, there is a problem of erroneously detecting the background when detecting the background. By counting the number of images with erroneous detection, the detection accuracy results for three sliding window sizes in the detection task are shown in Table 1. The detection accuracy calculation formula is as follows Eq. (1).

$$\mathrm{Acc} = \frac{1}{N}\sum_{N=1}^{N}\frac{T_N - f_N}{T_N} \tag{1}$$

where N is the number of images to be detected, T_N is the total number of segments after the N-th image is segmented, and f_N is the number of detection errors.

Table 1. Detection Accuracy Table

The size of the sliding window	Acc(%)
150 × 100	82.9
300 × 200	83.2
600 × 400	86

According to the detection results, it can be observed that with the increase in the size of the sliding window, the issue of incorrectly detecting the background has been somewhat alleviated, leading to an improvement in detection accuracy. However, as the sliding window size increases, certain features occurring on the specimen surface may be overlooked, resulting in potential errors in detection, Such errors are inevitable. Due to the potential errors, the detection accuracy formula defined equation in this article is also incorrect. The main reasons for the occurrence of false positive detections are twofold: firstly, the presence of rust falling in the background of the captured images, causing images that should belong to the background to include corrosion features; secondly, during the cutting process, some images may include the edges of features with minimal corrosion.

4 Conclusion

This paper proposes a sliding window method for detecting the degree of corrosion in profiled steel plates, achieving intelligent detection of corrosion levels. The research conclusions are as follows:

(1) In this paper, the sliding window method is employed as a tool to segment profiled steel plate images. The resulting image segments are fed into a corrosion level recognition network, utilizing the Swin-Transformer architecture, for the purpose of identifying corrosion levels. Different colors are utilized within the segmented regions to represent varying levels of corrosion. Following this, the identification results are amalgamated to enable intelligent detection of the corrosion levels present in profiled steel plates. Through this approach, the inherent limitations of manual detection are effectively alleviated.

(2) The detection accuracy of the method proposed in this paper is influenced by the size of the sliding window. Increasing the size of the sliding window may lead to the omission of some corrosion features, resulting in potential errors in detection accuracy. Addressing how to mitigate the issue of oversized sliding windows neglecting corrosion features and accurately selecting appropriate sliding window sizes will be key areas for future research. Therefore, it is imperative to conduct additional research aimed at identifying the optimal size for the sliding window, a critical step towards improving detection accuracy and minimizing errors. This endeavor will involve comprehensive experimentation and analysis to ascertain the ideal balance between window size and detection precision, thereby advancing the effectiveness of the proposed method in practical applications.

References

1. Ahmed, E., Badaruzzaman, W.H.W.: Finite element prediction on the structural performance of profiled steel sheet dry board structural composite system proposed as a disaster relief shelter. Constr. Build. Mater. **19**(4), 285–295 (2005)
2. Islam, S.Z., Ahamed, M., Islam, M.M.: Exploring thin shell profiled steel sheet roofing element for architectural and structural context. Khulna Univ. Stud. 259–70 (2022)
3. Majid, M.A., Wan Badaruzzaman, W.H.: Experimental evaluation of profiled steel sheet dry board wall panel system. Key Eng. Mater. **594–595**, 421–426 (2013)
4. Choi, K.-Y., Kim, S.: Morphological analysis and classification of types of surface corrosion damage by digital image processing. Corros. Sci. **47**(1), 1–15 (2005)

5. Koch, G.H., Brongers, M.P., Thompson, N.G.: Corrosion costs and preventive strategies in the United States. J. Endocrinol. (2002)
6. Křivý, V., Fabián, L.: Calculation of corrosion losses on weathering steel structures. Appl. Mech. Mater. **188**, 177–182 (2012)
7. Feliciano, F.F., Leta, F.R., Mainier, F.B.: Texture digital analysis for corrosion monitoring. Corros. Sci. **93**, 138–147 (2015)

Study on the Impact of Sudden Temperature Increase on the Hydrothermal Stability of Permafrost in Long-Term Permafrost Regions

Hui Ge[1(✉)], Chunxiang Guo[1], and Lei Zhang[2]

[1] School of Civil Engineering, Lanzhou Jiaotong University, Lanzhou 730070, China
645459556@qq.com

[2] Shaanxi Province Institute of Water Resources and Electric Power Investigation and Design, Xi'an 710001, China

Abstract. With the advancement of the "Western Development" strategy, there has been an increasing trend in engineering construction in the Qinghai-Tibet Plateau region. This study, based on a high-altitude airport project in the hinterland of the Qinghai-Tibet Plateau, aims to investigate the impact of sudden temperature increases on the hydrothermal stability of natural ground in permafrost regions. By establishing a coupled hydrothermal model and integrating field measurement data, the study analyzes the effects of temperature spikes on the hydrothermal behavior of permafrost ground. The research reveals that sudden temperature increases lead to elevated ground temperatures in permafrost regions, resulting in an increase in unfrozen water content and changes in frost heave. Through numerical simulations and comparative analysis, it is found that temperature spikes significantly affect the hydrothermal behavior of ground materials, potentially causing uneven settlement of ground materials and thereby impacting the safety of surface structures. These findings hold significant implications for engineering construction and infrastructure design in permafrost regions, providing theoretical support and technical guidance for future projects in the area.

Keywords: Thermo-hydrological coupling · Permafrost layer · sudden temperature increase · permafrost

1 Introduction

Frozen soil refers to a distinct type of geological formation found beneath the Earth's surface to a certain depth, where the temperature in the surrounding environment drops below 0 °C. Under these conditions, the moisture within the soil solidifies into ice, leading to the bonding of soil particles by ice formation, resulting in a unique type of lithosphere known as frozen soil [1]. Permafrost, also known as perennially frozen ground, experiences seasonal freezing and thawing primarily within the surface layers of the soil during the winter and summer seasons. Within permafrost, the interactions among soil particles, ice, and unfrozen water are influenced by external factors such as temperature. These interactions encompass processes such as movement, diffusion, and

B. Yuan et al. (Eds.): ICCASE 2024, SUCI, pp. 319–329, 2024.
https://doi.org/10.1007/978-3-031-76102-7_36

phase changes, ultimately leading to the phenomenon of freeze-thaw cycles within the soil [2]. In this process, the variations in moisture content and heat within the soil, and the consequent redistribution of stresses, are the primary factors contributing to frost heave and thaw settlement. The Qinghai-Tibet Plateau is one of the most sensitive and severely affected regions by temperature changes. Extreme high-temperature events can lead to the degradation of permafrost on the plateau, resulting in an increase in the thickness of the active layer. The active layer is highly sensitive to temperature fluctuations, and even short-term temperature increases can cause its thickness to expand.

Many scholars both domestically and internationally have conducted research on coupled hydrothermal models of frozen soil. Taber [3] employed benzene instead of water in experiments, proposing that moisture migration is the primary factor in soil frost heave. Their research refuted earlier understandings of frost heave mechanisms, paving the way for new directions in frost heave studies. Harlan [4] used an analogical approach to investigate the process of moisture migration in unsaturated soils and partially frozen soils. They proposed coupled equations between heat and moisture migration in fully frozen soils. Xu et al. [5] obtained an empirical expression for the relationship between unfrozen water content and ice content through experiments. Mao et al. [6] proposed a coupled calculation model for hydrothermal fields based on the non-steady-state temperature field, moisture field, and stress-strain field of subgrade, while Bai et al. [7] introduced the concept of solid-liquid ratio based on an empirical expression for unfrozen water content and established a hydrothermal contact equation. Cheng et al. [8] considering permeation, established a fully coupled mathematical model for freeze-thaw coupling based on the Harlan hydrothermal coupling model. Zhou et al. [9] based on the theory of frost heave, established a three-dimensional thermal-water-stress coupling model considering moisture migration through the three-dimensionalization of a rigid ice model.

Many scholars have conducted research on the hydrothermal changes of permafrost in the Qinghai-Tibet Plateau region, but there is relatively less research on the impact of climate change on permafrost. This study aims to investigate the influence of extreme high-temperature events on the hydrothermal dynamics of natural ground-based permafrost, providing data support for subsequent construction of surface structures.

2 The Establishment of the Water-Thermal-Mechanical Coupling Model

COMSOL Multiphysics is a multi-physics simulation software that excels in numerical computation of coupled multi-physics fields. This paper developed a fully coupled model of frozen soil-water-heat using the PDE module of COMSOL software. Through comparison with on-site ground temperatures, the effectiveness of the model and parameters was validated. Subsequently, predictions were made for the water, temperature, and displacement fields of the natural foundation over a period of 10 years of freeze-thaw cycles and 10 years following a sudden temperature increase due to global warming. This research provides theoretical support for the construction of buildings in perennially frozen soil areas.

During the freezing process of frozen soil, the interactions among soil, ice, and unfrozen water are complex. Therefore, it is necessary to neglect minor factors affecting freezing. For ease of computation, the following assumptions are made [10, 11]: (1) The internal soil of the model is homogeneous and continuous, exhibiting isotropy. (2) Soil particles and ice are considered incompressible. (3) During soil freezing, moisture migration occurs in the liquid phase and follows Darcy's law. (4) Only the effects of heat conduction are considered in the heat transfer process, while thermal convection is neglected. (5) The permeability coefficient remains consistent across different positions within the same soil layer.

During the freezing process, the latent heat released from the phase change of unfrozen water to ice is considered as the thermal source term in the system. The release of latent heat from phase change significantly affects the temperature field during the freezing process and is a critical factor in thermal equilibrium of the system. The governing equation for the temperature field is as follows:

$$\rho C(\theta)\frac{\partial T}{\partial t} = \lambda(\theta)\nabla^2 T + L \cdot \rho_I \frac{\partial \theta_I}{\partial t} \tag{1}$$

The equation is represented as follows: ρ is the density of the soil, ρ_I is the density of ice, T is the temperature of the soil, t is time, $C(\theta)$ is the specific heat capacity, $\lambda(\theta)$ is the thermal conductivity, L is the latent heat of phase change, which is taken as 334.56 kJ/kg.

Based on the principles of mass conservation and Darcy's law, incorporating the phase change of ice-water into the Richard's equation, the equation governing the water content field can be expressed as follows:

$$\frac{\partial \theta_u}{\partial t} + \frac{\rho_I}{\rho_w}\frac{\partial \theta_I}{\partial t} = \nabla[D(\theta_u)\nabla\theta_u + k(\theta_u)] \tag{2}$$

In the equation, θ_u represents the volumetric unfrozen water content, ρ_w denotes the density of water, θ_i stands for the volumetric ice content, $D(\theta_u)$ denotes the diffusion coefficient, and $k(\theta_u)$represents the diffusion coefficient in the direction of gravity.

Introducing the solid-liquid ratio $B_I(T)$ to relate the volumetric unfrozen water con tent to the volumetric ice content, the equation establishing the relationship between the water content field and the temperature field can be expressed as follows:

$$B_I = \frac{\theta_i}{\theta_u} = \begin{cases} 1.1\left(\frac{T}{T_f}\right)^B - 1 & T < T_f \\ 0 & T \geq T_f \end{cases} \tag{3}$$

where B_I represents the solid-liquid ratio, B is a constant related to the variation in soil properties and salt content, and T_f is the freezing temperature of the soil.

During the freezing process of soil, the strain of the soil includes thermal expansion and contraction, moisture migration, and the volume increase due to the phase change of water into ice, wherein the volume change resulting from water phase change into ice is approximately 9%. Considering that moisture migration during freezing is the main factor causing frost heave, the equation for soil strain is as follows:

$$\varepsilon = 0.09(\theta_0 + \Delta\theta - \theta_u) + \Delta\theta - n \tag{4}$$

The equation is represented as follows: where ε is the volumetric strain, θ_0 is the initial moisture content, $\Delta\theta$ is the moisture migration, and n is the porosity.

3 Numerical Model Calculation and Results Analysis

3.1 Model Parameters and Boundary Conditions

The engineering site is located in the Tuotuo River area of the Qinghai-Tibet Plateau. According to the geological column chart on-site, the natural permafrost layers consist of miscellaneous fill soil, silty clay, mudstone, and fully weathered sandstone. The geometric model of the natural permafrost layer in this area is simplified as shown in the Fig. 1, with a length of 40m and a depth of 20m. A water-thermal coupled model of the natural permafrost layer is established using COMSOL for calculation.

Initially, the temperature is set to −0.25 °C, and the initial moisture content is set to the measured moisture content of the soil. The temperature boundary condition follows Eq. (7). Without considering temperature rise, a calculation of 10 years of freeze-thaw cycles is performed to obtain the temperature distribution of the permafrost layer, which is then compared with the measured on-site ground temperature to validate the effectiveness of the model and parameters.

Using the ground temperature as the initial value of the temperature field, the temperature boundary condition follows Eq. (7). Considering temperature rise, the distribution of water-thermal properties of the natural permafrost layer is predicted for 10 years after the temperature rise.

Taking the temperature field obtained from 10 years of freeze-thaw cycles as the initial value, assuming a sudden temperature increase in the second summer, the simulation increases the temperature by 10 °C for one month before returning to normal. Freeze-thaw cycle calculations are performed for another 10 years, and compared with the normal temperature freeze-thaw cycle to analyze the effects of the temperature rise event on the temperature, moisture, and displacement fields of the natural permafrost layer.

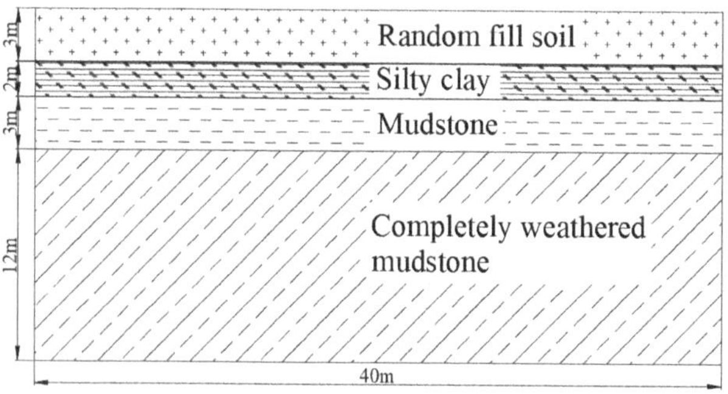

Fig. 1. Schematic Diagram of Natural Foundation Model.

Based on the field exploration and geotechnical test data, the water and thermal parameters of the model are summarized in Table 1.

Table 1. The values of the water and thermal parameters for the soil layers

Soil Layers	Density kg/m^3	Thermal conductivity KJ/(m·°C)	Specific heat capacity KJ/(kg·°C)	Freezing temperatures °C	Permeability coefficient m/s
Mixed fill soil	1900	1.45	0.9	−0.15	1.02×10^{-7}
Fine silty clay	1950	1.13	0.82	−0.51	5.5×10^{-8}
Mudstone	2000	1.72	1.05	−0.25	3.02×10^{-8}
Completely weathered mudstone	2200	1.75	1.07	−0.23	1.2×10^{-8}

The relationship between the elastic modulus and Poisson's ratio of the soil and temperature is expressed as follows:

$$\begin{aligned} E &= a_1 + b_1|T|^m \\ \mu &= a_2 + b_2|T| \end{aligned} \quad (5)$$

The parameters a_1, b_1, a_2, b_2 are experimental constants related to soil properties, with values shown in Table 2:

Table 2. Mechanical Parameters of a Permafrost Roadbed Structure

Soil Layers	a_1/Mpa	b_1	a_2	b_2
Mixed fill soil	61	53	0.35	−0.007
Fine silty clay	28	26	0.40	−0.008
Mudstone	28	26	0.38	−0.008
Completely weathered mudstone	30	26	0.37	−0.008

Considering the global warming factor, studies have shown that the annual average temperature increase in the Qinghai-Tibet Plateau region is 0.52 °C per decade [12]. Based on the meteorological data for this region in 2023, the upper boundary condition for temperature can be derived as follows:

$$T = -0.5 + 11.95 \sin\left(2\pi t + \frac{\pi}{2}\right) + \frac{0.52 * t}{10} \quad (6)$$

The formula features T as the soil temperature and t as time in civil engineering terminology.

The lower boundary condition is implemented in the form of heat flux, belonging to the third type of boundary conditions:

$$\frac{\partial T}{\partial z} = \xi = -0.047 \text{ W/m}^2 \quad (7)$$

The left and right sides are treated as adiabatic boundaries. For the water boundary, zero flux is applied at the left, right, and upper boundaries, while the lower boundary condition is as follows:

$$K\nabla H = q_\theta(x, y, t) \tag{8}$$

3.2 Temperature Field Analysis

Given an initial temperature of −0.25 °C and initial moisture content as per the soil test moisture content, the comparison between simulated values and on-site measured temperatures during summer for 10 years of freeze-thaw cycles is shown in Fig. 2. The simulated values are generally consistent with the on-site measured ground temperatures. The difference at 0.5 m depth is 0.3 °C, and the average difference below 1 m is only 0.1 °C. Therefore, the water-thermal coupling model and parameters in this study demonstrate effectiveness, enabling reasonably accurate predictive analysis.

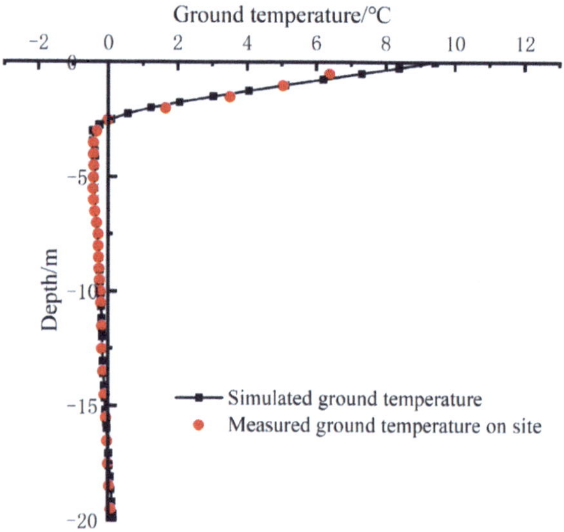

Fig. 2. Numerical simulation results are compared with in-situ measured ground temperatures.

The numerical simulation ground temperatures depicted in Fig. 2 are used as initial ground temperatures for predictive calculations. The temperature boundary conditions are altered to compute the temperature field, moisture field, and displacement field for two scenarios: considering climate warming over a 10-year freeze-thaw cycle and considering a temperature surge event followed by a 10-year freeze-thaw cycle. A comparative analysis of the results from both calculations is conducted to assess the impact of the temperature surge event on the soil's thermo-hydraulic properties. Figures 3 and 4 illustrate the distribution of temperature fields during winter and summer, respectively, after a 10-year freeze-thaw cycle under normal atmospheric conditions.

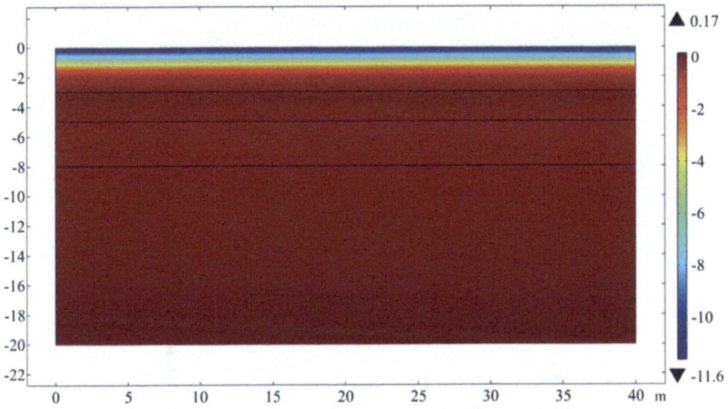

Fig. 3. The winter temperature field after a 10-year freeze-thaw cycle.

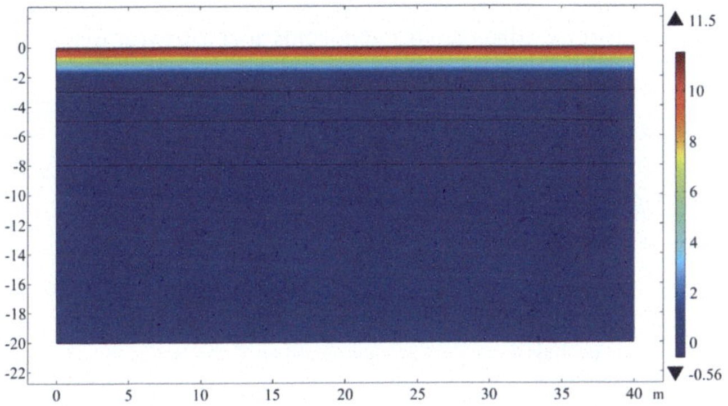

Fig. 4. The summer temperature field after a 10-year freeze-thaw cycle.

Figures 5 and 6 show the comparison of winter and summer temperature fields after a 10-year freeze-thaw cycle under normal temperature conditions and following a temperature spike event. In the range of 0 to 2.5 m, significant differences in winter ground temperatures are observed, with a maximum deviation of 0.7 °C, while differences in summer ground temperatures are smaller, reaching a maximum of only 0.3 °C. Below 5 m, the ground temperatures tend to converge. This discrepancy is attributed to the temperature spike event causing thawing of the permafrost, resulting in increased unfrozen water content and thickening of the active layer. During the freeze-thaw cycle following the temperature spike event, unfrozen water migrates towards the freezing front, leading to increased water content in the upper layer. During winter freezing, the phase change of water to ice releases latent heat, leading to an increase in temperatures in the upper layer.

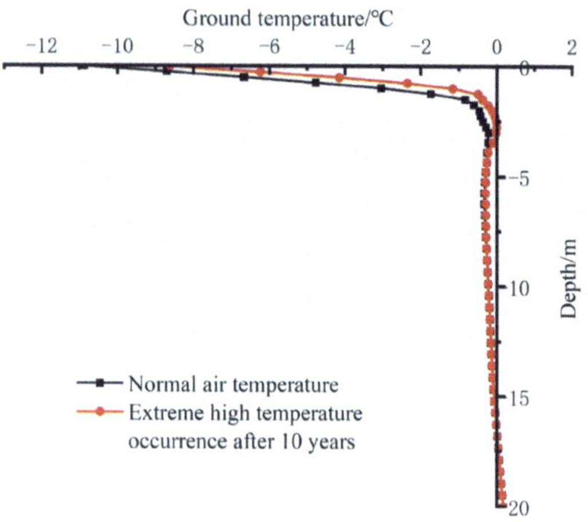

Fig. 5. Comparison of the winter temperature fields after a 10-year freeze-thaw cycle.

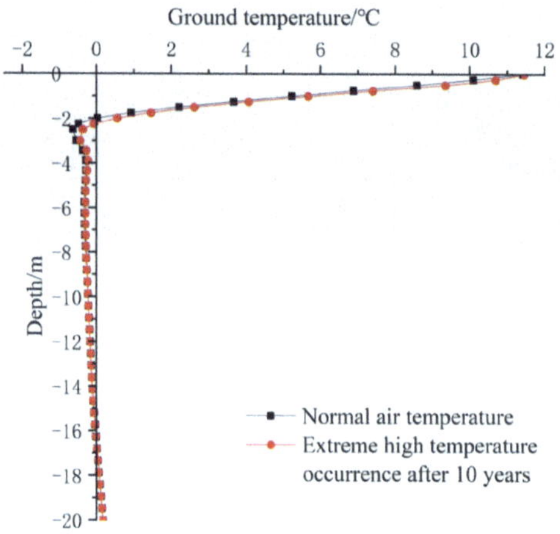

Fig. 6. Comparison of the summer temperature fields after a 10-year freeze-thaw cycle.

3.3 Water Field Analysis

Figures 7 and 8 compare the water fields after 10 years of normal winter and summer temperatures with those after 10 years of temperature spike events. After the temperature spike event, the unfrozen water content is greater than that under normal temperature conditions. At 2.6 m depth, the maximum difference in unfrozen water content during winter is 0.045, while at 5 m depth, the maximum difference during summer is 0.025. The difference in summer is smaller than that in winter. This is consistent with the temperature field analysis, as the temperature spike event leads to upward migration of water under the temperature gradient during freezing, resulting in increased water content in the upper soil layer, with a longer-lasting effect.

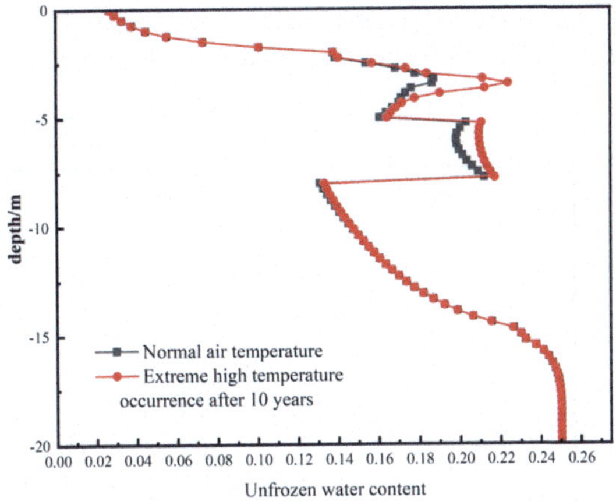

Fig. 7. Comparison of the water field after 10 years of freeze-thaw cycles during winter.

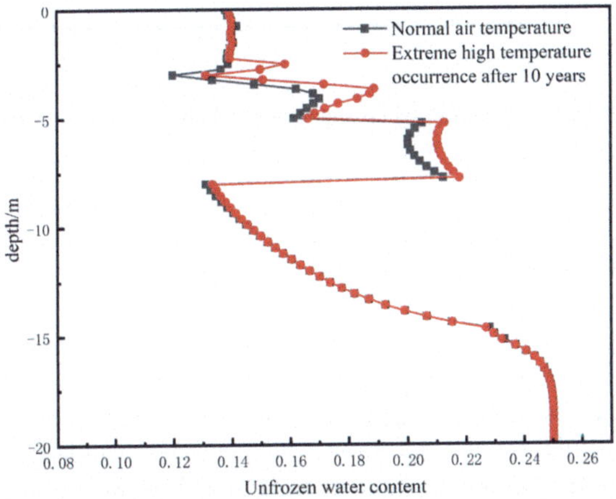

Fig. 8. Comparison of the water field after 10 years of freeze-thaw cycles during summer.

3.4 Displacement Field Analysis

Figure 9 presents a comparison of the displacement variations in the natural ground layer under normal temperature conditions and after a temperature spike event. From the figure, it can be observed that after the temperature spike event in the second year, the ground layer experiences frost heaving of 11.2 cm, which is 0.8 cm higher compared to the displacement under normal temperature conditions. Due to the increased moisture content in the surface layer, the frost heaving continues to increase during the freeze-thaw cycles, reaching a peak of 11.6 cm in the fifth year's winter. The difference in displacement compared to normal temperature conditions reaches 1.2 cm. With the consideration of climate warming effects, the frost heaving decreases after the fifth year, and simultaneously, the difference in displacement compared to normal temperature conditions also decreases. The non-continuous nature of the on-site soil layers means that the displacement changes caused by temperature spikes can result in uneven settlement in the natural permafrost layer on-site, thereby affecting the safe operation of surface structures.

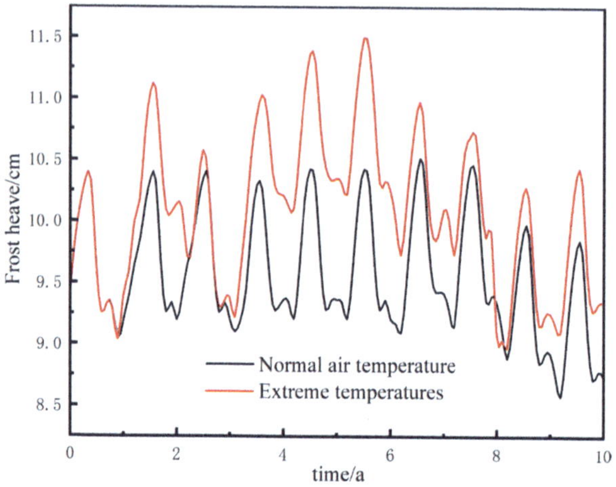

Fig. 9. Comparison of Displacement Changes Over 10 Years

4 Conclusion

Based on the coupled water-thermal equations, this study established a fully coupled water-thermal model of natural ground foundation. Utilizing the PDE module of the COMSOL software, the initial temperature values obtained from freeze-thaw cycles were compared with field-measured ground temperatures, validating the effectiveness and accuracy of the fully coupled water-thermal model in simulating the hydrothermal behavior of perennially frozen soil regions. Through predictive calculations comparing the effects of normal temperature conditions with those of sudden temperature changes on natural ground soil, the following conclusions were drawn:

(1) Ten years after the occurrence of a sudden temperature increase event, the temperature of the soil layer was higher than that under normal temperature conditions, with a significant temperature difference observed in the shallow layers of perennially frozen soil, reaching up to 0.7 °C in winter.
(2) The sudden temperature increase event resulted in an increase in unfrozen water content in the shallow layers of perennially frozen soil compared to normal temperature conditions, with the maximum difference observed in winter, reaching 0.045.
(3) The sudden temperature change event significantly affected the frost heaving, with the peak frost heaving reaching 11.6 cm, an increase of 1.2 cm compared to normal temperature conditions, and the effect persisted for a longer period. In practical engineering applications, the non-continuous nature of the foundation soil layer means that changes in frost heaving caused by sudden temperature changes can

lead to uneven settlement of the foundation, affecting the safety of above-ground structures.

(4) Sudden temperature change events can significantly affect the hydrothermal behavior of perennially frozen soil regions, leading to increased unfrozen water content and frost heaving, resulting in uneven settlement of soil layers and potentially impacting the safe operation of above-ground structures.

(5) In actual engineering design and construction processes, it is necessary to consider extreme climatic factors. During the design phase of projects in permafrost regions, insulation measures should be increased based on the frequency of extreme high-temperature events to enhance the hydrothermal stability of permafrost, thereby reducing the impact of uneven settlement. During the construction phase, it is advisable to avoid extreme high-temperature weather conditions and commence construction activities only after the hydrothermal performance of the permafrost layer stabilizes.

The study only examines the impact of sudden temperature increases on the hydrothermal stability of permafrost, without fully considering the broader aspects of climate change. Future research could explore the effects of extreme precipitation and extreme low temperatures on the hydrothermal stability of permafrost.

References

1. Ma, W., Wang, D.Y.: Mechanics of Frozen Soil. Science Press, Beijing (2014)
2. Wen, Z., Deng, Y.S., Feng, W.J.: Study on the mechanism of moisture migration in freezing soils: review and prospect. J. Glaciol. Geocryol. **45**(02), 588–598 (2023)
3. Taber, S.: The mechanics of frost heaving. J. Geol. **38**(4), 303–317 (1930)
4. Harlan, R.L.: Analysis of coupled heat-fluid transport in partially frozen soil. Water Resour. Res. **9**(5), 1314–1323 (1973)
5. Xu, X.Z., Deng, Y.S.: Experimental Study on Moisture Migration in Frozen Soil. Science Press, Beijing (1991)
6. Mao, X.S.: Study on the Coupling Effect of Water and Heat Field in Roadbeds in Permafrost Regions. Chang'an University (2004)
7. Bai, Q.B., Li, X., Tian, Y.H.: Study on coupled water and heat equations in permafrost and numerical simulation. J. Rock Soil Eng. **37**(S2), 131–136 (2015)
8. Cheng, H., Lin, J., Wang, B.: Mathematical model and test verification of seepage freezing in saturated sand layer. Sci. Technol. Eng **18**(12), 38–44 (2018)
9. Zhou, J., Li, Z.Y., Tang, Y.Q.: Three-dimensional coupled water-thermal-mechanical frozen-thawing model and its application in soft clay. J. Harbin Inst. Technol. **54**(02), 117–125 (2022)
10. He, M., Feng, X.P., Ning, L.: Improvement of coupled thermo-hydrological model for saturated frozen soil. Chin. J. Geotech. Eng. **40**(7), 1212–1220 (2018)
11. Li, Z., Jiang, X., Zhang, J.X., Zhao, Y., Wang, Y.Q., Fang, Y.X.: Numerical simulation of water-thermal-mechanical coupled freezing damage on slope protection of earth-rock dam. J. Hydroelectric Eng. 1–14 (2023)
12. Ming, Y., Zhang, M.Y.: Study on long-term stability of Qinghai-Tibet railway emban-kment. Cold Reg. Sci. Technol. **57**(2–3), 139–147 (2009)

Research on Seismic Performance of New Type Beam-Column Steel Joints with Double Loops

Xiao Xiao(✉), Baoming Guan, and Chao Ma

School of Civil Engineering, University of South China, Hengyang, China
brightxiao@163.com

Abstract. In recent years, box steel column and box steel beam have been used more and more widely in steel structure system because of their good overall stability and convenient construction. However, due to the influence of welding technology and the working environment of the construction site, it is difficult to ensure the quality of the weld at the joint of the beam and column. Therefore, in this paper, two new types of joints were proposed for steel structure connection: double ring hoop sleeve joints and double ring hoop π-shape joints, and two groups of components were designed to be analyzed by changing the ratio of the thickness of the double ring hoop and the size of the slotted hole of reaming-type beams, and the finite element simulation analysis of seismic performance was carried out by applying ABAQUS software. The results showed: the flexural strength of the joints increases with the increase of the ratio of the thickness of the double ring hoop to the columns; the two joints showed a better sliding effect when using the slotted holes type, which helps to improve the energy-consuming capacity of the structure. In conclusion, the double ring hoop sleeve joints and double ring hoop π-type joint have good force seismic performance, reflecting good seismic performance, and provided reference for the practical application of steel frame with new beam-column joints.

Keywords: double ring hoop sleeve connection · beam-column steel joints · numerical simulation · seismic performance · slotted-hole bolt

1 Introductory

The beam-column joints in steel structures are mostly connected by bolted connection, welded connection, end plate connection, top and bottom angle steel connection [1].

In recent years, scholars at home and abroad have proposed same new types of beam-column joints with different connection forms in terms of evaluation and improvement of mechanical properties, and have carried out theoretical analyses, experimental studies and numerical simulations. Feng et al. [2] In order to prevent the buckling of steel joint plates, a replaceable buckling-resistant connection steel frame joint was proposed, and the performance of the joints designed by this method met the design requirements. yi et al. [3] designed a new type of elliptical single-sided bolt which used in the joints of steel square tube columns and I-beams. This new bolt had a simple structure and is easy

B. Yuan et al. (Eds.): ICCASE 2024, SUCI, pp. 330–337, 2024.
https://doi.org/10.1007/978-3-031-76102-7_37

to install. Wei et al. [4] conducted a study on steel frame joints with different connection forms. The seismic performance of the bolted connection was compared to experimental data and proved to be the best. Kim et al. [5] developed an easy-to-construct beam-column connection through experimental and theoretical analyses, which also ensures the bending performance of the beams and columns. Through the reversed cyclic loading test and finite element simulation, the seismic performance of the joints and the influence of thickness and friction parameter on the performance of the joints were investigated.

Although there are many studies on new joints but none of them have studied the new joints with slotted hole bolts. In this paper, two new types of joints were proposed for application in steel structure based on box-type steel columns and box-type steel beams [6] and three groups of members as G1, G2 were designed respectively, and research on their seismic performance were carried out as well as the influence of thickness ratio, bolt-hole size on the joints were investigated, which provides a reference for the practical application of the engineering of the structural system of beam-column joints for steel frames.

2 Finite Element Method and Validation

In order to verify the applicability of the finite element analysis method on the new double hoop joints, the numerical simulation analysis of the XBL-3 specimen in the literature [7] were conducted and made comparison with the experimental results.

From Fig. 1, it can be seen that the moment-rotation curve calculated by finite element analysis is more consistent with the trend of the development of the moment-rotation curve obtained from the study in the paper, and the shapes and trends of the two are in good agreement and agree well with the experimental results.

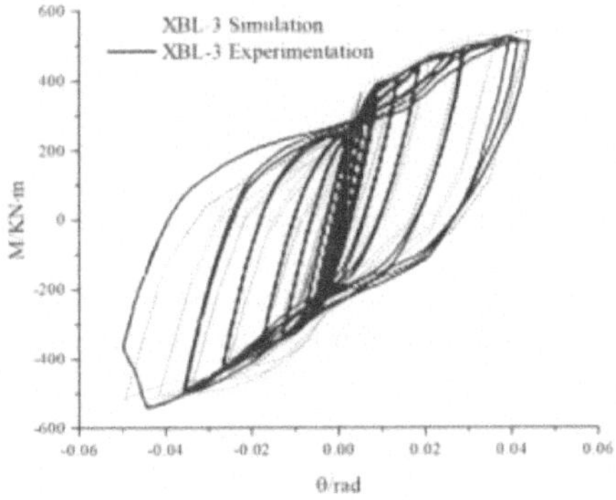

Fig. 1. Comparison of moment-rotation curves of specimen XBL-3.

3 Finite Element Analysis Model

3.1 Modeling

ABAQUS is internationally recognized as one of the most advanced large-scale finite element analysis software, it is specific powerful computational design simulation capabilities, with a large variety of cell types, material models and so on.

Based on the above design methodology ABAQUS software was used to analyze the seismic performance of the new beam-column joints with double ring hoop.

The joint model is shown in Figs. 2 and 3, slotted hole is shown in Fig. 4 (where A is the size of the enlarge hole and B is the radius of the original hole), and the design parameters were classified into three groups according to the thickness ratio, the size of slotted bolt holes, which are G1, G2, respectively, and the specific parameters are shown in Table 1.

Table 1. Design parameters for the specimens.

Subgroup No.	Component number	Thickness ratio X/Y	Reaming size/mm
G1	T1/AT1	1.2	0
	T2/AT2	1.4	0
	T3/AT3	1.6	0
	T4/AT4	1.8	0
	T5/AT5	2.0	0
G2	TH1/AH1	1.6	0
	TH2/AH2	1.6	2
	TH3/AH3	1.6	4

3.2 Material Constitutive Relation and Contact

The contact properties are selected as penalty function friction, and the friction coefficient is set to 0.3. The normal contact is set to hard contact, and the others are default settings, as shown in Table 2.

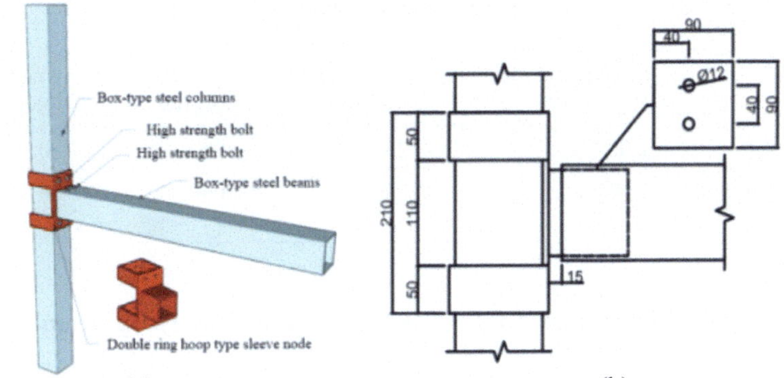

(a) (b)

Fig. 2. The joint with of double ring hoop sleeve: (a) Double ring hoop sleeve joint (b) Detail of the joint.

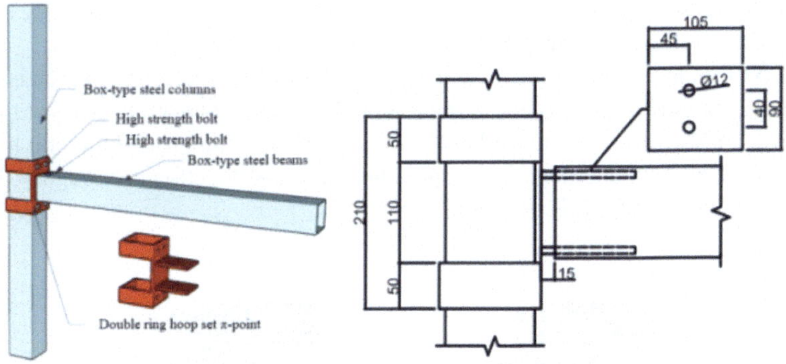

(a) (b)

Fig. 3. The joint with of π-type with double ring hoop: (a) Double ring hoop π-shape joint (b) Detail of joint.

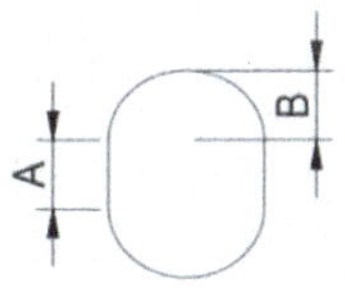

Fig. 4. The slotted hole for bolted connection with beam.

3.3 Loading Protocol

In this paper, the loading protocol recommended in the seismic code for steel structures (ANSI/AISC 341-16) is adopted with displacement controlled loading with a total of 13 cyclic loading groups.

Table 2. Material parameters.

Makings	Modulus of elasticity E/10^5 MPa	Yield strength σy/MPa	Tensile strength σu/MPa	Yield strain σy/%
Q235	2.03	235	400	0.11
M12 high strength bolts	2.06	940	1040	0.46

3.4 Simulation Results and Analysis of New Double Ring Hoop Beam-Column Joints

3.4.1 Failure Mode of Joints

The stress cloud diagram is shown in Fig. 5. Taking T1 and AT1 joints as an example, When the cyclic loading was over, failure mode of T1 was pulled away from the column surface for a certain distance, and obvious bulging and curvature deformation occurred in the weld between the double hoop and the socket, but the middle part of the end plate was almost not deformed, and the bolts at the joint were damaged by shear.

For AT1 under cyclic displacement loading, the central undeformed area is smaller, the stress is mainly concentrated in the double hoop and connecting plate at the upper and lower ends of the weld as well as the connection of joint around the bolt holes, the double hoop π-shape member stress spreading area and location are roughly the same as T1, but the spreading area is larger.

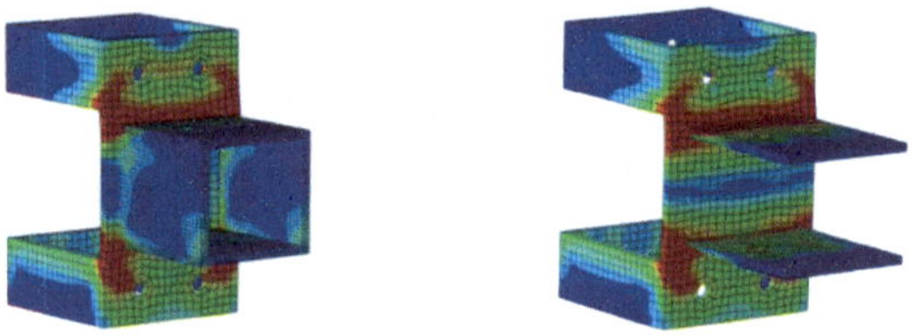

(a) (b)

Fig. 5. Stress at some joints: (a) T1 (b) AT1

3.4.2 Effect of Thickness Ratio on Seismic Performance of Joints

The parameters of thickness ratio are shown in Table 1, and the hysteresis curves, skeleton curves, obtained from its simulation analysis are shown in Fig. 6.

From Fig. 6, it can be seen that with the increase of the thickness ratio, when the thickness ratio is 1.6 and 2.0, the seismic performance of the T3 and T5 is greatly improved, the hysteresis curve changed significantly, the curve obviously becomes full and stiffness of joint has been greatly improved, the load capacity of the T3 increased by about 29.8% compared with that of T2, and load capacity of the T5 is increased by about 20.6% compared with that of T4.

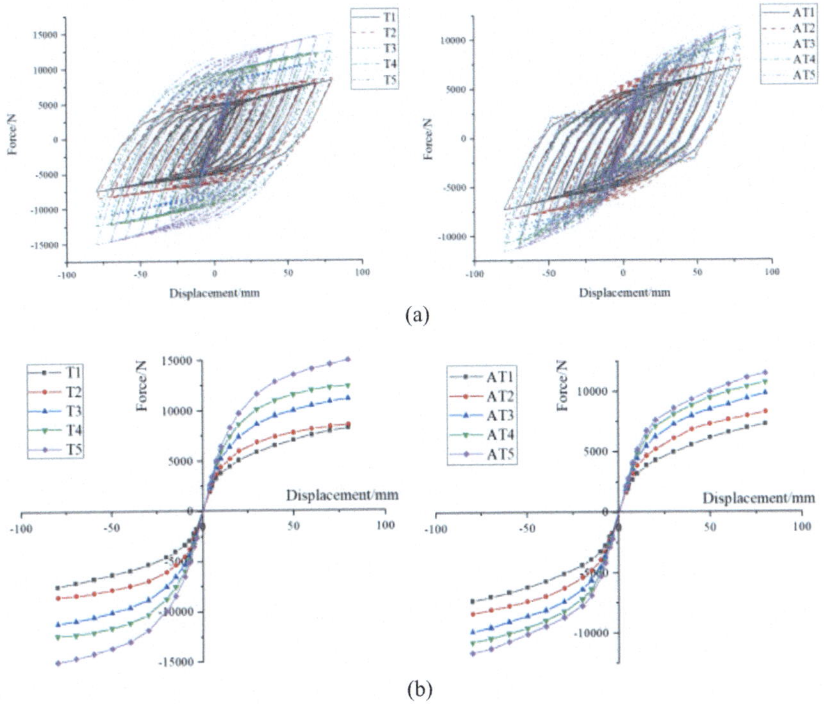

Fig. 6. Seismic performance analysis of joints under different thickness ratios: (a) Hysteresis curves of sleeve and π-shape for double ring hoops. (b) Skeleton curve of sleeve and π-shape for double ring hoops.

While for double ring hoop π-shape joint, it can be seen that the hysteresis curve of the double ring hoop π-shape joint is bow-shaped, and the seismic performance of the members enhanced with the increase of thickness ratio, and load carrying capacity of the AT3 increased by 17.9% compared with the AT2, and load carrying capacity enhanced by about 6–9% for every increase of 0.2 in the thickness ratio starting from the thickness ratio of 1.6.

3.4.3 Effect of Bolt Holes Size on Seismic Performance of Joints

The relevant parameters such as reaming size are shown in Table 1, and the hysteresis curve obtained from its simulation and analysis are shown in Fig. 7.

The effect of size of bolt hole was greater for the π-shape joint, where the bolt slip phenomenon was generated at the early stage of loading and the hysteresis curve was in the form of a parallelogram. For the sleeve joint, due to the larger rotational stiffness, the bolt slip phenomenon occurs at the later stage of loading, and the slip loads of the two are not the same, the slip load of the π-type joint of the double hoop is 1230 N, and that of the sleeve joint of the double hoop is 9742.5 N. The different amount of slip has an effect on the time point of the bolt contacting the wall of the hole resulting in the rise of stiffness, and the smaller the slip, the earlier stiffness rise.

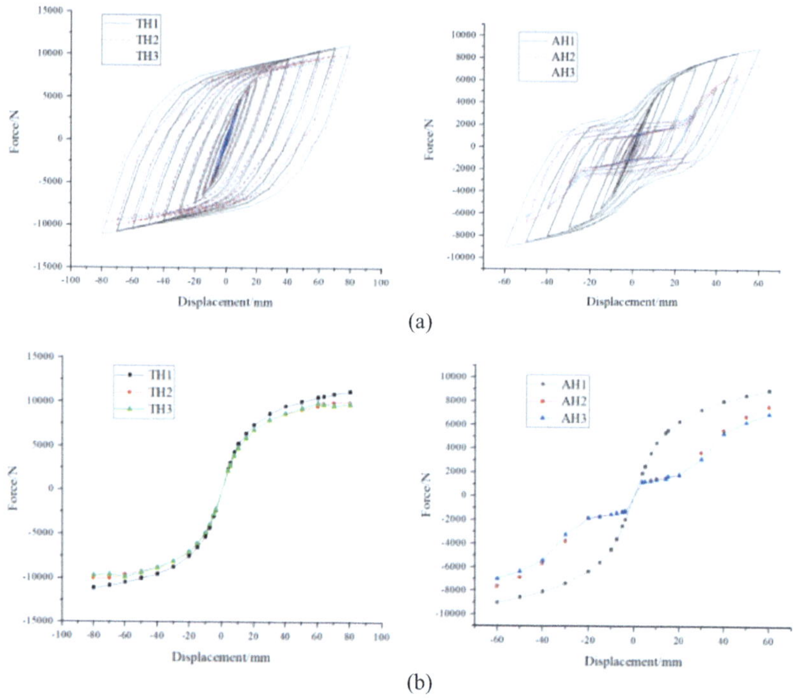

Fig. 7. Seismic performance analysis of joints under bolt-hole size: (a) Hysteresis curves of sleeve and π-shape for double ring hoops. (b) Skeleton curve of sleeve and π-shape for double ring hoops.

4 Conclusions

Two new types of beam-column joint were proposed in the paper and their seismic performance was simulated by ABAQUS finite element software, and the following conclusions can be obtained:

(1) The new double ring hoop sleeve joint and the new double ring hoop π-shape joint can be derived from the force displacement curve that the rotation reached 0.081 rad, which is much larger than the 0.03 rad of the U.S. standards, indicated that the joints have good seismic performance and ductile deformation capacity.
(2) Thickness ratio is the main influence factor of joint strength, stiffness, energy consumption and so on. With the increase of the thickness ratio of the two new types of joints, the force performance also increases, from the practical effective, the thickness ratio of 1.6 is the optimal thickness ratio to take the value of the higher efficiency, on the one hand, the performance has enough to protect.
(3) The increase in the size of bolt hole helps to improve the energy dissipation capacity of the joints. With the increase of the bolt hole size, the double hoop π joints showed better bolt slip, which helps to improve its energy-consuming capacity.

From the applicability point of view, the double ring hoop sleeve joints is more suitable for some steel structures with larger beam spans. For some steel structures that require energy consumption or a more complicated structures, the double ring hoop π-shape joint is more suitable for this environment due to it cost less under safety.

Further research is needed from the following aspects:

(1) This paper only studies the beam-column joints, and the following research can be extended from the beam-column joints to the whole frame. This provides a deeper understanding of how the new joints behave in the actual framework.
(2) Only the simulation analysis of beam-column joints cannot fully reflect the performance of such joints, so the research on the mechanical properties of such new joints needs to be further explored through tests.

Acknowledgments. The authors gratefully acknowledge the financial support provided by Natural science Foundation of Hunan provincial (grant No 2021JJ50015).

References

1. Hasan, J.M., Al-Deen, S., Ashraf, M.: Behaviour of top-seat double web angle connection produced from austenitic stainless steel. J. Constr. Steel Res. **155**, 460–479 (2019). https://doi.org/10.1016/j.jcsr.2018.12.015
2. Feng, Y.L., Wei, M.T., Chong, X., Jiang, Q., Zhu, Y.: Design and numerical analysis of steel frame joints with replaceable buckling-restrained links. J. Constr. Steel Res. **196**, 1–16 (2022). https://doi.org/10.1016/j.jcsr.2022.107415
3. Yi, W.T., et al.: Bearing Capacity of steel square tubular column i-shaped beam joints connected by new elliptical one-sided bolts. Int. J. Steel Struct. **22**(6), 1829–1840 (2022). https://doi.org/10.1007/S13296-022-00670-7
4. Wei, J.P., Qiao, H.Y., Guo, Y., Ke, Z.B., Bao, Y., Jiao, Z.A.: Cyclic loading test of earthquake-resilient steel frame joints with different connection forms. J. Constr. Steel Res. (2023). https://doi.org/10.1016/j.jcsr.2023.108130
5. Kim, S.D., Kim, S.S., Ju, Y.K.: Strength evaluation of beam–column connection in the weak axis of H-shaped column. Eng. Struct. **30**(6), 1699–1710 (2008). https://doi.org/10.1016/j.engstruct.2007.11.008
6. Cao, W.J., Zhao, J.Y., Wang, B., Yu, N.N.: Analysis of flexural strength of I-beam and box beam. Metall. Equipment **6**, 13–16 (2010). https://doi.org/10.3969/j.issn.1001-1269.2010.06.004
7. Li, S., Xu, T.T., Ding, S.H., Li, X.L., Guo, H.C., Liu, Y.H.: Seismic performance of a new assembled beam-column joint with cantilever beam splicing. J. Constr. Steel Res. **196**, 1–25 (2022). https://doi.org/10.1016/j.jcsr.2022.107432

Experimental Research on Liquefied Sand by Compaction Piles with Equal Energy and Equal Deformation in High Intensity Seismic Region

Yongkang Yang(✉), Wu Yang, Shuling Xie, Chunyan Feng, and Zhaoxiong Huang

Hainan Geological Bureau, Hainan Nonn-Ferrous Engineering Investigation and Design Institute, Haikou 570100, Hainan, China
289197277@qq.com

Abstract. Seismic precautionary intensity of start zone of Jiangdong New Area in Haikou is 8, and design basic acceleration of ground motion is 0.30g, which is a high intensity seismic region. Quaternary Holocene saturated liquefied sand layer is widely distributed in the zone, and the liquefaction grade of soil is mostly classified serious. According to the overall plan of Jiangdong New Area, the area is mainly built with medium and high-rise large-span office buildings, with large structural load. Reducing or eliminating sand liquefaction is a difficult problem for the engineering construction in this zone. The effect of compaction piles with equal energy and equal deformation to strengthen liquefied sand is remarkable. When the vertical reinforcement uses recycled aggregate, its economic benefits are very obvious. According to the topic "Compaction Piles with Equal energy and equal deformation Composite Foundation Treatment of Recycled Aggregate", 3 test areas are selected in the start zone of Jiangdong New Area, the test scheme is drawn up, the construction technical parameters of compaction piles with equal energy and equal deformation composite foundation are discussed, and the liquefaction evaluation parameters in this zone are studied. Through standard penetration test and heavy dynamic penetration test, the reinforcement effect before and after foundation treatment is compared and analyzed. The results show that this treatment method can improve the compactness of soil and effectively eliminate liquefaction. This work can be used for reference in geotechnical investigation, foundation engineering and ground treatment engineering in start zone of Jiangdong New Area.

Keywords: liquefied sand · High intensity seismic region · Liquefaction index

1 Introduction

Under the action of earthquake, dynamic load and other external forces, saturated sandy soil or silt loses shear strength, reduces effective stress, and increases pore water pressure, this phenomenon is called liquefaction. When the pore water pressure is greater than the weight of the soil, the soil particles are suspended, resulting in the phenomenon of sand

B. Yuan et al. (Eds.): ICCASE 2024, SUCI, pp. 338–347, 2024.
https://doi.org/10.1007/978-3-031-76102-7_38

blasting [1–5]. The investigation of earthquake damage shows that the liquefaction of saturated sand is the main cause of engineering disasters such as foundation failure, building damage, roadbed subsidence and landslide [6–11]. As a key area for the construction of China (Hainan) free trade pilot port, the start zone of Jiangdong New Area in Haikou is a high intensity earthquake region. It is necessary to take anti-liquefaction measures such as eliminating the liquefaction differential settlement completely, or eliminating the liquefaction differential settlement partially and taking measures for the foundation and upper structure.

According to the overall plan of Jiangdong New Area, the area is mainly built with medium and high-rise large-span office buildings, with large structural load. Considering the influence of site type and liquefied sand, the building foundation cost of this area is high, and the structural selection is also affected by it. The cost of recycled aggregate pile is low, and the effect of strengthening liquefied sand foundation is mainly manifested as follows: ① Improve the compaction of the soil around the pile and reduce the vibration density of the soil; ② pile shares the seismic horizontal shear stress and reduce the shear stress of the soil between piles [12]. We try to use pretreatment of the liquefied sand layer by compaction piles with equal energy and equal deformation, and evaluate the reinforcement effect, so as to provide a basis for the large-scale construction of the liquefied sand site in Jiangdong New Area.

2 Test Scheme

Based on the topic of "Compaction Piles with Equal energy and equal deformation Composite Foundation Treatment of Recycled Aggregate", site of the test scheme at Jiangdong New Area in Haikou is selected. The site project is planned to build seven 3–12-storey office buildings, with two basements as a whole. The excavation depth of the basement is 7 m. Seismic precautionary intensity of test section is 8, and design basic acceleration of ground motion is 0.30 g. According to the existing survey data, 3 test areas with liquefied sand layer thickness of 13–15 m are selected to carry out the treatment of compaction piles with equal energy and equal deformation, and the reinforcement effect is analysed.

2.1 Engineering Geological Condition and Hydrogeological Condition

The test site belongs to coastal accumulation plain geomorphic unit. Combined with field identification, in-situ test and laboratory test, the strata in the depth range of 20 m in the site are shown in Table 1.

Groundwater is found in all exploration holes. The buried depth of the stable groundwater level is 0.60–1.10 m, and the elevation is 1.82–2.20 m. According to the regional hydrogeological data, the range of water level in this area is about 3 m.

2.2 Construction Parameters

According to the current situation of the site, 3 test areas are selected, each with an area of 10 × 10 m. The top 0–7 m of each test area is the buried depth of the basement, and

Table 1. Columnar section of stratum

Soil layer	Age& Genesis	Min-Max Thickness	Avg Thickness	Legend
① Miscellaneous fill	Q_4^{ml}	1.5-2.6m	1.9m	
② Coarse sand	Q_4^{m}	1.5-3.5m	2.6m	c
②-1 Silty sand	Q_4^{m}	0.5-3.0m	1.4m	f
③ Gravelly sand	Q_4^{m}	6.0-11.0m	8.2m	l
④ Clay	Q_3^{m}	0.7-1.8m	1.2m	
④-1 Clay	Q_3^{m}	1.7-7.5m	3.6m	
④-2 Silty sand	Q_3^{m}	2.8-5.6m	3.7m	f

there is no liquefied sand after foundation pit excavation. The depth of ground treatment in test 1, 2 and 3 area are 7–15 m, 7–13 m and 7–15 m respectively.

Influenced by the complexity of geotechnical engineering, the construction parameters, construction technology, reinforcement effect and other aspects will be different according to different formation structures. According to regional experience, the square pile is used by compaction piles with equal energy and equal deformation. The diameter of the pile is 800 mm, and the spacing is 1800 mm.

The recycled aggregate adopts cheap construction waste, mainly including broken concrete, broken bricks, gravel, etc., which comes from the urban construction waste recycling. The recovery process is divided into the following two stages:

First of all, using machines to crush the construction waste, remove the iron material, through coarse crushing, medium crushing, to obtain semi-finished aggregate.

Secondly, the semi-finished aggregate is twice crushed and repeated until the finished recycled aggregate with particle size less than 20 cm is obtained.

With the construction waste filler as the carrier, the pile body plays the role of sharing the load in the stress process, and the clay soil plays the role of filling the pores. According to the experience, the filler is clay: construction waste = 2:1 (volume ratio). The volume of rammed fillers is controlled with pile diameter of 800 mm to ensure the consistency of test conditions, which in test 1, 2 and 3 area are 7.5, 6.5 and 7.5 m^3 respectively.

2.3 Construction Specification

The weight of 5.5 t rammer is lifted to 6 m, free to fall, and the hole is rammed with the energy of 3300 kN m. In the process of ramming, recycled aggregate is filled and compacted repeatedly. During vibration and compaction, the sand particles are rearranged and the compactness of the sand is increased. The construction process is shown in Fig. 1.

(1) hole forming by rammer
(2) construction of 560 mm diameter casing to design elevation
(3) layered filling recycled aggregate repeated compaction to form a bearing sphere

(4) Pull up the casing 2 m, expand the diameter to 800 mm by ramming. When the 3-hit penetration is less than 20 cm, the recycled aggregate is filled, so repeated
(5) Ramming fill to the design elevation

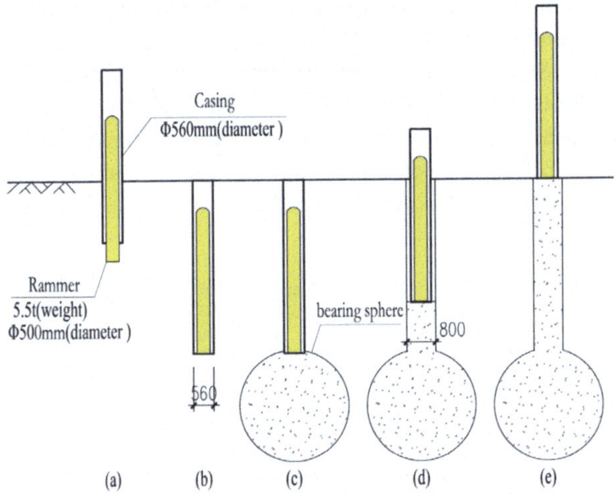

Fig. 1. Construction process

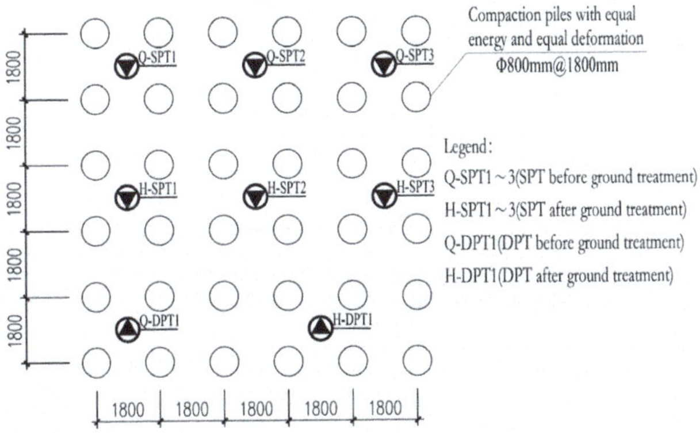

Fig. 2. SPT&DPT points in test 1 area

2.4 Experimental Parameters

In order to obtain the reinforcement effect of each test area, the standard penetration test and heavy dynamic penetration test are carried out before and after the ground treatment in the test 1, 2 and 3 area. 3 SPT holes and 1 heavy DPT holes with the depth of 20 m are arranged before and after ground treatment in each test area respectively, a total of 6 SPT holes and 2 heavy DPT holes, which is shown in Fig. 2. SPT is carried out every 1.0 m. The heavy DPT is carried out from the ground surface.

3 Comparative Analysis of Test

3.1 Comparative Analysis of Standard Penetration Test

In order to analyze the reinforcement effect of compaction pile with equal energy and equal deformation, a total of 18 drilling holes are arranged for standard penetration test before and after ground treatment. Q-SPT1 ~ 3 are completed before ground treatment in test 1 area, and H-SPT1 ~ 3 are completed after ground treatment. Q-SPT4 ~ 6 are completed before ground treatment in test 2 area, and H-SPT4 ~ 6 are completed after ground treatment. Q-SPT7 ~ 9 are completed before ground treatment in test 3 area, and H-SPT7 ~ 9 are completed after ground treatment. The typical N-h curves of test 1, 2 and 3 area are shown in Figs. 3, 4 and 5, respectively.

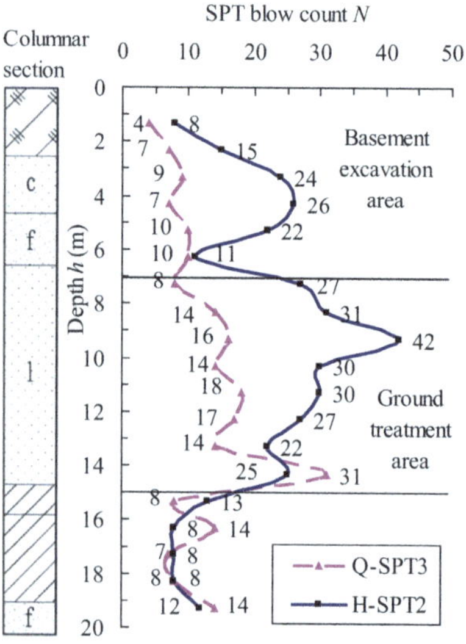

Fig. 3. Typical *N-h* curves of test 1 area

The comparison table of reinforcement effect of standard penetration test in each test area is shown in Table 2. Through calculation [13], the comparison chart of the average liquefaction index of the test 1,2,3 area before and after the ground treatment is shown in Fig. 6.

According to the test results, the average value of SPT blow count in the depth of 7–15 m after ground treatment is 29.5 in test 1 area. This value is 28.2 in the depth of 7–13 m in the test 2 area and 28.9 in the depth of 7–15 m in the test 3 area. The average values of SPT blow count of test 1, 2 and 3 area after ground treatment increased by 83, 82 and 101% respectively compared with those before ground treatment.

The average liquefaction index of Q-SPT1 ~ 3 is 40.78, and the liquefaction grade is serious. The corresponding index of H-SPT1 ~ 3 is 6.06, and the liquefaction grade is moderate. The average liquefaction index of Q-SPT4 ~ 6 is 42.00, and the liquefaction grade is serious. The corresponding index of H-SPT4 ~ 6 is 2.65, and the liquefaction

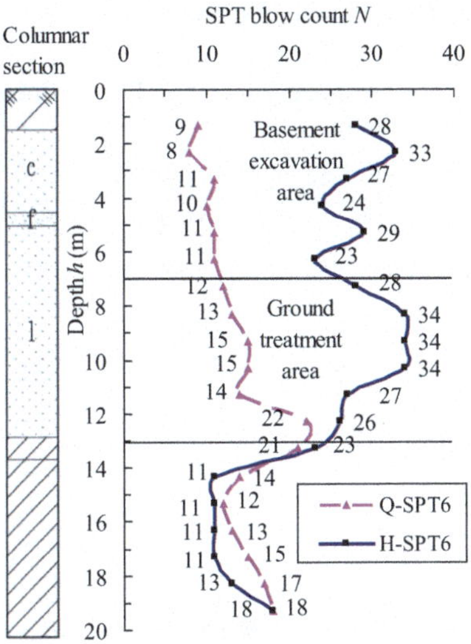

Fig. 4. Typical *N-h* curves of test 2 area

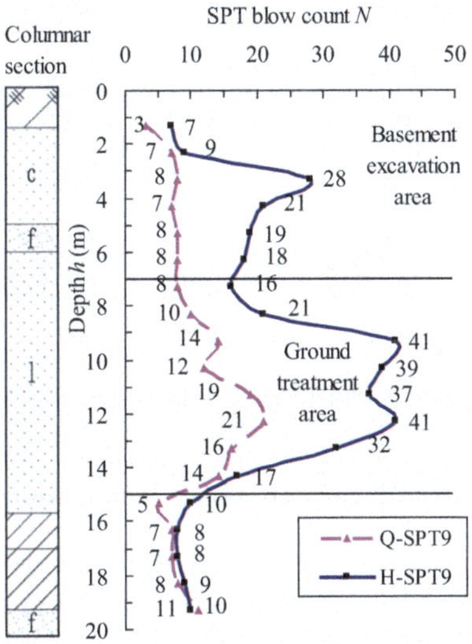

Fig. 5. Typical *N-h* curves of test 3 area

grade is light. The average liquefaction index of Q-SPT7 ~ 9 is 51.10, and the liquefaction grade is serious. The corresponding index of H-SPT7 ~ 9 is 4.91, and the liquefaction grade is light.

Table 2. Comparison of reinforcement effect of SPT

Name	Depth	Average value of SPT blow count		Increase percentage (%)
		Before	After	
Test 1 area	0–7 m	9.1	18.4	102
	7–15 m	16.1	29.5	83
Test 2 area	0–7 m	10.7	25.6	139
	7–13 m	15.5	28.2	82
Test 3 area	0–7 m	7.9	18.8	138
	7–15 m	14.4	28.9	101

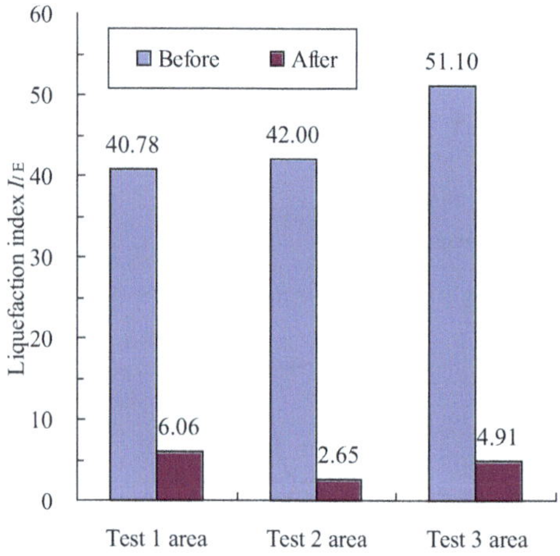

Fig. 6. Comparison of liquefaction index

3.2 Comparative Analysis of Heavy Dynamic Penetration Test

Before and after ground treatment, a total of 6 drilling holes are arranged for heavy dynamic penetration test. Q-DPT1 is completed before ground treatment in test 1 area, and H-DPT1 is completed after ground treatment. Q-DPT2 is completed before ground treatment in test 2 area, and H-DPT2 is completed after ground treatment. Q-DPT3 is completed before ground treatment in test 3 area, and H-DPT3 is completed after ground treatment. The typical $N_{63.5}$-h curves of test 1, 2 and 3 area are shown in Figs. 7, 8 and 9, respectively. The comparison table of reinforcement effect of heavy dynamic penetration test in each test area is shown in Table 3.

The average value of modified heavy DPT in the depth of 7–15 m after foundation treatment is 13.4 in test 1 area. This value is 13.6 in the depth of 7–13 m in the test 2 area and 9.1 in the depth of 7–15 m in the test 3 area. The average values of modified

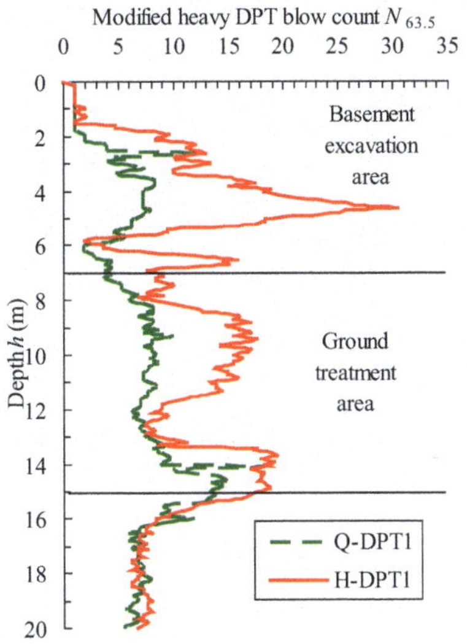

Fig. 7. Typical $N_{63.5}$-h curves of test 1 area

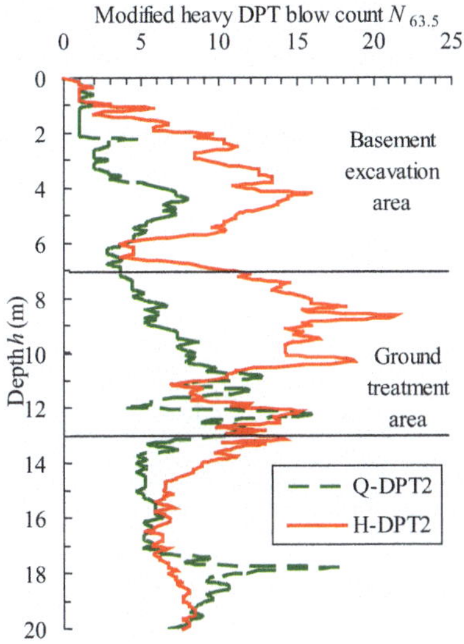

Fig. 8. Typical $N_{63.5}$-h curves of test 2 area

heavy DPT blow count of test 1, 2 and 3 area after ground treatment increased by 63, 74 and 30% respectively compared with those before ground treatment.

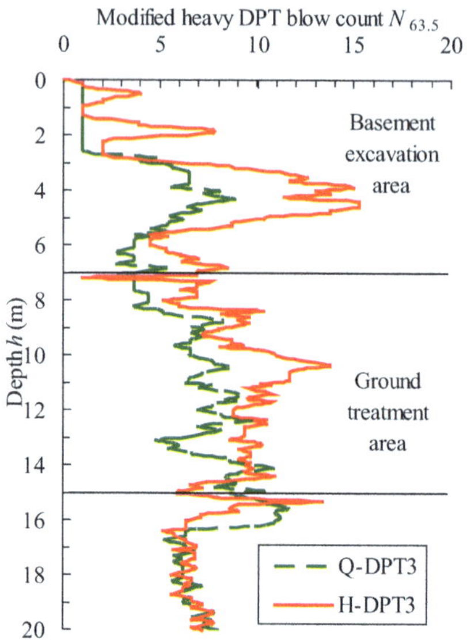

Fig. 9. Typical $N_{63.5}$-h curves of test 3 area

Table 3. Comparison of reinforcement effect of heavy DPT

Name	Depth	Average value of modified heavy DPT blow count		Increase percentage (%)
		Before	After	
Test 1 area	0–7 m	4.2	10.4	148
	7–15 m	8.2	13.4	63
Test 2 area	0–7 m	3.3	8.0	142
	7–13 m	7.8	13.6	74
Test 3 area	0–7 m	3.7	6.6	78
	7–15 m	7.0	9.1	30

4 Conclusions

Through standard penetration test and heavy dynamic penetration test, the tests of liquefied sand in high intensity seismic region by compaction piles with equal energy and equal deformation are researched, and the following conclusions are drawn:

After ground treatment, the average value of SPT blow count is increased by 82–101%, and the average value of modified heavy DPT blow count is increased by 30–74%. After ground treatment, the average liquefaction index decreased from 40.78–51.10 to 2.65–6.06, and the liquefaction grade changed from serious to light-moderate.

Compaction piles with equal energy and equal deformation can significantly increase the compactness of sand foundation, effectively reduce the liquefaction of sand. By adjusting the pile spacing, penetration degree and other construction parameters, the goal of eliminating liquefaction can be achieved.

The above construction scheme, parameters and test comparison results can provide reference for the determination of hole depth and hole distance in geotechnical engineering survey and the suggestion of foundation scheme, and also provide reference for the comparison of ground treatment engineering parameters before large area foundation reinforcement construction in Jiangdong New Area.

In this experimental study, only the diameter of pile of 800 mm, pile spacing of 1800 mm, fixed regeneration aggregate ratio (volume ratio of clay: construction waste = 2:1) were used to analyze the effect of reinforcing liquefied sand in high intensity seismic. According to the characteristics of different proposed projects in Jiangdong New Area, the following studies can be made:

(1) Study the anti-liquefaction effect through different diameter and spacing of the pile.
(2) Test through different materials, different ratios and different penetration degrees to study the reinforcement effect.

References

1. Sergeyev: Liquefaction and thixotropy of soil. Hydrogeol. Eng. Geol. **10**, 27–29 (1958)
2. Huang, W.: Investigations on stability of sand foundation and slopes against liquefaction. Water Resour. Hydropower Eng. **1**, 38–39 (1962)
3. Wang, W.S.: Study on pore water pressure of saturated sands due to vibration. J. Hydraul. Eng. **2**, 37–47 (1962)
4. Wang, W.S.: Distinction and interrelation between liquefaction, state of limit equilibrium and failure of soil mass. Chin. J. Geotech. Eng. **27**(1), 1–10 (2005)
5. Tang, L., et al.: Failure mechanism of pile foundations in liquefiable soils under seismic loading: status and challenge. Rock Soil Mech. **44**(10), 2979–2996 (2023)
6. Huang, Y., Hao, L.: Advances in failure mechanisms of pile foundations in liquefaction soils. J. Eng. Geol. **16**(2), 184–189 (2008)
7. Xu, G.H., Chen, Z.Y., Ren, Y.P., Liu, Z.X.: Recognition of the concept of sand liquefaction. Periodical Ocean Univ. China **53**(12), 138–145 (2023)
8. Chen, G.X., Jin, D.D., Chang, X.D., Li, X.J., Zhou, G.L.: Review of soil liquefaction characteristics during major earthquakes in recent twenty years and liquefaction susceptibility criteria for soils. Rock Soil Mech. **34**(10), 2737–2795 (2013)
9. Fujii, S., Isemoto, N., Satou, Y.: Investigation and analysis of a pile foundation damaged by liquefaction during the 1995 Hyogoken-Nambu earthquake. Soils Found. **38**(Special), 179–192 (1998)
10. Domenico, L., Subhamoy, B.: Modal analysis of pile-supported structures during seismic liquefaction. Earthq. Eng. Struct. Dyn. **43**, 119–138 (2014)
11. Bird, J.F., Bommer, J.J., Crowley, H., Pinho, R.: Modelling liquefaction-induced building damage in earthquake loss estimation. Soil Dyn. Earthq. Eng. **26**(1), 15–30 (2006)
12. Wang, J.Z.: The birth and development of piles with ram-compacted bearing sphere. Build. Struct. **38**(4), 118–119 (2008)
13. National Standard of the People's Republic of China: Code for Seismic Design of Buildings(GB50011-2010), pp. 21–22. China Architecture and Building Press, Beijing (2010)

Sensitivity Analysis of Factors Influencing Deformation Performance of Geosynthetic Reinforced Soil Abutment Based on Improved Grey Correlation Method

Hongchao Huang[1](✉), Haifang Wang[1], Tiekun Peng[2], and Yunsheng Zhu[2]

[1] CCCC Second Highway Consultants Co. Ltd., Wuhan 430056, Hubei, China
huanghongchao@ccccltd.cn

[2] Wuhan University of Technology, Wuhan 430063, Hubei, China

Abstract. In order to study the influence of different factors on the deformation performance of geosynthetic reinforced soil abutment, taking the reinforced soil abutment of a project as the prototype, the improved grey correlation method is used to analyse the influence of abutment height, reinforced spacing, reinforced length, reinforced material stiffness, bridge load and filling internal friction angle on the horizontal deformation sensitivity of reinforced earth abutment structure. The results show that the reinforced distance has the most significant effect on the horizontal deformation of reinforced soil abutment, and the calculation result of grey correlation degree is 0.932, followed by abutment height, bridge load and filling internal friction angle, respectively. The stiffness and length of reinforced materials have little influence on the deformation of abutment. In the structural design of reinforced soil abutment, more attention should be paid to the influence of reinforced spacing on the structure. The research results can provide an important reference for the design and construction of reinforced soil abutment.

Keywords: Geosynthetic reinforced soil abutment · Deformation performance · Grey correlation method · Sensitivity analysis

1 Introduction

Geosynthetic Reinforced soil-Integrated Bridge System (GRS-IBS) is the latest achievement in the application of reinforced soil technology to bridge abutment structures in the past decade. This structure was first proposed by the Federal Highway Administration (FHWA) in the United States to solve the replacement problem of hundreds of thousands of small and medium-sized bridges. This new type of structure is low-cost, easy to construct, has good deformation performance, and low maintenance costs [1]. As shown in Fig. 1, the GRS-IBS structure consists of three parts: (1) Reinforced soil foundation (RSF); (2) GRS abutment; (3) Integrated approach. RSF is composed of high-performance geotextile wrapped with well graded coarse particle fillers. As a supporting structure for GRS bridge piers, it can increase the load-bearing width and improve the

B. Yuan et al. (Eds.): ICCASE 2024, SUCI, pp. 348–361, 2024.
https://doi.org/10.1007/978-3-031-76102-7_39

bearing capacity of the piers. The wrapping design of geotextile allows the foundation to withstand a certain amount of water flow erosion. At the same time, the mesh characteristics of geotextile make it act as an anti-filter layer, reducing the erosion of water flow on the foundation. The GRS abutment is composed of a GRS composite composed of small spacing reinforced materials and well graded compacted fill layers (where the reinforcement spacing in the GRS composite does not exceed 0.3 m [2]), which is the direct load-bearing layer of the bridge load and the focus of structural analysis. Integrated approach is set up at the transition section between the main beam of the bridge and the roadbed. Integrated approach is composed of well graded dense backfill coarse particles and reinforced materials, which can effectively reduce differential settlement.

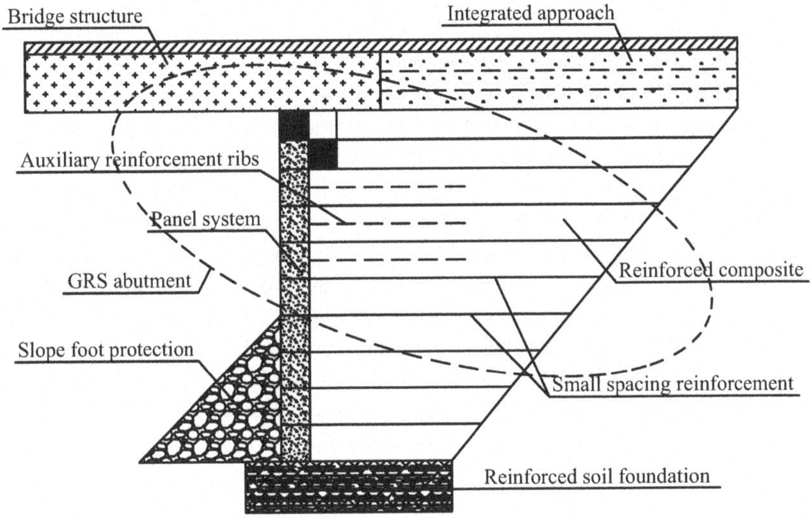

Fig. 1. Typical cross-section of GRS-IBS.

As a new type of bridge abutment structure, the mechanical performance of GRS-IBS under vehicle loads urgently needs to be studied. Many engineers and scholars have conducted extensive theoretical research and practice on GRS-IBS. The deformation of bridge piers is the most concerning aspect for engineers, and it is also a key focus of bridge pier structure design and on-site monitoring. At present, research on the deformation performance of GRS-IBS mainly focuses on two aspects: experimental and numerical simulation. Xu Chao et al. [3] constructed three two-dimensional abutment models and analyse the mechanical properties of reinforced soil structures and MSE structures under the same load. The experimental results confirmed that reinforced soil structures have excellent mechanical properties, and reinforced soil structures have excellent mechanical properties under small reinforcement spacing conditions. Jelušič et al. [4] present an experimental study of a geosynthetic reinforced soil abutment, and shows that the maximum strain was recorded below the sill, and was 50% higher than at the center of the bridge abutment. Xu Chao et al. [5] conducted plane-strain scaled model tests to investigate the deformation characteristics of GRS abutments subjected to vertical loads. Test results showed that increasing the setback distance could effectively reduce the deformations of the GRS abutment, but there existed an optimum setback distance. Muhammad Nouman Amjad Raja et al. [6] predict the settlement of the GRS abutments based on the novel hybrid artificial intelligence model, that is, ANN-HHO, and demonstrated that the ANN-HHO model predicted the settlement of GRS abutments with reasonable accuracy and yielded superior performance in comparison to counterpart models.

The above research indicates that design parameters such as reinforcement spacing, reinforcement length, and reinforcement stiffness have a significant impact on the deformation performance of bridge piers. Therefore, the influence of these factors must be considered in the structural design of bridge piers. However, GRS-IBS is a complex system composed of various single structures that are not obtainable, and the various parts of the structure (such as reinforcement, backfill soil, and blocks) are interconnected to jointly play their due mechanical performance roles. Therefore, it is necessary to analyse which factor has the greatest impact on the performance of the bridge abutment under multiple factors, that is, to determine the dominant factor through a certain method, and to sort the factors that affect the performance of the bridge abutment, in order to provide guidance for the construction of the bridge abutment. The current objective methods for multi factor correlation analysis include Technique for order preference by similarity to an ideal solution (TOPSIS), grey correlation method, and principal component method [7]. The weight selection of TOPSIS method is usually subjective, and the optimal and worst vectors will vary with the evaluation environment, leading to the lack of uniqueness in the evaluation results. The principal component method requires a large number of samples, the evaluation results are related to the sample size, and the evaluation results do not have uniqueness. The grey correlation analysis method can avoid the influence of weight subjectivity and sample size. Therefore, this article will use an improved grey correlation method to analyse the sensitivity of different factors to the deformation performance of reinforced soil abutment, and obtain a sensitivity sequence to guide the design and construction of reinforced soil abutment.

2 The Basic Principle and Improvement Methods of Grey Correlation Method

2.1 Basic Principles of Grey Correlation Method

Grey correlation analysis is an emerging method of multi factor analysis, which describes the degree of closeness between multiple factors and reference factors by comparing the geometric relationships of statistical sequences. The closer the geometric shapes between the sequence curves are, the greater the correlation between the corresponding multiple factors. The grey correlation method requires a small sample size, does not require the distribution pattern of the samples, is simple to calculate, and its analysis results are in good agreement with the results obtained from qualitative analysis; No special normalization is required for each data, and the original data can be directly analysed and calculated. The grey correlation method uses the grey correlation degree model for analysis and calculation, among which the Deng's grey correlation analysis model is the most classic [8], and its analysis process is shown in Fig. 2.

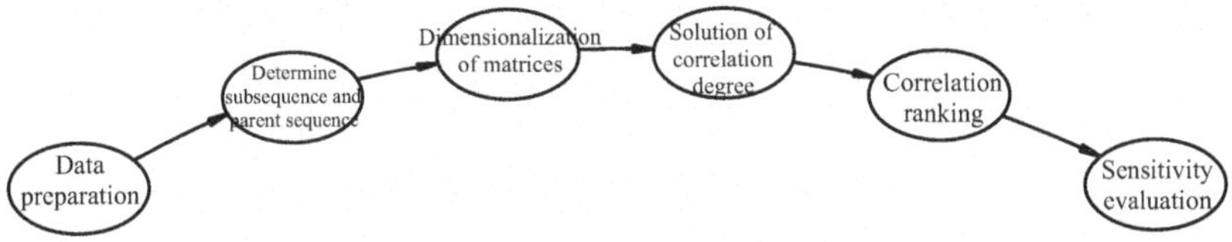

Fig. 2. Traditional grey relational model flowchart.

2.2 Principle and Calculation Steps of Improving Grey Correlation Method

The traditional grey correlation model has certain limitations in analysing correlations. When dealing with negative correlation factors in dimensionless processing, the calculation formula for positive correlation factors is uniformly used, thus ignoring the influence of negative correlation factors. This article adopts an improved grey correlation method [9], which uses a differentiation formula for dimensionless processing of positive and negative correlation factors in the sequence. The resolution coefficient ξ is dynamically taken to avoid outliers in the sequence factors dominating the overall system correlation. At the same time, the vector similarity method is selected to solve the weights of each sensitive factor to ensure absolute objectivity of the weights. The specific steps for conducting sensitivity analysis of bridge abutment parameters using the improved grey correlation method are as follows:

(1) Determine the sensitivity factor subsequence matrix and parent sequence matrix to obtain the grey correlation factor space.

Using sensitive factors such as abutment height, reinforcement spacing, reinforcement length, reinforcement material stiffness, bridge load, and internal friction angle of fill that have a significant impact on the horizontal deformation of the abutment as subsequence X, and the corresponding maximum horizontal displacement of the abutment retaining wall as the parent sequence Y, where:

$$X = [X_1, X_2, X_3, \cdots, X_n]^T Y = [Y_1, Y_2, Y_3, \cdots, Y_n]^T$$

Each sensitive factor in columns X and Y has several values:

$$X_i = [x_i(1), x_i(2), x_i(3), \cdots, x_i(n)]$$
$$Y_i = [y_i(1), y_i(2), y_i(3), \cdots, y_i(n)]$$

Rewrite columns X and Y into matrix form, as shown in Eqs. (1) and (2).

$$X = \begin{bmatrix} X_1 \\ X_2 \\ X_3 \\ \vdots \\ X_n \end{bmatrix} = \begin{bmatrix} x_1(1) & x_1(2) & x_1(3) & \cdots & x_1(n) \\ x_2(1) & x_2(2) & x_2(3) & \cdots & x_2(n) \\ x_3(1) & x_3(2) & x_3(3) & \cdots & x_3(n) \\ \vdots & \vdots & \vdots & \ddots & \vdots \\ x_n(1) & x_n(2) & x_n(3) & \cdots & x_n(n) \end{bmatrix} \tag{1}$$

$$Y = \begin{bmatrix} Y_1 \\ Y_2 \\ Y_3 \\ \vdots \\ Y_n \end{bmatrix} = \begin{bmatrix} y_1(1) & y_1(2) & y_1(3) & \cdots & y_1(n) \\ y_2(1) & y_2(2) & y_2(3) & \cdots & y_2(n) \\ y_3(1) & y_3(2) & y_3(3) & \cdots & y_3(n) \\ \vdots & \vdots & \vdots & \ddots & \vdots \\ y_n(1) & y_n(2) & y_n(3) & \cdots & y_n(n) \end{bmatrix} \tag{2}$$

(2) Dimensionalization of matrices

Using the interval relative value method, the values X and Y in matrices $x_i(j)$ and $y_i(j)$ are dimensionless using the following Eq. (3) or Eq. (4).

If the values $x_i(j)$ and $y_i(j)$ in the sensitivity factor are positively correlated, then:

$$\left.\begin{aligned} x_i'(j) &= \frac{x_i(j) - \min x_i}{\max x_i - \min x_i} \\ y_i'(j) &= \frac{y_i(j) - \min y_i}{\max y_i - \min y_i} \end{aligned}\right\} \tag{3}$$

If the values $x_i(j)$ and $y_i(j)$ in the sensitivity factor are negatively correlated, then:

$$\left.\begin{aligned} x_i'(j) &= \frac{\max x_i - x_i(j)}{\max x_i - \min x_i} \\ y_i'(j) &= \frac{\max y_i - y_i(j)}{\max y_i - \min y_i} \end{aligned}\right\} \tag{4}$$

In the formula, $\max x_i$ and $\min x_i$ are the maximum and minimum values of the parameter values corresponding to the i-th row in matrix X, respectively; $\max y_i$ and $\min y_i$ represents the maximum and minimum values of the parameter values corresponding to the i-th row in matrix Y.

(3) Establishment of correlation coefficient matrix

Solve the correlation coefficient between the matrices X' and Y' obtained through dimensionless processing. The correlation coefficient $l_i(j)$ between $x_i'(j)$ and $y_i'(j)$ in the two matrices can be solved by the following equation:

$$l_i(j) = \frac{\underset{i}{\min}\underset{j}{\min}\left|x_i'(j) - y_i'(j)\right| + \xi \underset{i}{\max}\underset{j}{\max}\left|x_i'(j) - y_i'(j)\right|}{\left|x_i'(j) - y_i'(j)\right| + \xi \underset{i}{\max}\underset{j}{\max}\left|x_i'(j) - y_i'(j)\right|} \tag{5}$$

In the formula, ξ is the resolution coefficient used to improve the significant difference between correlation coefficients. Generally speaking, $\xi \in [0, 1]$, in traditional grey correlation models, ξ is taken as 0.5.

The correlation coefficient matrix $l_i(j)$ is composed of various correlation coefficients L:

$$L = \begin{bmatrix} l_1(1) & l_1(2) & l_1(3) & \cdots & l_1(n) \\ l_2(1) & l_2(2) & l_2(3) & \cdots & l_2(n) \\ l_3(1) & l_3(2) & l_3(3) & \cdots & l_3(n) \\ \vdots & \vdots & \vdots & \ddots & \vdots \\ l_n(1) & l_n(2) & l_n(3) & \cdots & l_n(n) \end{bmatrix} \tag{6}$$

For the convenience of research, define new parameter $\Delta_{\max}$:

$$\Delta_{\max} = \max_i \max_j \left| x_i'(j) - y_i'(j) \right| \tag{7}$$

From Eq. (5), it can be seen that ξ, as the coefficient of $\Delta_{\max}$, represents the weight of $\Delta_{\max}$ in the correlation coefficient $l_i(j)$, which has a significant impact on the distribution of the correlation interval. Therefore, ξ cannot always be taken as 0.5, and should be carefully selected based on the actual situation. Now define two new parameters Δ_a and Y_Δ:

$$\Delta_a = \frac{1}{n \cdot n} \sum_{i=1}^{n} \sum_{j=1}^{n} \left| x_i'(j) - y_i'(j) \right| \tag{8}$$

$$Y_\Delta = \frac{\Delta_a}{\Delta_{\max}} \tag{9}$$

The resolution coefficient ξ can be calculated based on the following conditions:

$$\left.\begin{array}{l} Y_\Delta \le \xi \le 1.5Y_\Delta, \quad 0 \le Y_\Delta \le \frac{1}{3} \\ 1.5Y_\Delta < \xi \le 2Y_\Delta, \; Y_\Delta > \frac{1}{3} \end{array}\right\} \tag{10}$$

From Eq. (10), it can be seen that the value range of ξ is $[Y_\Delta, 2Y_\Delta]$. The value of ξ is influenced by Y_Δ and is limited within a certain interval to avoid outliers in the sequence that may affect the correlation degree.

(4) Grey correlation degree calculation

The traditional grey correlation model directly calculates the average value of each column of the obtained correlation coefficient matrix L, and can obtain the grey correlation degree of the corresponding factor in this row. This method does not consider the importance of various factors on the deformation of the bridge abutment, and there is a certain gap between it and the actual working conditions. The improved grey correlation method considers the importance of each sensitive factor on the deformation of the bridge abutment when calculating the grey correlation degree. That is, when calculating the average value of the final correlation coefficient, corresponding weight coefficients are assigned to each factor value to better fit the actual situation. The selection of weight coefficients is directly related to the reliability and authenticity of sensitivity evaluation results, and vector similarity method can be used for analysis. The basic principle of this method is to calculate the similarity between the feature vector composed of sample data of evaluation indicators and the corresponding ideal value indicator vector, and normalize the obtained similarity to obtain the weights of each indicator.

According to the relevant knowledge of matrix theory, any matrix can be represented in vector form. Represent the standardized matrices X' and Y' obtained through dimensionless processing in the previous text in vector form, which are the reference vector $X' = [X_1', X_2', X_3', \cdots, X_n']$ and the indicator vector $Y' = [Y_1', Y_2', Y_3', \cdots, Y_n']$, respectively. The Euclidean norm (length) of vectors X' and Y' is:

$$\|X'\| = \sqrt{x_1'^2 + x_2'^2 + x_3'^2 + \cdots + x_n'^2} \tag{11}$$

$$\|Y'\| = \sqrt{y_1'^2 + y_2'^2 + y_3'^2 + \cdots + y_n'^2} \tag{12}$$

The inner product of vectors X' and Y' is:

$$[X', Y'] = x_1'y_1' + x_2'y_2' + x_3'y_3' + \cdots + x_n' \tag{13}$$

The angle between vectors X' and Y' is:

$$\theta = \arccos\frac{[X', Y']}{\|X'\| \cdot \|Y'\|} \quad (0 \leq \theta \leq 180°) \tag{14}$$

The directional similarity β between vectors X' and Y' is:

$$\beta = 1 - \frac{\theta}{90°} \tag{15}$$

The norm similarity α between vectors X' and Y' is:

$$\alpha = \begin{cases} 1 - \frac{|\|X'\| - \|Y'\||}{\|X'\|} & \|Y'\| \leq 2\|X'\| \\ 0 & \|Y'\| > 2\|X'\| \end{cases} \tag{16}$$

The vector similarity γ between reference vector X' and indicator vector Y' is the product of the norm similarity α and directional similarity β of the two vectors:

$$\gamma = \alpha \cdot \beta \tag{17}$$

Calculate the similarity γ_i corresponding to each factor separately, and normalize according to Eq. (18) to obtain the weight k_i corresponding to each factor.

$$k_i = \frac{\gamma_i - \min \gamma_i}{\max \gamma_i - \min \gamma_i} \tag{18}$$

In the formula, max γ_i and min γ_i are the maximum and minimum values of vector similarity obtained by each factor, respectively.

Based on the above equation, the grey correlation degree ε_i of each factor is:

$$\varepsilon_i = \frac{1}{n}\sum_{j=1}^{n} k_i l_i(j) \tag{19}$$

Sort the grey correlation degree of each sensitive factor. The greater the correlation between factors, the greater the impact of this factor on the evaluation index (maximum horizontal displacement of reinforced soil abutment), indicating greater sensitivity; On the contrary, factors with lower correlation have lower sensitivity to evaluation indicators. Reinforced soil abutment design and construction personnel can carry out corresponding design and construction organization based on the grey correlation degree sequence.

3 Examples of Reinforced Soil Abutment Engineering

3.1 Project Overview

This article is based on a new construction project and plans to construct a small bridge with a prestressed concrete T-beam as the main beam at pile number K4 + 941. The original plan used a U-shaped expanded foundation for the bridge abutment, with a total length of 22 m. Now, a reinforced soil abutment structure is used instead of the original expanded foundation abutment. Figure 3 shows the numerical geometric model of the prototype bridge abutment modelled using Plaxis finite element software. According to Fig. 3, the height of the block from the top surface of the reinforced soil foundation to the top surface of the retaining wall is $H = 4$ m; The main beam has a length of $L_a = 11$ m and a height of $D_a = 1.0$ m (including a concrete beam height of 0.9 m and an asphalt layer thickness of 0.1 m), and the bridge span to height ratio is $R_{sd} = 2L_a/D_a = 22$; The length of the approach road is $L_b = 15$ m and the height is $L_c = 1.5$ m; To ensure uniform force distribution on the top surface of the bridge abutment, an L-shaped cap beam with a size of $d_a = d_b = 0.5$ m is installed on the top surface of the bridge abutment; The main reinforcement spacing is $S_v = 0.2$ m, and the reinforcement length is $L_r = 2.6$ m; A 1:2 excavation slope is left inside the bridge abutment, which is conducive to layered filling during construction and the stability of the bridge abutment after construction; The size of the abutment retaining wall blocks is 0.2 m × 0.2 m, with a total of 20 layers filled; Install auxiliary reinforcement bars in the area below the cap beam to improve the bearing capacity of this area, with a spacing of 0.1m between the reinforcement bars; The length of reinforced soil foundation is $B_r = 2.6$ m and the height is $B_a = 0.5$ m.

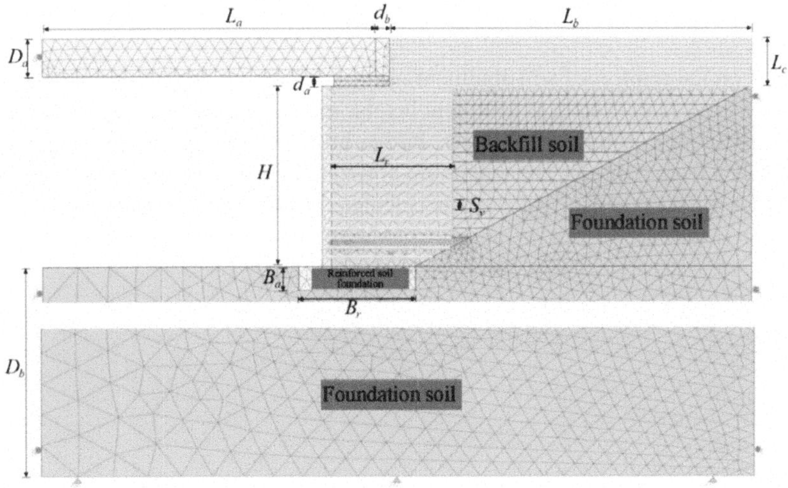

Fig. 3. Geometric model and grid division of abutment.

3.2 Selection of Abutment Parameters

The backfill soil of the bridge abutment is simulated using the advanced constitutive model of soil hardening [10] that comes with Plaxis software. The bridge abutment panel and geogrid are simulated using a linear elastic model, and the contact surface between structural elements is simulated using the interface element unique to the software. Select soil from the construction site for triaxial testing, and obtain the parameters of the hardening constitutive model of the backfill soil as shown in Table 1. The remaining parameter values refer to relevant literature and engineering experience values, as shown in Table 2.

Table 1. Parameters of soil hardening constitutive model.

Type	c[a] (kPa)	φ[b] (°)	ψ[c] (°)	E_{50}^{ref}[d] (kPa)	E_{ur}^{ref}[e] (kPa)	E_{oed}^{ref}[f] (kPa)	υ_{ur}[g]	m[h]	R_f[i]
Parameter values	22.8	33.55	3.55	31020	92829	31020	0.2	0.5	0.96

[a] Cohesive force.
[b] Internal friction angle.
[c] Dilation angle.
[d] Secant stiffness of standard triaxial drainage test.
[e] Tangent stiffness of lateral compression test.
[f] Unloading stiffness of lateral compression test.
[g] Poisson's ratio of unloading and loading.
[h] Stiffness stress related power exponent. $R_{sd} = 2L_a/D_a = 22$.
[i] Soil strength failure ratio.

Based on the reinforced soil abutment of the engineering project as the benchmark model, sensitive factors such as abutment height, reinforcement spacing, reinforcement length, reinforcement material stiffness, bridge load, and internal friction angle of the soil are taken from Table 2. The maximum horizontal displacement of the bridge abutment under different sensitive factors is calculated using finite element software. The calculation results are shown in Table 3.

Table 2. Material parameters of abutment model

Unit category	Parameter values
Panel	Linear elastic model. $E = 3 \times 10^7$ kPa. $\gamma = 12.5$ kN/m^3 Poisson's ratio ν=0
Geogrid	Linear elastic model. Tensile strength $T_u = 75.4$ kN/m Axial stiffness $EA = 600$ kN/m
Foundation soil	Mohr Coulomb model. Dry unit weight $\gamma_d = 15.2$ kN/m^3. Wet unit weight γ_t=18.65 kN/m^3. $c = 17.7$ kPa. φ=27°. $E = 30000$ kPa.ν=0.2
Contact surface between filler and geogrid	A linear elastic model following the Mohr Coulomb failure criterion $c = 8.6$ kPa. φ=40.4°
Contact surface between panel and geogrid	A linear elastic model following the Mohr Coulomb failure criterion $c = 7$ kPa. φ= 34°

Table 3. Calculation results of maximum horizontal displacement of bridge piers under different sensitive factors.

Sensitive factors	Values			
Abutment height (m)	4	5.6	7.2	8.8
Maximum horizontal displacement (mm)	0.5	5.5	8.7	10.5
Reinforcement spacing (m)	0.2	0.3	0.4	0.5
Maximum horizontal displacement (mm)	0.5	5.3	7.2	10.1
Reinforcement length (m)	2	2.8	3.2	4
Maximum horizontal displacement (mm)	1.4	1.125	0.8	0.4
Reinforcement material stiffness (kN/m)	300	600	900	1200
Maximum horizontal displacement (mm)	1.6	0.8	0.65	0.48
Bridge load (kN/m)	88.9	130.4	160.1	189.7
Maximum horizontal displacement (mm)	0.71	1.24	1.8	2.32
Internal friction angle of the soil (°)	35	40	45	51
Maximum horizontal displacement (mm)	1.2	0.93	0.77	0.6

4 Grey Correlation Degree Calculation and Sensitivity Evaluation

4.1 Grey Correlation Degree Calculation

According to the calculation results in Table 3, the sensitive factor values that affect the deformation performance of the bridge abutment are used as the reference matrix X, and the corresponding maximum horizontal displacement value of the bridge abutment is used as the indicator matrix Y, as shown below:

$$X = \begin{bmatrix} X_1 \\ X_2 \\ X_3 \\ X_4 \\ X_5 \\ X_6 \end{bmatrix} = \begin{bmatrix} 4 & 5.6 & 7.2 & 8.8 \\ 0.2 & 0.3 & 0.4 & 0.5 \\ 2 & 2.8 & 3.2 & 4 \\ 300 & 600 & 900 & 1200 \\ 88.9 & 130.4 & 160.1 & 189.7 \\ 35 & 40 & 45 & 51 \end{bmatrix}$$

$$Y = \begin{bmatrix} Y_1 \\ Y_2 \\ Y_3 \\ Y_4 \\ Y_5 \\ Y_6 \end{bmatrix} = \begin{bmatrix} 0.5 & 5.5 & 8.7 & 10.5 \\ 0.5 & 5.3 & 7.2 & 10.1 \\ 1.4 & 1.125 & 0.8 & 0.4 \\ 1.6 & 0.8 & 0.65 & 0.48 \\ 0.71 & 1.24 & 1.8 & 2.32 \\ 1.2 & 0.93 & 0.77 & 0.6 \end{bmatrix}$$

Perform dimensionless processing on the parameter values of positive and negative correlation factors according to Eqs. (3) and (4) respectively to obtain the standard reference matrix and indicator matrix, as shown below:

$$X' = \begin{bmatrix} 0 & 0.33 & 0.67 & 1 \\ 0 & 0.33 & 0.67 & 1 \\ 1 & 0.6 & 0.4 & 0 \\ 1 & 0.67 & 0.33 & 0 \\ 0 & 0.412 & 0.710 & 1 \\ 1 & 0.688 & 0.375 & 0 \end{bmatrix} \quad Y' = \begin{bmatrix} 0 & 0.50 & 0.82 & 1 \\ 0 & 0.50 & 0.698 & 1 \\ 0 & 0.275 & 0.6 & 1 \\ 0 & 0.714 & 0.848 & 1 \\ 0 & 0.329 & 0.677 & 1 \\ 0 & 0.450 & 0.716 & 1 \end{bmatrix}$$

According to Eqs. (7), (8), (9) and (10), it can be concluded that: $\Delta_{\max} = 1$. $\Delta_a = 0.374$. $Y_\Delta = 0.374/1 = 0.374 > 1/3$.

According to Eq. (10), the range of values for the resolution coefficient ξ is (0.561, 0.748). For ease of calculation, $\xi = 0.6$ is taken. Substituting ξ into Eq. (6) can obtain the correlation coefficients $l_i(j)$ corresponding to each factor. The correlation coefficient matrix L composed of correlation coefficients is as follows:

$$L = \begin{bmatrix} 1 & 0.779 & 0.800 & 1 \\ 1 & 0.779 & 0.955 & 1 \\ 0.375 & 0.824 & 0.532 & 0.375 \\ 0.375 & 0.758 & 0.483 & 0.375 \\ 1 & 0.793 & 0.703 & 1 \\ 0.375 & 0.716 & 0.637 & 0.375 \end{bmatrix}$$

According to Eqs. (11), (12), (13), (14), (15) and (16), the norm similarity α_i and directional similarity β_i of the reference vector X' and indicator vector Y' are calculated. The vector similarity γ_i can be obtained from Eq. (17), and then the weight k_i corresponding to each factor is normalized by Eq. (18). The calculation results are shown in Table 4.

Table 4. Calculation results of vector similarity.

Calculated value of norm similarity α_i		Calculated value of directional similarity β_i		Calculated value of vector similarity γ_i		Calculated value of the corresponding weight for each factor k_i	
α_1	0.889	β_1	0.913	γ_1	0.811	k_1	0.914
α_2	0.943	β_2	0.921	γ_2	0.870	k_2	0.999
α_3	0.841	β_3	0.211	γ_3	0.177	k_3	0.001
α_4	0.686	β_4	0.289	γ_4	0.198	k_4	0.030
α_5	0.908	β_5	0.871	γ_5	0.791	k_5	0.886
α_6	0.968	β_6	0.225	γ_6	0.218	k_6	0.059

According to the obtained correlation coefficient matrix L, the weight corresponding to each factor are input into Eq. (19) to calculate the grey correlation degree. The grey correlation degrees corresponding to each factor are shown in Table 5:

4.2 Sensitivity Evaluation and Its Impact on Design and Construction

According to the grey correlation calculation results of the above factors, it can be seen that the reinforcement spacing has the most significant impact on the deformation performance of the bridge abutment, followed by the abutment height, bridge load, and internal friction angle of the soil. However, the stiffness and length of the reinforcement material have a relatively small impact on the deformation of the abutment. Therefore, in the structural design of reinforced soil abutment, the reinforcement spacing, the abutment height, and the bridge load are the most important factors to consider. These three factors

Table 5. Grey correlation degree of various factors.

Relational factors	Abutment height	Reinforcement spacing	Reinforcement length	Reinforcement material stiffness	Bridge load	Internal friction angle of the soil
Grey correlation degree	0.818	0.932	0.001	0.0149	0.7743	0.310

greatly affect the deformation performance and stability of the abutment. When the deformation and stability of the designed abutment do not meet the requirements, the influence of these three factors should be considered first, and parameters such as the reinforcement spacing and the abutment height should be redefined according to the design requirements.

The sensitivity of the internal friction angle of the soil is weaker than that of the reinforcement spacing, abutment height, and bridge load. However, as a physical parameter of GRS-IBS, it has a certain impact on the mechanical properties of the reinforced composite, so it cannot be ignored in the design of the abutment. The influence of stiffness and length of reinforced materials on the deformation of bridge abutment is minimal compared to other factors, but the stiffness and length of reinforced materials, as two parameters of reinforced materials, can also affect the physical properties of reinforced composites to a certain extent. If the reinforcement stiffness is too small, it will cause significant deformation of the reinforcement material itself, thereby increasing the deformation of the bridge abutment. If the reinforcement length is too small, it will reduce the friction force of the reinforcement material and cannot suppress the development of the fracture surface. Therefore, when designing the bridge abutment, it is also necessary to choose appropriate material stiffness and reinforcement length.

From a construction perspective, the main sensitive factors such as reinforcement spacing, abutment height, and bridge load will have a significant impact on the deformation of reinforced soil abutments during the construction process. The secondary sensitive factors such as the internal friction angle of the soil, the stiffness and length of the reinforced material will affect the long-term creep of the reinforced soil abutment. Therefore, using the grey correlation calculation results of each factor is beneficial for evaluating construction risks and guiding construction in the most economical and efficient way.

5 Conclusions

By improved grey correlation method, the sensitivity analysis of the factors affecting the maximum horizontal deformation of reinforced soil abutment can lead to the following conclusions:

(1) The grey correlation degree calculation value of the reinforcement spacing is 0.932, which is the highest value among all sensitive factors, indicating that its impact on the deformation of reinforced soil abutment is the greatest. The ranking results of sensitive factors based on grey correlation degree are: reinforcement spacing, abutment height, bridge load, internal friction angle of soil, reinforcement material stiffness, and reinforcement length. This sequence can guide the design and construction of reinforced soil bridge abutment structures.
(2) The improved grey correlation method can accurately and quantitatively analyse the influence of various sensitive parameters on the horizontal deformation of reinforced soil abutment, and can clearly display the main and secondary factors that affect the deformation performance of reinforced soil abutment. It is a reliable and practical analysis method.
(3) In addition to evaluating the sensitivity of the influencing factors on the horizontal deformation of reinforced soil abutment, the improved grey correlation method can also be used for the analysis of the influencing factors on the stability and consolidation time of reinforced soil abutment. It can also be extended to analysis other types of civil engineering structures, which will be the direction of further research in the future.

References

1. Adams, M., Nicks, J., Stabile, T.: Geosynthetic reinforced soil integrated bridge system interim implementation guide. Federal Highway Administration (US) (2012)
2. Han, J., Jiang, Y., Xu, C.: Recent advances in geosynthetic reinforced retaining walls for highway applications. Front. Struct. Civ. Eng. **12**, 239–247 (2018)
3. Chao, X., Min-min, L.: Differences in performance and evaluation methods between geosynthetic reinforced soil structure and mechanically stabilized earth structure. J. Yangtze River Sci. Res. Instit. **36**, 1–7 (2019)
4. Jelušič, P., Žlender, B.: Experimental study of a geosynthetic-reinforced soil bridge abutment. Geosynth. Int. **28**, 479–490 (2021)
5. Chao, X., et al.: Scaled model tests investigating deformation characteristics of geosynthetic reinforced soil (GRS) abutments under vertical loads. Materials (Basel, Switzerland) **16**, 4601 (2023)
6. Raja, M.N.A., Jaffar, S.T.A., Bardhan, A., Shukla, S.K.: Predicting and validating the load-settlement behavior of large-scale geosynthetic-reinforced soil abutments using hybrid intelligent modeling. J. Rock Mech. Geotech. Eng. **15**(3), 773–788 (2023). https://doi.org/10.1016/j.jrmge.2022.04.012
7. Xiaofen, Y., Dai, F.: A review of multi index comprehensive evaluation methods. Stat. Decision **3**, 119–121 (2004)
8. Julong, D.: Basic Methods of Grey System. Huazhong University of Technology Press, Wuhan (1987)
9. Huang Anping, Sensitivity analysis of factors affecting the stability of high fill slopes. Lanzhou University of Technology (2020)
10. Schanz, T., Vermeer, P.A., Bonnier, P.G.: The hardening soil model: Formulation and verification. In: Brinkgreve, R.B.J. (ed.) Beyond 2000 in Computational Geotechnics, pp. 281–296. Routledge (2019). https://doi.org/10.1201/9781315138206-27

Experimental Study on the Response Characteristics of Typical Beam Bridges to Near-Fault Ground Motions

Haoyu Xie[1,2,3](✉), Jianlin Li[1,2], and Yuxin Mao[1,2]

[1] CMCT Research and Design Institute Co., Ltd., Chongqing 400067, China
{xiehaoyu,lijianlin,maoyuxin}@cmhk.com
[2] State Key Laboratory of Bridge Safety and Resilience, Chongqing 400067, China
[3] School of Civil Engineering, Chongqing University, Chongqing 400045, China

Abstract. The design response spectrum in the current bridge seismic design specifications is mainly based on far-fault observation records and cannot accurately reflect the ground motion characteristics and structural seismic response characteristics in the near-fault area. Therefore, the seismic design of bridges in the near-fault area may be unsafe. An experimental study was conducted based on shaking table, in order to study and verify the response characteristics and rules of typical beam bridges under near-field vibrations with the input of both horizontal and vertical ground motion accelerations. The result shows that for a simply supported beam bridge, the biggest feature of the horizontal effect of near-field vibrations is to amplify the displacement of the superstructure and the shear deformation of the supports. The impact on the stress at the pier bottom and the acceleration of the superstructure is less than the impact on the displacement. The main mode of beam bridge and the peak ground motion acceleration range when the bearing enters into stretch state under vertical ground motions are also figured out in the experiment.

Keywords: Beam bridges · Near-fault ground motion · Experimental study · Shaking table test

1 Introduction

Near-fault earthquakes, also known as near-field earthquakes, are generally believed to occur in areas where the fault distance is less than 20 km. Their time-frequency domain characteristics and impact on the seismic response of engineering structures have become a major issue in recent years, as one of the most popular topics in seismic engineering [1, 2]. Investigations of earthquake damage caused by many intensive earthquakes in history have shown that due to the existence of rupture directional effects, hanging wall effects, sliding effects, strong long-period components, and significant vertical ground motion effects, the earthquake response and damage of engineering structures in the near-fault area is much more serious. Recently, a large number of bridges have been built, among which beam bridges are a very common structural form [3–5]. In the past

B. Yuan et al. (Eds.): ICCASE 2024, SUCI, pp. 362–371, 2024.
https://doi.org/10.1007/978-3-031-76102-7_40

design and calculation of beam bridges, the effect of near-fault earthquake motion on the bridge was often ignored. In recent years, a large amount of near-fault seismic motion data has been recorded, enriching the seismic data and accelerating the research on the response of bridge structures under near-fault earthquakes. Earthquake is a complex random process, which contains many uncertain factors. Some studies have recognized the difference between near-fault earthquakes and mid- and far-fault earthquakes and have also selected representative earthquake records in their studies to calculate structures. Pu [6] established the acceleration spectrum function based on the structural period, structural damping and ground motion pulse period, and further developed the displacement spectrum of the highly damped system by introducing damping correction factors for the acceleration and displacement spectra. Guo [7] discussed the seismic damage characteristics of a seven-span high-speed railway simply supported girder bridge under near-fault earthquake excitation with obvious velocity pulse and high-frequency vibration characteristics. Jiang [8] analyzed the applicability of high-speed railway systems under near-fault earthquakes and concluded that high-speed railway piers are difficult to yield and the shear bearing capacity of the bridge piers is insufficient under near-fault earthquakes. Zhou [9] analyzed the impact of near-fault pulse-type ground motion on simply supported bridges of high-speed railways and believed that the response of simply supported bridges of high-speed railways depends on the acceleration intensity parameters and has a strong linear relationship with the spectral intensity parameters corresponding to the basic period.

Responses and comparisons have led to some useful conclusions, but there are still many unknown areas waiting for us to explore. For example, when studying the influencing factors of the structure's near-fault seismic response, since the near-fault ground motion itself is very complex and has many influencing factors, existing research has shown that both the horizontal and vertical components of the near-fault ground motion are related to those of the mid- and far-fault earthquakes. However, there is a lack of near-fault seismic design response spectrum and a lack of research on its impact on the response of simply supported girder bridges, especially a lack of shaking table test research. The paper conducts a shaking table test on the possible response characteristics of a typical simply supported girder bridge under the action of near-field vibrations, inputs near-field and far-field vibrations vertically and horizontally, compares the response results, and summarizes the rules.

2 Project Overview and Test Model

A 3 $\times$ 13 m simply supported hollow plate structure girder bridge is selected as the research object. The upper structure is a 70 cm standard hollow plate beam, 9.25 m wide; the lower structure is a 5 m high double-column pier (after river excavation); the bearing model is a circular plate rubber bearing. The bearing model is GYZ 150 $\times$ 42, with 16 bearings arranged on the bridge platform and 32 bearings arranged on the piers.

The similarity relationship of the shaking table scale model test is determined based on the dimensional analysis method, as shown in the Table 1.

The model was scaled down from one span in the actual project, in which the double-column pier was simplified to a single-column type, and the main purpose was to study the

Table 1. Structural dynamic model test similarity coefficient

Type of physical quantity	Physical quantity	Similar constant symbols	Relational expression	Design similarity constant
Material properties	σ	S_σ	S_σ	1
	ε	S_ε	1	1
	E	S_E	$S_E = S_\sigma$	1
	ρ	S_ρ	$S_\rho = S_\sigma/(S_a \cdot S_l)$	1
Geometric properties	l	S_l	S_l	1/3
	A	S_A	$S_A = S_l^2$	1/9
	I	S_I	$S_I = S_l^4$	1/81
Load	F	S_F	$S_F = S_E \cdot S_l^2$	1/9
	q	S_q	$S_q = S_E \cdot S_l$	1/3
	p	S_p	$S_p = S_E$	1
	M	S_M	$S_M = S_E \cdot S_l^3$	1/27
Dynamic properties	m	S_m	$S_m = S_\rho \cdot S_l^3$	1/27
	k	S_k	$S_k = S_E \cdot S_l$	1/3
	c	S_c	$S_c = S_m/S_t$	1/9
	t	S_t	S_t	1/3
	T	S_T	$S_T = S_t$	1/3
	x	S_x	$S_x = S_l$	1/3
	v	S_v	$S_v = S_x/S_t$	1
	a	S_a	$S_a = S_x/S_t^2$	3

seismic response characteristics of the bridge in the longitudinal and vertical directions. Design drawings and actual models are as Figs. 1 and 2 below.

3 Input Ground Montion and Working Condition

Historical time-series and artificial time-series [10] are both selected for the shaking table test in order to find out the characteristics of deformation and acceleration response to the target structure. Furthermore, bridge longitude working conditions and vertical working conditions are conducted to obtain the corresponding response patterns, especially focusing on the conditions and failure modes of tensile failure of bearings under intensive and extreme vertical near-fault earthquakes.

The input ground motions include El-Centro waves (typical far-fault natural wave), Chi-Chi waves (typical near-fault natural wave), artificial waves fitted according to the

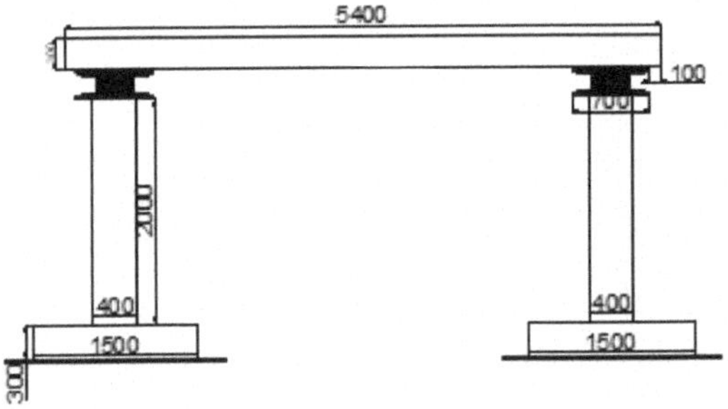

Fig. 1. Schematic diagram of the bridge of the shaking table experimental model (mm)

Fig. 2. Photos of scale model of vibration table installation

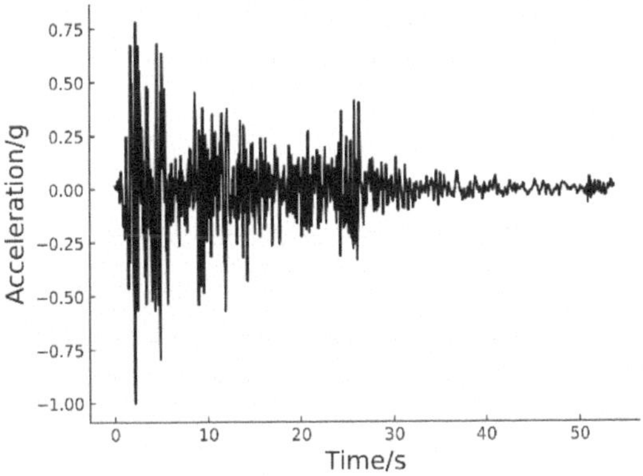

Fig. 3. Time history of El-Centro

standard response spectrum (standard spectrum), and artificial waves fitted according to the modified response spectrum (modified spectrum). A total of 4 seismic acceleration time histories, after unifying the PGA to 0.1 g, the time history curves and response spectra of the 4 input seismic motions are shown in Figs. 3, 4, 5, 6, 7, 8, 9 and 10. The response spectra fitting methodology is based on the simulation methodology considering iterative correlation and phase spectra.

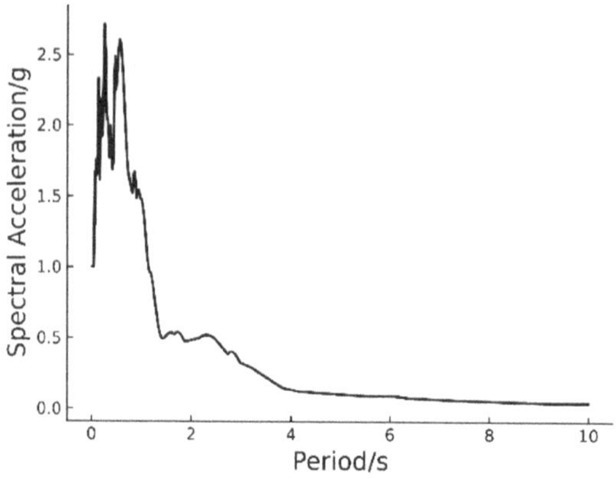

Fig. 4. Response spectrum of El-Centro

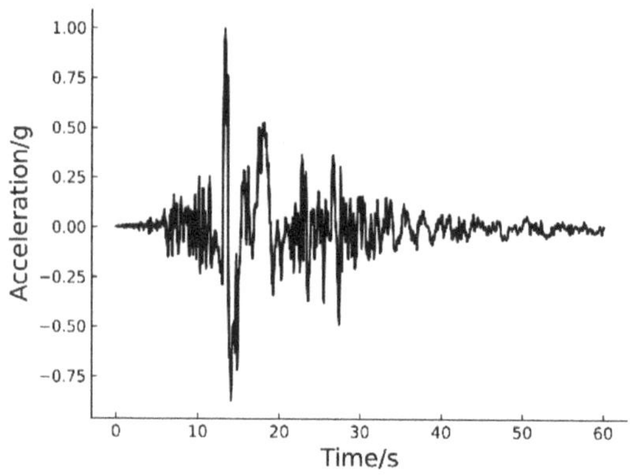

Fig. 5. Time history of Chi-Chi

The shaking table test includes bridge-longitude working conditions, vertical working conditions, and bridge longitude + vertical coupled working conditions, a total of 44 working conditions.

4 Experimental Results and Comparison

4.1 Horizontal Response

There are two longitudinal bridge-direction displacement sensors arranged on the beam, corresponding to the top positions of the two bridge piers. The maximum longitudinal displacements of the beam under different peak accelerations are shown in Table 2. It can be seen that when the peak acceleration of 0.3 g is input to the longitudinal bridge ground motion, the longitudinal displacement of the beam under the action of near-field vibration exceeds the far-field vibration by an average of 144% (2.44 times). And under the condition of 0.6 g peak acceleration input to the longitudinal bridge ground motion,

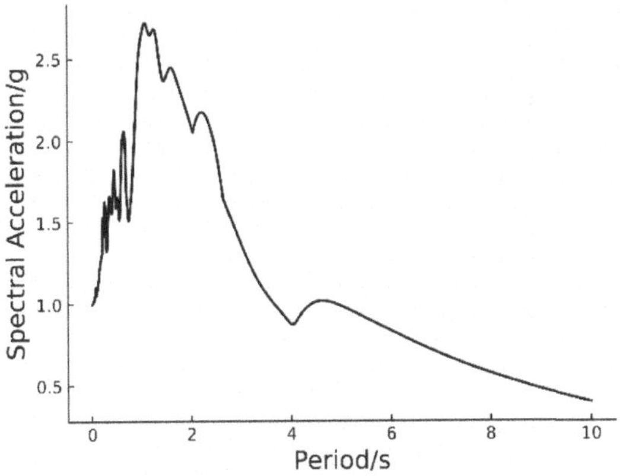

Fig. 6. Response spectrum of Chi-Chi

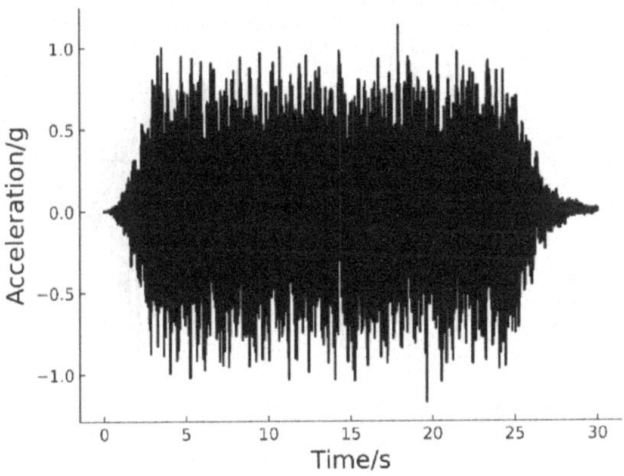

Fig. 7. Time history of artificial ground motion (standard spectrum)

the longitudinal displacement of the beam under the action of near-field vibration exceeds the far-field vibration by an average of 139% (2.39 times).

There are two longitudinal bridge-direction acceleration sensors arranged on the beam, corresponding to the top positions of the two bridge piers. The maximum longitudinal accelerations of the beam under different peak accelerations are shown in Table 3. It can be seen that when the peak acceleration of 0.3 g is input to the longitudinal bridge ground motion, the longitudinal acceleration of the beam under the action of near-field vibration exceeds the far-field vibration by an average of 5%, which is far less different than the displacement results of the same working condition. The 0.6 g condition shares the similar conclusion.

There are 8 strain gauges arranged in the longitudinal reinforcements of the same section at the bottom of the pier, corresponding to the longitudinal and transverse bridge directions respectively. When 0.3 g peak acceleration is input into the longitudinal bridge

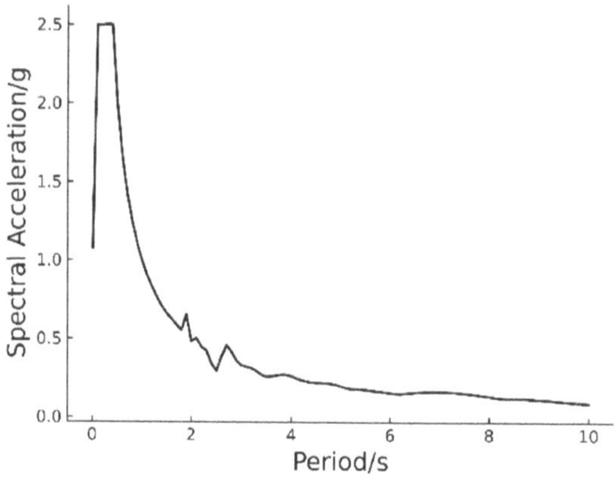

Fig. 8. Response spectrum of artificial ground motion (standard spectrum)

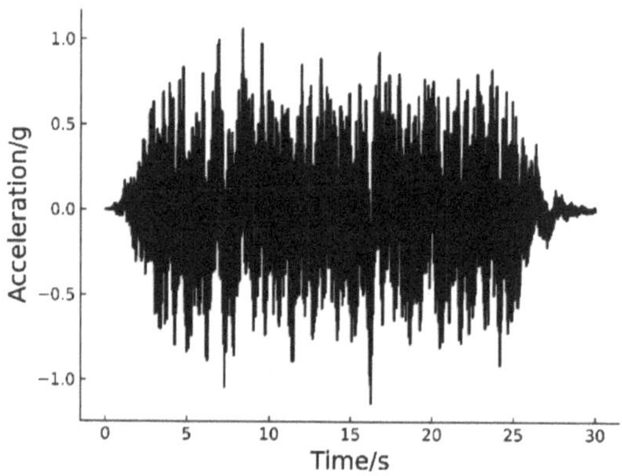

Fig. 9. Time history of artificial ground motion (modified spectrum)

ground motion condition, the law of strain at the pier bottom and the law of displacement and acceleration of the upper structure under the action of near-field vibration are obviously different. The reason is that the control mode of the beam bridge that affects the bending moment at the bottom of the pier corresponds to a shorter period (high frequency) and is not affected by the strong long-period component of near-field vibrations. The 0.6 g condition shares the similar conclusion.

4.2 Vertical Response

There are two vertical bridge-direction displacement sensors arranged on the beam, corresponding to the top positions of the two bridge piers. The maximum vertical displacements of the beam under different peak accelerations are shown in Table 4. It can be found that under vertical earthquake conditions, the vertical displacement of the beam body under the action of near-fault vibration exceeds the far-fault vibration by about 10% on average. Because the frequency of the vertical main mode of the beam bridge is

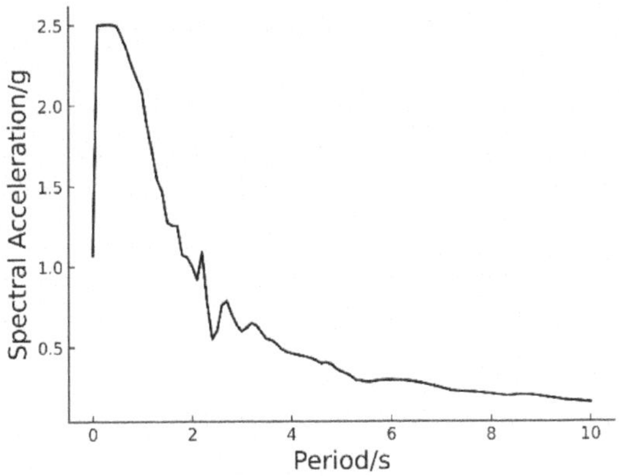

Fig. 10. Response spectrum of artificial ground motion (modified spectrum)

Table 2. Statistical table of maximum beam displacement under longitudinal bridge direction conditions

Ground Motion	El-Central	Chi-Chi	Artificial (standard)	Artificial (modified)
Maximum Displacement under 0.3 g	9.8 mm	31.7 mm	14.4 mm	27.3 mm
Maximum Displacement under 0.6 g	19.3 mm	63.0 mm	30.2 mm	55.1 mm

Table 3. Statistical table of maximum beam acceleration under longitudinal bridge direction conditions

Ground Motion	El-Central	Chi-Chi	Artificial (standard)	Artificial (modified)
Maximum Acceleration under 0.3 g	0.52 g	0.49 g	0.36 g	0.43 g
Maximum Acceleration under 0.6 g	1.03 g	1.04 g	0.79 g	0.85 g

much higher than that of the isolation bearing, the impact of vertical near-field vibrations under the same PGA is not as prominent as the horizontal impact.

There are two vertical direction acceleration sensors arranged on the beam, corresponding to the top positions of the two bridge piers. The maximum vertical accelerations of the beam under different peak accelerations are shown in Table 5. The average vertical acceleration of the beam under the action of near-fault vibration is close to the results of far-fault vibration, and there is no big difference. And the bearing begins to be stretched between 0.6 g and 0.9 g.

Table 4. Statistical table of maximum beam displacement under vertical bridge direction conditions

Ground Motion	El-Central	Chi-Chi	Artificial (standard)	Artificial (modified)
Maximum Displacement under 0.3 g	7.1 mm	13.3 mm	17.0 mm	15.6 mm
Maximum Displacement under 0.6 g	19.3 mm	30.1 mm	36.6 mm	33.8 mm
Maximum Displacement under 0.9 g	30.7 mm	43.8 mm	56.0 mm	57.9 mm
Maximum Displacement under 1.2 g	41.7 mm	56.6 mm	75.0 mm	68.1 mm

Table 5. Statistical table of maximum beam acceleration under vertical bridge direction conditions

Ground Motion	El-Central	Chi-Chi	Artificial (standard)	Artificial (modified)
Maximum Acceleration under 0.3 g	0.38 g	0.36 g	0.29 g	0.32 g
Maximum Acceleration under 0.6 g	0.76 g	0.63 g	0.62 g	0.65 g
Maximum Acceleration under 0.9 g	1.18 g	0.94 g	0.92 g	1.01 g
Maximum Acceleration under 1.2 g	1.63 g	1.35 g	1.21 g	1.27 g

5 Conclusion

A shaking table test study was conducted on the longitudinal and vertical response characteristics of a typical beam bridge subjected to near-fault ground motions, and the main conclusions are as follows. 1) The horizontal design acceleration response spectrum given in my country's current industry standard is not suitable for the seismic design of bridges under horizontal earthquakes near fault areas, which is considered unsafe. 2) For a simply supported girder bridge, comparing the design response spectrum to modified response spectrum near-fault, the magnification of the bearing displacement and pier displacement is greater than the pier bottom bending moment and pier bottom shear force. 3) The most significant feature of the horizontal effect of near-field ground motions is to amplify the displacement of the superstructure and the shear deformation of the supports. The effect on the stress at the pier bottom and the acceleration of the superstructure is less than the effect on the displacement. 4) For a simply supported beam bridge, when the near-fault vertical ground motion acts between PGA 0.6 g and 0.9 g, the support may enter a tension state, and its main control mode is the vertical vibration of the superstructure.

For the further study, it is planned to conduct more research on different bridge types to near-fault earthquakes relying on numerical simulation and model testing, further summarize the response characteristics, and provide a basis for the seismic design methodology of bridges in the near-fault areas.

Acknowledgments. The work was financially supported by grant No. CSTB2023NSCQ-MSX0775 of Chongqing Natural Science Foundation and grant No. 12032008 of National Natural Science Foundation of China.

References

1. Bray, J.D., Rodriguez-Marek, A.: Characterization of forward-directivity ground motions in the near-fault region. Soil Dyn. Earthq. Eng. **24**(11), 815–828 (2004)
2. Tang, G., Yu, W., Xie, H., et al.: Research progress on dynamic response of bridge structures under near-field earthquakes. J. Highw. Transport. Technol. **39**(01), 60–68 (2023)
3. Zong, Z., Lin, Y., Lin, J., et al.: Research advances of across-fault ground motions and their effects on bridge structures. China J. Highw. Transport **36**(01), 80–96 (2023). (in Chinese)
4. Mavroeidis, G.P., Papageorgiou, A.S.: A mathematical representation of near-fault ground motions. Bull. Seismol. Soc. Am. **93**(3), 1099–1131 (2003)
5. Zhong, J., Jiang, L., Pang, Y., et al.: Near-fault seismic risk assessment of simply supported bridges. Earthq. Spectra **36**(4), 1645–1669 (2020)
6. Pu, W., Wu, M., Huang, B., et al.: Quantification of response spectra of pulse-like near-fault ground motions. Soil Dyn. Earthq. Eng. **104**, 117–130 (2018)
7. Guo, W., Gao, X., Hu, P., et al.: Seismic damage features of high-speed railway simply supported bridge–track system under near-fault earthquake. Adv. Struct. Eng. **23**(8), 1573–1586 (2020)
8. Jiang, L., Yu, J., Zhou, W., et al.: Applicability analysis of high-speed railway system under the action of near-fault ground motion. Soil Dyn. Earthq. Eng. **139**, 106289 (2020)
9. Zhou, T., Jiang, L., Xiang, P., et al.: Effects of near-fault pulse-type ground motions on high-speed railway simply supported bridge and pulse parameter analysis. Bull. Earthq. Eng. **2022**, 1–26 (2022)
10. Xie, H., Zheng, W., Zhang, W., et al.: Response spectra-compatible artificial ground motion simulation considering iterative correlation and phase spectra. Acta Seismol. Sin. **42**(3), 341–348 (2020). (in Chinese)

Optimization of Anti-Seepage System for an Asphalt Concrete Core Rockfill Dam on a Deep Overburden Foundation

Jie Lu[1], Detan Liu[2,3], Jin Yang[2], Guoqiang Lei[1], Haoxuan Li[3], Zhenzhong Shen[3], and Yan He[3](✉)

[1] Sichuan Datang International Ganzi Hydroelectric Co., Ltd., Kangding 626001, China
[2] Datang Hydropower Science & Technology Research Institute Co., Ltd., Chengdu 610074, China
[3] College of Water Conservancy and Hydropower Engineering, Hohai University, Nanjing 210098, China
1902010101@hhu.edu.cn

Abstract. Seepage control is the key to building dams and reservoirs on deep overburden foundation. A reasonable seepage control scheme can not only effectively reduce the leakage in the reservoir area, but also help reduce the project investment. Taking an asphalt concrete core rockfill dam on a deep overburden foundation as an example, according to the topographic and geological characteristics of the dam site area and the seepage control scheme, a three-dimensional finite element model for dam seepage analysis was established. The saturated unsaturated seepage theory was used to calculate and analyze the seepage field, seepage flow and seepage gradient of the reservoir area under the design seepage control scheme, and the design seepage control scheme is optimized by changing the depth and length of the anti-seepage curtain. The results show that the curtain depth at the right bank dam foundation, left bank dam abutment and right bank dam abutment should be shortened by 10 m, and the curtain length of the right bank scheme should be extended by about 50 m. The optimized seepage control scheme can not only effectively control the leakage in the reservoir area, but also reduce the construction difficulty and project cost.

Keywords: deep overburden · asphalt concrete core rockfill dam · anti seepage curtain · optimization of anti-seepage system

1 Introduction

In order to realize the national strategic goal of "carbon peak, carbon neutral" and comply with the transformation and upgrading of energy structure, efficient and clean hydro energy resources will become an important member of China's energy development strategy [1]. China's northwest and southwest regions are rich in hydropower resources, but the geological conditions are complex, and the development and utilization of hydropower resources often require the construction of water conservancy

B. Yuan et al. (Eds.): ICCASE 2024, SUCI, pp. 372–386, 2024.
https://doi.org/10.1007/978-3-031-76102-7_41

projects on the foundation of the deep cover layer. Deep cover layer has loose structure and strong local permeability, which is easy to form concentrated seepage channels [2], which adversely affects the seepage safety and structural stability of the dam, therefore, the construction of dams on the deep cover layer needs to analyze the characteristics of seepage field at the base of the dam, and select a reasonable anti-seepage system arrangement scheme to prevent infiltration damages at the base of the dam [3].

Curtain grouting for the construction of dams on the deep cover is usually used for seepage control measures[4], in the actual project by the geological conditions and project cost and other factors, the need to carry out anti-seepage program optimization research, curtain layout form and design size need to be adapted to local conditions, in order to ensure that the reservoir has good seepage control performance while reducing the cost of the project[5]. At present, many scholars have carried out research on anti-seepage system optimization, Arifin, Muhammad Zainal et al. [6] carried out 3 alternatives, extended curtain grouting, cut off wall and upstream blanket respectively and the result indicated that, the installation curtain grouting was the most effective alternative to reduce seepage; He Yunxiao et al. [7] applied FLUENT to simulate the impact of different anti-seepage curtain depth on the seepage volume of the base of the dam, and came up with a reasonable depth of curtain; Deng Lunxin et al. [8] combined the finite element and composite optimization algorithm to optimize the anti-seepage curtain grouting parameters; Liu Jie et al. [3] set up a three-dimensional finite element model of a sand and gravel dam with a deep overburden layer to compare the anti-seepage system of different dam site areas with the anti-seepage program, and then compared the anti-seepage schemes of different dam site areas. The three-dimensional finite element model of a gravel dam on a deep overburden was established, and the seepage characteristics of the dam site area with different anti-seepage schemes were compared, and a reasonable anti-seepage arrangement was obtained. The above studies provide references for the optimization of dam anti-seepage system and calculation of seepage characteristics.

A hydropower station hub project dam site area geological conditions are complex, the riverbed part of the alluvial cover layer is deep, the structure level is complex, the reservoir water through the two sides of the mountain to the downstream bypass seepage is larger, in order to prevent the occurrence of uneven settlement of the dam foundation, infiltration deformation and other problems, it is necessary to carry out a study of the depth of its anti-seepage curtain and the length of extension of the curtain on both sides of the preferred choice. This paper combines the topographic and geological conditions of the reservoir hub project, establishes the three-dimensional finite element mesh of the reservoir dam based on the equivalent continuous medium model, calculates the change rule of seepage field during the operation period of the reservoir according to the theory of saturated-unsaturated seepage calculation, and analyzes the trend of infiltration gradient and seepage in the key parts of the reservoir such as the dam body, dam base and shoulder under the different anti-seepage schemes, and obtains the reasonable anti-seepage arrangement scheme. The results of the study can provide a reference for the setting of similar anti-seepage systems for asphalt concrete core walls using both anti-seepage walls and anti-seepage curtains.

2 Basic Theory of Saturated-Unsaturated Seepage Flow

2.1 Basic Differential Equations for Seepage Calculation

The basic differential equation for unsaturated seepage is derived by the same method as for saturated seepage under the assumption that Darcy's law also applies to unsaturated seepage, where the coefficient of permeability is no longer a constant but a function of the degree of saturation of the soil, and the basic differential equation for seepage is as follows [9]:

$$\frac{\partial}{\partial x_i}\left[k_{ij}^s k_r(h_c)\frac{\partial h_c}{\partial x_j} + k_{i3}^s k_r(h_c)\right] - Q = [C(h_c) + \beta S_s]\frac{\partial h_c}{\partial t} \tag{1}$$

where, x_i, x_j are the coordinate values (i, j = 1, 2, 3); k_{ij}^s is the saturated permeability coefficient tensor, m/s; $k_r(h_c)$ is the relative permeability ($0 \leq k_r(h_c) \leq 1$); h_c is the pressure head, m; β is the saturated-unsaturated selection constant in the interval [0,1]; Q is the source-sink term, m^3/s; S_s is the elasticity of the water storage rate, m-1; and $C(h_c)$ is the specific volume of water, m-1.

2.2 Infiltration Flow Calculation

The seepage q through any cross section is calculated as follows:

$$q = -\iint_s k_n \frac{\partial h}{\partial n} \mathrm{d}S \tag{2}$$

where h is the seepage head, m; n is the unit vector normal to the section; S is the overflow section; k_n is the permeability coefficient in the n direction.

Take any eight nodes hexahedron seepage calculation unit shown in Fig. 1, select the center section abcda for the overflow section S, S_x, S_y, S_z respectively for the section S in the YOZ, ZOX, XOY plane of projection, then calculate the infiltration flow rate through the unit in the cross section of the formula is as follows:

$$q = -k_x \frac{\partial h}{\partial x} S_x - k_y \frac{\partial h}{\partial y} S_y - k_z \frac{\partial h}{\partial z} S_z \tag{3}$$

Seepage calculation and analysis are carried out based on the basic theory of seepage, Darcy's law and the theory of equivalent continuous media suitable for engineering applications.

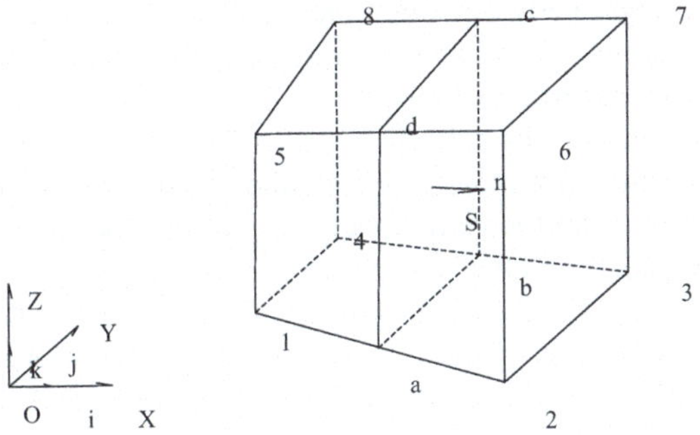

Fig. 1. Schematic diagram of the infiltration flow calculation unit.

3 Engineering Geology of the Dam Area and Computational Modeling

3.1 Project Overview

A hydropower station dam site area ground elevation of 2491 m–2495 m, the hydropower station to power generation, the normal storage level of 2545 m, the dead water level of 2541 m, the total reservoir capacity of 158 million m^3, regulating the reservoir capacity of 0.2 billion m^3. Reservoir retaining structures for the construction of asphalt concrete core wall rockfill dams built on a deep cover layer, the top of the dam elevation of 2549 m, the width of the top of the dam is 10 m, dam crest length of 385 m. The lowest excavation elevation of the dam foundation is 2480 m, the maximum dam height is 69 m, the typical section of the dam is shown in Fig. 2.

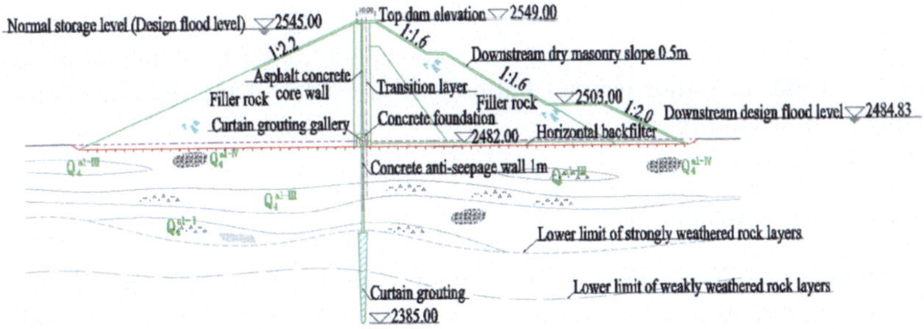

Fig. 2. Typical profile of a reservoir dam (m).

Reservoir hub area geological conditions are complex, according to the exploration borehole data, the dam site area riverbed alluvial cover thickness of 17.70 m–63.30 m, the overall presentation of the center of the river along the thicker layer of accumulation, to the sides of the gradual thinning. The structure of the overburden is dominated by sandy pebble gravel layer, between which there are local gravel-containing medium-coarse sand, mud-containing gravel powder and fine sand lenses, with different granular compositions and complex structural levels.

The dam seepage control adopts linear, crushed asphalt concrete core wall, the core wall is gradually thickened from top to bottom, the maximum thickness of the bottom of the wall is 1.7 m. The dam foundation seepage control adopts concrete anti-seepage wall, anti-seepage curtain and other comprehensive measures, anti-seepage curtain is laid along the foundation of the core wall at the concrete base (both sides) and under the concrete anti-seepage wall (riverbed), and curtain grouting holes are set up at the shoulders of both sides of the dam to both sides of the river, and seepage control of the dam foundation and both sides anti-seepage curtains are controlled according to the depth of 5 m below the 5 Lu line.

3.2 Finite Element Model

The model coordinate system is selected according to the following requirements: take the right dam end pile B1 point on the dam axis as the origin of the model coordinates; take the X-axis as the upstream and downstream direction, the upstream pointing to the downstream is positive, perpendicular to the dam axis; the Y-axis is the direction of the dam axis, pointing to the left bank from the right bank is positive; the Z-axis is the vertical direction, the upward direction is positive, and the elevation is consistent with the upstream direction. The upstream boundary is 600 m above the foot of the upstream slope, and the downstream boundary is 340 m beyond the foot of the downstream dam. The left bank boundary is 640 m to the left of the diversion power generation system, and the right bank boundary is 580 m to the right of the right shoulder of the dam. The bottom boundary is intercepted about 185 m below the elevation of bottom of the curtain curtain of the riverbed (elevation 2200.00 m). According to the topography of the dam site area and the building arrangement, 16 control sections are selected to form a three-dimensional super-cell grid, and 68513 grid nodes and 66083 cells are encrypted and generated according to the "control section super-cell finite element automatic dissecting method", and the three-dimensional finite element model of the reservoir dam after dissecting is shown in Fig. 3.

3.3 Calculation Parameters

According to the results of indoor tests and field water injection experiments, and combined with the preliminary ground investigation data and design data, the dam body and reservoir rock materials are divided into 12 parameter zones according to permeability, see Table 1.

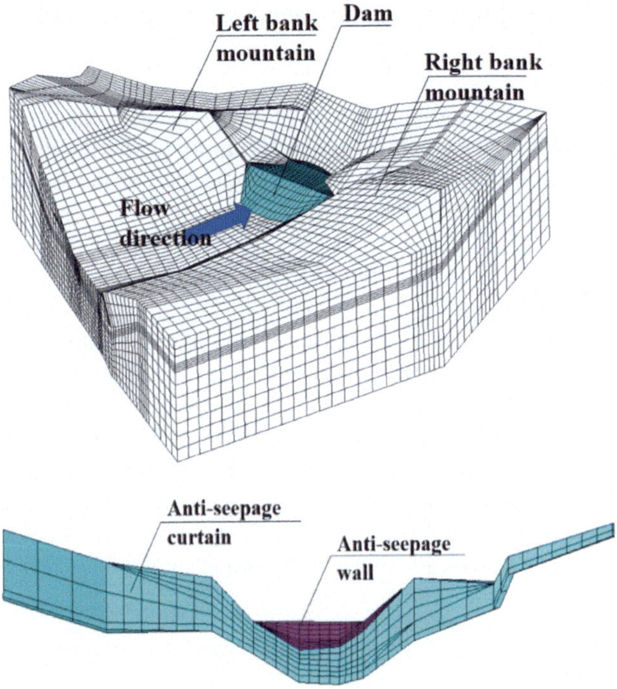

Fig. 3. Schematic diagram of meshing of the 3D finite element model.

Table 1. Dam materials and rock permeability coefficients.

Material partition	Permeability coefficient (cm/s)	Material partition	Permeability coefficient (cm/s)
Asphalt concrete core wall	1.0×10^{-8}	Curtain	3.0×10^{-5}
Transition material	1.0×10^{-2}	Moderately permeable	1.2×10^{-3}
Main pile of stone	5.0×10^{-1}	Weakly permeable	1.0×10^{-4}
Downstream rock pile	6.0×10^{-1}	Relatively impermeable	3.3×10^{-5}
Anti-filtration material	1.0×10^{-2}	A sandy pebble gravel layer with drifted stones	2.5×10^{-2}
Concrete anti-seepage walls	1.0×10^{-7}	Gravelly medium to coarse sand	5.0×10^{-2}

4 Design Anti-Seepage Program Seepage Field Calculation Results and Analysis

The maximum depth of grouting curtain under the design anti-seepage program is about 61 m, the right bank curtain grouting hole is 200 m long, and the left bank is 430 m long. Calculation takes the water level in normal storage period, the upstream water level 2545.00 m, the downstream water level 2484.83 m.

4.1 Seepage Field

Figure 4 shows the contour map of groundwater level in normal storage level condition of the anti-seepage system using the design scheme, the seepage field trend of the dam site area during the operation period is basically in line with the trend of topographic relief in the area of model calculation, the groundwater level of the shoulder of the dam on both sides of the bank is higher, and it gradually decreases to the center of the riverbed, and the water in the reservoir is seeped downstream through the dam body, the dam foundation, and the anti-seepage curtain on both sides of the shoulder of the dam, as well as the mountain outside the range of the seepage control.

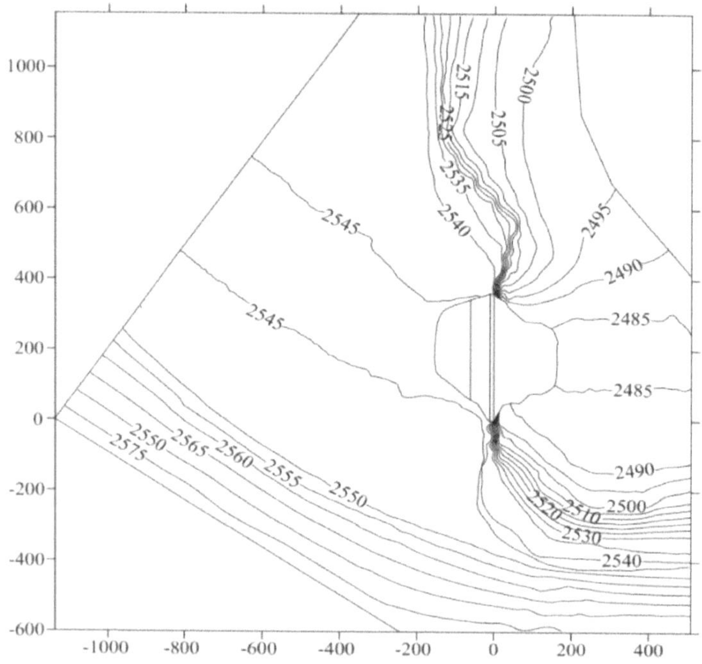

Fig. 4. Contour map of groundwater level during the operation period of hydropower station (m).

From the distribution of groundwater potential in the dam profile in Fig. 5, the infiltration surface forms a sudden drop downstream of the asphalt concrete core wall, and the anti-seepage system effectively abates the head, and the percentage of abated head is defined as follows [10]:

$$M = \frac{H_{\text{uphead}} - H_{\text{downhead}}}{H_{\text{up}} - H_{\text{down}}} \times 100\% \tag{4}$$

where: H_{uphead} is the head on the upstream face of the core wall; H_{downhead} is the head on the downstream face of the core wall; H_{up} is the upstream water level; H_{down} is the downstream water level.

In the case of normal water storage level, the anti-seepage system composed of core wall, anti-seepage wall and anti-seepage curtain cuts the head by 55.36 m, accounting for 92.01% of the total head, which shows that the anti-seepage system composed of asphalt concrete core wall, anti-seepage wall and anti-seepage curtain has a significant seepage control effect.

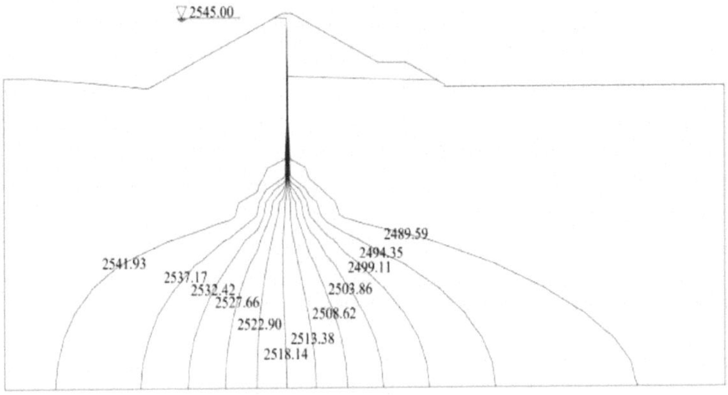

Fig. 5. Distribution of groundwater level potential in typical section of dam body (m).

4.2 Leakage and Infiltration Slope Drop

Take the seepage calculation section along the direction of the seepage control system arrangement, as shown in Fig. 6, divide the section into six calculation areas, respectively, for the left shoulder and right shoulder, the dam foundation and embankment, according to the infiltration coefficient and area of each part of the dam and the mountain body to calculate the infiltration flow in each part.

Table 2 shows the amount of seepage in each calculation area during the normal storage level operation period. As can be seen from Fig. 7, most of the reservoir leakage is generated by the left and right bank foundation and shoulder bypass seepage, and the infiltration flow of the embankment and riverbed foundation accounts for a smaller proportion, only 8.1% of the total leakage. The left shoulder leakage is the largest, accounting for 30.9% of the total leakage in the reservoir area, and the right shoulder leakage accounts for 25.6% of the total leakage. This is due to the asymmetry of the left

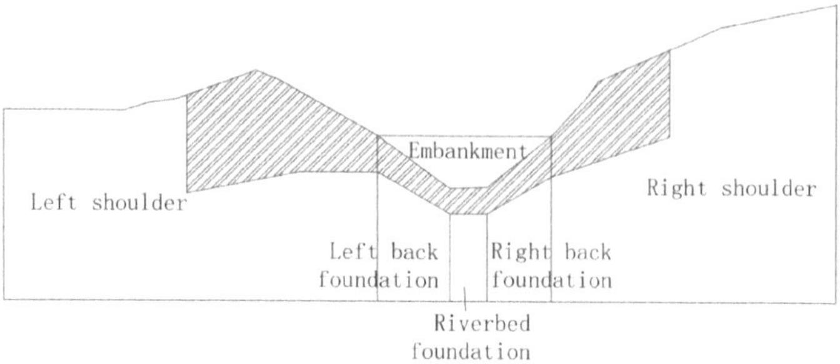

Fig. 6. Schematic diagram of calculated infiltration flow zoning.

and right bank topography, the seepage field of the dam site area is also asymmetrical, and the right shoulder has a higher overall water table and a smaller range of bypassing seepage, so the bypassing seepage of the left shoulder is more obvious compared to that of the right shoulder. From the calculation results, it can be seen that the seepage caused by the seepage around the two banks of the mountain is larger, so it is necessary to explore the extension of the two banks of the curtain grouting range can reduce the amount of reservoir seepage. Combined with the actual project, the study of the sensitivity of the different curtain depth schemes on the seepage control effect of the reservoir area, to reduce the difficulty of the construction and the cost of the project at the same time to achieve a better seepage control effect.

Table 2. Leakage in each calculation area during the normal storage level operation period (m^3/d).

Serial number	Part Name	Leakage
1	Left shoulder	5130.43
2	Left bank foundation	2699.14
3	Riverbed foundation	1331.42
4	Embankment	22.89
5	Right bank foundation	3149.28
6	Right shoulder	4250.2
7	Total	16583.36

Table 3 gives the calculated values of infiltration slope drop of the dam body, dam foundation and bedrock of the mountain in each major part under the basic design scheme, in which the infiltration slope drop of the dam body is larger, the infiltration slope drop of the asphalt concrete core wall of the dam body reaches 66.78, the infiltration slope drop of the dam foundation anti-seepage wall and the anti-seepage curtain are 55.36 and 11.07 respectively, and the anti-seepage system has a better seepage blocking effect, and the infiltration slope drop of the rest of the parts are smaller. The infiltration slope

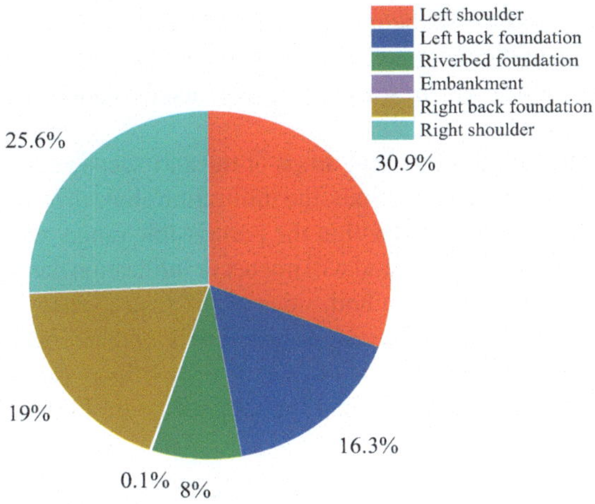

Fig. 7. Percentage of leakage in each calculation partition.

drop of all parts is within the range of allowable infiltration slope drop, which meets the requirement of infiltration stability.

Table 3. Infiltration slope drops in various parts of the reservoir area under the design scenario.

Calculation site	Slope infiltration	Allowable infiltration slope
Dam foundation anti-seepage curtain	11.07	15 ~ 20
Concrete anti-seepage wall	55.36	80
Left shoulder curtain	7.05	15 ~ 20
Right shoulder curtain	8.86	15 ~ 20
Core wall	66.78	100
Riverbed foundation	0.26	1
Left shoulder bedrock	0.35	1
Right shoulder bedrock	0.23	1

5 Comparison and Optimization of Different Seepage Control Schemes

The original design of the curtain grouting hole on both sides of the right bank is 200 m long, the left bank is 430 m long, on the basis of the design program, respectively, both sides of the dam shoulder curtain length extension of 50 m (Scheme 1), extension of 100 m (Scheme 2); design program seepage control curtain depth 5 Lu line below 5 m, on the basis of the dam base and both sides of the shoulder seepage control curtain deepened by 10 m as a whole (Scheme 3), deepened by 20 m (Scheme 4), shortened by 10 m (Scheme 5). Five schemes were calculated to compare the seepage characteristics of different seepage control schemes, according to which the dam seepage control system arrangement is optimized.

Figure 8 shows the infiltration slope drops of the anti-seepage curtains under different seepage control arrangement schemes, the infiltration slope drops of the anti-seepage curtains under the five schemes are within the permissible range, which meet the basic requirements of infiltration stability and will not occur infiltration damage. When extending the length of curtain grouting on both sides of the river (Scheme 1, Scheme 2), the permeability slope drop of the anti-seepage curtain will increase, because extending the length of curtain grouting on both sides of the river will increase the seepage diameter, the water retaining area increases, and the seepage control effect is also enhanced.

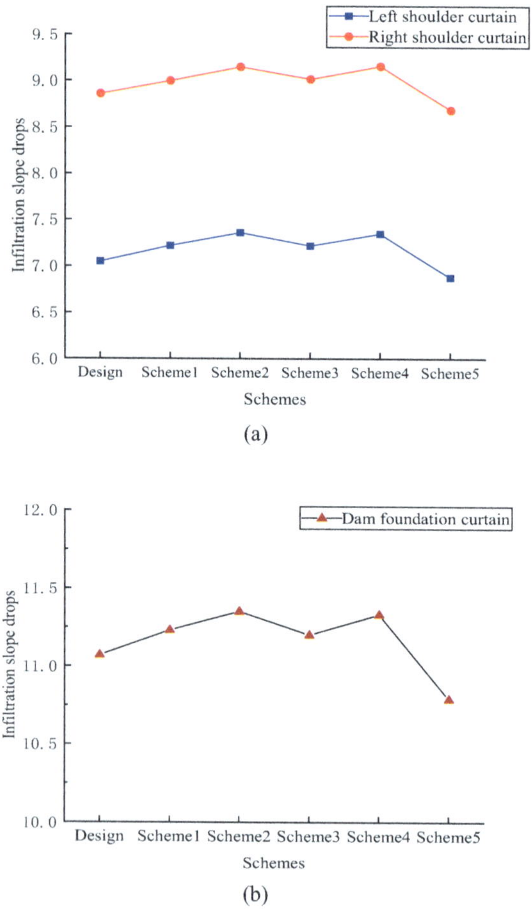

Fig. 8. Curtain infiltration slope drops of dam for different schemes. (a) is the changes in curtain infiltration slope drops on both sides of the dam shoulder. (b) is the changes in curtain infiltration slope drops at dam foundation.

When deepen curtain grouting depth (Scheme 3, Scheme 4), the permeability slope drop with curtain depth deepening and increase, riverbed dam base curtain and both sides of the shoulder curtain permeability slope drop change rule is consistent; on the contrary, shrinking.

Shortening the depth of curtain grouting (Scheme 5) decreases the curtain's ability to hold back water and the infiltration slope drop decreases as a result.

In order to more intuitively compare the seepage control effect of different seepage control arrangement schemes, quantify the seepage control curtain length, depth changes on the degree of influence on the seepage control effect of the hub area, were calculated under the five schemes under the infiltration flow rate and total infiltration flow rate of the various parts of the calculation results are shown in Table 4.

Figure 9 gives the change rate β of leakage and total leakage in various parts of the hub area under different seepage control schemes compared with the basic design scheme, taking $|\beta| = 5\%$ as the control standard, if the change rate of leakage is more than 5%, then it is considered that the change of seepage control scheme has a significant effect on the seepage of the reservoir area.

As seen in Fig. 9(a), when the curtain grouting holes on both banks are extended by 50 m (Scheme 1), the total seepage in the reservoir area is reduced by 1.42%, among which the change of seepage in the right shoulder part is more obvious, and the rate of change of this part β is -5.06%, and the seepage volume is reduced by 215.10 m^3/d. This is due to the fact that after the extension of the seepage curtains, the truncation of some of the moderately permeable layers of the rock in the mountain body results in the decrease of its infiltration flow, and the decrease of the right dam shoulder bypass seepage. The seepage around the right dam shoulder is reduced, so extending the length of curtain grouting on the right bank can reduce the leakage of the right dam shoulder more significantly. Extending the length of the curtain grouting hole has a small effect on the leakage of the left shoulder and the dam foundation, and the change rate of the leakage does not exceed 5%.

On the basis of Scheme 1, the length of curtain grouting holes on both sides of the river continues to be extended by 50 m (Scheme 2), then the total leakage rate of the reservoir area changes from -1.42% to -2.48%, of which the leakage rate at the right shoulder changes from -5.06% to -7.29%, compared with the leakage reduced by 2.23%, and the remaining parts of the rate of change is not more than 5%, so it can be seen that the continuation of extension of curtain grouting lengths on both sides of the river has significantly improved the seepage control effect on the hills and riverbed dam foundation. It can be seen that continuing to extend the length of curtain grouting on both sides of the mountain and the riverbed dam foundation seepage control effect enhancement is no longer obvious.

From Fig. 9(b), the overall deepening of the anti-seepage curtain 10 m (Scheme 3) only reduce the total leakage of 0.19%, the anti-seepage curtain deepened 20 m (Scheme 4) only reduce the total leakage of 0.41%, which shows that the deepening of the depth of the anti-seepage curtain is not a significant enhancement of the seepage control effect of the reservoir area, which is mainly due to the original design of the anti-seepage curtain at the bottom of the depth of the weakly permeable layer ($q \leq 5Lu$) below the 5 m, the bedrock permeability coefficient has been similar to the permeability coefficient of the curtain. The permeability coefficient of the bedrock here is already similar to that of the anti-seepage curtain, and the anti-seepage curtain has formed a relatively closed seepage control system with the foundation bedrock, so continuing to deepen the depth of the anti-seepage curtain can not significantly improve the seepage control effect [11]. If the depth of anti-seepage curtain is shortened by 10 m (Scheme 5), the total seepage in the reservoir area increases by 4.86%, and the seepage of the dam base on the left bank and the dam base on the riverbed increases by 5.44% and 8.63% respectively, which is more than the control rate of change of 5%, which is mainly due to the larger permeability coefficient of the rock layer above the line of 5 Lu, and the closure of the anti-seepage system will be destroyed if the depth of anti-seepage curtain is shortened and the centralized leakage will be easy to occur.

Table 4. Infiltration flow by component under different impermeability schemes (m^3/d).

Schemes	Embankment	Left bank foundation	Right bank foundation	Left shoulder	Right shoulder	Riverbed foundation	Total
Design	22.89	2699.14	3149.28	5130.43	4250.20	1331.42	16583.36
Scheme 1	22.87	2676.67	3159.65	5122.66	4035.10	1330.56	16347.51
Scheme 2	22.93	2696.54	3167.42	5011.20	3940.32	1333.15	16171.56
Scheme 3	22.89	2698.25	3142.26	5119.20	4239.37	1329.88	16551.85
Scheme 4	22.91	2697.26	3139.50	5111.02	4222.45	1321.56	16514.70
Scheme 5	22.80	2846.01	3303.08	5347.07	4423.91	1446.34	17389.21

In summary, considering the influence of reservoir area on the infiltration stability and seepage amount in different seepage control system optimization schemes, the following suggestions are made for the optimization of seepage control system of this hydropower station: considering the reduction of the construction difficulty and cost of the project, shorten the depth of curtain at the right bank dam base, left bank dam shoulder and right bank dam shoulder by 10 m, and maintain the original design of seepage control curtain depth at the riverbed dam base and the left bank dam base; extend the length of the seepage control curtain at the left bank dam shoulder for a smaller improvement of the seepage control effect. The seepage control curtain length on the left shoulder has little improvement on the seepage control effect. The length of the anti-seepage curtain on the left shoulder is maintained at the original design length, and the extension of the length of the curtain grouting hole on the right bank by 50 m improves the effect of anti-seepage on the right bank, so the extension of the anti-seepage curtain on the right bank by 50 m is considered.

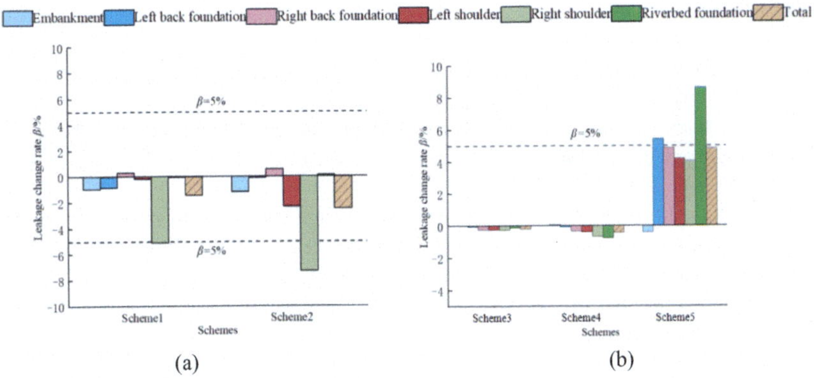

Fig. 9. Leakage change rate of various parts in different seepage control schemes. (a) represents the leakage change rates by varying the length of both sides curtain. (b) is the leakage change rates by varying the depth of curtain

6 Conclusion

This paper establishes a three-dimensional seepage finite element calculation model of a hydropower station hub project area on a deep overburden foundation, calculates the seepage field, infiltration slope drop and seepage leakage in the reservoir area under the design of seepage control program. Based on these, the optimization of anti-seepage system has been carried out, which calculates and compares the changes of seepage characteristics in the reservoir area under the different lengths and depths of the seepage control curtains. Conclusions are as follows:

(1) Under the design of seepage control program, all parts of the reservoir area meet the requirements of permeability stability, and the seepage control system cuts the water head by a total of 55.36 m under the normal water storage level, accounting for 92.01% of the total water head, and the seepage control effect is remarkable. The total seepage volume of the reservoir area is 16583.36 m^3/d, of which the proportion of seepage volume of the dam body and riverbed dam base is very small, while the geological conditions of the mountains on both sides of the river are poorer, and there are more fissures between the mountains, and the seepage volume is larger.
(2) The optimization of seepage control system shows that, in order to reduce the seepage around the right bank mountain, the curtain grouting hole on the right bank needs to be extended outward by 50 m, while the curtain depth can be shortened by 10m at the base of the right bank dam, the shoulder of the left bank dam and the shoulder of the right bank dam, so that the seepage amount of the reservoir area can be effectively controlled while reducing the construction difficulty and the cost of the project.
(3) For similar asphalt concrete core walls with combined anti-seepage walls and anti-seepage curtains, anti-seepage systems are advisable to consider the double optimization of the depth and length of the anti-seepage curtains at the same time.

References

1. Zhou, J., Du, X., Zhou, X.: Research on hydropower development strategy for new power system. J. Hydropower Gener. **41**(07), 106–115 (2022)
2. Wang, Z., Mao, H., Shin, J., et al.: Study on the influence of localized strongly permeable layer on seepage in deep and thick overburden. J. Water Resour. Water Eng. **30**(04), 159–166 (2019)
3. Liu, J., Zhao, J., Shen, Z., et al.: Analysis of seepage characteristics and seepage control measures in gravel dams over deep cover. Hydropower Energy Sci. **29**(08), 62–65 (2011)
4. Waddell, R.W., Bou-Sleiman, C.: Understanding modern dam foundation grouting and cut-off wall fundamentals. In: Association of State Dam Safety Officials Annual Conference Proceedings (2021)
5. Wang, S., Shen, Z., Ji, Y., et al.: Optimization of seepage control system of a diversion power station over a deep cover. J. Water Resour. Water Eng. **33**(01), 144–151+158 (2022)
6. Arifin, M.Z., Purwana, Y.M., Dananjaya, R.H.: Seepage and piping control of earth fill dam: (case study of Pidekso dam Indonesia). Lecture Notes Civ. Eng. **225**, 311–322 (2023)
7. He, Y.. Zhao, G.: Influence of seepage control curtain depth on seepage flow in concrete dam foundation. China Rural Water Conservancy Hydropower (09), 119–121+129 (2013)
8. Lunxin, Z., Qiao, Y., Shengli, Z., et al.: Optimization of seepage control curtain scheme for deep-thick gravel dam foundation. J. Three Gorges Univ. (Nat. Sci. Ed.) **37**(03), 44–47 (2015)
9. Ren, J., Shen, Z., Ji Yang, CYu.: Back analysis of the 3D seepage problem and its engineering applications. Environ. Earth Sci. **75**(2), 113 (2016). https://doi.org/10.1007/s12665-015-4837-1
10. Zhang, W., Shen, Z., Chen, G., et al.: Analysis of three-dimensional seepage field and evaluation of seepage control effect in reservoir area under the geological condition of karst development. J. Water Res. Water Eng. **31**(05), 218–224+234 (2020)
11. Xu, Z., Cao, C., Li, K., et al.: Analysis of local seepage control for seepage control in the upper reservoir of a pumped storage power station. J. Appl. Mech. (2018)

Numerical Study on Deformation Characteristics and Influencing Factors of Super-High Filled Subgrade

Guijun Wan[1], Jiangxin Liu[2(✉)], Chunduo Liu[1], Jing Yang[1], Xudong Wang[1], and Peng Li[2]

[1] China Overseas Construction Co., Ltd., Shenzhen 518000, China
[2] Ministry of Transport, Research Institute of Highway, Beijing 100088, China
liu.jx@rioh.cn

Abstract. Super-high filled subgrades are a critical component in highway construction, especially in mountainous or uneven terrains where establishing a stable and even road surface presents significant engineering challenges. The stability and deformation behavior of such structures under the influence of various factors are of paramount importance for ensuring the longevity and safety of the transportation network. Based on the numerical simulations, the effects of filling height, slope grade, and the wide platform on the deformation characteristics of a super-high filled highway subgrade are discussed. The results showed that with the increase of the filling height, both the settlement and the horizontal displacement of the subgrade gradually increased. Due to the natural ground is inclined, the subgrade settlement and horizontal displacement exhibited asymmetric. After the subgrade filling was completed, the overall settlement tended to be smaller on the left and larger on the right, with the maximum settlement occurring at the thicker filling area; while the horizontal displacement decreased from the top to the bottom of the embankment, and the maximum horizontal displacement occurred within a certain distance to the left of the centerline of the road surface. The position of the maximum horizontal displacement at the top of each embankment level developed gradually from left to right. The reduction of slope grade and the setting of a wide platform at the base of subgrade both contribute to reducing the settlement and horizontal displacement, and the influence of slope grade on the maximum horizontal displacement of the subgrade is greater than its effect on the maximum settlement.

Keywords: Super-high fill subgrade · Settlement · Horizontal displacement · Slope grade · Wide platform

1 Introduction

The construction and operational stability of super-high fill subgrades are pivotal concerns in highway engineering, posing significant challenges in terms of safety, sustainability, and functionality [1, 2]. The issue of subgrade settlement notably affects traffic

B. Yuan et al. (Eds.): ICCASE 2024, SUCI, pp. 387–399, 2024.
https://doi.org/10.1007/978-3-031-76102-7_42

safety and comfort, making the understanding and mitigation of deformation a crucial element in the engineering of high fill subgrades [3, 4]. Despite the abundant research on settlement and deformation of high embankments, existing studies primarily focus on general and high embankment scenarios, leaving a noticeable gap in the specific challenges and solutions pertinent to super-high fill subgrades that often exceed the hundred-meter mark [5].

Pioneering work by B. Indraratna et al. examined the settlement and lateral deformation over soft foundations with the finite element method, exploring ground capacity enhancement measures [6]. Similarly, research by Liu et al. and Li et al. compared settlement characteristics between high and general embankments and investigated the effects of dynamic filling construction processes, respectively [7, 8]. Further, Chen et al. and Yao et al. delved into deformation mechanisms on soft foundations and the post-construction settlement attributes of rock-soil mixed fill high embankments, providing quantitative analyses and modeling approaches [9, 10]. Despite these contributions, most have not distinctly tackled the nuanced behaviors and remediation strategies specific to super-high fill subgrades. Moreover, Zhao et al.'s predictive modeling for post-construction settlement has opened avenues for forward-looking studies yet remains to be specifically adapted to super-high fill contexts [11].

Therefore, a detailed understanding of the deformation characteristics, particularly the nuanced influences of factors like filling height, slope grade, and platform width on super-high fill subgrades, remains underexplored. Recognizing this deficiency, the present study aims to explicitly analyze the deformation characteristics and influencing factors specific to super-high fill subgrades. Drawing on the foundation laid by previous research, this research employs FLAC3D numerical software within the ambit of a practical engineering project, providing a targeted examination of these structures. This investigation seeks not only to enrich the corpus of knowledge on super-high fill subgrades but also to propose practical guidelines for their design and construction, directly contributing to safer and more robust highway infrastructures.

2 Methodology and Modelling

2.1 Methodology

Detailed numerical simulations in construction process with FLAC 3D software, with robust performance in analyzing the mechanical behavior of geotechnical structures, to replicate real-world conditions as closely as possible are performed, focusing on the deformation of the embankment in a mountainous terrain. It is of great necessity to accurately predict the geotechnical behavior of the subgrade under the dynamic conditions of construction and usage, incorporating the material parameters and geometrical specifics elucidated. To simulate the complexities of the embankment construction process, the model incorporates both the initial geostress field and the subsequent filling stages, using progressively activated embankment levels and calculating displacements and velocities to reflect the sequential construction stages and analyze the potential impacts on the subgrade stability.

2.2 Model Creation

The super-high fill subgrade located in mountainous area is constructed along a valley with sloping terrain. The natural ground has a lateral slope ranging approximately from 1:2.6 to 1:1.3. The road width is 24.5 m, and there is a filled area with width of 104.5 m on the left of the road. The subgrade and the filling area were filled simultaneously. The maximum height at the center of the subgrade is 63.6 m, with the right side's maximum slope height being 114.4 m. There are 13 levels of embankments from the top down, with each of the first 12 levels having a height of 8 m and the 13th level having a height of 10 m. A retaining wall is installed at the base of the slope to consolidate the slope; the wall's height is 10 m with a foundation depth of 1.6 m. There is original soft soil distributed locally at the foundation, which is replaced with rock fragments. The first level has a slope of 1:1.5, the second to fifth levels have slopes of 1:1.75, the sixth to eleventh levels have slopes of 1:2, the twelfth level has a slope of 1:1.5, and the thirteenth level has a slope of 1:1.75. Terraces are set between each level; there is an 8-m-wide terrace at the bottom of the seventh level slope, a 52.7-m-wide terrace at the bottom of the eleventh level slope, and other levels have terraces with a width of 2 m at the bottom of the slopes. The fillers comes from nearby tunnel muck, mainly composed of moderately weathered interbedded sandstone and slate, with a tested single axial saturated compressive strength ranging from 29.5 MPa to 39.8 MPa, which is considered as relatively hard rock.

Based on the actual dimensions, a numerical simulation model is constructed in Fig. 1.

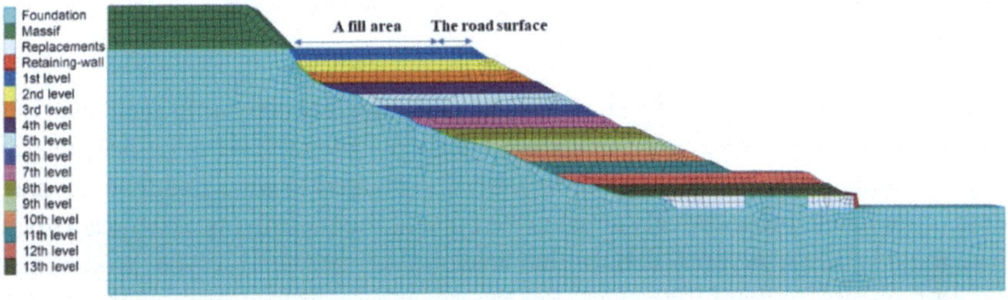

Fig. 1. Schematic of numerical model

2.3 Model Creation

The model material parameters are listed in Table 1. K is the bulk modulus and G is the shear modulus. It is therefore necessary to convert the elastic modulus E and Poisson's ratio μ into the bulk and shear moduli using formulas below:

$$K = \frac{E}{3(1-2\mu)} \quad G = \frac{E}{2(1+\mu)}$$

Table 1. Physical and mechanical parameters of materials

Category	Density (kg/m^3)	Elastic Modulus (GPa)	Poisson's Ratio	Tensile Strength (MPa)	Cohesion (MPa)	Friction Angle (°)
Massif	2200	15	0.2	2	2	40
Foundation	2200	15	0.2	2	2	40
Replacements	2200	15	0.2	2	2	40
Retaining-wall	2400	15	0.2	2.5	2.5	45
Embankment	2150	5	0.25	0.5	0.5	30

2.4 Boundary Condition and Initial Stress

During calculation, the bottom of the model is a fixed, the four sides of the model are roller-supported, and the surface of the model is freely deformed.

Before the filling operation, the embankment grid is assigned as nulls, while the grids for the mountain, foundation, soft soil replacement at the base, and retaining walls are assigned the corresponding materials. First, the modulus and cohesion parameters for the mountain, foundation, soft soil replacement at the base, and retaining walls are assigned infinite values for solving to ensure that the elements do not yield under the action of gravity. Then their parameters are set to the real values for calculation to further obtain the initial geostress field.

2.5 Filling Process Simulation

The filling process is dynamically simulated by progressively activating each level of the embankment using the model null command. The displacements and velocities generated during the initial stress calculation process are first cleared to zero, and then the staged loading is applied to progressively activate the embankment elements. After each level of embankment is filled, the grid cells for that level are activated and solved, producing the calculation results for that filling stage. For the super-high embankment, there are a total of 13 levels, thus, 13 activations and solution calculations are required. By dynamically simulating the filling stage by stage, it is possible to better analyze the patterns of how the settlement and horizontal displacement of the embankment evolve during the filling process.

3 Results and Analysis of Subgrade Deformation

3.1 Subgrade Settlement

During the progressive embankment construction from the 13th level to the 1st level, the maximum subgrade settlement with the filling process is shown in Fig. 2. It can be observed that the maximum settlement gradually increases with the height of the embankment. Upon completion of the filling, the maximum settlement is 313.5 mm, and with the central fill height of the subgrade being 63.6 m, the settlement ratio is approximately 313.5 mm/63.6 m = 4.9‰.

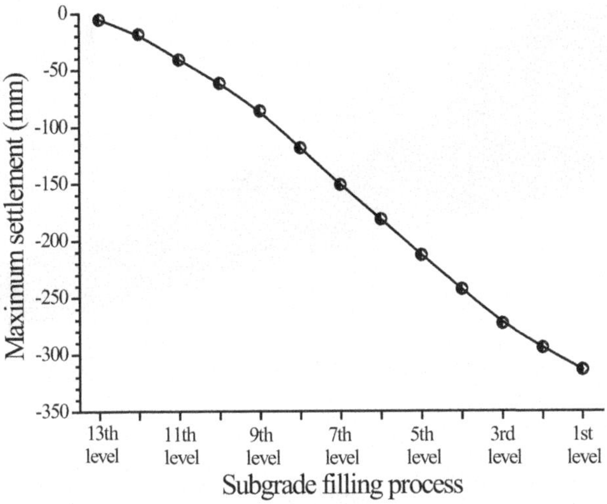

Fig. 2. Embankment maximum settlement in filling process

In general, when an embankment is constructed on flat ground, the settlement typically shows a pattern where the center of the subgrade settles more than the sides, with a symmetrical type along the centerline of the embankment. However, the pattern of subgrade settlement varies due to the ground slope. The settlement nephogram after the completion of filling is shown in Fig. 3, and the settlement curves of the embankment from the 1st to the 13th level are drawn in Fig. 4.

According to Figs. 3 and 4, before filling, the foundation was bedrock, resulting in minimal foundation settlement. Therefore, the subgrade settlement is primarily due to the embankment's own settlement. Due to the natural ground slope being higher on the left and lower on the right, the overall settlement is smaller on the left and larger on the right, showing asymmetric on both sides outside the centerline of the road surface. The maximum settlement is 313.5 mm, occurring in areas with thicker filling, specifically from the right side slope of the 1st to 5th level embankment. Vertically, from the top to the bottom of the embankment, the settlement gradually decreases to almost zero as it nears the bedrock foundation. The settlement at the right shoulder of the road is 286.2 mm, and on the left side, it is 233.7 mm, resulting in an uneven settlement rate of (286.2–233.7) mm/24.5 m = 2.1‰.

3.2 Lateral Displacement of Subgrade

The maximum lateral displacement of the subgrade in filling process is indicated in Fig. 5. As the height of the embankment increases, the maximum lateral displacement progressively increases. After the filling, the maximum horizontal displacement is 129.0mm, which gives a ratio of maximum lateral displacement to maximum settlement of 0.41.

When constructing an embankment on flat ground, the horizontal displacement typically shows the pattern of minimal horizontal displacement at the center of the embankment, symmetry greater displacement at the toe of either side. Similarly, due to the

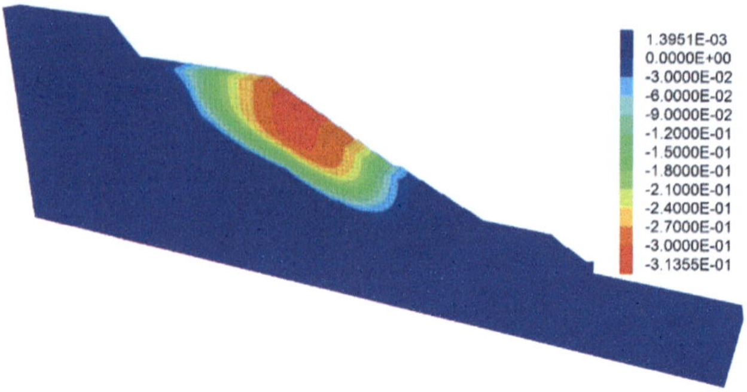

Fig. 3. Settlement nephogram when subgrade filling is completed (Unit: m)

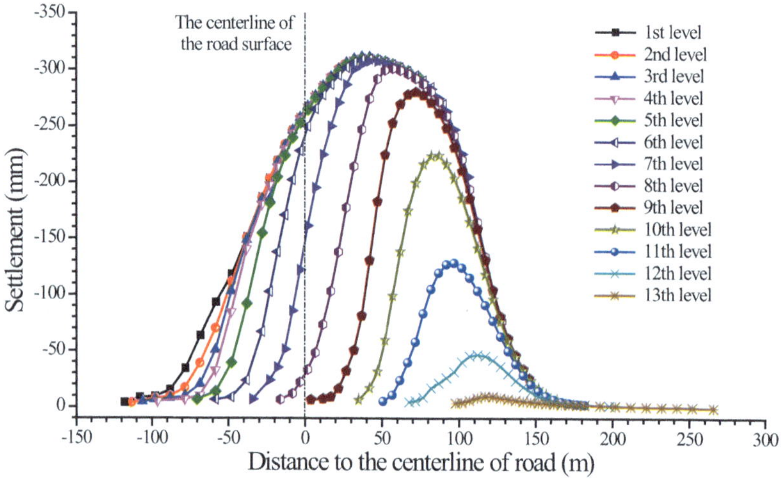

Fig. 4. Settlement curve of embankment top when each level filling is completed

original ground being a slope, the pattern of the base's horizontal displacement is different as shown in Fig. 6. The horizontal displacement curves from level 1 to 13 after filling completion are drawn in Fig. 7.

Figures 6 and 7 indicate that the foundation's horizontal displacement is minimal; the subgrade's horizontal displacement is mainly caused by the embankment's own displacement. On either side of the centerline, the horizontal displacement also exhibits asymmetric. The maximum horizontal displacement 129.0mm is primarily located at the top of level 1, with the maximum horizontal displacement area within a 50m range to the left of the centerline. From level 1 to level 13, horizontal displacement progressively decreases; there is a slight increase in horizontal displacement at the slope of levels 7 to 9, which then gradually decreases to a minimum 11.6mm at level 13 embankment. From level 1 to level 13, the position of the maximum horizontal displacement at the top of each level progressively shifts from left to right. From the top of level 1 embankment downward to the natural ground, horizontal displacement gradually decreases to around 10-20mm. The horizontal displacement at the vertical downward path of level 11 embankment is stable, around 20mm. This indicates that the 52.7m wide platform established at the base of level 11 embankment effectively acted as a counter-pressure measure for stabilizing the slope for levels 12 and 13 embankments. Near 5m outside the retaining wall, the foundation's horizontal displacement value tends to be zero.

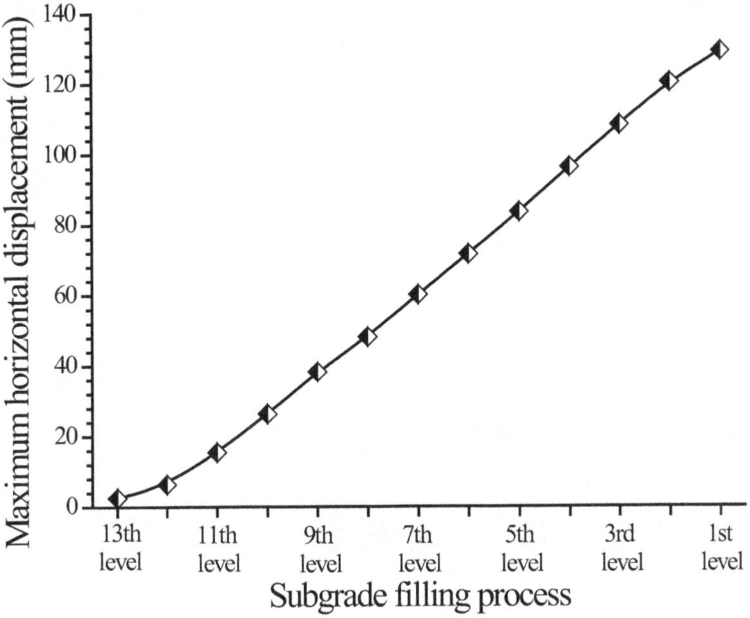

Fig. 5. Embankment maximum horizontal displacement in filling process

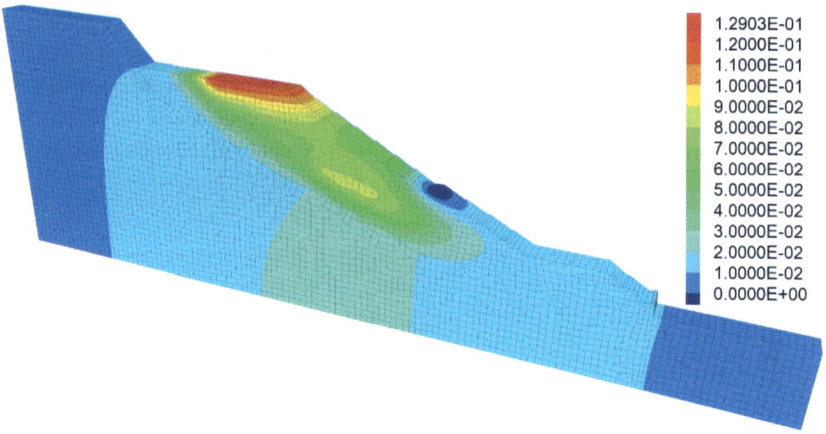

Fig. 6. Horizontal displacement nephogram when subgrade filling is completed (Unit: m)

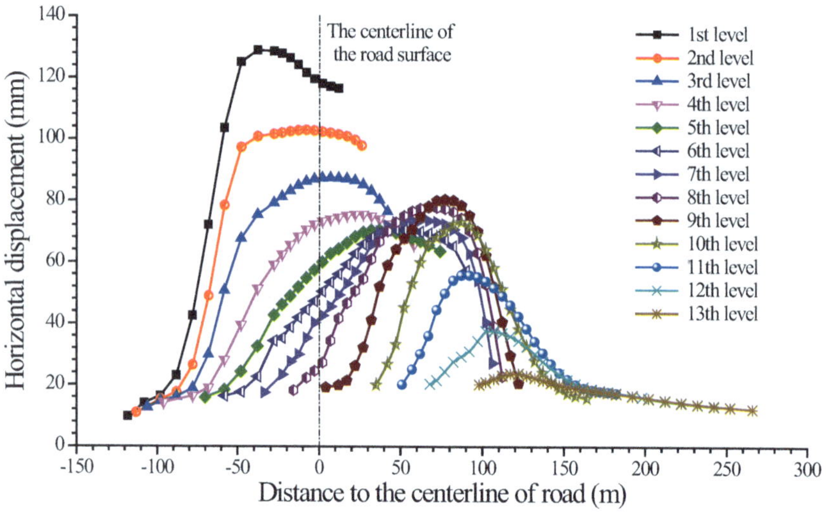

Fig. 7. Horizontal displacement curve of embankment top when each level filling is completed

3.3 Horizontal Displacement of Ground Below Slope Toe

A curve depicting the horizontal displacement of the ground below the slope toe with depth during the filling process is illustrated in Fig. 8. It is observed that the horizontal displacement gradually increases as the embankment is progressively constructed. The horizontal displacement decreases and stabilizes as it goes deeper. Within a depth range of 0–10 m, due to the lateral support provided by the retaining wall, the horizontal displacement sharply decreases with depth. In the depth range of 10–20 m, where there is no support from a retaining wall, the rate of decrease in horizontal displacement slows down. Below a depth of 20 m, the horizontal displacement gradually becomes stable.

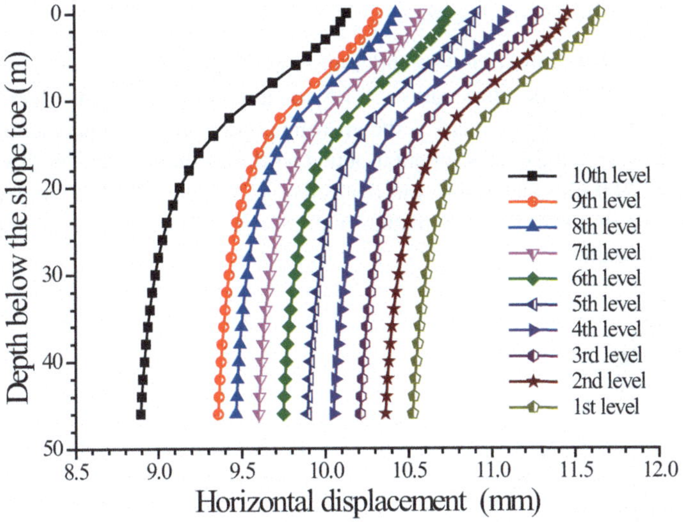

Fig. 8. Ground horizontal displacement along depth below the slope toe

4 Influencing Factors and Discussions

4.1 Slope Grade

The actual conditions are as follows: the slope grade of the first-level is 1:1.5, the second to fifth-level is 1:1.75, the sixth to eleventh-level is 1:2, the twelfth-level is 1:1.5, and the thirteenth-level is 1:1.75. Subsequently, all slope grades of the 13 levels of embankment are set to three different conditions: 1:1.5, 1:1.75, and 1:2, and corresponding models are established with all other conditions remain unchanged, then a comparative analysis is conducted between different slope grades and the actual condition.

(1) Maximum settlement

The maximum settlements of different slopes are shown in Fig. 9. The gentler the slope of the subgrade, the smaller the maximum settlement. When the slope ratio is 1:1.5, the maximum settlement is 328.6mm; when the slope ratio is 1:2, the maximum settlement is 302.7mm. However, in actual working condition and with the slope ratio of 1:1.75, the difference is very small, at 313.5mm and 317.0mm respectively. It is evident that changes in slope ratio have a certain impact on the maximum settlement of the subgrade.

(2) Maximum horizontal displacement

Similarly, Fig. 10 shows that the gentler the slope, the smaller the maximum horizontal displacement. When the slope ratio is 1:1.5, the maximum horizontal displacement is 142.5 mm; when the slope ratio is 1:2, the maximum horizontal displacement 121.0 mm. The maximum horizontal displacement in actual working condition and with a slope ratio of 1:1.75 are 129.0 mm and 134.4 mm respectively. It is clear that the impact of slope grade on the maximum horizontal displacement of the subgrade is greater than on the maximum settlement.

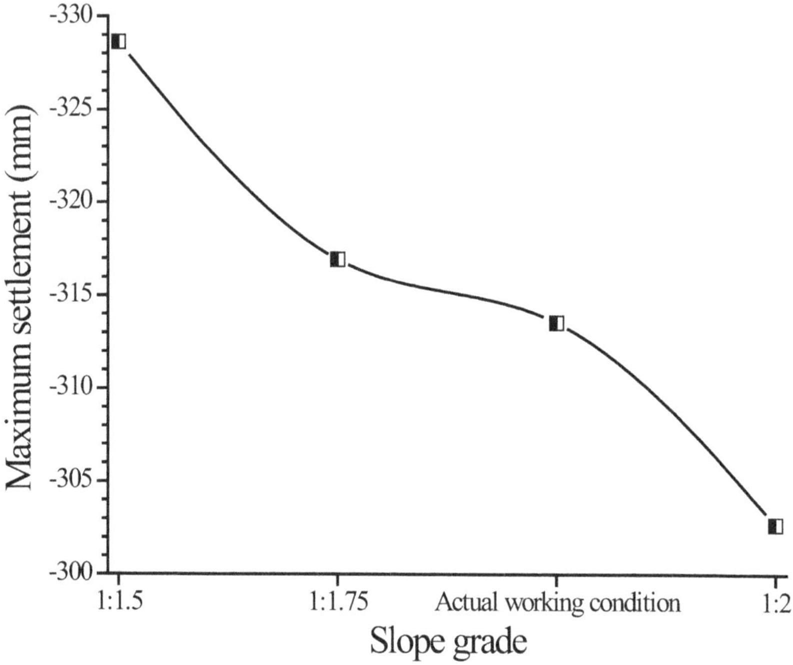

Fig. 9. Maximum settlement of subgrade under different slope grades

4.2 Wide Platform

In actual condition, a wide platform with a width of 52.7 m is set at the base of the 11th level embankment slope. This has now been changed to a regular platform with a width of 2 m with all other conditions remain unchanged.

(1) Maximum displacement

When there is a wide platform in the actual condition, the maximum settlement of the subgrade after filling is 313.5 mm, and the maximum horizontal displacement is 129.0 mm. Without the wide platform, they increase to 343.6 mm and 141.3 mm.

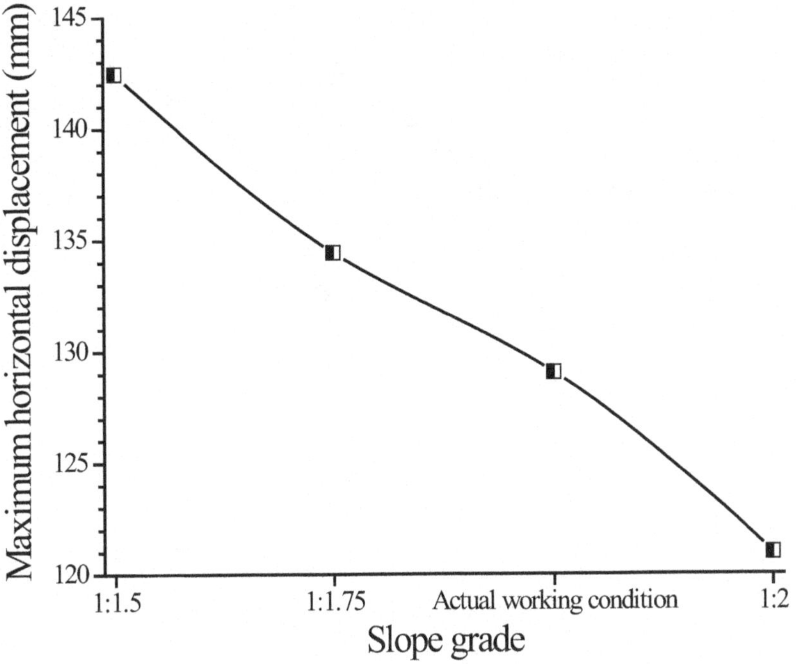

Fig. 10. Maximum horizontal displacement of subgrade under different slope grades

Compared with the wide platform, there is an increase in both the maximum settlement by 30.1 mm and the maximum horizontal displacement by 12.3 mm.

(2) Horizontal displacement of ground

The horizontal displacement of ground below the slope toe with depth for both with and without a wide platform are illustrated in Fig. 11. It can be observed that both situations exhibit a similar pattern where the value of horizontal displacement gradually decreases and tends to stabilize as the depth increases. With a wide platform in actual conditions, the horizontal displacement at the toe is 11.6 mm, and it stabilizes to 10.5 mm with increasing depth. Without the wide platform, the horizontal displacement at the toe is 12.4 mm, and it stabilizes to 11.3 mm with increasing depth, which indicating a relatively minor change.

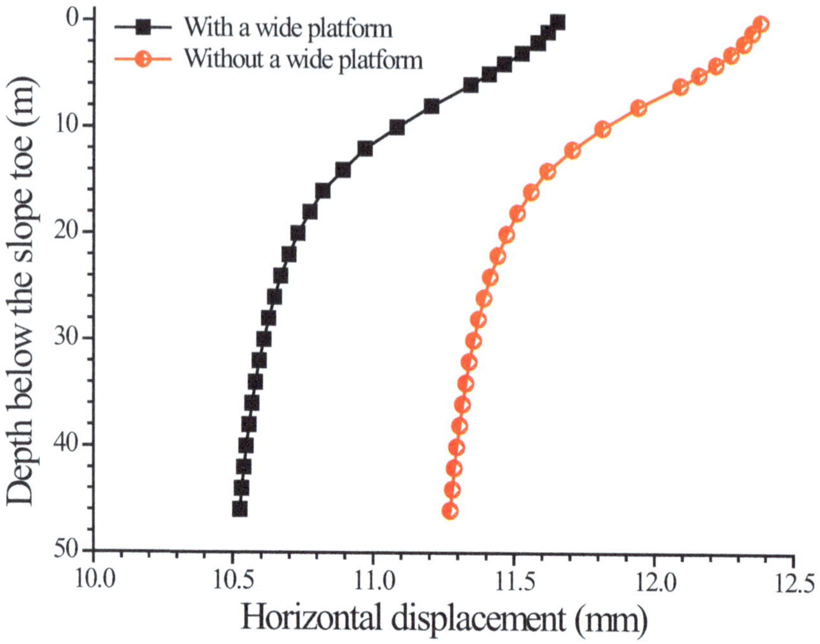

Fig. 11. Horizontal displacement of ground with or without a wide platform

5 Conclusions

This study on super-high subgrades has yielded pivotal insights into their deformation characteristics and influencing factors, summarized as follows:

1. An increase in fill height is directly correlated with an increase in both vertical and horizontal displacements, exhibiting asymmetry due to natural ground unevenness. Notably, the maximum settlement observed was 313.5 mm (a 4.9‰ rate), and the peak horizontal displacement of 129.0 mm was primarily within 50 m to the left of the road centerline, indicating a lateral progression of peak displacement across embankment levels.
2. Modifying the slope grade effectively reduces both vertical and horizontal deformations, with a more pronounced effect on horizontal displacement. This underscores slope optimization's relevance in subgrade design for deformation mitigation.
3. The presence of a toe wall significantly reduces horizontal displacement by imposing a sharp decrease with depth, and a wide platform at the embankment base effectively diminishes both vertical and horizontal displacements, indicating their preventive roles against excessive subgrade deformation.

These findings provide critical guidance for the design and construction of super-high subgrades, emphasizing the need to consider fill height, structural reinforcements, and slope optimization. However, the study's limitations, including the reliance on specific assumptions and conditions, suggest the necessity for further research to broaden these conclusions' applicability to diverse engineering contexts.

References

1. Zeng, A.: Discussion on construction technology and disease prevention measures of highway high fill subgrade of expressway. Chin. Overseas Architect. **4**, 194–196 (2020). https://doi.org/10.19940/j.cnki.1008-0422.2020.04.062
2. Xu, Y., Zhuang, Q., Xu, P., et al.: Study on settlement characteristics and influencing factors of high fill subgrade in loess gully area. Technol. Highw. Transport **39**(5), 1–9 (2023). https://doi.org/10.13607/j.cnki.gljt.2023.05.001
3. Nakamura, H., Yokota, S., Fujioka, K., Yokohama, S., Miura, S.: Evaluation of settlement on the construction of the high embankment using ashitaka loam. Doboku Gakkai Ronbunshuu C **70**(1), 125–134 (2014)
4. Qu, T., Xu, Y., Zhang, J.: Research on numerical simulation of influencing factor of high fill subgrade deformation. Northern Commun. **4**, 73–78 (2016). https://doi.org/10.15996/j.cnki.bfjt.2016.04.019
5. Zhu, Y., Yang, X.: Analysis of Foundation Deformation and Slope Stability of High fill engineering. China Architecture Publishing & Media Co., Ltd., Beijing (2020)
6. Indraratna, B., And, A.S.B., Sivaneswaran, N.: Analysis of settlement and lateral deformation of soft clay foundation beneath two full-scale embankments. Int. J. Numer. Anal. Meth. Geomech. **21**, 599–618 (1997)
7. Liu, M., Jiang, X., Chen, X., et al. : Analysis of settlement deformation characteristics of layered buried high embankment. J. Transport. Eng. Inform. **18**(1), 61–67, 76 (2020)
8. Li, X., Li, X.: Study on settlement deformation characteristics of high fill subgrade along a highway. Subgrade Eng. **6**, 202–207 (2023). https://doi.org/10.13379/j.issn.1003-8825.202303003
9. Chen, T., Yang, W.: Deformation and stability analysis on high embankments of soft ground in mountain slopes. J. Municipal Technol. **37**(3), 45–48 (2019)
10. Yao, X., Que, Y., Sun, T., et al.: Post-construction settlement characteristics of high embankment filled with soil-rock mixtures in v-shaped valley. J. Nanchang Univ. (Eng. Technol.) **44**(3), 265–272 (2022). https://doi.org/10.13764/j.cnki.ncdg.2022.03.007
11. Zhao, Z., Zhu, Y., Ye, S.: Analysis on settlement deformation law and influencing factors of loess high fill foundation. J. Architect. Civ. Eng. **40**(3), 161–169 (2023). https://doi.org/10.19815/j.jace.2022.01037

Analysis and Discussion on Investigation and Design of Karst Bridge Pile Foundation in Shijing Area of Guangzhou

Jinlong Yin[1,2,3], Yang Xiang[4], Zhu Liang[1,2,3](✉), Weiping Peng[1,2,3], and Yanping Jin[1,2,3]

[1] Guangzhou Urban Planning and Design Survey Research Institute Co., Ltd., Guangzhou 510060, China
499324556@qq.com
[2] Guangzhou Collaborative Innovation Center of Natural Resources Planning and Marine Technology, Guangzhou 510060, China
[3] Guangdong Enterprise Key Laboratory for Urban Sensing, Monitoring and Early Warning, Guangzhou 510060, China
[4] School of Hydraulic and Electric Power, Heilongjiang University, Harbin 150080, China

Abstract. Guangzhou Shijing area is a karst development area. In recent years, the Guangzhou Baiyun Station integrated transportation hub has carried out many municipal bridge projects in the Guangzhou Shijing area. Karst collapses and ground subsidence accidents caused by many engineering construction causes have resulted in different degrees of economic losses. The main reasons are the lack of research on the characteristics of karst development in the Guangzhou Shijing area and the lack of pertinence for the investigation and design of the karst bridge pile foundation. The law of karst development in the Shijing area of Guangzhou is studied through many on-site geological surveys. Combining with the actual geological situation, this paper analyzes and discusses the key points of the karst bridge pile foundation investigation and design. And summarizes the investigation and design process, aiming to provide a reference for the future investigation and design of the karst bridge pile foundation in the Guangzhou Shijing area, to reduce the occurrence of safety accidents.

Keywords: Shijing area of Guangzhou · Karst · Bridge pile foundation · Survey and design

1 Introduction

With the rapid development of urbanization in our country, the problem of karst collapse in karst areas is becoming more and more serious. Among them, the problem of karst ground collapse induced by pile foundation construction has been ignored. The design of bridge pile foundations in the karst area generally adopts an end-bearing pile, which has higher requirements for the bearing layer of the pile end. If there are karst caves under the bearing layer, the stability of the karst cave roof is the key to the pile

B. Yuan et al. (Eds.): ICCASE 2024, SUCI, pp. 400–407, 2024.
https://doi.org/10.1007/978-3-031-76102-7_43

foundation design. The stability of the karst cave roof mainly focuses on the study of roof failure mechanism and qualitative, semi-quantitative, and quantitative analysis [1, 2]. The engineering Geology manual simplifies the karst cave roof into beam, plate, arch, and other models, and attached simple estimation formulas for each simplified model, which can be used for approximate analysis of the stability of the karst cave roof [3–5]. However, according to the requirements of the existing design code, the bridge pile foundation has a higher requirement on the thickness of the karst cave roof. In the past, most of the research was based on empirical formulas or theories for karst bridge pile foundation design, and the characteristics of karst development were ignored. How to find out the actual geological conditions and carry out targeted design according to the actual geological conditions is the key to the investigation and design of bridge pile foundations.

In this paper, the Shijing area of Guangzhou is taken as the research area. The karst development law of the Shijing area of Guangzhou is studied through many on-site geological surveys. Combined with the actual geological conditions, this paper analyzes and discusses the key points of the karst bridge pile foundation investigation and design, and summarizes the investigation and design process.

2 General Situation of the Study Area

2.1 Geological Environmental Conditions

Guangzhou Shijing area is located in the Baiyun District of Guangzhou. It is a plain area with relatively flat terrain and a developed water system. The upper soil layer is mainly filled with soil, sand, and silty clay layer, and the underlying bedrock is mainly limestone of the Hutian Group of the Carboniferous system and carbonaceous limestone of the Qixia Formation of the Permian system, which has geological conditions for karst development.

2.2 Characteristics of Karst Development

Through a lot of geological surveys in the Shijing area of Guangzhou, 18,789 geological boreholes in the study area were collected. According to statistics and analysis, the hole finding rate of karst boreholes in the Guangzhou Shijing area is 34.8%, and the linear karst rate is 14.14%, revealing that karst caves and soil caves of 2 to 12 layers are often associated with karst caves, and it is a strong karst development area.

The top buried depth of the karst cave is 6.60–69.30 m, the drilled hole height is 0.10–20.80 m, and the average hole height is 2.92 m. There is no obvious regularity in the distribution of karst caves, which is disordered.

The filling types of karst caves can be divided into no filling, half filling, and full filling, and the filling materials are mostly soft and fluid-plastic silty clay, loose sand or silty clay sand limestone debris, etc., which are relatively soft. The main phenomena in drilling are water leakage, soft soil layer, and self-weight sinking of drilling tools.

According to statistics, 77.6% of the cave roof thickness is less than the height of the cave body, the cave roof thickness is thin, and the minimum thickness of the roof is only 0.10 m. Under the action of external forces, the roof may collapse, and the ground will collapse, affecting the site stability and construction safety, and the karst stability is poor, and there are great security risks.

3 Karst Exploration Means

The key to the design of a bridge pile foundation in a karst area is to find out the geological condition of the lower part. Only using geological drilling and geophysical exploration, can we comprehensively find out the development of karst, to carry out targeted design.

3.1 Geological Drilling Technology

Geological drilling is one of the most commonly used technologies for bridge pile foundation geological investigation in karst areas. This method is almost unaffected by geology and has the characteristics of high accuracy. The process of geological drilling in karst areas, should be based on drilling standards, procedures, and purposes, make overall planning and standardization of the investigation process, record the changes of drilling in real-time, and lay the foundation for subsequent construction.

3.2 Geophysical Exploration Technology

Geophysical exploration technology mainly includes the seismic exploration method, radioactive exploration method, tube wave test method, direct current method alternating current method, etc. Although it has the advantages of large information data acquisition, simple operation, strong adaptability, wide scope, and high efficiency, this technology is highly susceptible to interference by other factors, resulting in the impact of exploration results. In the exploration process, exploration data should be analyzed based on actual karst geology to further improve the accuracy of exploration results [4–7].

4 Key Points of Bridge Pile Foundation Design in Karst Area

4.1 Calculation of Safe Thickness of Pile End-Bearing Layer

In the design process of the bridge pile foundation, to ensure the stability of the bridge pile foundation, the safety thickness of the pile bottom holding layer should be calculated. The current common calculation methods are as follows [3]:

1. The calculation is based on the bending moment control of the karst cave roof

When the karst cave roof rock stratum has the characteristics of large span, high strength, thick layer force, and relative integrity, the main control object is the bending moment. Therefore, in the design process of the bridge pile foundation, the stress conditions of the beam plate should be analyzed first, and on this basis, the minimum design bearing layer plate thickness is calculated, the calculation formula is as follows:

$$H = \sqrt{\frac{6M}{b\sigma}} \tag{1}$$

where M is the bending moment, the unit is kN•m; σ is 1/8 of the allowable compressive strength of limestone, in kPa; b is the width of the beam and plate, in m.

When using this method for calculation, the roof rock is relatively complete, and the calculation process can also be divided into fixed-end beams, cantilever beams, and simply supported beams. If the cave roof is relatively complete, but there are cracks at the bottom of the pile foundation when the cave roof is relatively complete around the pile foundation, but there are cracks at the base of the pile foundation, the cantilever beam can be used as the basis for calculation; When the bottom of the pile foundation is relatively complete, but there are cracks around the roof of the cave, the simple supported beam can be used as the basis for calculation; When the bottom of the pile foundation and the roof of the cave are relatively complete, the calculation can be based on the fixed end beam.

2. The calculation is based on the control of shear stress

When the roof stratum of a karst cave in a karst area is relatively complete, the span of the cave is small and the strength of rock mass is large, the safe thickness of the pile bottom bearing layer can be calculated based on shear stress control. The calculation formula is mainly as follows:

$$H = \sqrt{\frac{4f_s}{S}} \tag{2}$$

where, fs is the shear force at the support, in kN; S is 1/12 of the allowable compressive strength of the limestone, in kPa.

4.2 Foundation Form Selection

1. Friction pile

   When the soil layer above the karst cave is thin, the friction provided by it is relatively small, the roof thickness is small and the distribution of the karst cave is shallow, so the rock-socketed pile cannot be used for construction [4]. To ensure the stability of the bridge foundation, the bridge pile foundation should pass through the karst cave group. In the design process of the bridge pile foundation, the structure of the friction pile should be optimized based on the shear strength of the rock and the results of the pile test, and the cohesiveness between pile concrete and surrounding rock should be controlled. When the soil layer of the bridge construction site is thick and the beaded karst cave under the soil layer is bottomless, the friction pile group can be used to design, and the pile foundation need not pass through the karst cave.
2. Rock-socketed pile

   (1) Single-row pile and group pile

   In the construction process of the bridge project, if it has the characteristics of large pile foundation width, aperture deviation, and the same foundation condition, it can choose the pile group and single row pile structure for design. Under this condition, the axial lateral stability of the single pile and the allowable bearing capacity of the axial compression is relatively high, which meets the requirements of the bridge structure codes and standards.

(2) Single column and single pile foundation

Single column and single pile foundations are mainly suitable for bridge construction with small widths and small aperture. In the construction process of bridge engineering, if the distribution of karst caves in the lower pier is dense, the scope is deep, and the single column and single pile foundation should be selected for the design.

(3) Multi-pile foundation

The karst of the original pile site is strongly developed, and the beaded karst cave is bottomless. If there is complete bedrock in the adjacent area, the pile location can be adjusted, the number of piles can be increased, the pile diameter can be reduced, and the cap and the ground beam can be connected through the ground beam. By increasing the number of piles, multiple pile caps can be formed to reduce the load requirements of each pile; Reducing the pile diameter can reduce the requirements of rock entry and pile end-bearing layer. In the karst development area, by adjusting the pile location, pile number, and pile diameter, to adapt to different geological conditions and ensure the requirements of the upper load and the lower pile foundation bearing layer.

5 Engineering Examples

5.1 Project Overview

A bridge in Shijing, Guangzhou is located in a strong karst development area. The bridge adopts a 13 × 30 m span combination, pile diameter of 1.80 m, single pile bearing capacity requirement of 15000 kN, and single column and single pile. Figure 1 shows the situation of karst collapse at the engineering site.

Fig. 1. Collapse site

5.2 Survey Results

According to the site investigation results, the super large karst cave was revealed at mile 2 + 446, which was a multi-layer karst cave. The maximum height of the cave was

20.80 m, which was an unfilled karst cave. The thickness of the upper covering layer was only 17.20 m, mainly composed of loose sand and silty clay, and the thickness of the cave roof was only 2.30 m. In Mileage 2 + 540, beaded karst caves were revealed, with up to 12 layers of dissolved soil caves, distributed in the depth of 14.80–71.30 m, and the size of karst caves 0.40–9.70 m. Most of them were unfilled or half-filled karst caves. The upper covering layer was only 14.80 m thick, mainly consisting of silt, loose sand, and silty clay layers. According to the geological investigation on site, the geological structure has a great influence on the stability and safety of bridge pile foundation construction.

5.3 Key Points of Design

According to the site survey data, the pile foundation construction at milepost 2 + 446 has great risks, and the site construction may cause ground collapse at any time. The site is no longer suitable for the bridge pile foundation holding layer. Through the exploration of the adjacent area, combined with the tube wave test, it is found that the complete rock stratum meets the load layer of the pile foundation, so two 1.50 m single piles are used, according to the design of rock-socketed piles, and connected through the ground beam to support the load of the upper pier column jointly.

The pile sites at 2 + 540 miles are beaded karst caves, and the exploration in the adjacent areas, combined with the tube wave test, are also beaded karst caves and the pile sites have no adjustment position, which can not be designed according to the rock-socketed pile, so the friction pile design is adopted. According to the survey results report, the standard value of soil friction resistance on the side of the pile of moderately weathered rock formation is 320 kPa, and that of the breezed rock formation is 650 kPa. Only the side friction resistance of the rock formation pile is calculated, excluding the side friction resistance and pile tip resistance of the upper soil layer, which are used as the bearing capacity reserve. Through calculation, the pile length is finally determined to be 59.00 m. The pile end rock socketed is 0.50 m, and the pile end to the cave roof is not less than 2 m, although according to the friction pile design, the rock-socketed pile final hole is clear.

Through the above investigation and design, the pile foundation construction is smooth, through the pile foundation detection, the operation is good.

5.4 Key Points of Construction

In the process of bridge pile foundation construction in a karst area, if there are a large number of karst caves in the area, pile foundation investigation and design should analyze the number and distribution characteristics of karst caves, and treat karst caves based on normative requirements and standards. The treatment methods of karst caves are mainly as follows:

1. Conventional punching method. In the process of treating small cracks, dissolved trenches, dissolved grooves, karst caves, etc., usually only need to use conventional punching methods for processing. For the area where the thickness of the cave roof is thin, small strokes are used to slowly rush into it to ensure the stability of the soil mass.

2. Karst cave dumping and filling method. When the height of a single cave is not more than 5 m, the cave dumping method can be used for treatment. If the cave is large and needs full filling treatment, the whole bag of cement, clay, and stone should be put into the punching process to form mud wall protection inside the cave and ensure the safety of drilling. Suppose the cave is half-filled or has no filling and there is a sudden drop in the slurry level in the protection cylinder during punching. In that case, the clay and schist (volume ratio of 1:1) and the appropriate amount of cement should be immediately put into the drilling hole, and the punching treatment should be carried out again. After the liquid level is stable, it should be left for a period to form a stable slurry wall on the surface of the cave. After many backfills and impacts. Ensure that there is no slurry leakage problem, and you can continue construction.
3. Guard tube follow-up method. If the karst cave is covered with a sand gravel layer or silt soil, a multi-layer karst cave shows a beaded shape, the karst cave is large and has other characteristics, you can use the protection tube follow-up method for construction. Usually, the protection tube is a steel plate, the thickness is greater than 10 mm, and the diameter should be 10–30 cm larger than the pile diameter. Before opening the hole, a steel casing should be driven into the cave first, and the casing should be connected to the hole while punching the hole. When passing through the development section of the cave or the cover layer that is easy to collapse, the casing should be treated with vibration or static pressure sinking. Under normal circumstances, the multi-layer beaded karst cave can be treated with two or three layers of protection, and the general covering layer can be treated with a single layer of protection.

6 Conclusions

In summary, the law of karst development in the Shijing area of Guangzhou is studied through many geological surveys. Using comprehensive investigation, the geological conditions and karst development in the pile site area are ascertained. According to the actual geological conditions, the key points of the karst bridge pile foundation investigation and design are analyzed and discussed, and the pile foundation design is carried out reasonably by adjusting pile location, connecting multiple pile caps, and comparing foundation forms. According to the size and filling of karst caves, appropriate construction measures should be used to ensure the safe formation of holes and piles, to improve the overall construction quality of Bridges, and to provide a reference for the future investigation and design of karst bridge pile foundations in Shijing area of Guangzhou.

Acknowledgments. This research was funded by the Collaborative Innovation Center for Natural Resources Planning and Marine Technology of Guangzhou (No. 2023B04J0301, No. 2023B04J0326), the Key-Area Research and Development Program of Guangdong Province (No. 2020B0101130009), the Guangdong Enterprise Key Laboratory for Urban Sensing, Monitoring and Early Warning (No. 2020B121202019), the Science and Technology Foundation of Guangzhou Urban Planning & Design Survey Research Institute (Grant No. RDI2210204020).

References

1. Farhangi, V., Karakouzian, M.: Design of bridge foundations using reinforced micropiles. In: Proceedings of the International Road Federation Global R2T Conference & Expo, 19–22. Las Vegas, NV, USA (2019)
2. Sheng, M., et al.: Bearing behavior of pile foundation in karst region: Physical model test and finite element analysis. Applied Rheology **34**(1), 20230115 (2024)
3. Tharp, T.M.: Design against collapse of karst caverns. In: Karst Geohazards, pp. 397–406. Routledge (2018)
4. Byle, M.J., Mackey, R.: Grouting karst to support driven pile foundation for a truss rail bridge. Geotechnical Engineering for Transportation Projects, 1915–1924 (2004)
5. Kang, Y., Geng, Z., Chen, L., Zhu, Y., Li, K., Liu, B.: Ground reinforcement method for closely spaced overlapping tunnels passing beneath high-speed railway bridge. J. Geotech. Geoenviron. Eng. **148**(12), 05022007 (2022)
6. Lu, Y., Nie, C., Zhou, D., Shi, L.: Research on programmatic multi-attribute decision-making problem: An example of bridge pile foundation project in karst area. PLoS ONE **18**(12), e0295296 (2023)
7. Gotman, N.Z., Evdokimov, A.G.: Determination of additional load on the bridge foundation pile under karst deformation. J. Phys. Conf. Series IOP Pub. **1928**(1), 012058 (2021)

A Study on the Force Characteristics of Wooden Arch Corridor Bridges in Gansu

Junren Yuan, Yanhai Liu(✉), and Aijun Li

Lanzhou Jiaotong University, Lanzhou, Gansu 730070, China
1657808489@qq.com, 1059278331@qq.com

Abstract. Baling Bridge in Weiyuan, Gansu Province is a unique combination system bridge of outrigger timber beam and through timber arch. The design mode and mechanical characteristics of this combination type of bridge have major significance in the study of traditional timber bridge. In this paper, Midas/civil finite element model of main arch structure is established, the deformation and internal force changes of joints under elastic and rigid connection are analyzed, and the main mechanical characteristics of the bridge are studied. The results show that: Baring bridge is mainly stressed by the through wooden arch system, and the extender beam system enhances the bearing capacity of the bridge. The stronger the joint stiffness is, the better the overall rigidity of the structure is, and the overall vertical displacement of the bridge is obviously improved.

Keywords: wooden arcade bridges · structural form · nodal stiffness · mechanical properties

1 Introduction

Research on wooden bridge is coming into public view. Due to the late discovery of no protective measures and the poor durability of wooden structure Bridges, there are very few existing types of wooden beam Bridges with extended arm. From the perspective of bridge engineering, modern scientific and technological methods are applied to study wooden beam Bridges with extended arm to provide references for the research of wooden structure Bridges.

In recent years, the research on the traditional wooden bridge is mainly focused on the "Bianshui Hongqiao" and "Fujian Zhejiang wooden arch covered bridge". At present, there is no relevant literature on the structure of the outrigger wooden beam bridge. Moreover, it is not certain from the structure that it was improved by BianShuiHong Bridge or Minzhe wooden arch Bridge, but it has a certain connection with these two kinds of Bridges. In the literature [2], the Baling Bridge in Weiyuan is regarded as a variant wooden arch bridge, whose structure is the combination of the outrigger wooden beam bridge in Gansu and the wooden arch bridge in Fujian and Zhejiang, and the design mode of the Bian Shui Hongqiao Bridge and the wooden arch bridge in Fujian and Zhejiang is simplified, and its mechanical characteristics are analyzed. Literature [11] measured the Baling Bridge in Weiyuan on site, obtained the structural dimensions

B. Yuan et al. (Eds.): ICCASE 2024, SUCI, pp. 408–419, 2024.
https://doi.org/10.1007/978-3-031-76102-7_44

of the bridge, analyzed the force transfer mechanism of the bridge, and calculated the internal force and deformation law of the bridge under dead load and live load. Literature [11] SAP2000 finite element software to model the main arch of a typical wooden arch corridor bridge under construction in Jing Ning County, southern Zhejiang Province, and analyzes the stress and deformation of the components under different working conditions, and obtains that the main arch components can meet the relevant index requirements of the specification. Literature [6–11] established a simplified model of Fujian Zhejiang wooden arch corridor bridge with finite element software, and replaced the two hinged arches of the secondary system in the wooden arch bridge with an arc. Through the calculation, analysis and comparison of the original bridge model and the simplified model, it is concluded that the mechanical characteristics of the two structural models are very similar. Fujian Zhejiang wooden arch corridor bridge is an arch structure, which belongs to a wooden arch bridge.

2 The Design Model of the Baling Bridge Introduction to the Baling Bridge

Fig. 1. Weiyuan Baling Bridge

Ba Ling Bridge is located in Weiyuan County, Gansu Province, south of the Qingyuan River, the bridge is 40 m long, span of 27.1 m, 15.4 m high, 4.8 m wide, 14 rows of 64 columns (including the head house 8 columns) [5], the bridge deck and the bottom of the bridge are in each row of 10 stout logs, and is classified as 11 groups from the two sides of the abutments at the bottom of the gradient.

2.1 The Structural Characteristics of the Baring Bridge

Figure 2 gives a three-dimensional illustration of the Baling Bridge, Baling Bridge has two systems, respectively, for two groups of a flat and two sloping three-section rod (commonly known as three-section seedling) and four groups of layers out of the extension arm beams.

The first system is situated between the second system and the reach girder system, along the longitudinal direction of the bridge consists of two longer diagonal saplings

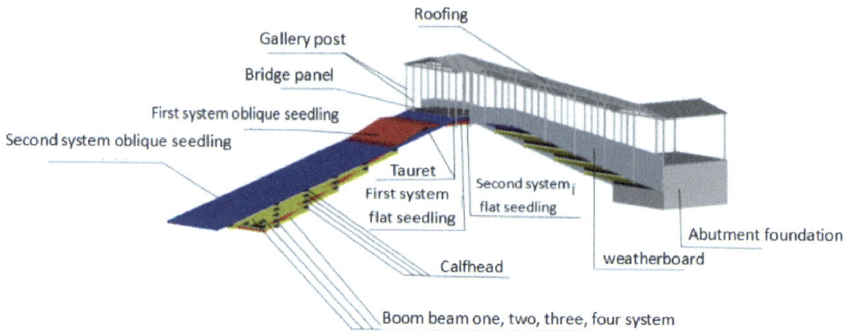

Fig. 2. Weiyuan Baling Bridge 3D model drawing

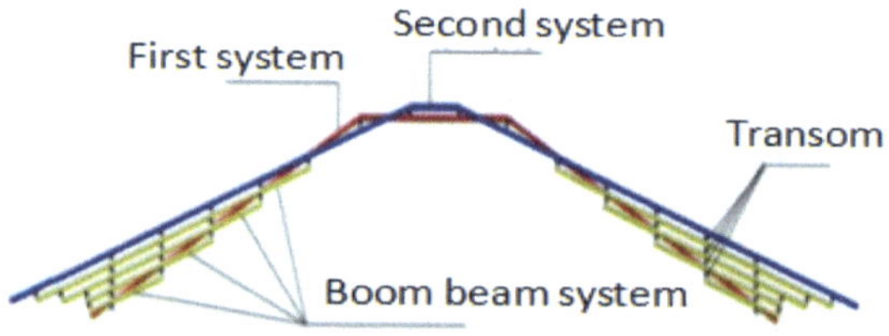

Fig. 3. 2D modeling diagram of the Weiyuan Baling Bridge

and one flat sapling (Fig. 3), two diagonal saplings are respectively picked out from the innermost position of the abutment at the end of the girder, and the angle of inclination is slightly larger than that of the second system and the reach girder system in the horizontal direction, the diagonal saplings and the flat saplings are connected through the crossbeam in the form of tenon and mortise. The telescopic girder system consists of four layers of telescopic girders picked out between the first and second systems at the abutment position, picked out from the bottom to the top, with the length of the bars increasing from one layer to the next, and each layer of the system connected to each other by means of a crossbeam.

The crossbeam is located between each layer of the system, which forms an organic link between the systems of the main arch structure, enhances the stability of the overall structure of the bridge, and makes the transverse load distribution more uniform.

3 Calculation of the Dimensions of the Baling Bridge

Baling Bridge net span of 27.65 m, the bridge along the longitudinal direction of the bridge is divided into equal 13 points the specific structure as Fig. 4.

As shown in Fig. 5, and the clear span of the bridge is Lp, the bridge is divided into equal lengths along the longitudinal direction; the length of the first system of diagonal seedlings is, the L_2, horizontally at an angle of β; the length of the second system of diagonal seedlings is, the, horizontally at an angle of β; the bottom-up lengths of the extended-arm beam system are, respectively, the L_3, L_4, L_5, L_6, horizontally at an angle

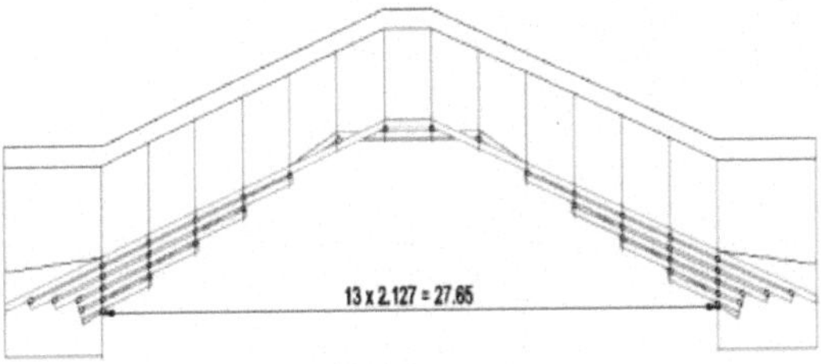

Fig. 4. Elevation of Bailing Bridge

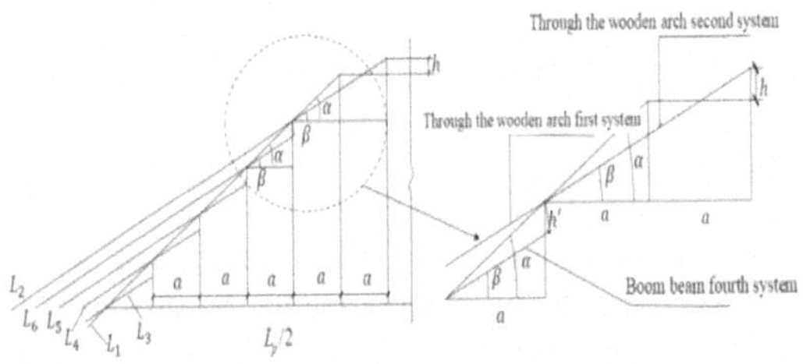

Fig. 5. Bailing bridge set up rod design details

of β; the axial distance between the first system and the second system of flat seedlings is h, the axial distance between the first system and the end of the extended arm beam system is h'.

From the geometrical relations of the above arrangement it follows that.

$$2\tan\beta - \tan\alpha = \frac{h}{a} \tag{1}$$

$$\tan\alpha - \tan\beta = \frac{h\prime}{a \cdot \cos\beta} \tag{2}$$

Each member of the cross-wood arch bridge structure is very tightly connected, and the span and member size of the bridge are determined according to the above arrangement method, and the structure is formed and the vector height is determined [4]. With the net span of the bridge as the fixed value and the equal length as the fixed value, if the diameter of each round rod and the size of the beam are determined, the trial algorithm can be used to find the above formula, so that the length of each rod can be obtained:

$$L_1 = \frac{5a}{\cos\alpha} \tag{3}$$

$$L_2 = \frac{h + 5a \cdot \tan\alpha}{\tan\beta \cdot \cos\beta} \tag{4}$$

$$L_3 = \frac{\frac{h\prime}{\cos\beta} + a \cdot \tan\alpha}{\tan\beta \cdot \cos\beta} + \frac{B}{2} \tag{5}$$

$$L_4 = \frac{\frac{h\prime}{\cos\beta} + 2a \cdot \tan\alpha}{\tan\beta \cdot \cos\beta} + \frac{B}{2} \tag{6}$$

$$L_5 = \frac{\frac{h\prime}{\cos\beta} + 3a \cdot \tan\alpha}{\tan\beta \cdot \cos\beta} + \frac{B}{2} \tag{7}$$

$$L_6 = \frac{\frac{h\prime}{\cos\beta} + 4a \cdot \tan\alpha}{\tan\beta \cdot \cos\beta} + \frac{B}{2} \tag{8}$$

4 Finite Element Modeling of the Baling Bridge

4.1 Node Constraint Simulation

In the finite element simulation of bridge, the beam is regarded as an independent member, the connection between the beam member and the end joint of each member is regarded as a full hinge, and the connection between the beam and the middle part is regarded as a half hinge. The connection between the beam and the system of Baling Bridge has a certain inclination Angle, and the force is mainly transverse grain compression and shear, which is different from the connection of ancient wooden structure which is mainly compression. Therefore, in this paper, two methods of rigid connection and elastic connection are used to simulate the mortise and tenon connection of Baling bridge.

Connection of beams to systems.

(1) Rigid connections: The connections between the second system and the crossbeam, between the crossbeam and the reach beam, and between the crossbeam and the first system are modeled using rigidity in elastic connections.
(2) Flexible connection.

The elastic connection of the bridge components is as follows: between the beam end and the beam system because the beam end and the beam unable occur relative linear displacement, the larger linear stiffness of the elastic connection is used to constrain the linear displacement in 3 directions, S_{Dx}, S_{Dy}, S_{Dz} are taken to 1×10^5 kN/m, Rx, Ry, Rz are taken to 1500 kN·m/rad. The connection of the inclined beam rib and the beam of the same system adopts the "only pressure" spring unit simulation, S_{Dx} takes 1×10^5 kN/m, the beam and the system are interlaced with each other through the concave tenon, and an elastic connection of S_{Dx} is added in both, S_{Dx} takes 0 kN/m, S_{Dy}, S_{Dz} takes 80 kN/m, and S_{DX} takes 0 kN/m. S_{Dy} and S_{Dz} at the joint between the top beam of the second system and the flat rib of the first system are taken to be 100 kN/m, and Rx, Ry and Rz are taken to be 1500 kNm/rad.

4.2 Material Selection

The finite element model is established by MIDAS/CIVIL software. The wood rod used in the timber structure is simulated by beam element (beam element). According to the

(GB50005-2017), the strength grade of Chinese fir belongs to TC11A, the longitudinal compressive strength is 10 MPa, the design value of bending strength is 11 MPa, the elastic modulus is 9000 MPa, the Poisson ratio is 0.25, and the bulk reweight is 2.95 kN/m^3.

4.3 Dimensions of the Individual Bars of the Baring Bridge

Baring Bridge Dimensions. Lp = 27.65 m. $\alpha = 2.127$ m, with the first system, the second system and the boom system all of log diameter d = 0.15 m, beams of square timber side lengths B = 0.10.

From Eqs. (1) and (2).

$$\alpha = 23.5^\circ \quad \beta = 16^\circ.$$

From Eqs. (3), (4), (5), (6), (7), (8).

$L_1 = 11.600$ m, $L_2 = 17.865$ m, $L_3 = 2.309$ m, $L_4 = 5.665$ m. $L_5 = 9.020$ m, $L_6 = 12.375$ m.

4.4 Bridge Loads

① Constant load [6].

$$S_k = \mu s \tag{9}$$

$$S_G = V\gamma = \gamma \sum n \frac{\Pi D^2}{4} 1 \tag{10}$$

Of which. S_k -roofing roles. S_G -Structural gravity. μ -Surface distributed loads. s -Roof area. D-Diameter of logs. γ -Cedarwood heavy; V-Volume of the component, n-Number of components.

Formulas (9) and (10) determine the self-weight of the verandah on Baling Bridge, convert the verandah load into column load, and act on the node. It can be calculated that the load on the eave column is 7.62 KN and the load on the gold column is 9.28 KN.

② Live load.

The crowd load is equivalent to the uniformly distributed load of 3.5 KN/m^2, in which 2.5 KN/m is taken from the inclined seedling of the second system and 2 KN/m from the flat seedling of the first system, and is loaded on the beam unit (Fig. 6).

4.5 Finite Element Modeling

5 Analysis of the Results of the Finite Element Model of Baling Bridge

The analysis models include: elastic connection model, node articulation model; The analysis condition is divided into full span uniform loading and half span uniform loading (Fig. 7).

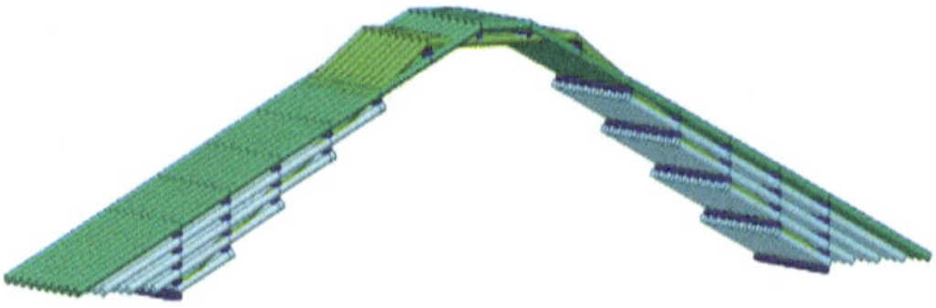

Fig. 6. Finite element model of Bailing Bridge

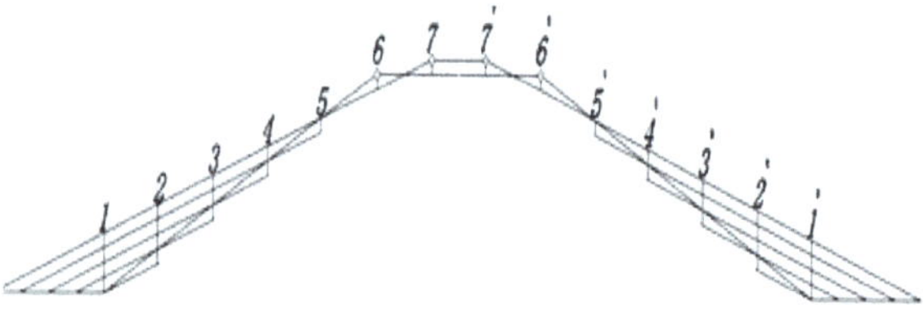

Fig. 7. Bailing Bridge node numbering diagram

5.1 Effect of Node Stiffness on the Deformation of the Baring Bridge

Figure 8 shows the vertical deflection value of the main structure of the bridge under the action of bridge dead weight + corridor house load simulated by finite element model. It can be concluded that the joint stiffness simulated by elastic connection has a larger deflection at the time point and reaches the maximum at the position No. 4, while the deflection is smaller at the position No. 5 when simulated by rigid connection.

5.2 Influence of Node Connection Stiffness on Internal Forces in Corridor Bridges

5.2.1 Bending Moments Under Constant and Full Span Uniform Loads for Different Node Connection Stiffness Models

As can be seen from Figs. 9 and 10, different node connections lead to different stress conditions of covered Bridges. The joint connection mode is elastic connection, the maximum bending moment of the corridor bridge is at the position of nodes 3 to 4, and the nodes are elastic connection. Under symmetric load, the maximum bending moment of the bridge is at the position of 1/4 span, which conforms to the form of arch stress. When the joint connection mode is rigid and the load is symmetrical, the maximum bending moment is at positions 1 and 2, so the maximum bending moment appears at the bridge abutment position.

In the case of elastic connection and rigid connection, the maximum bending moment is generated in the first system, and the maximum internal force varies with the joint stiffness, as shown in Table 1. (Figs. 11 and 12)

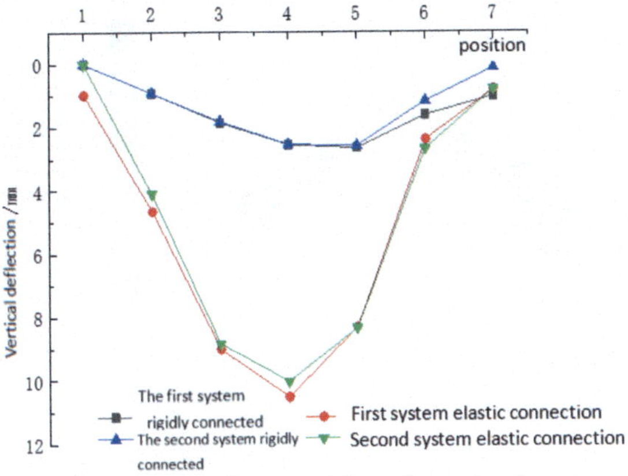

(a) Node displacement diagram of through-wood arch system

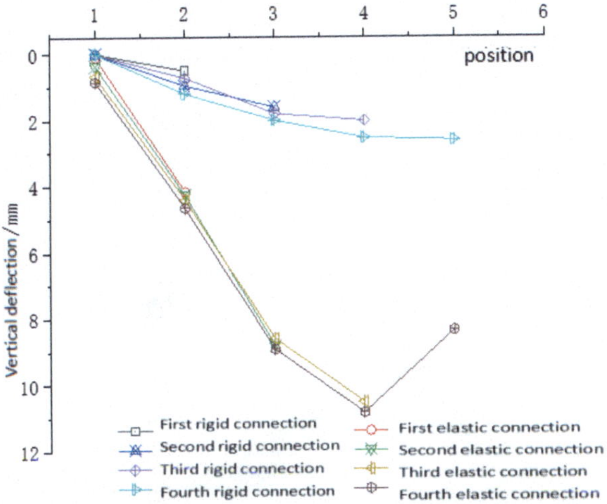

(b) Node displacement diagram of extended arm beam system

Fig. 8. Node displacement diagram of Bailing Bridge

Fig. 9. Bending moment variation under elastic connection conditions

Fig. 10. Bending moment variation under rigid connection conditions

Table 1. Maximum bending moment values for different connection (kN·m)

	Rod		
Connection method	Three-section seedlings The first system	Three-section seedlings The second system	Outrigger beam system
Resilient connection	3.3	2.3	2.6
Rigid connection	2.1	1.2	1.2

Fig. 11. Shear force variation under elastic connection conditions

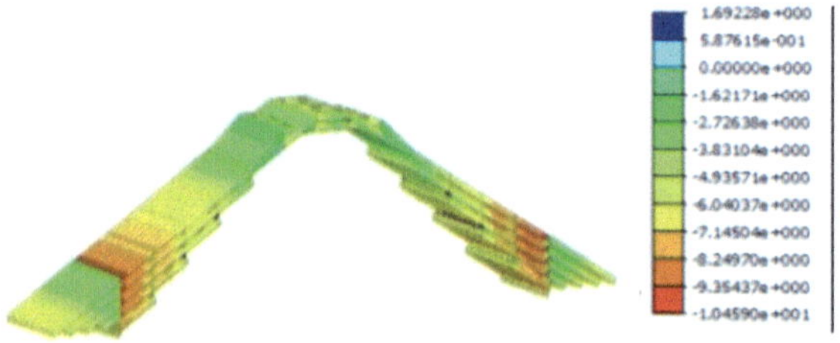

Fig. 12. Shear force variation under rigid connection conditions

Table 2. The maximum shear value of each system with different connection methods

	Rod		
Connection method	Three-section seedlings The first system	Three-section seedlings The second system	Outrigger beam system
Resilient connection	3.0	2.4	2.7
Rigid connection	1.6	1.3	2.6

5.2.2 Shear Forces Under Constant and Full Span Uniform Loads for Different Node Connection Stiffness Models

Table 2 shows the maximum shear force of each system in the finite element calculation model. When the joint connection is rigid, the maximum shear force generated by the member of each system is larger than that when the elastic connection is adopted (Table 3 and Figs. 13 and 14).

5.2.3 Axial Forces Under Constant and Full Span Uniform Loads for Different Node Connection Stiffness Models

Fig. 13. Variation of axial force under elastic connection conditions

Fig. 14. Axial force variation under rigid connection conditions

From Figs. 1, 2 and 3, it can be seen that the joint connection adopts elastic connection, and the axial force borne by the rods of the first system of the transversal arch and the extender beam system is smaller than that borne by the rigid connection, and

Table 3. Coefficients for the modification factor

Connection method	Rod		
	Three-section seedlings The first system	Three-section seedlings The second system	Outrigger beam system
Resilient connection	4.1	−31.1	18.2
Rigid connection	12.7	−23.1	−3.9

the axial force changes more stably when the elastic connection is made. When the joint connection mode is rigid, the axial force on the bar will be sudden change, which requires higher construction of the joint.

6 Conclusion

(1) Under full collapse uniform load, different joint connection modes have a great influence on the deflection change of wood arcade bridge. In the actual manufacturing process of corridor bridge, the joint connection stiffness should be ensured as much as possible, and the overall displacement of corridor bridge structure can be significantly improved.
(2) The internal force changes of each member of the bridge under different connection stiffness under dead load and live load are calculated, and the bending moment, shear force and axial force of the joint increase with smaller joint stiffness.
(3) Through the analysis of structural internal force deformation, the main load-bearing structure is the through timber system, and the overhanging beam system assists the first system to jointly bear the upper load. The weaker the connection stiffness of the system, the more obvious the arch effect.

References

1. Tang, H.: History of Science and Technology in China-Bridges. China Construction Industry Press, Beijing (2010)
2. Tang, H.: Wooden arch bridges in China. China Construction Industry Press, Beijing (2010)
3. Tang, H.: Ancient Bridges of China. China Construction Industry Press, Beijing (2010)
4. Yang, S., Research on stacked girder arch bridges. Highway (12), pp. 1–6 (2002)
5. Yang, Y.: Experimental design research based on the wooden arch corridor bridge in Taishun, Zhejiang. Nanjing ArtInstitute (2021)
6. Fang, X.: Structural design and analysis of Bianshui rainbow bridge type wooden arch bridge. Fujian Agriculture and Forestry University (2014)
7. Liu, J.: Research on force behavior of wooden arch bridges in Fujian and Zhejiang. Fuzhou University (2011)
8. Zhang, X., Wang, B.: Flood response analysis of wooden arch corridor bridge based on numerical simulation. Urban Road and Bridge and Flood Prevention (05), 243–248+29 (2020). https://doi.org/10.16799/j.cnki.csdqyfh.2020.05.069

9. Ou, J.: Finite element analysis of force mechanism of wooden arch corridor bridge. Zhejiang University (2014)
10. Zhang, L.: Model test research on force mechanism of wooden arch corridor bridge. Zhejiang University (2014)
11. Sun, W.: Experimental design research based on the wooden arch corridor bridge in Taishun, Zhejiang. Nanjing Art Institute (2021)

Prediction of the Creep Behavior of Ultra-High Performance Concrete Beams

Ting Liu[1,2], Xiaohui Yan[1], and Xuli Gu[3](✉)

[1] Department of Municipal and Ecological Engineering, Shanghai Urban Construction Vocational College, Shanghai, China
liuting@succ.edu.cn, 406779136@qq.com
[2] Prefabricated Building Research Institute of Anhui Province, Hefei, China
[3] Digital Construction Institute, Shanghai Urban Construction Vocational College, Shanghai, China
351513046@qq.com

Abstract. Ultra-high performance concrete (UHPC) is a new cement-based composite material that offers ultra-high compressive strength (120 MPa), excellent toughness, and excellent durability. It is a new cement-based material with excellent performance, which usually uses different particle size gradations of cementing materials (cement, silica ash, fly ash, mineral powder, etc.), fine aggregates and steel fibers to achieve the closest packing at an ultra-low water-binder ratio (about 0.2). These components and characteristics significantly reduce the size and dead weight of UHPC components, and significantly improve the seismic resistance and seawater corrosion resistance. The creep of concrete is a kind of non-elastic deformation which depends on load and is related to time, and will increase with the extension of load time. The characteristics of concrete creep models widely used in the world such as ACI 209R-92, CEB-FIP 2010, B3, GL2000, AASHTO and NFP18-710-2016 are analyzed in this paper. On this basis, the finite element analysis is carried out with the above model, and the comparison and analysis are made with the test values. Furthermore, the reasonable selection of creep model of UHPC is proposed.

Keywords: ultra-high performance concrete · creep · predict model · creep deflection

1 Introduction

Due to its convenient materials, relatively low cost, and excellent durability, fire resistance, plasticity, and integrity, concrete structures have gradually become one of the most important building structures at home and abroad, and are widely used in the field of civil engineering. The continuous expansion of the application field of concrete structures has led to an increasing demand in civil engineering for lightweight, large-span, environmentally-friendly, and highly durable performance [1, 2]. Conventional building materials and structures are no longer able to fully meet these evolving needs. In order to fulfill the developmental requirements, there has been an increasing application of

B. Yuan et al. (Eds.): ICCASE 2024, SUCI, pp. 420–430, 2024.
https://doi.org/10.1007/978-3-031-76102-7_45

ultra-high performance concrete (UHPC) characterized by a low water-cement ratio, high strength, exceptional impermeability, and superior corrosion resistance [3].

Research findings have indicated [4–6] that creep is an inherent characteristic of UHPC, significantly impacting the load-bearing capacity, stress distribution, deformation behavior, and crack control of structures constructed with this material. Failure to accurately assess its influence on structures [6] may lead to substantial deviations in future design and application of UHPC structures, posing significant safety risks and potential for severe injuries and property damage.

Presently, concrete structures' long-term performance has been extensively investigated by scholars both domestically and internationally. Bazant et al. [7, 8] developed the B3 model, a predictive model for concrete shrinkage and creep, in conjunction with the Kelvin chain model based on concrete aging theory. They established a finite element model to analyze the time-dependent behavior of prestressed concrete bridges and utilized this model to investigate the long-term deflection of KB prestressed box girders. Kistek et al. conducted an analysis on the impact of shear hysteresis on the long-term performance of concrete box girders and discovered its significant influence on calculating concrete creep deformation and stress redistribution resulting from structural system changes during construction. Fang et al. [9] proposed a method for analyzing the long-term structural performance of reinforced concrete box beams, considering the influence of reinforcement. Liu et al. [10] investigated the collaborative working performance of prestressed concrete simply supported beam-track systems on ballastless tracks, incorporating concrete creep and temperature effects using a beam-rail system model. Cao et al. [11], based on long-term performance tests of reinforced concrete continuous beams, analyzed the impact of stress redistribution due to creep by employing a sectional transformation method for long-term deflection deformation in box girder structures. Reference [12] considered the effects of concrete shrinkage creep and steel relaxation, establishing a three-dimensional finite element model to analyze the time-dependent behavior of prestressed concrete beams and study their influence on long-term structural performance. Based on long-term load tests conducted on two-span prestressed continuous beams, Cao et al. [13] proposed a method for calculating the long-term deflection of continuous beams that takes into account stress redistribution. Pan et al. [14], through shrinkage and creep tests on high-performance concrete, developed a modified model for shrinkage and creep of high-performance concrete and performed an uncertainty analysis on the long-term performance of Su-Tong Bridge's continuous rigid frame based on this modified model. The results indicated that the effects of shrinkage and creep uncertainties should be considered when analyzing the long-term behavior of concrete [15, 16]. Despite significant research being done regarding the long-term performance of ordinary and high-performance concrete, there is relatively little research concerning UHPC beam's behavior under prolonged loading conditions.

Due to the intricate nature of concrete creep, there exist multiple prediction methods for considering different factors, leading to significant disparities among them. Consequently, it becomes imperative for researchers to carefully select an appropriate creep model and accurately determine the creep coefficient of UHPC in long-term deformation calculations. This article aims to conduct a comparative analysis of widely utilized creep prediction models, namely CEB-FIP, ACI 209R-92, B3 model, B4 model, GL 2000 model, AASHTO model, and NFP18-710-2016 model. These models will be incorporated as constitutive models for concrete creep into a time-dependent finite element analysis program developed for secondary purposes. Subsequently, the program will be employed to simulate the long-term deformation of UHPC beams mentioned in existing literature. By comparing the computed results with experimental findings, recommendations regarding appropriate selection of creep prediction models for UHPC will be proposed.

2 Creep Model

Since Hatt's initial observation of concrete creep in 1907 [17], extensive research has been conducted by scholars both domestically and internationally, leading to the proposal of empirical and semi-empirical models based on experimental data and theoretical studies [18–21]. Currently, widely accepted creep models include CEB-FIP, ACI 209R-92, B3/B4 model, GL 2000 model, AASHTO model, and NFP18-710-2016 model [21, 22].

Concrete creep is commonly characterized by the creep coefficient, which represents the ratio of creep strain to initial strain. Various models have been proposed for concrete creep: (1) superposition method combining basic and drying creep, such as the B3 and B4 model; (2) superposition method considering recoverable and non-recoverable creep, like CEB-FIP MC 1978 [17, 30, 31]; (3) holistic description methods, including European CEB-FIP MC 1990, CEB-FIP MC 2010, American ACI 209R-92, GL2000 model, NFP18-710-2016 model and the models based on viscoelastic rheological theory. Details of creep models are given in Table 1.

The models employed in this study are all empirical or semi-theoretical and semi-empirical formulas, taking into consideration various influencing factors (as presented in Table 2 [21, 22]). Currently, the CEB-FIP MC1990 model is adopted by the current highway specification in China (JTG 3362-2018), while the European standard EN 1992-2 model is utilized by the current construction standard GB 50010. Additionally, for structures made of active powder concrete, the technical regulations DBJ 43/T 325 adopt the French standard NFP18-710-2016 model. It should be noted that both EN 1992-2 and CEB-FIP MC2010 models employ identical calculation methods.

According to Table 2, the GL2000 model, CEB-FIP model, AASHTO model, and NFP18-710-2016 model demonstrate the minimum parameter requirements for predicting concrete creep. It is important to note that these prediction models have distinct applicability ranges due to their coefficients being derived from experimental data. The long-term performance of UHPC is more susceptible to factors such as environmental temperature and humidity, loading age, and curing methods due to its inherent material characteristics including a low water-cement ratio.

3 Procedure for the Secondary Development of UMAT Subroutines

Based on the existing researches, this article utilizes the user subroutine UMAT provided by the general finite element software Abaqus for secondary development, implemented using the Fortran programming language. The program is utilized to conduct a finite

Table 1. Details of creep models

Creep model	Creep coefficient $\phi(t, t_0)$	Category
B3 [20, 27]	$\phi(t, t_0) = E_c[C_0(t, t_0) + C_d(t, t_0, \tau)]$	basic and drying creep
B4 [28]	$\phi(t, t_0) = E_c[C_0(t, t_0) + C_d(t, t_0, \tau)]$	basic and drying creep
CEB-FIP MC 1990 [24]	$\phi(t, t_0) = \phi_{RH}\beta(f_{cm})\beta(t_0)\beta_c(t - t_0)$	holistic description methods
CEB-FIP MC 2010 [25]	$\phi(t, t_0) = \phi_{RH}\beta(f_{cm})\beta(t_0)\beta_c(t - t_0)$	holistic description methods
ACI-209R-92 [26]	$\phi(t, t_0) = \frac{(t-t_0)^{0.6}}{10+(t-t_0)^{0.6}}\phi(\infty, t_0)$ $\phi(\infty, t_0) = 2.35\gamma_{c,t_0}\gamma_{c,RH}\gamma_{c,vs}\gamma_{c,M}\gamma_{c,\psi}\gamma_{c,\alpha}$	holistic description methods
GL 2000 [23]	$\phi(t, t_0) = \Phi(t_c)[U_1(t, t_0)+U_2(t, t_0)]$	holistic description methods
AASHTO LRFD [29]	$\phi(t, t_0) = 1.9k_s k_{hc} k_f k_{td} t_i^{-0.118}$	holistic description methods
NFP18-710-2016 [32]	$\phi = \phi_\infty \frac{(t-t_0)^{0.6}}{(t-t_0)^{0.6}+10}$	holistic description methods

Note: coefficients regarding in the formulas refer to relevant literatures

Table 2. Relevant factors considered in commonly utilized creep models

Relevant factors	B3	B4	CEB-FIP (1990/2010)	ACI-209R-92	GL 2000	AASHTO LRFD	NFP18-710-2016
Cement varieties	√	√	√	√	√	×	×
Cement content	√	√	×	√	×	×	×
Water-cement ratio	√	√	×	×	×	×	×
Fine aggregate /coarse aggregate	×	×	×	√	×	×	×
Slumps	×	×	×	√	×	×	×
Aggregate /Cement Weight	√	√	×	×	×	×	×

(*continued*)

Table 2. (*continued*)

Relevant factors	B3	B4	CEB-FIP (1990/2010)	ACI-209R-92	GL 2000	AASHTO LRFD	NFP18-710-2016
Air content	×	×	×	√	×	×	×
Cement density	×	×	×	√	×	×	×
Water content	√	√	×	×	×	×	×
Relative humidity	√	√	√	√	√	√	×
Modulus of elasticity for 28-day concrete	√	√	√	×	×	×	×
Elastic modulus of concrete under loading conditions	√	√	×	√	√	√	×
28-day compressive strength of concrete	√	√	√	√	√	√	×
Concrete age at the time of loading	√	√	√	√	√	√	√
Curing age	√	√	√	√	√	√	×
Volume-Surface ratio	√	√	√	√	√	√	×
Curing method	√	√	×	√	×	×	√

element analysis on UHPC beams, providing data for comparative analysis of creep model prediction.

The user material subroutine is a Fortran program interface provided by Abaqus, enabling users to define their own material properties and utilize material models not included in the Abaqus material library. The user material subroutine UMAT facilitates seamless communication with the main solver program of Abaqus through its well-defined interface [33].

The essence of UMAT lies in providing the Jacobian matrix that defines the material constitutive model, representing the rate ($\partial \Delta \sigma / \partial \Delta \varepsilon$) of change between stress increments and corresponding strain increments, and updating the stress for ABAQUS main program. For instance, given the results (σ_n, ε_n) at step N, when ABAQUS main program provides a strain increment $d\varepsilon_{n+1}$, UMAT calculates new stress σ_{n+1} based on the provided Jacobian matrix DDSDDE.

The UMAT subroutine functions as an interface within the ABAQUS main program, being invoked at integration points of elements. At the onset of each incremental step, the main program accesses UMAT through its designated interface. The initial values of essential variables at the current integration point are transferred to corresponding variables in UMAT. Subsequently, UMAT calculates the Jacobian matrix for material integration points of the element and updates stress tensor and state variables accordingly. Ultimately, it returns these updated variable values back to the main program via the interface. It is evident that a dynamic interaction and collaborative process exist between the ABAQUS main program and UMAT subroutine for seamless data transfer.

The UMAT subroutine is implemented in the Fortran programming language, adhering to a fundamental programming approach as outlined below.

(1) The main program transmits stress $\sigma(t)$, strain $\varepsilon(t)$, time increment Δt, and total strain increment DSTRAIN to the UMAT subroutine.
(2) The state variables are extracted from STATEV and utilized in the elastic-plastic analysis conducted through KSTEP for value assessment.
(3) When KSTEP = 1, perform elasticity analysis to obtain the incremental elastic strain DSTRAN and form the elastic Jacobian matrix DDSDDE. Then, calculate the stress-dependent term according to Eq. (1).

$$\sigma(t + \Delta t) = \sigma(t) + \Delta\sigma(t) = \sigma(t) + DDSDDE * DSTRAN \quad (1)$$

(4) When KSTEP = 2, a nonlinear analysis is conducted to obtain the elastic strain and the strain increment DSTRAN caused by concrete shrinkage and creep, resulting in the formation of the elastic Jacobian matrix DDSDDE. The strain increment and stress increment are obtained through iterative calculations, while the time-dependent stress is computed based on Eq. (2).

$$\sigma(t + \Delta t) = \sigma(t) + \Delta\sigma(t) = \sigma(t) + DDSDDE * (DSTRAN - \Delta\sigma(t)) \quad (2)$$

(5) Compute the updated Jacobian matrix DDSDDE based on Eq. (1) and Eq. (2).
(6) Generate a new array of state variables STATEV within the subroutine.
(7) Transfer the time-dependent stress back to the ABAQUS main program from the subroutine.

4 Comparison Between Experimental Results and Finite Element Analysis Calculations

The program incorporates multiple calculation model modules for concrete creep, encompassing widely used models such as CEB-FIP, ACI 209R-92, B3 model, GL 2000 model, AASHTO model, and NFP18-710-2016 model. To demonstrate the variations in application of these creep prediction models, all six models were employed to forecast the creep deformations of four 360-day UHPC beams described in Reference [34], followed by comparative analysis with experimental data.

The study in Reference [34] conducted an experiment of four I-shaped UHPC beams on 360-day long-term behaviors. Three-point loading was applied to the beams. Specific design parameters and experimental results are shown in Table 3.

Table 3. Details of the beams

Beams	Cross-sectional dimensions (mm)	Length (mm)	Steel-fiber percentage (%)	Sustained loading (kN)	Temperature (°C)	Humidity (%)	Deformation (mm)
UHPCB-1	120*240	3200	0.00	18.94	20 ± 2	60 ± 5	1.16
UHPCB-2	120*240	3200	0.50	20.79	20 ± 2	60 ± 5	0.91
UHPCB-3	120*240	3200	1.50	22.63	20 ± 2	60 ± 5	0.76
UHPCB-4	120*240	3200	2.55	24.92	20 ± 2	60 ± 5	0.65

The experiment mainly considered the factor influencing the steel-fiber content. Based on the six different creep prediction models, a predictive analysis on the long-term deformation of the test beams is conducted by GL2000, CEB-FIP, ACI 209R-92, B3, AASHTO, and NFP18-710-2016 creep prediction models. Table 4 presents the ratios of analytical midspan deflections to experimental midspan results.

Table 4. Ratios of calculated midspan deflections to experimental results for UHPC beams

Source	Beams	ACI 209-92	CEB-FIP 2010	GL 2000	B3	AASHTO LRFD 2017	NFP18-710-2016
Experimental studies (360 days)	UHPCB-1	1.370	1.165	1.165	1.742	1.433	1.092
	UHPCB-2	1.373	1.167	1.167	1.731	1.416	1.102
	UHPCB-3	1.366	1.125	1.125	1.751	1.328	1.086
	UHPCB-4	1.389	1.134	1.134	1.737	1.434	1.100
	Mean/Standard deviation	1.420/0.025	1.148/0.018	1.374/0.009	1.740/0.008	1.403/0.044	1.095/0.007

Note: The values in the table are the ratios of analytical midspan deflections to experimental midspan results

Table 4 summarized the ratio of midspan deflections calculated by ACI 209R-92, CEB-FIP 2010, GL 2000, B3, AASHTO LRFD 2017, and NFP 18-710-2016 to the r experimental results. The six methods yields mean ratios of 1.420, 1.148, 1.374, 1.740, 1.403 and 1.095 respectively, with corresponding standard deviations of 0.025, 0.018, 0.009, 0.008, 0.044 and 0.007. Notably, the results obtained from NFP 18-710-2016 aligned closely to the deflections of the experimental UHPC beams. In contrast, the other existing standards (ACI209-92, CEB-FIP 2010, GL 2000, B3 and AASHTO LRFD 2017) yielded unconservative prediction on the long-term deflections of UHPC beams.

5 Conclusions

This paper analyzes the characteristics of concrete creep models widely used in the world such as ACI 209R-92, CEB-FIP 2010, B3, GL2000, AASHTO and NFP18-710-2016. Then the finite element analysis is carried out with the above models, and the comparison and analysis are made with the test values. Furthermore, the reasonable selection of creep model of UHPC is proposed. The main conclusions are:

(1) Due to the complexity of concrete creep, different models incorporate diverse mechanisms and parameter selections. In general, the ACI 209R-92 model considers ten influential factors including relative humidity, component size, compressive strength and elastic modulus of concrete at 28 days, compressive strength and elastic modulus of concrete during loading, cement type, curing method, loading age period, and slump. Compared to the ACI 209R-92 model, the CEB-FIP 2010 model omits the influence of compressive strength during loading as well as curing method and slump. The B3 model emphasizes concrete's inherent characteristics such as water-cement ratio and aggregate-to-cement content ratio. The GL2000 model primarily takes into account factors like component size, compressive strength and elastic modulus of concrete at 28 days, compressive strength and elastic modulus of concrete during loading, and loading age period. Similarly to GL2000, the AASHTO LRFD 2017 model also considers these aspects while the NFP 18-710-2016 model solely focuses on differences in loading age period and curing methods for concrete. According to the characteristics of UHPC, its long-term performance is significantly influenced by environmental temperature and humidity, loading age, and curing methods.

(2) Through a comparative analysis between the results of FEA and experiment, it is evident that the B3 model exhibits an average error of 74.0% in predicting long-term deformations over a period of 360 days when compared to experimental values. In contrast, the ACI 209R-92, AASHTO LRFD 2017, and GL2000 creep prediction models demonstrate average errors of 42.0%, 40.3%, and 37.4% respectively for the same duration. Conversely, the CEB-FIP 2010 model displays an error rate of only 14.8% when compared to experimental values over a period of 360 days; whereas the NFP18-710-2016 model maintains an error within or below 10%. Ultimately, the B3 model predicts long-term deformation with an error rate as low as 9.1% when compared to experimental values over a period of 360 days. The analysis suggests that both CEB-FIP 2010 and NFP18-710-2016 models are more suitable for calculating the long-term deformations of UHPC beams.

(3) It is important to note that due to the multitude of factors influencing the creep behavior of UHPC, there exists no universally accepted theory or calculation formula for predicting its creep. Different creep prediction models exhibit significant variations in error when estimating long-term deformation of the same structural element. While experimental methods can accurately account for the influence of UHPC composition, simulating the actual placement environment and conducting lengthy and costly creep tests pose challenges. Therefore, during the design phase, it is recommended to select a creep prediction model that considers relevant influencing factors based on the specific characteristics of the component itself in order to achieve more accurate predictions of long-term deformations caused by creep. In the

advanced stages of research, it is feasible to commence with considering the creep behavior principles of UHPC materials and integrate long-term performance tests and theoretical analysis of various components to propose a simplified calculation method for predicting their long-term deformation.

Acknowledgments. The authors acknowledge the financial support of Prefabricated Building Research Institute of Anhui Province (No. AHZPY2022KF02).

References

1. Wu, Z.W., Lian, H.Z.: High-performance concrete. China Railway Publishing House, Beijing (1999). ISBN: 9787113034580
2. Feng, N.Q.: High-performance concrete structures. Machinery Industry Press, Beijing (2004). ISBN: 9787111132943
3. Peng, G.F., Yang, J., Gao, Y.X., Wang, B.: Factors influencing the compressive strength of ultra-high performance concrete with coarse aggregates. J. North China Inst. Water Conserv. Hydroelect. Power (Natural Science Edition) **33**, 5–9 (2012). https://doi.org/10.19760/j.ncwu.zk.2012.06.002
4. Li, Y., Guo, Z.G.: How to consider the effect of creep and shrinkage in reinforced concrete tall building structure design. Industrial Construction **25**, 40–46 (1995). https://doi.org/10.13204/j.gyjz1995.04.009
5. Xi, X.: Experimental study on mechanical properties of steel-reinforced concrete under long-term load after high temperature exposure. Hunan University (2013). https://doi.org/10.7666/d.Y2355891
6. Chen, K.L.: Bridge collapse accidents in the Republic of Palau. Foreign Roads **18**, 31–33 (1998). https://doi.org/10.14048/j.issn.1671-2579.1998.03.009
7. Bazant, Z.P., Yu, Q.A., Li, G.H.: Excessive long-term deflections of prestressed box girders. I: Record-span Bridge in Palau and other Paradigms. J. Struct. Eng. **138**, 676–686 (2012). https://doi.org/10.1061/(ASCE)ST.1943-541X.0000487
8. Bazant, Z.P., Yu, Q.A., Li, G.H.: Excessive long-term deflections of prestressed box girders. II: Numerical Analysis and Lessons Learned. J. Struct. Eng. **138**, 687–696 (2012). https://doi.org/10.1061/(ASCE)ST.1943-541X.0000375
9. Fang, Z., Wu, Q.F., Cao, G.H.: Bar simulation method for analysis on long-term behavior of reinforced concrete box girder. J. China Railway Soc. **34**(8), 87–93 (2012). https://doi.org/10.3969/j.issn.1001-8360.2012.08.014
10. Liu, P., Zheng, Z.H., Yu, Z.W.: Cooperative work of longitudinal slab ballastless track prestressed concrete simply supported box grider under concrete creep and a temperature gradient. Structures **27**, 559–569 (2020). https://doi.org/10.3969/j.issn.1006-2106.2019.07.005
11. Cao, G.H.: Short-term and long-term mechanical properties of concrete box beams. Hunai University (2007). https://doi.org/10.7666/d.y1209542
12. Ebrahimi, M.H.R., Rahai, A.: Long-term behavior of a prestressed concrete bridge with corrugated steel webs. J. Bridg. Eng. **27**, 1–9 (2022). https://doi.org/10.1061/(ASCE)BE.1943-5592.0001801
13. Cao, G.H., Zhang, W., Hu, J.X.: Long-term performance of prestressed concrete continuous box girders in the natural environment. Can. J. Civ. Eng. **47**, 856–864 (2020). https://doi.org/10.1139/cjce-2018-0447

14. Pan, Z.F., Lyu, Z.T., Liu, Z.: Uncertainty analysis of shrinkage and creep effect of continuous rigid frame of Sutong bridge. Engineering Mechanics **26**, 67–73 (2009). https://doi.org/10.1061/(ASCE)BE.1943-5592.0000147
15. Yang, I.H.: Uncertainty and sensitivity analysis of time-dependent effects in concrete structures. Eng. Struct. **29**, 1366–1374 (2007). https://doi.org/10.1016/j.engstruct.2006.07.015
16. Henrik, M., Zdenek, P.B.: Uncertainty analysis of creep and shrinkage effects in concrete structures. ACI Journal Proceeding **80**, 116–127 (1983). https://doi.org/10.14359/10467
17. Bazant, Z.P.: Prediction of concrete creep and shrinkage: past, present and future. Nucl. Eng. Des. **203**, 27–38 (2001). https://doi.org/10.1023/A:1020031701950
18. Hubler, M.H., Wendner, R., Bazant, Z.P.: Statistical justification of Model B4 for drying and autogenous shrinkage of concrete and comparisons to other models. Mater. Struct. **48**, 797–814 (2015). https://doi.org/10.1617/s11527-014-0516-z
19. Wendner, R., Hubler, M.H., Bazant, Z.P.: Statistical justification of model B4 for multi-decade concrete creep using laboratory and bridge databases and comparisons to other models. Mater. Struct. **48**, 815–833 (2015). https://doi.org/10.1617/s11527-014-0486-1
20. Bazant, Z.P., Baweja, S.: Creep and shrinkage prediction model for analysis and design of concrete structures: model B3. Mater. Struct. **28**, 357–365 (1995). https://doi.org/10.1007/BF02473152
21. Goel, R., Kumar, R., Paul, D.K.: Comparative study of various creep and shrinkage prediction models for concrete. J. Mater. Civ. Eng. **19**, 249–260 (2007). https://doi.org/10.1061/(ASCE)0899-1561(2007)19:3(249)
22. Bazant, Z.P., Li, G.H.: Unbiased statistical comparison of creep and shrinkage prediction models. ACI Materials Journal **105**, 610–621 (2008). https://doi.org/10.14359/20203
23. Gardner, N.J., Lockman, M.J.: Design provisions for drying shrinkage and creep of normal-strength concrete. ACI Materials Journal **98**, 159–167 (2001). https://doi.org/10.14359/10199
24. BéTON, C.E.-I.D.: CEB-FIP model code 1990: Design code. Thomas Telford Publishing, London (1993)
25. CODE, C-F.M.: Design code for concrete structures. International Federation for Structural Concrete (fib). Lausanne, London (2010)
26. Rath, B., Deo, S., Ramtekkar, G.: Modification of ACI 209R–92 Concrete Shrinkage Model for Partial Replacement of Cement with Fly Ash and Sand with Pond Ash. Adv. Civil Eng. Mat. **9**, 602–620 (2020). https://doi.org/10.1520/ACEM20200036
27. Zdenek, P.B., Joong-koo, K.: Improved prediction model for time-dependent deformations of concrete: Part 1-Shrinkage. Mater. Struct. **24**, 327–345 (1991). https://doi.org/10.1007/BF02472066
28. Roman, W., Mija, H.H., Bažant, Z.P.: The B4 Model for Multi-decade Creep and Shrinkage Prediction. CONCREEP-9 (2013). https://doi.org/10.1061/9780784413111.051
29. American Association of State Highway and Transportation Officials: AASHTO LRFD Bridge Design Specifications 8th. LRFD-8, Newyork (2017)
30. Bazant, Z.P., Li, G.H.: Comprehensive database on concrete creep and shrinkage. ACI Mater. J. **105**, 635–637 (2008). https://doi.org/10.1002/tal.478
31. Hedegaard, B.: Creep and shrinkage modeling of concrete using solidification theory. J. Mater. Civ. Eng. **32**, 1–13 (2020). https://doi.org/10.1061/(ASCE)MT.1943-5533.0003256
32. Director General of AFNOR: French standard NF P 18-710 National addition to Eurocode 2 — Design of concrete structures: specific rules for Ultra-High Performance Fibre-Reinforced Concrete (UHPFRC) . Association Francaise de Normalisation (AFNOR-French standard institute), French (2016)

33. Bazant, Z.P., Baweja, S.: Creep and shrinkage prediction model for analysis and design of concrete structures–model B3. Materials & Structures. **28**, 357–365(1995). https://doi.org/10.1007/bf02486204
34. Barbos, G.: Long-term behavior of ultra-high performance concrete (UHPC) bended beams. Procedia Technol. **22**, 203–210 (2016). https://doi.org/10.1016/J.PROTCY.2016.01.045

Load Carrying Capacity Analysis of Continuous Girder Bridge Based on Load Test

Yongkang Mao[1], Xiaojun Ning[1](✉), and Dejun Mao[2]

[1] Faculty of Civil Engineering and Architecture, Kunming University of Science and Technology, Kunming 650500, China
adn648@outlook.com

[2] School of Architecture and Civil Engineering, Kunming University, Kunming 650214, China

Abstract. In order to accurately assess the load carrying capacity and actual working condition of a newly constructed continuous girder bridge, the load test of a newly constructed three-span prestressed variable cross-section continuous girder bridge is taken as the engineering background, and the check coefficient of the static load test, relative residuals and other parameters are used to judge whether the indexes such as the strength and stiffness of the structure satisfy the requirements or not. The results show that the deflection and strain calibration coefficients are less than 1 and the relative residual value is not more than 20% when the loading efficiency is between 0.85 and 1.05, indicating that there is surplus of structural stiffness and the deformation can be recovered after unloading, and the structure is in an elastic working state.

Keywords: continuous girder bridge · load test · calibration coefficient · bearing capacity

1 Introduction

With the continuous development of transportation industry, China's unique topography has created the image of a big bridge country. Continuous girder bridge with uniform internal force distribution and good overall performance is the most widely used bridge type, which occupies a large share in China's bridge construction market, and it is of great significance to fully grasp its stress performance and working condition for the development of bridge industry. Load test is the most direct and effective way to check the performance and working condition of the bridge [1], for the newly opened bridge must be produced to identify the nature of the completion of the static and dynamic load test for the scientific evaluation of the bridge structure of the strength, stiffness, load-bearing capacity and structural performance of the structure as a whole to provide first-hand information, and at the same time, also for the evaluation of the quality of the project's construction, the reliability of the design and the reasonableness of the bridge as well as the completion of the acceptance of providing a reliable basis [2]. It also provides a reliable basis for evaluating the construction quality, design reliability and reasonableness as well as the acceptance of the completion of the bridge [2].

B. Yuan et al. (Eds.): ICCASE 2024, SUCI, pp. 431–440, 2024.
https://doi.org/10.1007/978-3-031-76102-7_46

Load test is divided into static load test and dynamic load test, static load test is mainly through the measurement of bridge structure in the static load under the action of each control section of the stress and structural deformation, so as to determine the actual working condition of the bridge and the design of the desired value is consistent with the method, the dynamic load test is mainly measured under the action of the environmental loads and traffic loads and the characteristics of the self-oscillation and forced vibration. There are many engineering cases where load tests have been used to assess the load carrying capacity of bridges. Jia Yi et al. used load tests to evaluate the load carrying capacity of Goldfish Bridge [3] and Shunqi River Bridge [4]; Liu Jinfu [5] used in-situ static load tests to evaluate the working performance and actual load carrying capacity of a pre-stressed concrete continuous girder bridge that has been in service for 20 years. Zheng Xu et al. [6] proposed a rapid load test method based on single-vehicle mobile loading and a method based on virtual loading of influence lines that can quickly evaluate the load carrying capacity of small and medium-span bridges.

In this paper, a three-span prestressed concrete continuous girder bridge is subjected to static load test to analyze the difference between its strain, deflection and other parameters and the calculated values, and the parameters such as calibration coefficient are used to judge whether the strength and stiffness of the bridge meet the actual requirements and whether the construction quality is up to the standard. It provides theoretical basis for the design, construction, monitoring and maintenance of the same type of bridge in the future.

2 Project Overview

The ratio of girder height to span diameter is 1/17.2, the ratio of girder height to pivot point girder height is 1/2.07; the thickness of the web plate varies from pivot point to span from 0.75 m to 0.45 m; the lower edge of the bottom plate is 1.8 parabola, the thickness of the span is 30 cm, and the thickness at the pivot point is 80 cm. The main pier adopts hollow thin-walled pier, the outer outline of which is rectangle, and it is poured with C40 concrete. As shown in Fig. 1.

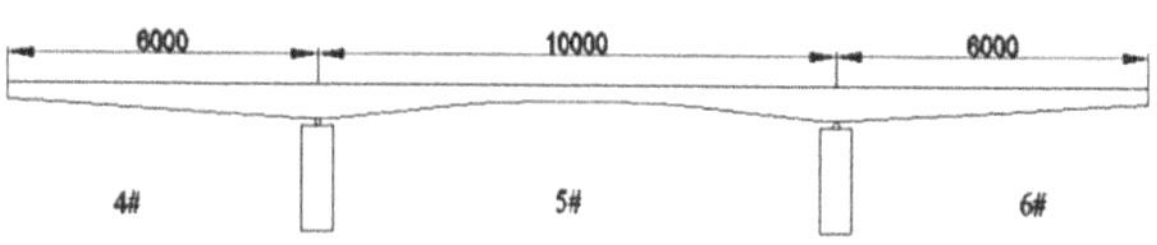

Fig. 1. Elevation of the bridge (unit: cm)

3 Finite Element Analysis

The Midas/Civil bridge finite element analysis software was used as the computational tool for this test bridge. As shown in Fig. 2. The following assumptions are proposed for modeling:

(1) The materials used in the bridge, concrete and steel reinforcement, are ideal elastic materials, and the modulus of elasticity is constant;
(2) The elastic modulus of box girder C55 concrete is taken as 3.55×104 MPa; the elastic modulus of pier C40 concrete is taken as 3.25×104 MPa, and the cross-sectional deformation of its members conforms to the assumption of flat section;
(3) The influence of bridge deck pavement on bridge force is not considered, but the influence of the weight of bridge deck pavement on the structure's self-oscillation frequency is considered.

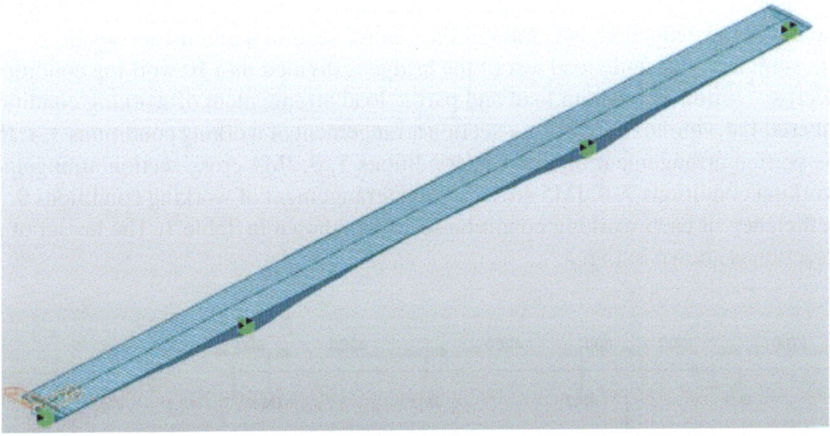

Fig. 2. Finite element model

4 Static Load Test Program

4.1 Loading Vehicles

A total of six three-axle vehicles of about 460 kN were used in this test, and the wheelbases of the front and rear axles of the test vehicles are shown in Fig. 3.

4.2 Selection of Working Conditions

According to the most representative and most unfavorable stress principle to determine the test conditions and test section. The control sections JM1, JM2, JM3, JM4, JM5 were taken for longitudinal direction according to the most unfavorable position, and

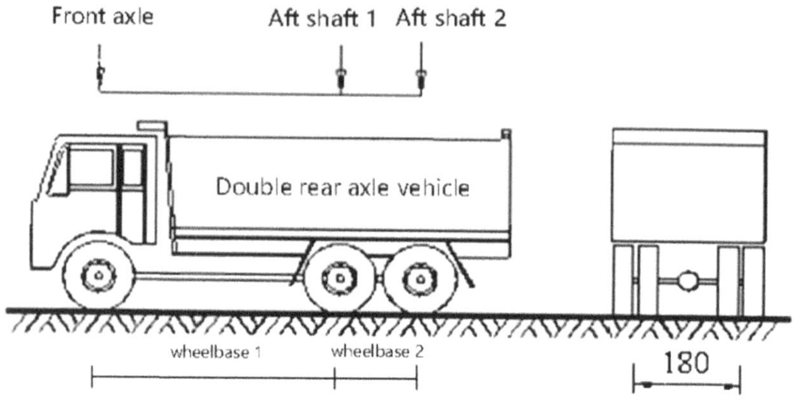

Fig. 3. Schematic diagram of the loaded vehicle (unit: cm)

the transverse direction was divided into partial load and medium load, and the test content was the deflection and strain of the control section under the test load.

In summary, the static load test of the bridge is divided into 10 working conditions, JM1 cross-section of medium load and partial load arrangement of working conditions numbered 1, 2, similarly, JM2 cross-section arrangement of working conditions 3, 4, JM3 cross-section arrangement of working conditions 5, 6, JM4 cross-section arrangement of working conditions 7, 8, JM5 cross-section arrangement of working conditions 9, 10, the efficiency of each working condition loads are shown in Table 1. The layout of the test section is shown in Fig. 4.

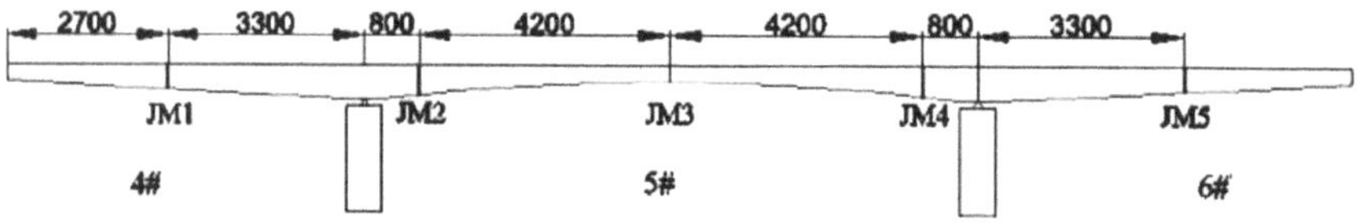

Fig. 4. Layout of test sections (unit: cm)

4.3 Principles of Loading Calculations

The number and loading position of the vehicles to be loaded in the test under each working condition will be equivalent to the principle of the following equation according to the most adverse effect value generated by the live load of the design standard [7]:

$$\eta = \frac{S_s}{S(1+\mu)}$$

where: η- Static test load efficiency coefficient; S_s- Calculated value of displacement or force of the detection part under the action of test load; S- Calculated value of displacement or force of the test part under the action of the standard live load of the design standard (excluding the impact action); μ- Impact coefficient for design.

According to the provisions of the "Highway Bridge Load Test Regulations" (JTG /T J21-01-2015) [8] (hereinafter referred to as the "Test Regulations"), for the delivery (completion) acceptance load test, the static load test load efficiency coefficient should be between 0.85–1.05.

Table 1. Load efficiency (KN·m)

working load	Conditions1, 2, 9, 10	Conditions3, 4, 7, 8	Conditions5, 6
Highway-I effect	16579.1	−33890.4	17425.8
Loaded vehicle effect	17489.4	−32321.7	17958.8
Load efficiency	1.05	0.95	1.03

Note: Impact factor $\mu = 0.05$ has been factored in

4.4 Measurement Point Arrangement and Test Method

4.4.1 Strain Gauge

Bridge JM1, JM2, JM3, JM4, JM5 cross-section strain measurement points are arranged in the box girder exterior (web and bottom plate), JM1, JM3 and JM5 cross-section are arranged 7 measurement points (3 measurement points at the bottom, 2 measurement points at the left and right webs), JM2 and JM4 cross-section are arranged 11 points (3 measurement points at the bottom, 4 measurement points at the left and right webs), a total of 43 strain measurement points. In total, there are 43 strain measurement points; the testing method adopts resistance strain gauges with a resistance value of 120 Ω, a sensitivity factor of 2.08, and a marking distance of 100 mm, which are matched with the DH3819 static strain testing system for data acquisition.

4.4.2 Deflection Point

The longitudinal deflection measurement points of the bridge along the bridge are arranged on the left and right sides of the sidewalk, 5 deflection measurement points are arranged on the side spans and 4 measurement points are arranged on the middle spans, the specific arrangement is shown in Fig. 5; the testing method adopts the precision level meter, together with the indium steel ruler for testing.

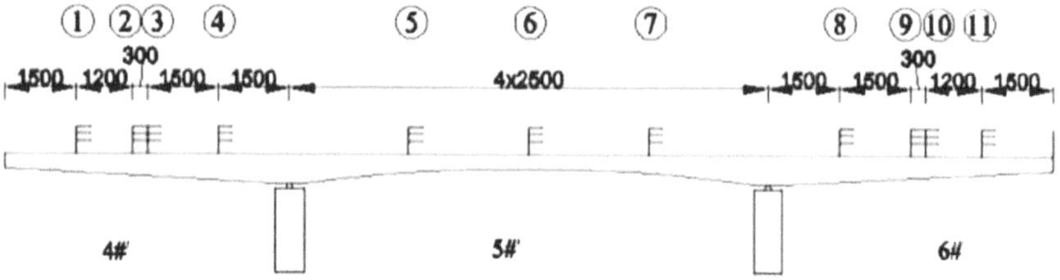

Fig. 5. Longitudinal bridge deflection arrangement (unit: cm)

4.5 Analysis of Results

Due to the symmetry of the two side spans of the bridge, the results obtained are similar, and usually the bias load is more reflective of the most unfavorable loading conditions, therefore, limited to the space only analyze the test results of conditions 1, 3 and 5.

4.5.1 Deflection

Table 2. Deflection evaluation index

conditions	numbering	1	2	3	4	5	6	7	8	9	10	11
1	Calibration factor	0.58	0.54	0.53	0.51	0.48	0.46	0.43	/	/	/	/
	relative residue	2.75	3.42	4.41	5.17	6.39	3.54	4.88	/	/	/	/
5	Calibration factor	0.51	0.45	0.46	0.49	0.54	0.57	0.56	0.50	0.46	0.45	0.50
	relative residue	7.07	7.59	7.56	8.65	4.65	3.84	5.46	7.47	5.97	8.75	7.07

As shown in Table 2, under the test load, the deflection calibration coefficient of the bridge ranges from 0.43 to 0.57, and the maximum measured deflection increment is 16.69 mm, which is in accordance with the specification requirements. After the unloading of each loading condition, the measured relative residual deflection of each

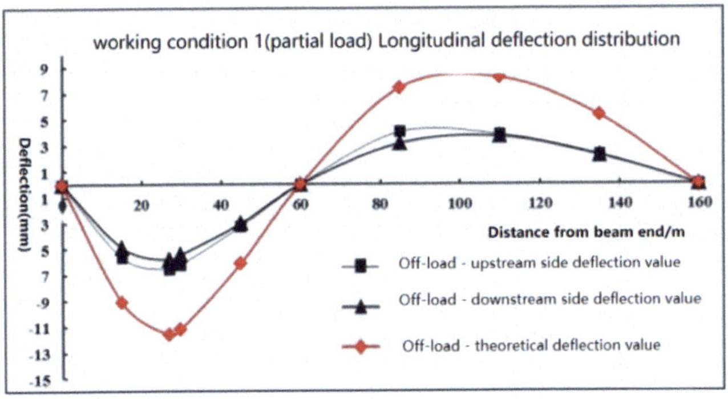

Fig. 6. Longitudinal deflection along the bridge for Case 1

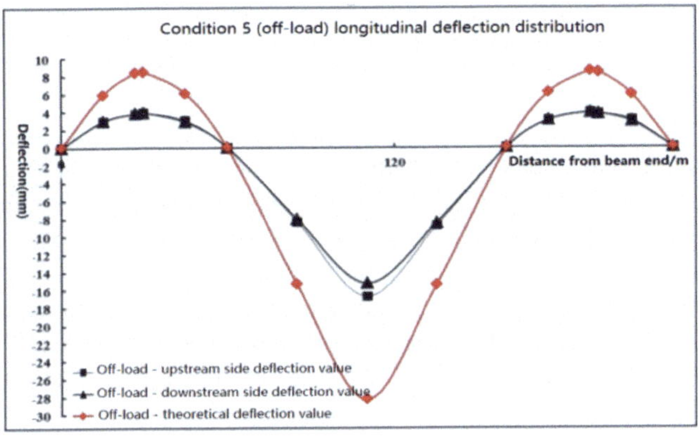

Fig. 7. Longitudinal deflection along the bridge for Case 5

control point (maximum 8.75%) is less than the limit value (20%) stipulated in the "Specification for Testing and Evaluating the Bearing Capacity of Highway Bridges (JTG/TJ21–2011) [9] (hereinafter referred to as the "Evaluation Specification").

From Figs. 6 and 7, it can be seen that the measured deflection values are smaller than the theoretical values, and the deflection trend of each control section along the longitudinal distribution curve is smooth and continuous, which is basically consistent with the theoretical calculation curve, indicating that the longitudinal force transmission of the bridge is normal, and the overall structural performance of the bridge span is normal.

4.5.2 Strains

Strain at the Bottom of the Beam. The comparative analysis of strain test results and theoretically calculated values for each strain test section under static load test conditions 1, 3 and 5 is shown in Table 3.

Under the test load, the measured strains of JM1, JM2 and JM3 sections were generally normal, and the strain calibration coefficients were in the range of 0.56–0.77,

Table 3. Beam bottom strain

working condition	measurement point	fe(με)	fp(με)	fs(με)	η(fe/fs)	(fp/ft) %
1	3#	52	0	72	0.72	0
	4#	48	1	72	0.67	2.04
	5#	44	2	72	0.61	4.35
3	3#	−27	−1	−35	0.77	3.57
	4#	−24	0	−35	0.69	0
	5#	−22	−1	−35	0.63	4.35
5	3#	63	2	99	0.64	3.08
	4#	58	1	99	0.59	1.69
	5#	55	1	99	0.56	1.79

Note: *fe*-Measured elastic strain; *fp*-Measured residual strain; *fs-calculated value*

which were generally lower than or in the normal values of the common bridge calibration coefficients recommended by the Ministry of Transportation and Communications' "Test Regulations" (the strain calibration coefficients of the concrete strains of the prestressed concrete bridges were in the range of 0.6–0.9), due to the actual strength of the materials being higher, the importance of the calculation theories or the Simplified calculation schema is biased towards safety, etc. may be the reason for the small calibration coefficient. The above results show that: the measured relative residual deformation of each control strain measurement point is less than the limit value (20%) specified in the "Assessment Regulations", and the maximum relative residual strain is 4.35%.

Web Strain. After averaging the left and right side strains of the beam web, the measured strain distribution along the beam height is shown in Fig. 8.

By analyzing the strain distribution along the girder height, it can be seen that the strain varies normally with the girder height, the theoretical calculation of the neutral axis position from the bottom of the girder JM1–JM3 were 123.33 cm, 242.27 cm, 106.58 cm, respectively, and the measured neutral axis from the bottom of the girder JM1–JM3 were 150.62 cm, 267.45 cm, 114.62 cm, respectively, and the reason for the deviation is possibly The reason for the large size may be related to the concrete bridge deck pavement, concrete guardrail and other participation in the force, which is a normal situation.

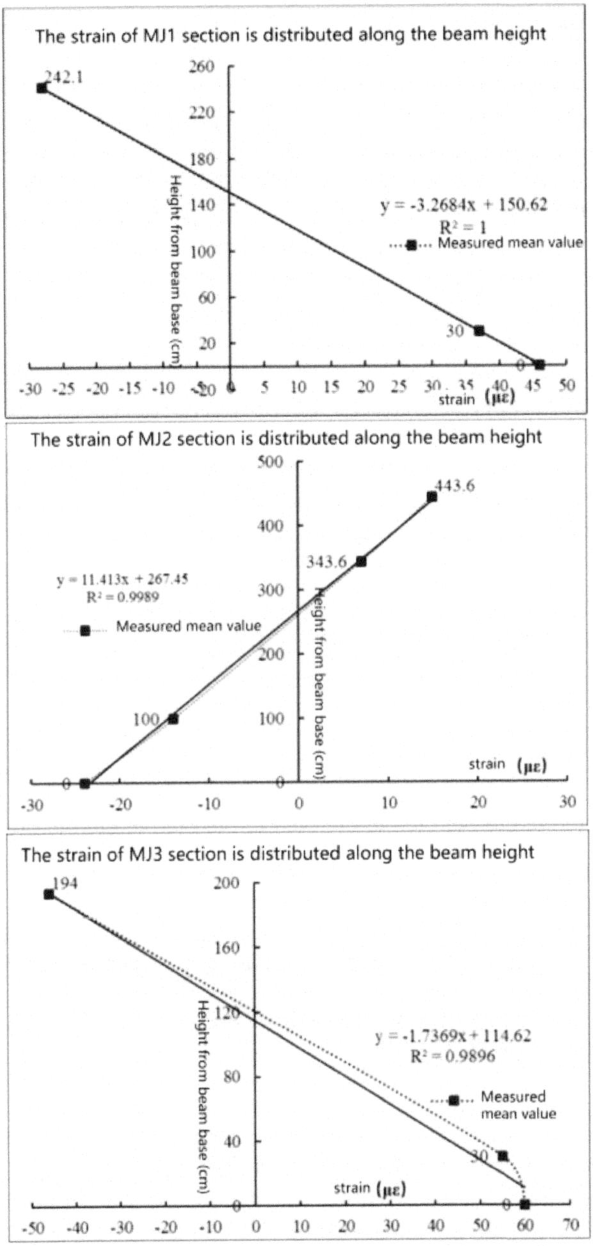

Fig. 8. JM1, JM2 and JM3 beam side measured strain-beam height distribution diagram

5 Conclusion

(1) Under the equivalent static load with load efficiency not less than 0.85, the relative residuals of strain and deflection measurement points are less than 20%, indicating that the structure is in an elastic state;
(2) The deflection and strain calibration coefficients of the measuring point are less than 1, and the beam does not crack under the load of each test condition, and the maximum value of the measured deflection does not exceed the design value under the test load, and the beam is completely restored after unloading. The strength, stiffness and crack resistance of the bridge superstructure meet the design load (Highway-Class I) and code requirements.
(3) Through the static load test of the structure, the original data of future maintenance and management are provided, and a theoretical basis is provided for the design and construction of such bridges in the future.
(4) Static load tests often require traffic interruptions and are not economical, so in the future the load test procedure should be further optimized or rapid evaluation methods such as calculation coefficients should be used.

Acknowledgments. Scientific Research Project of Introducing Talents in Kunming College (No. XJ20220031); Yunnan Province "Three Regions" Science and Technology Talents Project in 2023 (ZX20230237); Yunnan Province Science and Technology Specialist Program in 2023 (ZX20230180).

References

1. Rongfeng, C.H.E.N., Dan, S.O.N.G., Peng, G.A.O.: A review of load tests on highway bridges. Transportation Standardization **10**, 28–30 (2006)
2. Huang. L., Shi, X., Xu, L.: Dynamic and static load tests and structural evaluation of large-span continuous girder bridges. J. Shijiazhuang Rail. Univ. (Natural Science Edition) **25**(3), 24–29 (2012)
3. Yi, Y., Wang, Y., Liao, P.: Experimental study on bridge-forming load of Goldfish Bridge. Railway Construction **6**, 33–35 (2016)
4. Yi, Y., Lu, L., Ji, Y., et al.: Load test and bearing capacity assessment of Shunqihe Bridge. J. Build. Sci. Eng. **34**(6), 104–109 (2017)
5. Lio, J.: Research on static load test of prestressed concrete continuous girder bridge in service for 20 years. Bridge Construction **43**(5), 75–80 (2013)
6. Zheng, X., Yi, T., Yang, D., et al.: Research on rapid detection and virtual evaluation method of bearing capacity of small and medium-span bridges. Engineering Mechanics, 1–10
7. Liu, H., Zhu, Y., Yang, J.: Experimental study on static dynamic characteristics of cast-in-place continuous box girder bridge. National Urban Highway Society 16th (2007) Annual Conference: Transportation Science and Technology, pp. 60–62 (2007)
8. JTG/T J21-01-2015 Load test specification for highway bridges. People's Transportation Press, Beijing (2015)
9. JTG/T J21-2011 Specification for testing and evaluating the bearing capacity of highway bridges. People's Transportation Press, Beijing (2011)

The Influence of Mixed Fibers on the Bending Performance of SRSSC Beams Analyzed Through Finite Element Analysis

Zhixin Peng and Changchun Pei(✉)

Department of Civil and Hydraulic Engineering, Yanbian University, Yanji, China
peicc@ybu.edu.cn

Abstract. In this study, we analyzed the influence of glass-polypropylene fiber hybridization at different ratios on the flexural performance of steel reinforced sea sand concrete (SRSSC) beams. Through simulated experiments, we observed specimen failure modes and the load-deflection relationship, analyzed the effects of fiber incorporation on cracking, yielding, ultimate load, and deflection of the beams, elucidated the patterns of different fiber ratios on beam flexural stiffness, and revised the current code's load-carrying capacity calculation formulas based on experimental data. The results indicate that incorporating glass fibers alone and a glass-polypropylene hybrid fiber ratio of 1 can significantly enhance the flexural performance of SRSSC beams. These findings provide a theoretical basis for the selection and design of concrete materials in coastal and island construction.

Keywords: steel · seawater-sand concrete · hybrid fibers · bending performance · finite element analysis

1 Introduction

With the development of the construction industry, there is a significant consumption of natural resources, and some resources are facing the problems of depletion or exhaustion, especially for the construction of coastal cities and islands. Resources such as freshwater and river sand are severely scarce, which increases the construction period and cost [1]. Seawater Sea-sand Concrete (SSC) has similar properties to Normal Concrete (NC) and thus shows good application prospects in offshore and island construction [2]. SSC has high early strength [3], but its later strength is inferior to NC, and it is prone to brittle failure [4, 5].

In order to improve the mechanical properties of SSC materials, people usually add fibers to the concrete, which leads to a good improvement in various basic properties of SSC. For example, Huang Kaiji [6] improved the microstructure of the concrete and enhanced the erosion resistance of SSC by adding a mix of polypropylene and basalt fibers. Xu Aiyan et al. [7] delayed the erosion of OH- ions inside by adding carbon-basalt fibers, thereby improving the compressive strength. Bai Song et al. [8] improved various basic properties of SSC by incorporating glass-polypropylene fibers.

B. Yuan et al. (Eds.): ICCASE 2024, SUCI, pp. 441–446, 2024.
https://doi.org/10.1007/978-3-031-76102-7_47

In the research of SSC, most studies focus on the mechanical properties of SSC, while there is a gap in the research of Steel Reinforced Seawater Sea-sand Concrete (SRSSC) components. To fill this research gap, the ANSYS finite element simulation method was used for the first time, and the influence of different proportions of glass-polypropylene hybrid fiber on the bending performance of SRSSC beams was studied by changing the fiber incorporation. This research provides a theoretical basis for the subsequent study of SSC concrete engineering structures.

2 Research Scheme Design

2.1 Research Scheme

In this study, 6 beams (180 mm × 240 mm × 1600 mm) spanned 1300 mm were designed with a w/c ratio of 0.48 and a sand content of 28.5% (seawater and sea sand). The fiber mixing ratios of each beam and the mechanical performance parameters of the concrete are shown in Table 1. Longitudinal reinforcement included 4 HRB400 18 mm rebars, stirrups of HPB300 6.5 mm rebars at 100 mm intervals, and a Q345 I14 hot-rolled I-beam. The performance parameters of the steel section and the rebar are presented in Table 2. Protective layer thicknesses were 20 mm for rebar and 50 mm for concrete. Both rebar and steel were rust-treated. A four-point bending load method was used with displacement control in the finite element analysis.

According to literature review [9–11], the addition of glass fiber (GF) and polypropylene fiber (PPF) at a volume fraction of 0.1% to 0.3% can effectively improve the performance of concrete; when GF and PPF are hybridized, it combines the crack arrestment and toughening effects of both, enhancing various performance indicators of concrete such as tensile and compressive strengths. Therefore, this study selects GF and PPF at a volume fraction of 0.3% to investigate their performance improvement effects on SRSSC beams. The dimensions of the beam, the positions of the loading points, and the arrangement of the reinforcement in the cross-section of the beam are shown in Fig. 1.

Table 1. Experimental Beam Parameter Design

Experimental ID	Specimen ID	Fiber Volume Fraction (%)		Cuboid Compressive Strength (N/mm^2)	Tensile Strength (N/mm^2)	Elastic Modulus (10^4 N/mm^2)
		GF	PPF			
1	SRSSCB1	0	0	28.09	3.09	3.98
2	SRSSCB2	0.3	0	34.51	3.46	4.39
3	SRSSCB3	0	0.3	30.83	3.25	3.7
4	SRSSCB4	0.1	0.2	30.76	3.24	3.91
5	SRSSCB5	0.15	0.15	36.25	3.55	4.4
6	SRSSCB6	0.2	0.1	31.7	3.3	4.18

Table 2. Steel Material Performance Table

Steel Classifiction/Performance	Elastic Modulus (N/mm^2)	Yield Strength (N/mm^2)	Ultimate Tensile Strength (N/mm^2)
HRB400 longitudinal bars (HPB300 stirrups)	2.00×10^5 (2.10×10^5)	400 (300)	540 (420)

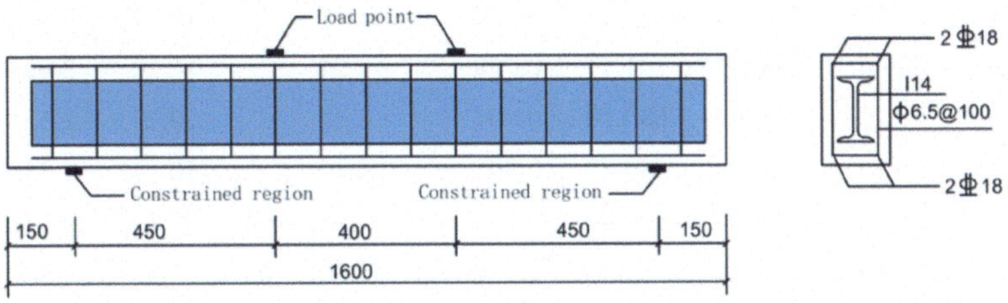

Fig. 1. Schematic Diagram of Beam Specimen Configuration (Unit: mm)

2.2 Modeling Process

Due to property differences among steel section, rebar, and seawater-seasand concrete, this study used separate modeling. Concrete, steel, and rebar were represented by SOLID65, SOLID45, and LINK8 elements. Rigid pads (50 mm × 100 mm) were positioned on beam tops and bottoms. Concrete's Poisson's ratio was 0.2; rebar and steel's, 0.3. Multilinear kinematic hardening models were applied for steel and rebar. Von Mises yield criterion was used for FE simulation with 50 load steps and 50 equilibrium iterations. Element details are in the Fig. 2.

(a) Concrete Unit (b) Profile Steel Unit (c) Rebar Unit

Fig. 2. Steel-Concrete Composite Beam Model

3 Result Analysis

3.1 Analysis of Characteristic Point Parameters

From the Table 3, it is observed that with the addition of fibers, the ultimate carrying capacity of SRSSC beams has generally increased. Specifically, beams 2 and 5 experienced a significant enhancement, with improvements of 4.7% and 5.9% respectively.

Table 3. Characteristic point loads, midspan deflection parameters

Specimen ID	P_{cr}/kN	P_y/kN	P_u/kN	P_y/Pu	f_y/mm	f_u/mm	U (f_u/f_y)
SRSSCB1	66.69	532.28	662.21	0.804	4.18	11.23	2.687
SRSSCB2	70.55	555.40	693.18	0.801	4.21	10.48	2.489
SRSSCB3	65.43	534.44	675.67	0.791	4.18	10.91	2.610
SRSSCB4	67.21	539.41	675.23	0.799	4.19	10.97	2.618
SRSSCB5	71.44	560.61	701.41	0.799	4.21	10.20	2.423
SRSSCB6	69.20	546.79	680.10	0.804	4.20	10.73	2.555
Mean	68.23	545.34	680.35	0.798	4.19	10.77	2.570
Variance	5.53	131.88	195.99	2.29E−05	1.90E−04	0.14	9.14E−03
Standard Deviation	2.35	11.48	14.00	4.78E−03	1.38E−02	0.37	9.56E−02

However, the P_{cr} of beam 3 decreased, which is attributed to the more pronounced agglomeration effect of polypropylene fibers in concrete. Additionally, after the incorporation of fibers, the ductility coefficient u of the beams has decreased. From the ratio of P_{cr} to P_u, it is known that the beams have a safety reserve of nearly 20%.

3.2 Analysis of Load-Deflection Curves at Midspan

As shown in the Fig. 3, the failure process of the SRSSC beam is divided into three stages: elasticity, service with cracks, and ultimate failure. The cracking and yield loads are found to be within the ranges of 0.097 to $0.102P_u$ and 0.791 to $0.804P_u$, respectively. With the incorporation of fibers, there is an enhancement in the ultimate load-bearing capacity, particularly for beam number 2 and 5, which show a significant increase.

3.3 Analysis of Cross-Sectional Load-Bearing Capacity

According to the Chinese standard YB9082-2006, the formula for calculating the ultimate load capacity is as shown below, where M_u is the simulated value, M'_u is the calculated value, and the results can be found in the Table 4 below.

$$M_{u'} \leq M_{by}^{ss} + M_{bu}^{rc};\ \text{Among which } M_{by}^{ss} = \gamma_s W_{ss} f_{ssy},\ M_{bu}^{rc} = A_s f_{sy} \gamma h_{b0}$$

From the Table 4, it is observed that the ratio of M_u to M'_u is greater than 1, and the data dispersion is relatively small. This indicates that the calculated results are on the safe side, with errors all less than 10%. It is feasible to use the standard formula for calculations.

Revise the formula to include α1 as the fiber influence factor (set to 1 when no fibers are added) and α2 as the concrete brittleness influence factor:

$$M_u = \alpha_1 (M_{by}^{ss} + \alpha_2 M_{bu}^{rc})$$

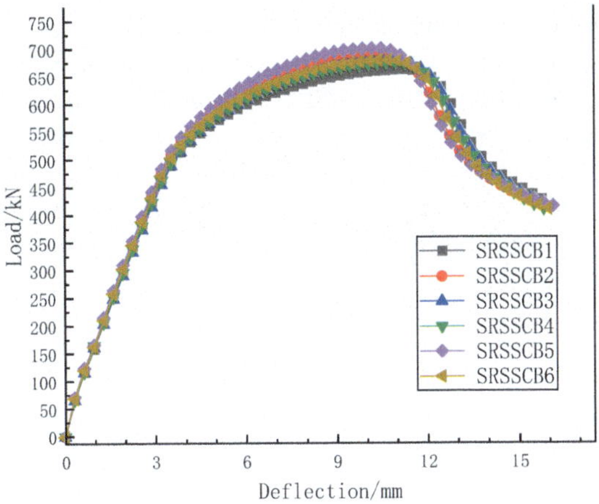

Fig. 3. Specimen Load-Deflection Curve

Table 4. Comparison of Simulated and Calculated Bending Moment Values

Specimen ID	M_u/kN·m	M'_u/kN·m	M_u/M'_u
SRSSCB1	148.997	140.196	1.063
SRSSCB2	155.966	142.120	1.097
SRSSCB3	152.026	141.115	1.077
SRSSCB4	151.927	141.094	1.077
SRSSCB5	157.817	142.524	1.107
SRSSCB6	153.023	141.374	1.082
Mean	153.293	141.404	1.084
Variance	9.922	0.681	2.55E−04
Standard Deviation	3.150	0.825	0.016

By substituting the parameters, the values of $\alpha 1$ were obtained as 1.032, 1.013, 1.013, 1.041, and 1.018, respectively, and the value of $\alpha 2$ was found to be 1.098.

4 Conclusions

This paper examines the effect of glass-polypropylene hybrid fibers on the flexural performance of SRSSC beams, with the following conclusions:

1. The incorporation of glass-polypropylene fibers leads to a certain enhancement in the load-bearing capacity of SRSSC beams. Notably, when glass fibers (GF) are added alone or in a ratio of GF to PPF of 1:1, the increase in load-bearing capacity is more pronounced. The critical load (P_{cr}) is increased by 5.79% and 7.12% respectively, and the yield load (P_y) is enhanced by 4.34% and 5.32% respectively. The ultimate load (P_u)

also sees an increase of 4.68% and 5.92% respectively. However, with the addition of fibers, there is a reduction in the ductility coefficient (μ) of the SRSSC beams.
2. The load-bearing and failure process of SRSSC beams is similar to that of conventional concrete beams, and it can be divided into three stages: the elastic stage (0.097 to 0.102 P_u), the stage with crack propagation (0.791 to 0.804 P_u), and the failure stage.
3. According to the statistical analysis of the results, the improvement in the flexural load-carrying capacity of SRSSC beams due to the addition of glass-polypropylene fibers is limited and not significant.

References

1. Yang, S.T., et al.: Research progress on material and structural performance of marine aggregate concrete. J. Ocean Univ. China (Natural Science Edition) (10), 11–19 (2023). https://doi.org/10.16441/j.cnki.hdxb.20230028
2. Vidhyadharan, V.P., Johny, A.: Utilization of seawater and sea sand in concrete for the sustainability of natural resources. Materials Today: Proceedings (2023). https://doi.org/10.1016/j.matpr.2023.03.559
3. Hanif, A.: Applicability and limitations of using seawater and sea sand in concrete. Materials Today: Proceedings (2023). https://doi.org/10.1016/j.matpr.2023.05.339
4. Mashayekhi, A., Hassanli, R., Zhuge, Y., Ma, X., Chow, C.W.: Synergistic effects of fiber hybridization on the mechanical performance of seawater sea-sand concrete. Constr. Build. Mater. **416**, 135087 (2024). https://doi.org/10.1016/j.conbuildmat.2024.135087
5. Pan, D., Yaseen, S.A., Chen, K., Niu, D., Leung, C.K.Y., Li, Z.: Study of the influence of seawater and sea sand on the mechanical and microstructural properties of concrete. J. Build. Eng. **42**, 103006 (2021). https://doi.org/10.1016/j.jobe.2021.103006
6. Huang, X.K.: Research on the erosion resistance of hybrid fiber sea sand concrete [Master's thesis, Hebei University of Science and Technology] (2022). https://doi.org/10.27107/d.cnki.ghbku.2022.000704
7. Xu, A.Y., Du, Y.X., Pan, L.J.T., Zhu, D.J.: Degradation mechanism and life prediction of tensile performance of hybrid carbon-basalt fiber in sea water and sea sand concrete. J. Composite Mat. 1–11 (2024). https://doi.org/10.13801/j.cnki.fhclxb.20231206.001
8. Bai, S., Xu, J.S., Xin, M., Lu, Z.Q., Wang, Y.: Analysis of basic mechanical properties of fiber-reinforced sea water and sea sand concrete. J. Lanzhou Ins. Technol. **05**, 16–19 (2023)
9. Su, W.W.: Study on the physical and mechanical properties of fiber-reinforced recycled coarse aggregate concrete [Master's thesis, Guangxi University] (2021). https://doi.org/10.27034/d.cnki.ggxiu.2021.000285
10. Xu, H.F.: Research on the basic mechanical properties of hybrid fiber reinforced recycled concrete [Master's thesis, Xi'an University of Architecture and Technology] (2016). https://doi.org/10.27393/d.cnki.gxazu.2016.000302
11. Yao, L., Pei, C.H.: Experimental study on the basic mechanical properties of hybrid fiber reinforced recycled concrete. Shanxi Architecture (35), 117–119 (2016). https://doi.org/10.13719/j.cnki.cn14-1279/tu.2016.35.062

Experimental Study on the Bonding Performance of Rarite Radiation Shielding Concrete with Anticorrosion Coated Steel Bars

Kangle Yue(✉)

Colleague of Civil Engineering, Lanzhou Jiaotong University, Lanzhou 730000, China
1342391225@qq.com

Abstract. Barite concrete can absorb X-rays and γ-rays and has good shielding performance, which is important for the development of nuclear energy. In this paper, C30 concrete is formulated, and six sets of center-drawn specimens are used to study the effects of factors such as rebar diameter and rebar anticorrosive coating on the bond performance of barite concrete and anticorrosive-coated steel bars. It was found that the anticorrosive coating of the steel reinforcement leads to a decrease in bond performance, and the smaller the diameter, the greater the effect. These studies can provide some data reference for the design and construction of barite anticorrosion concrete structures.

Keywords: Barite Radiation Shielding Concrete · Bonding performances · Centre pull-out test

1 Introduction

As a clean energy source, nuclear power is mainly used to generate electricity, which is an important energy component for realizing the goal of "carbon peaking and carbon neutrality". As of February 24, 2023, China has 53 nuclear power plants in operation, the third largest in the world. This shows that nuclear power is developing vigorously in China. However, while vigorously developing nuclear power, the safety of nuclear facilities should not be ignored.

Barite concrete whose aggregate is mainly barite and its chemical composition is barium sulfate. It has its own high density, stable chemical properties, can absorb X-rays and γ-rays, and can play a good anti-radiation performance [1–3]. Therefore, barite concrete can be used in nuclear facilities. At present, the research of scholars on barite concrete mainly focuses on the aggregate properties, proportion, mechanical properties and shielding properties of barite concrete [1–3].

As for reinforced concrete, good bonding properties are the basis for the normal operation of reinforced concrete structural members. Therefore, the importance of bond properties for reinforced concrete structures is self-evident. For the bond performance between reinforced concrete, scholars have long ago used beam test and pull-out test to carry out research. It was found that factors such as concrete strength [4, 8], rebar diameter [5, 7], bond length [6, 8], and rebar coating [7] would affect the bond performance.

B. Yuan et al. (Eds.): ICCASE 2024, SUCI, pp. 447–452, 2024.
https://doi.org/10.1007/978-3-031-76102-7_48

However, little research has been conducted on the bonding performance of barite concrete with corrosion-resistant coated steel bars. Therefore, this paper designs 6 groups (3 specimens in each group) of center-drawn specimens and adopts the center-drawn test to study the bond performance between anticorrosive-coated reinforcement bars and barite radiopaque concrete.

2 Experimental Materials and Mixing Ratios

Barite concrete in this experiment consisted of barite crushed stone, barite sand, PO 42.5 normal silicate cement, S2 grade fly ash and tap water. The content of barium sulfate is 93.1%, the diameter of barite crushed stone is 5-20mm, the apparent density is 3.99kg/cm3, the diameter of barite sand is 0-5mm, the apparent density is 4.05kg/cm3. This experiment in accordance with the experimental requirements for the preparation of concrete strength grade C30, concrete fit such as the following Table 1, to produce a set of 100 × 100 × 100 mm compressive test block, standard curing 28 days to test the compressive strength of 36.64 Mpa, in line with the specification of the "standard test method for mechanical properties of ordinary concrete," the requirements. Concrete density is greater than 2800 kg/m^3, in line with the "barite anti-radiation concrete application technical specifications" specification requirements.

In this experiment, HRB400 grade rebar with diameters of 12 mm, 16 mm and 20 mm were used. The floating rust of the rebar was treated and then ordinary anticorrosive paint was applied.

3 Center Pull-Out Test

3.1 Experimental Design

In this experiment, drawing on the ordinary reinforced concrete center pull test, a total of 6 groups (3 specimens per group) of center pull specimens were designed to investigate the bonding properties between barite radiopaque concrete and anticorrosive coated reinforcement bars, taking into account the strength of the concrete (C30), the diameters of the reinforcement bars (12 mm, 16 mm, and 20 mm), and the length of the bond of 5 D (D is the diameter of the reinforcement bars), and, for the pull test specimen that probed for anticorrosive coatings, the The bonded section of the rebar was coated with anti-corrosion coating. The size of the center pullout specimen was 150 mm × 150 mm × 150 mm, and a PVC pipe with a diameter slightly larger than that of the rebar was used to control the bond length of the rebar in the concrete. The specimens were made according to the mix ratio and cured under standard conditions for 28 days for the next experiment.

3.2 Loading Methods

The center pullout test was carried out on the HUALONG universal testing machine in the Mechanics Laboratory of Lanzhou Jiaotong University, as shown in Fig. 1. The displacement method was used to control the whole loading device, and the loading rate

Table 1. Radiation Shielding Concrete Proportions

Series	W/C	sand rate (%)	Water (kg/m^3)	Cement (kg/m^3)	Coal ash (kg/m^3)	Coarse aggregate (kg/m^3)	Fine aggregate (kg/m^3)
C30	0.48	35	200	374.99	41.67	1692.46	911.32

was 0.01 mm/s. Three center-drawing specimens were measured in each batch, and the loading was stopped when the displacement of the loading section was 20 mm or the specimen was damaged. The loading device is shown in Fig. 1.

Fig. 1. Drawing device diagram

4 Analysis of Experimental Results and Influencing Factors

4.1 Results of the Experiment

Equation (1) was used to calculate the ultimate bond strength between barite radiant barrier concrete and steel reinforcement, and the results are shown in Table 2

$$\tau_{max} = \frac{F}{\pi dl} \tag{1}$$

where: F is the ultimate pullout load, kN; τmax is the ultimate bond strength, Mpa; d is the diameter of the steel bar, mm; l is the length of the reinforced concrete bond section, mm.

Figure 3 shows the bond-slip curves of barite radiopaque concrete and steel reinforcement, and it can be seen that the bond-slip curves of barite radiopaque concrete and steel reinforcement are basically the same as those of ordinary concrete and steel reinforcement. In which the pullout damage can be roughly divided into three stages, namely, the rising stage, the falling stage and the residual stage, while the splitting damage has only the rising stage. For the ordinary center-drawing specimen with splitting damage, the concrete between the ribs of the reinforcement was crushed, while for the coated center-drawing specimen, the concrete between the ribs of the reinforcement was more intact, but the anticorrosion coating was damaged (Fig. 2).

Table 2. Details of specimens for pull-out test

Series	Average ultimate bond stress (MPa)	Mean ultimate displacement of the loaded end (mm)	Destruction pattern
C30–16-5D	15.55	5.87	S
C30–16-5DT	12.58	4.49	S
C30–20-5DT	11.51	4.11	S
C30–12-5D	16.40	4.20	P
C30–20-5D	12.58	5.32	S
C30–12-5DT	10.16	5.33	P

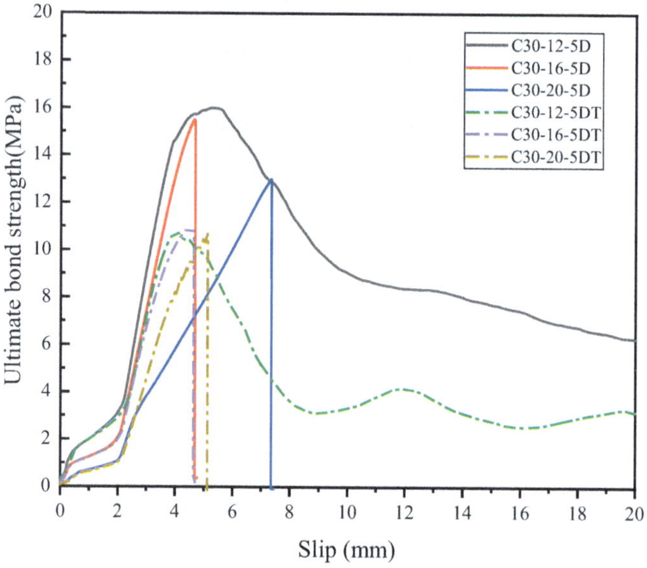

Fig. 2. Bond-slip curves with different influencing factors

4.2 Analysis of Factors Affecting Bonding Performance

Figure 3 shows that the ultimate bond strength of the specimens with the same bond length (5D) and concrete strength (C30), etc., decreased by 23.3%, 22.3%, and 8.5%, respectively, as the diameter of the reinforcement bars was changed from 12 mm to 16 mm and then to 20 mm. It can be seen that the ultimate bond strength decreased significantly after applying ordinary anticorrosion coating on the bond section. In the case of constant concrete strength, the larger the diameter of the rebar, the smaller the effect of anticorrosion coating on the ultimate bond strength. The larger the diameter of the reinforcement bar, the larger the rib-to-rib spacing of the reinforcement bar, after applying the anticorrosion coating, due to the same thickness of the coating, the smaller the effect of the coating on the rib-to-rib spacing of the reinforcement bar, and therefore the smaller the effect on the ultimate bond strength.

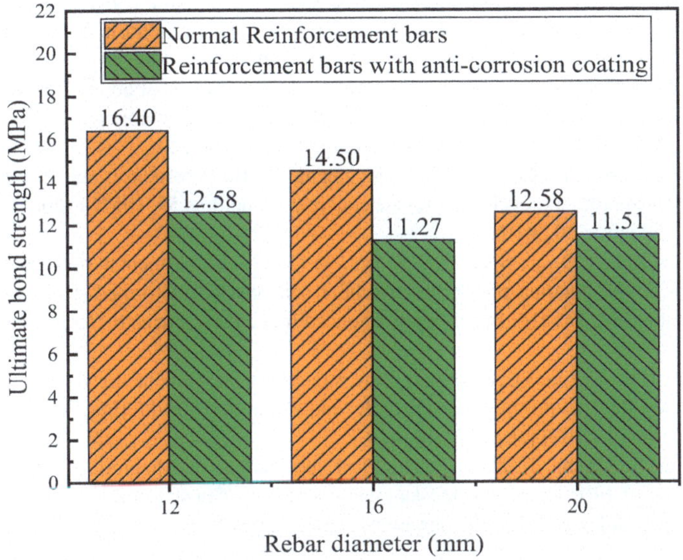

Fig. 3. Ultimate bonding strength

5 Conclusions

1. The density of barite concrete in this study is 3221 kg/m^3, which is greater than 2800 kg/m^3. The cubic compressive strength is 36.64 MPa, which are in accordance with the specification.
2. The anti-corrosion coating causes a reduction in the bond between the reinforcement and the barite concrete, the smaller the diameter of the reinforcement, the more the bond is reduced.

3. Ordinary anti-corrosion coatings were damaged after the completion of the test loading, so in the actual project, the anti-corrosion coatings used not only need to meet the requirements of the anti-corrosion function, but also need to meet the strength requirements, especially in the barite concrete structure.

References

1. Lee, S.Y., Daugherty, A.M., Broton, D.J.: Assessing aggregates for radiation-shielding concrete. Concrete int. **35**(5), 31–38 (2013). https://www.concrete.org/publications/internationalconcreteabstractsportal/m/details/id/51685628
2. Azeez, M.O., Ahmad, S., Al-Dulaijan, S.U., Maslehuddin, M., Naqvi, A.A.: Radiation shielding performance of heavy-weight concrete mixtures. Constr. Build. Mater. **224**, 284–291 (2019). https://doi.org/10.1016/j.conbuildmat.2019.07.077
3. Mostofinejad, D., Reisi, M., Shirani, A.: Mix design effective parameters on γ-ray attenuation coefficient and strength of normal and heavyweight concrete. Constr. Build. Mater. **28**(1), 224–229 (2012). https://doi.org/10.1016/j.conbuildmat.2011.08.043
4. Fathifazl, G., Razaqpur, A.G., Isgor, O.B., Abbas, A., Fournier, B., Foo, S.: Bond performance of deformed steel bars in concrete produced with coarse recycled concrete aggregate. Can. J. Civ. Eng. **39**(2), 128–139 (2012). https://doi.org/10.1139/l11-120
5. Yu, Y., Xu, J., Zhou, L., Liao, Z., Chen, W., Zheng, Y.: Bond responses and anchorage length of GFRP bar in precast recycled aggregate concrete. Structures **59**, 105761 (2024). https://doi.org/10.1016/j.istruc.2023.105761
6. Bai, Y. H., Zhang, D. Y., Shen, K., & Yan, Y. (2023). Experimental study and numerical simulation on the bonding properties of rubber concrete and deformed steel bar. Construct. Build. Mater. **383**, 131416. https://www.sciencedirect.com/science/article/pii/S0950061823011297
7. Zhang, W., Liu, Z., Liu, Q., Zhang, W.: Investigation of bond behavior between waterborne epoxy-coated steel reinforcing bars (WECR) and concrete. J. Build. Eng. **76**, 107328 (2023). https://doi.org/10.1016/j.jobe.2023.107328
8. Liu, S., Bai, C., Zhang, J., Zhao, J., Hu, Q.: Experimental and theoretical study on bonding performance of FRP bars-Recycled aggregate concrete. Constr. Build. Mater. **361**, 129614 (2022). https://doi.org/10.1016/j.conbuildmat.2022.129614

Experimental Study of the Shear Behaviour of the Contact Surface Between a Frozen Chalk Structure and a Foundation Under the Influence of Roughness

Chen Cheng(✉) and Dai-jun Jang

Lanzhou Jiaotong University, Lanzhou, Gansu 10732, China
1069357873@qq.com

Abstract. Bearing capacity of a pile foundation depends mainly on the magnitude of the pile-side freezing force. The magnitude of the pile-side freezing force is affected by a variety of factors, among which the roughness of the pile body is an important influencing factor. In this paper, the freezing strength problem of the pile-soil contact surface is transformed into the shearing strength problem of the frozen soil-concrete contact surface. The shear characteristics of the permafrost-concrete contact surface are investigated by a large-scale straight shear test. The results show that the peak shear strength at the contact surface increases with increasing roughness. The increase in contact surface roughness has a more pronounced effect on cohesion, while the effect on the angle of internal friction is relatively small.

Keywords: frozen silt · Roughness · mechanical properties · Large-scale direct shear test · Shear strength

1 Introduction

In recent years, the development policy for the western and northeastern regions has led to the construction of numerous cold-zone structures, which has resulted in a range of issues related to the interface between permafrost and structures. In engineering foundations in cold regions, temperature changes can cause complex phase change conditions within the soil. This leads to a more complex study of the soil-foundation contact surface than in normal-temperature soils. The problem of contact between permafrost and various structures cannot ignore roughness as one of the important factors [1, 2]. The bearing capacity of foundations for buildings in cold regions is significantly influenced by the level of roughness present in the foundation surfaces.

A significant number of both domestic and foreign scholars have utilised the results of straight shear tests to model the freezing strength of soil-structure contact surfaces. These tests were conducted on a range of soil-concrete binary specimens of varying sizes and materials. The objective of these studies was to analyse the mechanical properties of the contact surfaces. In addition, the effect of roughness on the contact surface shear characteristics was also investigated. Pan Yi-ming [3] conducted modified straight shear

B. Yuan et al. (Eds.): ICCASE 2024, SUCI, pp. 453–458, 2024.
https://doi.org/10.1007/978-3-031-76102-7_49

tests on permafrost-concrete specimens with varying roughness. The results showed that an increase in roughness leads to a significant increase in contact surface shear strength, primarily due to an increase in adhesive cohesion. The effect on the angle of internal friction is minimal. Zhang Ming-yi et al. [4] conducted straight shear tests on clayey soil and concrete slabs with varying roughness. The results show that the adhesive force at the contact surface increases with roughness and is close to the cohesive force of the clayey soil itself. Additionally, the shear strength of the contact surface increases significantly with roughness. In the context of the interpretation of phenomena associated with damage to test blocks: Felix and Cheng Hao et al. [5, 6] found that as roughness increases, the shear damage surface moves gradually upward from the contact surface into the upper soil. With regard to the reflection of the effect of roughness on the stress-strain curve of the contact surface: Chen Jun-huah et al. [7] adapted the sand-filling method to apply to regularly toothed contact surfaces. Their experimental results showed that the shear stress-displacement curves of clayey soils gradually transitioned from softening to hardening phenomena as surface roughness increased. Wang Tian-liang et al. [8] discovered that the shear stress-displacement curves of permafrost and normal-temperature soil are significantly different, exhibiting abrupt brittle damage. Additionally, the effect of roughness on shear strength is more pronounced at negative temperatures.

The present paper takes the loess in the Lanzhou area as the research object and creates different roughness frozen soil-concrete specimens. The shear strength law of the contact surface under different roughness and different normal stresses under negative temperature is analysed. Furthermore, the results are summarised so that the results of this test may be referenced in engineering practice in the frozen soil area.

2 Experimental Procedure

2.1 Test Procedure

In this paper, the three-dimensional problem of soil contact with the foundation surface is simplified to a planar problem of contact between permafrost and concrete. In order to reduce the influence of the direct shear specimen size effect on the results, a large-scale direct shear apparatus is used to conduct direct shear tests on the permafrost-concrete dichotomy specimen block. Figure 1 shows the device.

Fig. 1. High tension controlled straight shear

2.2 Specimen Preparation and Test Programme

The dimensions of the concrete test block employed in this experiment are 200 mm in length, 200 mm in width, and 100 mm in height. Concrete specimens with different rough surfaces are shown in Fig. 2

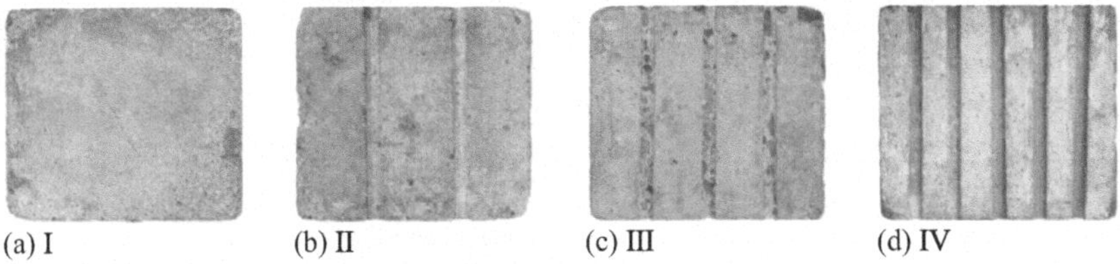

Fig. 2. Specimens with different concrete surface roughness

The loess used was extracted from Gaolian County in Lanzhou City. Its liquid limit is 27%, its plastic limit is 13.5%, its maximum dry density is 1.63 g·cm^{-3}, and its optimum moisture content is 15.5%.

The shear strength of the contact surface between permafrost and structures has a curvilinear distribution with water content. Viloyav et al. [9] demonstrated that the limited water content is between 80% to 90% of the saturated water content. Therefore, in this paper, we conducted the test using a sample of chalky soil with a water content of 22%.

Due to the greater regularity of the surface grooves on concrete specimens, this research employs the sand filling method as an evaluation technique for the structure's roughness. Check the equation below:

$$\mathrm{R} = \frac{V}{S} \tag{1}$$

where R represents roughness, V represents the total volume of sand filled on the surface, and S represents the surface area of the concrete specimen.

Spread the configured soil samples in layers on concrete test blocks, compact them sufficiently and test after freezing for 24 h.

3 Analysis and Discussion of Test Results

3.1 Characteristics of Shear Stress-Displacement Curve

By analyzing the results of the direct shear test at −3 °C on the contact surface of frozen chalk and concrete, we obtained the shear stress-displacement relationship under different normal pressures, as shown in Fig. 3. The findings indicate that the shear stress-shear displacement curves of frozen chalk concrete exhibit a trend that can be divided into four distinct stages: I Pressure-dense deformation stage. II stage of elastic deformation. III Plastic deformation stage. IV Slip damage stage.

Figure 4 shows the variation of shear strength with normal pressure for different roughness, under the same conditions of roughness. Shear strength increases as the normal pressure increases. The data points exhibited a linear relationship, as demonstrated by the fitted curves in Fig. 4, with correlation coefficients exceeding 0.97.

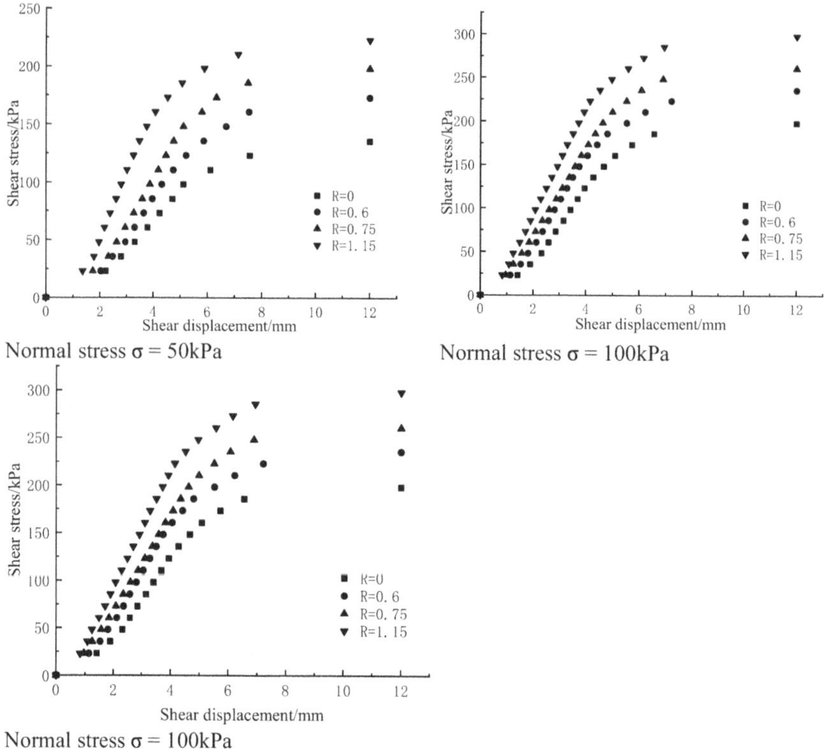

Fig. 3. Shear stress-shear displacement test values at frozen chalk-concrete interface

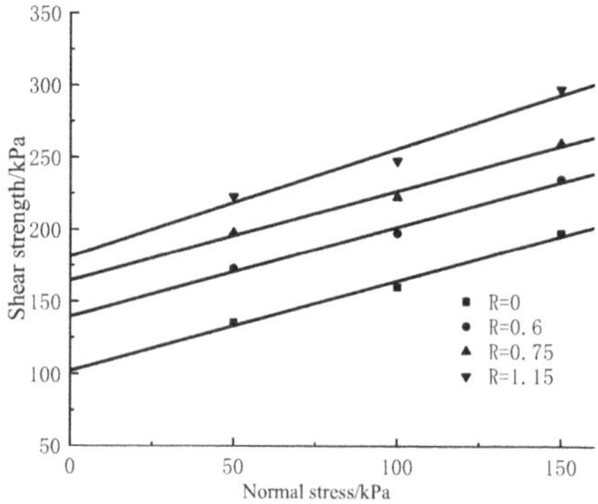

Fig. 4. Plot of normal stress versus shear strength under the influence of roughness

3.2 The Influence of Surface Roughness on Shear Strength Parameters

As the roughness at the contact surface remains constant, the cohesion and angle of internal friction at the contact surface of the permafrost-concrete specimens exhibit a linear relationship with increasing roughness. In Fig. 5, it can be observed that the cohesion growth rate is larger than the growth rate of the angle of internal friction in areas where the latter has a small growth rate.

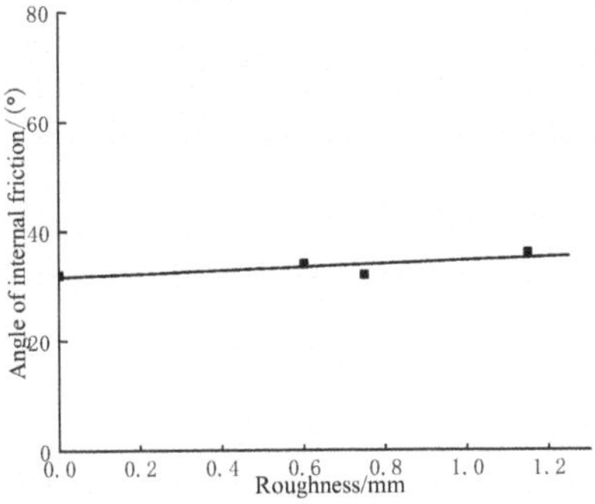

(a) Plot of roughness versus angle of internal friction

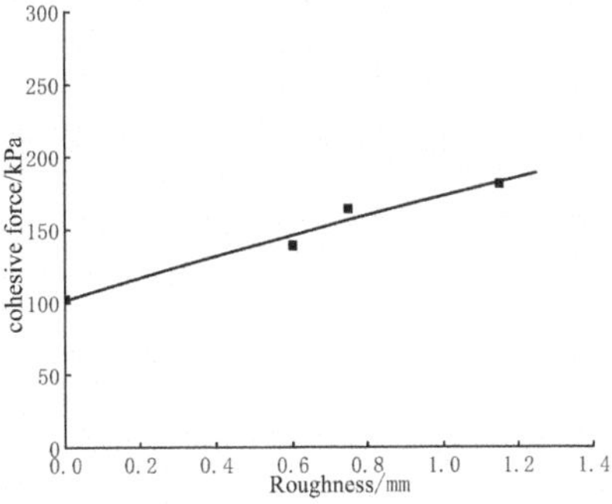

(b) Roughness versus Cohesion Graph

Fig. 5. Roughness vs. shear strength parameters

Shear damage along the contact surface of the specimen is primarily caused by the overcoming of friction and cohesion forces.

The friction force has two main components: the freezing force between the soil particles and the concrete on the contact surface, and the friction force between the soil embedded in the interior of the groove and the soil on the upper part of the contact surface, as well as the cementation force of the ice crystals in the soil.

The bonding force comprises three components: adhesion between soil particles and the concrete surface, adhesion between soil particles inside the recess and soil in the upper part of the recess, and cementation of ice crystals inside the soil at the recess.

4 Conclusion

(1) The increase in normal pressure at the frozen chalk-concrete interface increases the shear strength of the contact surface. As the roughness of the contact surface increases, the shear strength of the contact surface increases and gradually approaches the shear strength of the soil body. Increases in normal stress and roughness have varying degrees of improvement in contact surface adhesion and internal friction angle.

(2) The findings of this study demonstrate that, in practical engineering applications, a pile foundation with a corrugated base surface exhibits superior bearing capacity compared to a smooth indication. Furthermore, it can be concluded that modifying the surface roughness of the base surface is an effective method for enhancing the bearing capacity of the foundation. Nevertheless, the boundary value of the contact surface roughness was not subjected to a comprehensive investigation in this paper. Consequently, further research could be conducted to expand the range and gradient of roughness, with the objective of identifying the contact surface shear strength that corresponds to the maximum critical roughness.

References

1. Experimental study on mechanical characteristics of soil-structure interface. Chinese J. Geotech. Eng. **23**(4), 431-435 (2001). https://doi.org/10.3321/j.issn:1000-4548.2001.04.010
2. Experimental study on monotonic behavior of interface between soil and structure. Chinese J. Geotech. Eng. **26**(1), 21–25 (2004). https://doi.org/10.3321/j.issn:1000-4548.2004.01.003
3. Yiming, P.: Experimental Study on Shear Mechanical Properties of Frozen Soil-Concrete Contact. Jilin University, ChangChun (2020)
4. Zhang, M.Y., Bai, X.Y., Gao, Q., et al.: Indoor experimental study on force mechanism of pile-soil interface in clayey soil. Geotechnical **38**(08), 2167–2174 (2017). https://doi.org/10.16285/j.rsm.2017.08.002
5. Feligha, M., Hammoud, F., Belachia, M., et al.: Experimental investigation of frictional behavior between cohesive soils and solid materials using direct shear apparatus. Geotech. Geol. Eng. **34**(2), 567–578 (2016). https://doi.org/10.1007/s10706-015-9966-5
6. Cheng, H., Zeng, G., Zhou, M., Wang, S., Zhang, J., Wang, J.: Modeling of peak shear strength of clay-concrete contact surface considering the effect of roughness. Eng. Sci. Technol. **53**(04), 168–177 (2021). https://doi.org/10.15961/j.jsuese.202000574
7. Chen, J.H., Zhang, J.S., Li, K.: Mechanical property test of cohesive soil-structure contact surface considering roughness. J. Sichuan Univ. (Eng. Sci. Ed.) **47**(04), 22–30 (2015). JournalArticle/5b3bd04bc095d70f009176ad
8. Wang, T., Wang, H., Wang, O., He, Y., Feng, Z.: Experimental study on shear strength characteristics of silt and groove structural surface. J. Beijing Jiaotong Univ. **43**(3), 115 (2019). https://doi.org/10.11860/j.issn.1673-0291.20180179
9. Vialov, C.C.: Rheology of permafrost. Liu, J., Liu, Y. (et al.) (Translated). China Railway Press, Beijing (2005)

Study on the Influence of Geo-Studio Water Level Elevation on High Embankment Slope Stability

Koupeng Lv and Ming Li(✉)

School of Civil Engineering, University of South China, HengYang, China
1121446131@qq.com, 1826853441@qq.com

Abstract. The research of the reservoir bank slope under the condition of water level change has been very mature, but the stability state of the high embankment slope under the condition of water level change in the middle reaches of Xiangjiang River is very few. In this paper, a high embankment slope of the middle reaches of Xiangjiang River was selected as the research object. Based on the Geo-Studio software to analyze the steady water level of the high embankment slope at 25 m, 31 m and 37 m, the research results obtained the change of the infiltration line in different water levels and the stability of the high embankment slope. The results show that the higher the static water level, the more stable the slope is, the slope stability is in the most unfavorable state during the water level decline, and the faster the decline speed, the faster the safety factor drops, the more likely to occur instability.

Keywords: stability of embankment slope · water level change · seepage field · finite element simulation

1 Introduction

At present, the majority of experts and scholars at home and abroad have conducted extensive research on the change law of slope stability under the water level change [1–8], but there are few studies on the middle flow embankment of the Xiangjiang River. The author will carry out research in this aspect to make up for the deficiencies in this field and facilitate the application and promotion of engineering personnel in the construction of embankments. This paper analyzes the evolution of the seepage field during the water level rise and fall, and explores the mechanism of the water level change on the stability of the high embankment slope.

2 The Landslide Profile

The landslide area is located in the middle reaches of Xiangjiang River and the middle of Hengyang Basin, the surface cover is thick, the head layer is artificial filled soil, cohesive clay, weak permeability and thickness of 3.0 m; the second layer of silty clay has weak

B. Yuan et al. (Eds.): ICCASE 2024, SUCI, pp. 459–470, 2024.
https://doi.org/10.1007/978-3-031-76102-7_50

permeability; and silt, with low dry strength, toughness, medium permeability, and landslide, as shown in Fig. 1; the fourth layer is pebble with silt with strong permeability; the bedrock of the sliding bottom is strongly weathered mudstone with controlled thickness of 1.4 m. The water level in the study area is kept at 25 m in the year round, the water level can reach 31 m in rainy season and the highest 37 m during flood. Landslide status as shown, the landslide slope elevation of 38 m, slope is about 50 m wide, riprap and primary slope landslide will be completely exposed to the embankment, in the rainfall and water level rise directly into the soil layer, silt under the infiltration of water and water is very easy to further, instability in the rainy season flood comes will pose a serious threat to the stability of the high embankment, high embankment collapse, will cause incalculable losses.

Fig. 1. Land landslide of high embankment slope.

The most direct cause of the landslide is the change of water level caused by rainfall. In the initial stage, the water level is low, the soil saturation is small, the substrate suction is large, the pore water pressure is small, the soil effective is large, and the high embankment is in a stable state. With the increase of water level, after the soil saturation, the physical and mechanical properties of the landslide decrease; as the water level falls, the empty surface of the high embankment slope increases at the effective and severe decrease, which causes additional drag force on the one hand, infiltration damage in the soil, and the superposition of various factors leads to the slope instability.

3 Numerical Simulation Analysis

3.1 Theory of Computation

3.1.1 Theory of Shear Strength of Unsaturated Soil

The soil of high embankment slope usually changes with rainfall and water level, making the saturation state of soil in the state of saturation and unsaturated. In this case, the theory of unsaturated soil is more appropriate to study the stability of the shear strength theory of high embankment slope. For unsaturated soils, the substrate suction must be introduced as a new variable. For this reason, the bilateral quantity theory of shear strength in unsaturated soil is proposed by Fredlund et al.

$$\tau = c' + (\sigma - u_a)\tan\phi' + (u_a - u_w)\tan\phi^b \tag{1}$$

In the formula: c' *and* ϕ' is the effective stress strength parameter; σ is the normal total stress; u_w is the pore water pressure; $(u_a - u_w)$ is the matrix suction, ϕ^b is the capillary suction formed by the presence of the gas phase; which is a function of the substrate suction.

3.1.2 Soil-Water Characteristic Curve

The soil-water characteristic curve is the fundamental constitutive relationship in unsaturated soil mechanics, which describes the relationship between matrix suction and soil moisture content. In order to describe the influence of matrix suction on the stability of high embankment slope, water-soil characteristic curve can be used for analysis. At present, the characteristic equations used to describe the soil-water characteristic curve mainly include:

(1) The Fredlund & Xing Eq.

Fredlund to deduce the soil-water characteristic curve expression for any soil:

$$\frac{\theta}{\theta_s} = F(\psi) = C(\psi)\frac{1}{\left\{ln\left[e + \left(\frac{\psi}{a}\right)^b\right]\right\}^c} \tag{2}$$

$$C(\psi) = 1 - \frac{\ln(1 + \psi/\psi_r)}{\ln(1 + 10^6/\psi_r)} \tag{3}$$

In the formula: a is the parameter of the intake value; b is related to the dehydration rate; c is the parameter of the residual water content function; ψ is the matrix suction; ψ_r is the matrix suction force corresponding to the θ_r; θ_s is the saturated volume water content. Fredlund & Xing indicates that when the correction coefficient = 1, the effect is very small. The formula is: *Curve shape is minimally affected by*$C(\psi)$.

$$\theta_w = \frac{\theta_s}{\left\{ln\left[e + (\psi/a)^b\right]\right\}^c} \tag{4}$$

(2) Van Genuchten Eq.

Van Genuchten The relationship between the moisture content of the unsaturated soil and the matrix suction:

$$\frac{\theta - \theta_r}{\theta_s - \theta_r} = F(\psi) = \frac{1}{\left[1 + (\psi/\text{a})^b\right]^{\left(1-\frac{1}{b}\right)}} \tag{5}$$

The fitting parameters are a, b, and the symbolic meaning is the same as before.

3.1.3 The Saturated-Unsaturated Seepage Control Equation

$$\frac{\partial}{\partial x}\left(k_x\frac{\partial H}{\partial x}\right)+\frac{\partial}{\partial y}\left(k_y\frac{\partial H}{\partial y}\right)+Q=m_w\rho_w g\frac{\partial h}{\partial t} \tag{6}$$

In the formula: k_x, k_y is the permeability coefficient in x, y direction; Q is the applied boundary flow rate; ρ_w is the density of water; G is the gravity acceleration; The m_w is the specific water bulk weight; the m_w is defined as the negative value of the partial derivative of the volume water content θ_w on the matrix suction $(u_a - u_w)$, namely:

$$m_w = -\frac{\partial\theta_w}{\partial(u_a - u_w)} \tag{7}$$

3.2 Model Building and Parameter Setting

In order to further analyze the stability of high embankment slope and study the specific influence of water level change and seepage on the stability of high embankment slope, Geo-Studio is used to analyze different high embankment slope, different rainfall types and rainfall intensity, different water level and water level change rate. Geo-Studio software was used to establish a two-dimensional model of S1 high embankment slope (Fig. 2), divide the grid (Fig. 3), and import the parameters to calculate the numerical simulation of different conditions. The grid divided 3031 nodes and 2913 cells. The model boundary conditions are the lateral and bottom vertical direction constraints, with free above.

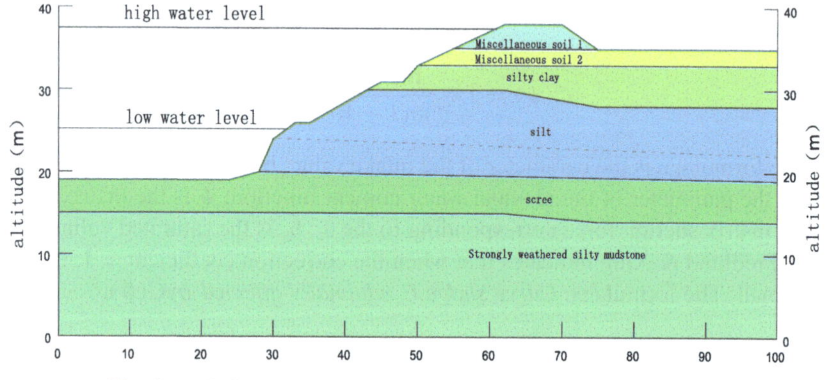

Fig. 2. A 2-dimensional model diagram of the high embankment.

Soil-water characteristic curve (SWCC) is an essential parameter. The relationship between matrix suction and volume water content of each soil layer is shown in Fig. 4. The physical and mechanical properties of the soil layer are shown in Table 1.

3.3 Working Condition Setting

Combined to the actual situation, in order to simulate the influence of different water levels, different water level rising rate and falling rate on high pore water pressure,

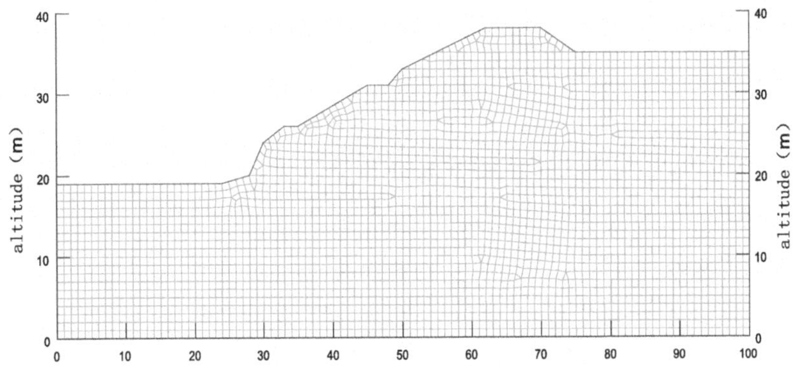

Fig. 3. Grid division diagram of the high embankment.

Table 1. Selection of rock and soil layer parameters.

Stratum	Natural density kN/m^3	Modulus of elasticity/GPa	Poisson ratio	Internal friction angle /(°) Natural state, and the saturated state		Cohesion /MPa Natural state, and the saturated state	
Miscellaneous soil 1	18.8	7.90	0.25	17.6	15.4	38.5	32.5
Miscellaneous soil 2	18.1	5.81	0.31	20.0	17.8	28.0	22.6
silty clay	19.5	6.30	0.33	22.3	20.1	29.0	23.1
silt	19.5	6.52	0.32	18.8	16.8	15.4	14.6
scree	20.5	50.00	0.21	35.0	29.6	/	/
Strong weathering mudstone	21	100.00	0.22	/	/	/	/

displacement and safety factor, consider the following six working conditions shown in the table, among which three different static water levels are set:

Working condition 1: the water level is 25 m (conventional water level);
Working condition 2: the water level is 31 m (the water level in rainy season);
Working condition 3: water level 37 m (flood level);
Set 3 dynamic water levels:
Case 4: the water level rises from 25 m to 4 m/d to 37 m and drops from 4 m/d to 25 m from the 60th day after the change,

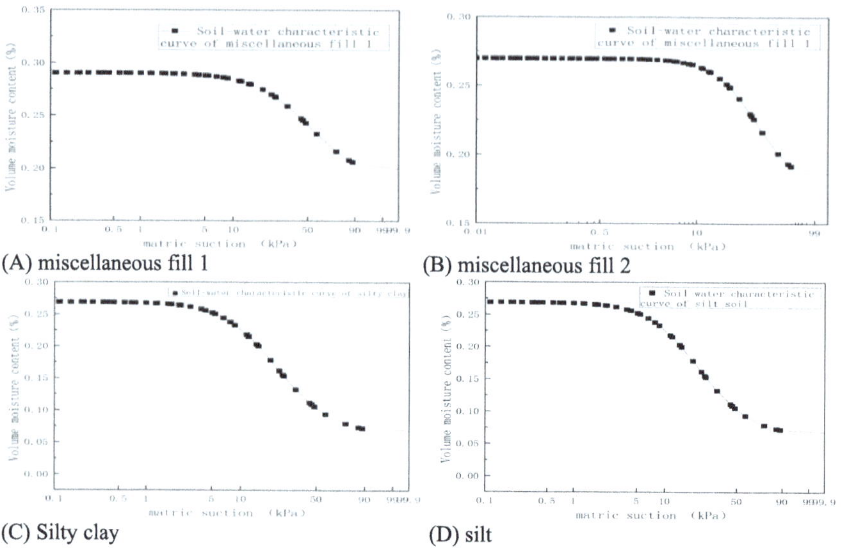

Fig. 4. Soil-water characteristic curve of each soil layer sample.

Case 5: the water level rises from 25 m at 2 m/d to 37 m and drops to 25 m at 2 m/d starting on the 60th day after the change;
Case 6: the water level elevation rises from 25 m to 0.5 m/d to 37 m from the 60th day to 0.5 m/d, and the water level elevation changes with time as shown in the Fig. 5.

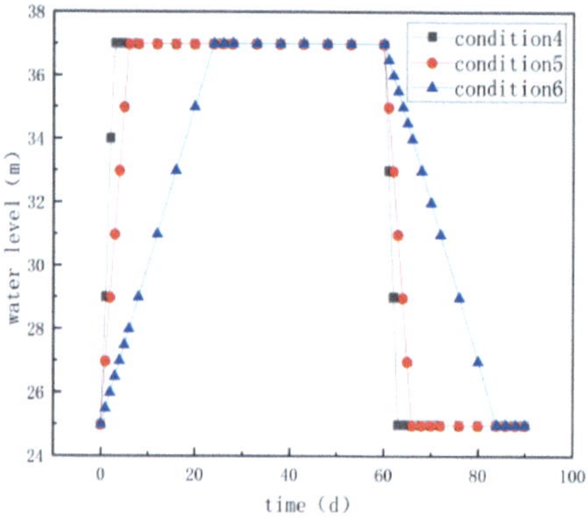

Fig. 5. Plot of water level elevation change over time.

3.3.1 Analysis of the Stability of High Embankment by Different Water Level

The simplified Bishop method was used to calculate the safety factor in SLOPE/W, setting the top and tip position of the sliding surface, respectively. The safety factor of the slope under different static water level working conditions is shown in Fig. 6. The safety factors of the water level of 25 m, 31 m and 37 m are 1.117,1.218 and 1.513 respectively. It can be seen that the higher the water level, the greater the safety factor, and the safety factors of the water level of 31 m and 37 m increase by 9% and 35% respectively on the basis of the water level of 25 m.

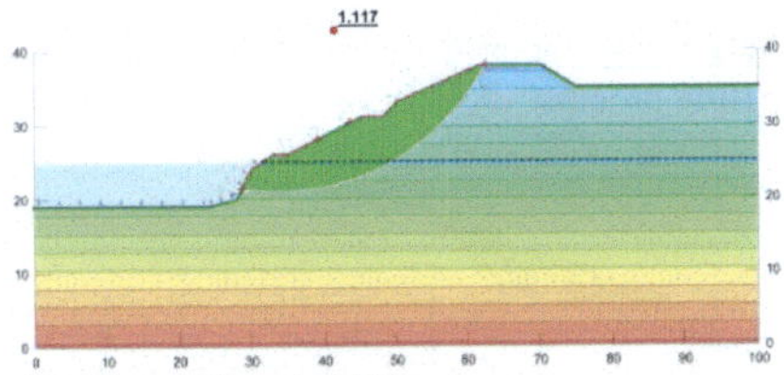

(A) S1 working condition 1

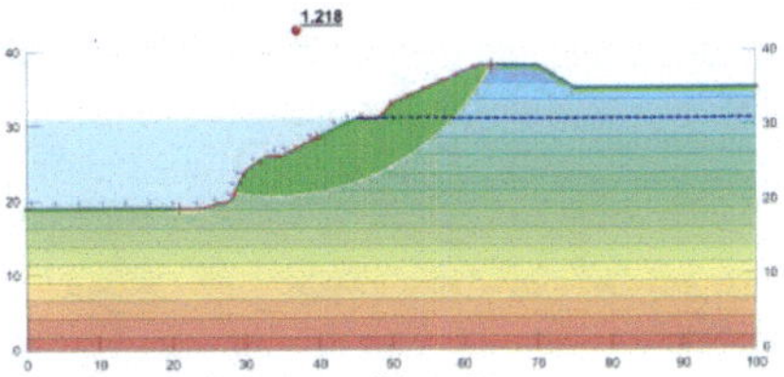

(B) S1 working condition 2

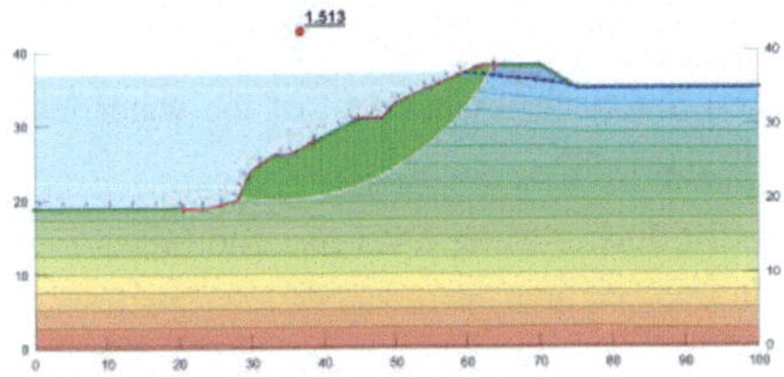

(C) S1 working condition 3

Fig. 6. Safety coefficient of different water levels of high embankment slope.

3.3.2 Analysis of Stability of High Embankment by Different Rate of Water Level

The Figs. 7 and 8 above shows the change of the infiltration line in the bank slope at different time nodes when the water level rise speed is 4 m/d, 2 m/d and 0.5 m/d, respectively. As can be seen from the figure, the infiltration line will change with the change of the water level, and the slower the water level changes, the smaller the distance between the infiltration lines, the smaller the radian of the infiltration line, and the lag effect is not obvious. The faster the water level changes, the greater the spacing between the infiltration lines, the radian of the infiltration line will become larger, and the more obvious the lag effect will be. It can also be learned from the figure that, during the rise of water level, the infiltration line shows a concave shape and the water level is flush with the highest end of the infiltration line, the infiltration line is convex and the water level is flush with the lowest end of the infiltration line.

In SLOPE/W, the shear entrance and exit of the sliding surface are set at the top and foot of the slope. The variation of the safety factor of the high embankment slopes is shown in Fig. 9. During the rise of the high embankment water level, The safety factor of high embankment slope will increase with rising water level, And the faster the water level rises, The higher the safety factor of the high embankment slope increases, When the water level no longer rises, After reaching the highest water level, The safety factor gradually decreases over time and finally tends to a stable value, The faster the water

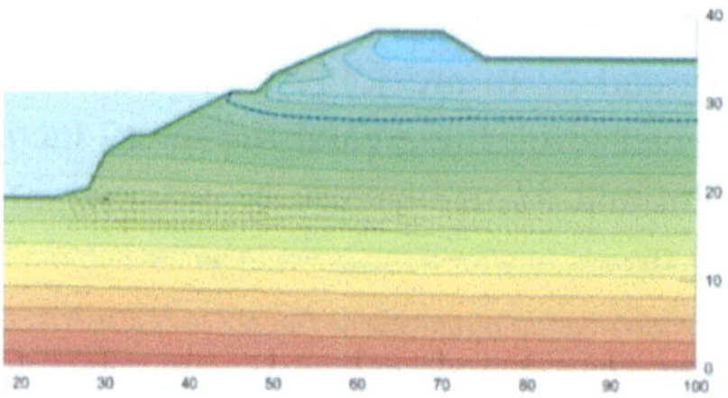

(A) Inflow diagram during the water level rise process

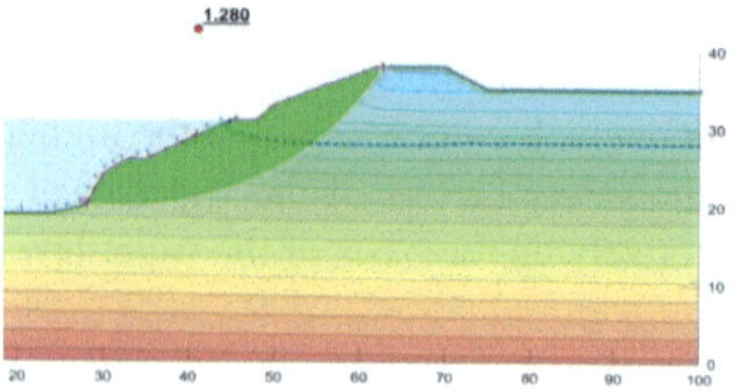

(B) Safety factor diagram of the water level rising process

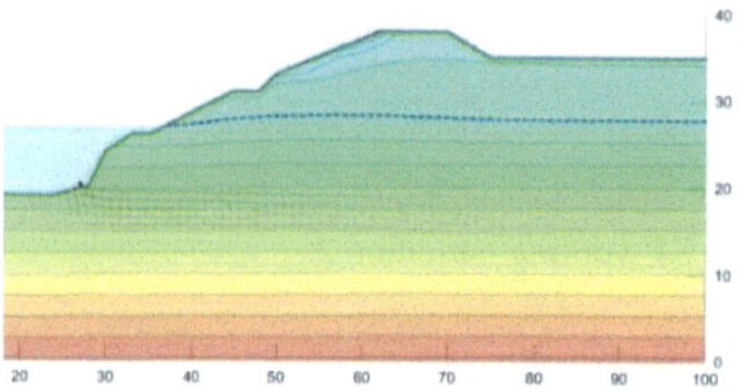

(C) Seepage diagram of water level descent process

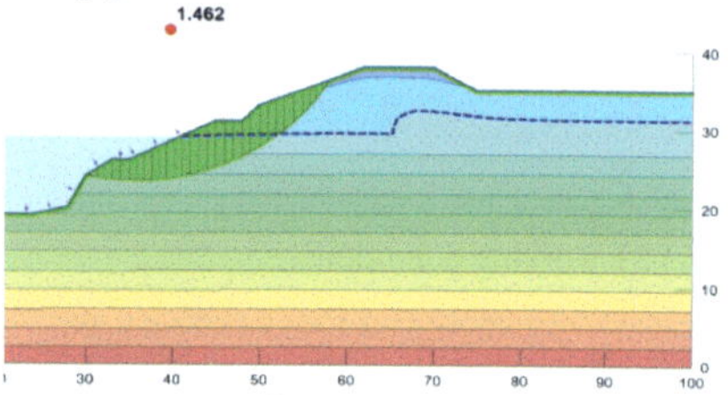

(D) Safety factor diagram of the water level drop process

Fig. 7. Diagram of seepage and safety factor during water level rise and fall.

level rises, The safety factor will also be greater after stability, The slower the water level rises, After stability, the safety factor will be smaller; In the course of the falling water level, The safety factor of the bank slope decreases as the water level drops, The faster the water level drops, The faster the safety factor is lowered, The slower the water level drops, The slower the safety factor is lower, When the water level drops to a minimum, The gradually rising safety factor will eventually lead to a stable value, And the faster the drop, the smaller the final safety factor is.

3.4 Analysis of Computational Results

3.4.1 Analysis of Conditions 1,2 and 3

The higher the water level, the effective severe reduction of the soil above the infiltration line under the action of the river water. The river water infiltration has the effect on the soil. In the stability of the high embankment slope, the decrease will have double effect: First, the gravity component along the slope decreases, and the stability of the high embankment slope, which is not conducive to the stability of the high embankment slope. When the water level rises, the gravity of the water itself will give a positive pressure to the slope, which is conducive to the stability of the high embankment slope and increase the safety factor.

3.4.2 Analysis of the Results of Working Condition 4-S1,5-S1 and 6-S1

During the water level, The slope internal seepage field has a lag effect, And the faster the rise, the more obvious the lag effect, And simultaneously produce a dynamic water pressure pointing from the outside of the slope to the inside of the slope body, The faster

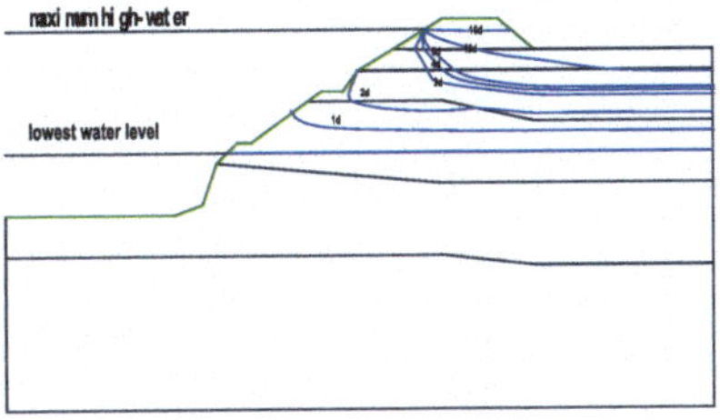

(A) Change of four water level rising infiltration line under working condition

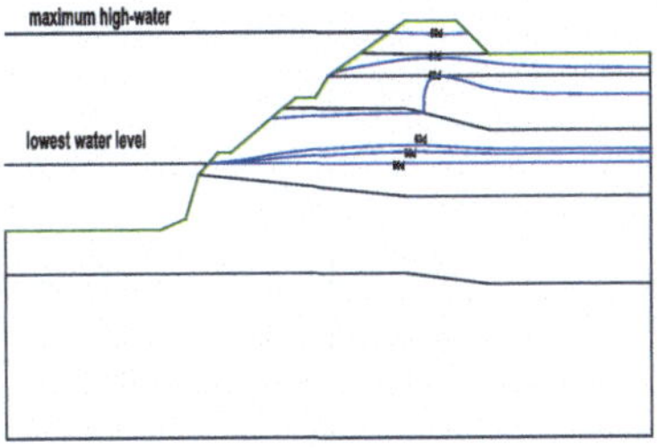

(B) Change diagram of four water level drop infiltration line in working condition

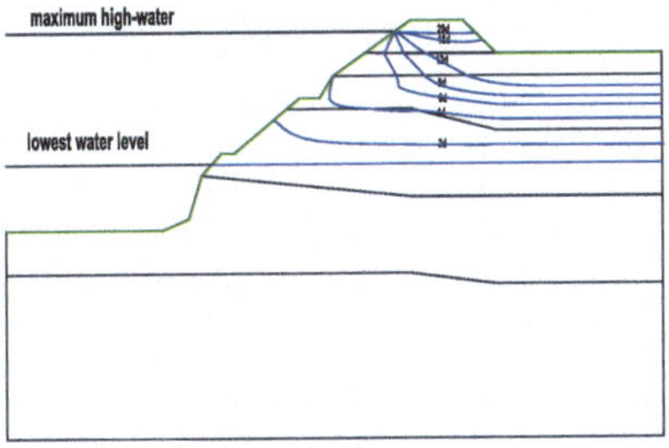

(C) Change diagram of water level rising infiltration line under working condition 5

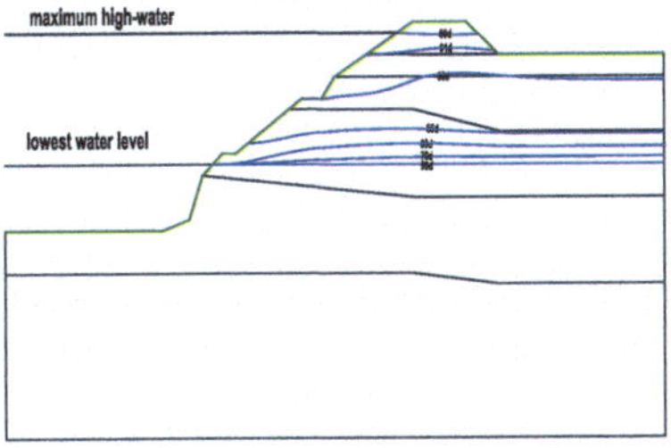

(D) Change diagram of water level drop infiltration line in working condition 5

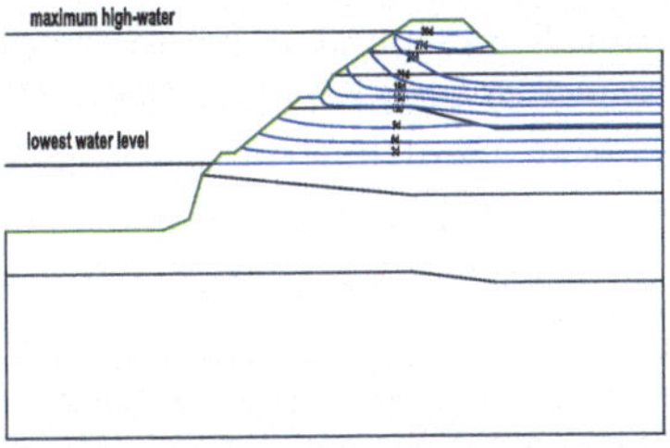

(E) Change diagram of six water level rising infiltration line under working condition

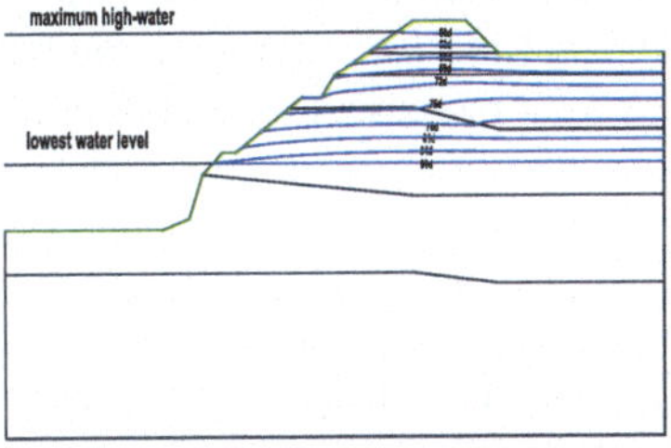

(F) Change diagram of six water level drop infiltration line in working condition

Fig. 8. Change diagram of water level rising and falling infiltration line under different working conditions.

the water level rises, The greater the dynamic water pressure is, Benefit to the stability of high embankment slopes, The safety factor increases; The rising water level also forms a head difference from inside the soil, The water level rise shows that the water seepage direction is pointing to the inside of the slope body, Particles in the soil will be pushed by the seepage force of water, Thus making the structure inside the soil more compact, And the faster the water level rises, The difference will also increase, This phenomenon is also subsequently more obvious; Due to the decrease of the gravity component along the inclined plane, Reduced the sliding force of the rock and soil mass, At the same time, the sliding resistance of the sliding body is decreased; to sum up, The stability of the high embankment slope, And the faster the water level rises, The faster the safety

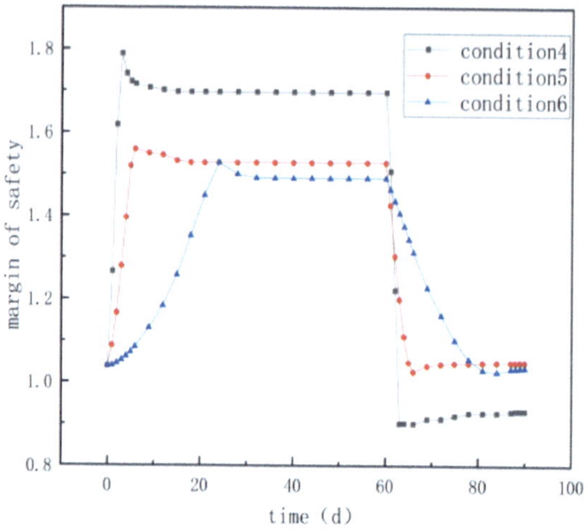

Fig. 9. Change diagram of high embankment safety coefficient over time under different working conditions.

factor increases. The water level rises to the highest, and the safety factor reaches its maximum.

As the water level gradually stabilized, the dynamic water pressure also disappear, under the permeability, the slope moisture will gradually increase, the infiltration line below the soil reaches saturation, the bulk of the soil from natural severe to saturated severe, weight increase, the saturation of the soil substrate suction close to zero, lower shear strength, sliding resistance, and the other disadvantages of high embankment slope stability weakened, safety coefficient stability after gradually reduced, stable value is greater than the initial value of the safety factor. And the faster the rise speed, the greater the safety factor will be after the final stability.

During the water level drop, The slope with positive water pressure is in the air state with the decline of water level, Not conducive to the stability of the slope; leap, The water level will form a head difference from the soil inside, Water will seep from inside the slope to outside the slope, The resulting osmotic pressure pointing to the outside of the slope body, The seepage direction is consistent with the landslide direction, The force of water can move particles in the soil, And make the soil become more relaxed, Not conducive to the stability of high embankment slopes, The faster the decline is made, This phenomenon will also become more apparent, This is the main reason for the lower safety factor in the early stage of the decline, When the water level drops to a minimum, The safety coefficient of the high embankment slope also reaches the lowest value.

In the water level is stable after a period of time, the high embankment slope soil inside the seepage field gradually stabilized, the head difference will be more and more small until disappear, the water permeability pressure reduced until disappear, the stability of the slope relative to the lowest water level, the upper slope infiltration line back to the original saturation, the matrix suction increases, strengthen the mechanical properties of soil, soil shear strength, skid resistance increases, is conducive to the stability of the high embankment slope, slope is the safety factor after the water level dropped to the lowest point. High embankment slope after water soaking for a long time, and restore the natural state, equivalent to a dry and wet cycle, slope in the soil mechanical properties, mainly lead to internal friction Angle and cohesion drop, compared to the initial the water level of 25 m its stability, and due to the water level changes fast slope soaking time is relatively longer, the faster the decline, it eventually tends to stabilize after the safety factor will be lower.

4 Conclusion

As can be learned from working Conditions 1,2 and 3, the higher the water level, the more stable the slope, the greater the safety factor will be.

It can be concluded by working Conditions 4,5 and 6, As the water level rises, The more stable the slope, And the faster the water level rises, The more that we have increased the stability, The water level is most stable at the highest point, The safety factor reaches the highest level; When the water level reaches the highest water level, Stability and then gradually decreases, 1 Finally finally the stabilization, The stability value of the safety factor is larger than the initial value; And the faster the water level rises, The safety factor is relatively large after stability; During the descent of high embankments, Lower stability of high embankment slope, The lower the safety factor will also be, The faster the descent rate is, The more unstable it is, The faster the safety factor drops, When the water level drops to a minimum, The stability of the bank slope reaches the lowest level; After the water level is stabilized, The slope body gradually returns to a stable value, However, the stability value of the final safety factor is lower than the original safety factor value of the same water level, And the stability value of the final safety factor with a faster decline speed is slightly lower than the stability value of the slower safety factor.

When the bank slope is most prone to instability, the water level drops, and the faster the decline speed, the more likely it is to instability.

It is concluded that the instability of high embankment slope can most easily occur in the process of water level sudden drop, and the main factor of the instability is the internal seepage of the slope, which has great practical significance for the prevention and treatment of embankment.

Expectation. In this paper, the water level change affects the stability of the high embankment slope, but there are many factors affecting the stability of the high embankment slope. The subsequent research can combine other influencing factors to set more complex working conditions.

References

1. Guo, Z.H., Zhou, C.B., Sheng, Q., et al.: Effect of reservoir water level change on slope stability. J. Rock Soil Mech. **26**(S2), 29–32 (2005)
2. Yang, H.: Slope stability analysis under the action of reservoir water seepage. J. Water Conservancy Plann. Des. (01), 110–113 + 158 (2018)

3. Yan, J.F., Zhu, H.H., Shi, B., et al.: Experimental study on the stability of soil slope model under water level change based on DOFS. J. Eng. Geol. **22**(06), 1263–1270 (2014)
4. Stinkang. Numerical analysis of the stability of binary structure slope under the condition of reservoir water level rise. Kunming: Kunming University of Science and Technology (2020)
5. Zhang, L.T.: Study on the Mechanism of Bank Slope Instability Induced by Reservoir Water Combined with Rainfall. Fuzhou University, Fuzhou (2017)
6. Lei, Z.: Study on Deformation Mechanism and Prediction of Bank Slope Based on DFOS. Nanjing University, Nanjing (2020)
7. Zhao, P., Zhang, H., Yu, C., et al.: Effect of water level change and tower load on the slope stability of the transmission line at the source end of hydropower station. Water Conservancy Hydropower Technol. (Chinese and English) (2024)
8. Liu, W., Xu, C., Hu, S.T, et al.: Slope reliability analysis of J considering spatial variation of unsaturated permeability coefficient under rainfall and reservoir level. J. Civ. Env. Eng. (Chinese and English) (2023)

Correction to: Research on Intelligent Vibration Parameters of Assembled Lightweight T-Beam Members

Zuoqian Xu, Zhitian Wang, and Dading Sun

Correction to:
Chapter 19 in: B. Yuan et al. (Eds.): ***Structural Safety and Ground Improvement on Bridge*****, SUCI,**
https://doi.org/10.1007/978-3-031-76102-7_19

In the original version of the chapter 19, the name of the chapter author has been corrected as "Zuoqian Xu".

The updated version of this chapter can be found at
https://doi.org/10.1007/978-3-031-76102-7_19

B. Yuan et al. (Eds.): ICCASE 2024, SUCI, p. C1, 2026.
https://doi.org/10.1007/978-3-031-76102-7_51

Author Index

B. Yuan et al. (Eds.): ICCASE 2024, SUCI, pp. 471–473, 2024.
https://doi.org/10.1007/978-3-031-76102-7

The manufacturer's authorised representative in the EU is Springer Nature Customer Service Centre GmbH, Europaplatz 3, 69115 Heidelberg, Germany. If you have any concerns regarding our products, please contact ProductSafety@springernature.com

Printed and bound by CPI Group (UK) Ltd, Croydon, CR0 4YY
01/10/2026
02225911-0012